SMART FOOD INDUSTRY
The Blockchain for Sustainable Engineering
Volume II - Current Status, Future Foods, and Global Issues

I0028913

Editors

Eduardo Jacob-Lopes

Food Science and Technology Department
Federal University of Santa Maria
Santa Maria, RS, Brazil

Leila Queiroz Zepka

Food Science and Technology Department
Federal University of Santa Maria
Santa Maria, RS, Brazil

Mariany Costa Deprá

Food Science and Technology Department
Federal University of Santa Maria
Santa Maria, RS, Brazil

CRC Press
Taylor & Francis Group
Boca Raton London New York

CRC Press is an imprint of the
Taylor & Francis Group, an **informa** business

A SCIENCE PUBLISHERS BOOK

Cover image taken from Shutterstock.

First edition published 2024
by CRC Press
2385 NW Executive Center Drive, Suite 320, Boca Raton FL 33431

and by CRC Press
4 Park Square, Milton Park, Abingdon, Oxon, OX14 4RN

© 2024 Eduardo Jacob-Lopes, Leila Queiroz Zepka and Mariany Costa Deprá

CRC Press is an imprint of Taylor & Francis Group, LLC

Library of Congress Cataloging-in-Publication Data (applied for)

ISBN: 978-1-032-13865-7 (hbk)
ISBN: 978-1-032-13866-4 (pbk)
ISBN: 978-1-003-23117-2 (ebk)

DOI: 10.1201/9781003231172

Typeset in Times New Roman
by Radiant Productions

Preface

Food is fuel for human life on Earth. However, the base for all life depends on a clean and healthy planet.

We move past supermarket shelves and often look for our favorite food products. In a nutshell, most of our decisions are based on convenience, taste, and price. However, unfortunately, sustainability is not always at the forefront of our choices. Even for those of us who wish to live planet-friendly lives.

However, the same fuel that powers humanity also drives it towards global environmental collapse. Currently, it is estimated that the environmental burdens associated with food production are the crucial contributing factor for the ecosystem to approach its planetary limits. Sustainability is no longer a simple trend and has become imperative in the food production chain and industry. There is a resounding call to do good—not just avoid evil.

Notoriously, the food industry is far from sustainable. Nevertheless, options to address these challenges do exist, but they need to be carefully considered. The food industry must strive to go beyond the traditional "do no harm" imperative. While some slight progression—from pure evil to necessary evil—may have been understood in the past, there is a growing expectation and enthusiasm for regenerative and nature-positive food production.

In this transition towards sustainability, we seek to place the food industry in a prominent spotlight. Instead of just being the story rogue, in this new narrative, the food industry—in its most innovative form—takes on the role of a solution, which can reduce food loss and waste, and provide new foods for the future, while becoming environmentally friendly.

Therefore, a more sustainable future applied to the food industries can be achieved, but getting there will not be easy. That is because addressing one sustainability concern can cause another to pop up, leaving food industries playing a never-ending game of "whack-a-mole". Therefore, to win this game, it will be necessary to understand what all the gaps can be filled, considering the levels of commitment of decision-makers, beyond the effectiveness in facing future challenges. After all, the future is uncertain, but to act now, we need to have a good sense of what the world might look like under potentially different paths.

In light of this, this book, *Smart Food Industry: The Blockchain for Sustainable Engineering: Volume II: Current Status, Future Foods, and Global Issues*, reviews the literature and frameworks of sustainable food engineering, aiming to fill the knowledge gap about the future of the food industry. Divided into three parts, this book discusses the (i) status of sustainable food industry; (ii) next generation and future technology for sustainable foods; and (iii) policy, social, economic, and environmental aspects in food industries. Here, the assembled chapters draw on stakeholder input to present a kind of sustainability compass, comprised of a comprehensive set of metrics for food industry assessments. Thus, this book proposes new concepts and strategies to face future sustainability challenges that are on the horizon that can affect and impact future generations.

Eduardo Jacob-Lopes
Leila Queiroz Zepka
Mariany Costa Deprá

Contents

Part I
Status of Sustainable Food Industry

1

The Food System with Optimum Nutrition Vision

Ayten Aylin Tas[1] and *Sedef Nehir El*[2,*]

1. Introduction

Ten years ago, the focus of the global nutrition targets was on providing nutritious, sufficient and safe food for all people and eliminating hunger by 2030. There is now strong evidence that these targets will not be achieved by 2025 globally and in most countries worldwide (Global Nutrition Report 2021). The world is at a critical point, and bold actions are needed to address significant causes of food insecurity and malnutrition, especially when the world population is projected to reach 10 billion by 2050 (UN Department of Economic and Social Affairs 2017).

Despite increased interest in nutrition in recent years, the progress in achieving the sustainable development goals (SDSs) and reducing food insecurity and malnutrition has been insufficient. Globally, we continue to experience the triple burden of malnutrition in the form of (1) impaired child growth manifested as stunting, (2) micronutrient deficiencies and (3) the growing epidemic of diet-related non-communicable diseases (NCDs) linked to overweight and obesity (Global Panel on Agriculture and Food Systems for Nutrition 2020). The following is where we stand currently regarding the state of global malnutrition, hunger and health:

- The number of undernourished people continued to increase, and between 720 and 811 million people in the world faced hunger in 2020. More than half of those (418 million) were in Asia, and more than one-third (282 million) were in Africa.

- Nearly one in three people (2.37 billion) did not have access to adequate food in 2020 (this equates to an increase of almost 320 million people in just one year).

- Two billion people were deficient in essential micronutrients, especially iron, zinc, vitamin A and iodine, and suffered from "hidden hunger".

- 149.0 million children under five years of age were stunted, 49.5 million were wasted, and 40.1 million were overweight.

[1] Department of Health Professions, Faculty of Health and Education, Manchester Metropolitan University, Manchester, United Kingdom.
[2] Food Engineering Department, Nutrition Section, Ege University, Izmir, Türkiye.
* Corresponding author: sedef.el@ege.edu.tr
Ayten Aylin Tas, ORCID ID: 0000-0001-5642-939X
Sedef Nehir El, ORCID ID: 0000-0002-2996-0537

- 772 million adults were obese, and 2.2 billion (40.8% of all women and 40.4% of men) were overweight. 39 million children under the age of five were overweight or obese in 2020. Over 340 million children and adolescents aged 5–19 were overweight or obese in 2016.

- Globally one in every five deaths in adults was associated with unhealthy diet. Among NCDs, cardiovascular diseases accounted for 17.9 million deaths, followed by cancers (9 million) and diabetes mellitus (1.6 million) (FAO 2017, WHO 2020, FAO IFAD UNICEF WFP and WHO 2021, Global Nutrition Report 2021, WHO/EURO 2021).

The Covid-19 pandemic that emerged in 2019 (still prevailing as of 2022) impeded millions of people's food security and nutrition status. Its cumulative effects on global nutrition and diets are still a significant concern, particularly for the vulnerable (Global Panel on Agriculture and Food Systems for Nutrition 2020). Dietz and Pryor (2022) stated that Covid-19 added another pandemic of food insecurity to already existing ones, which manifest themselves as obesity, undernutrition and climate change (collectively known as "the global syndemic"). Covid-19 revealed the weaknesses in the food systems (FSs) and introduced further challenges by adding an estimated 83 to 132 million more hungry people to existing figures (FAO IFAD UNICEF WFP and WHO 2021). Decisions, strategies and targets made on nutrition, food security and public health before the outbreak had to be reviewed and reconsidered. We are now at a turning point where we must address the threats that create vulnerabilities in our FSs, with the responsibility of taking and implementing swift action to eliminate/mitigate them before it is too late. The challenges caused by the major drivers of food insecurity and malnutrition (i.e., conflicts, climate change and economic adversity) can be overcome by the following strategies (FAO IFAD UNICEF WFP and WHO 2021):

(1) Reduce conflict in affected areas by integrating peacebuilding policies and encouraging socio-economic development

(2) Reduce the negative effects of variable and extreme climate conditions

(3) Build resilience against economic adversity

(4) Increase the affordability of nutritious foods by intervening along food supply chains

(5) Reduce extreme poverty and structural inequalities by ensuring that interventions are not only pro-poor and inclusive but also empowering women and youth

(6) Promote healthier food environments that encourage the consumption of healthy and safe foods with a lower impact on the environment.

The FS includes the essential elements that affect our food throughout its journey from farm to fork—such as people, institutions, activities, processes and infrastructures. It involves all stages of the value chain—from growing and harvesting fresh produce to processing, packaging, transporting, selling, preparation/cooking and disposal of waste food (UNEP 2016). The Food and Agricultural Organization (FAO) defined a sustainable FS as "the one that contributes to food security and nutrition for all in such a way that the economic, social, cultural and environmental bases to generate food security and nutrition for future generations are not compromised" (HLPE 2017). Indeed, the "food systems" approach requires us to think about broader issues such as environmentally sustainable livelihoods and planetary health.

Unfortunately, today's FSs are exacerbating risk to our and planetary health and therefore are no longer fit for purpose. This situation necessitates changing how FSs are managed, governed and utilised (Global Panel on Agriculture and Food Systems for Nutrition 2020). Existing FSs need to be transformed to provide adequate and affordable food for all with the involvement of all interested actors, drivers and stakeholders (von Braun et al. 2021). Transformation should at the same time allow FSs that do not drive climate change, biodiversity loss, soil degradation and pollution (UNEP 2016).

The Scientific Group of the UN Food Systems Summit described five action tracks (areas) as recommendations for transforming FSs to align them with the objectives of the 2030 agenda and

SDGs and for reshaping relevant policies and strategies with the consideration of the impact of Covid-19 on food security and nutrition (von Braun et al. 2021, Neufeld et al. 2021). These action tracks are described as follows:

(1) Enable everyone to have access to nutritious, healthy and safe food

(2) Promote and encourage sustainable consumption patterns (including efforts to reduce food waste)

(3) Promote food production methods that produce sufficient food for all and with the least impact on the environment and sustainability

(4) Reduce inequalities by increasing income and wealth and expanding inclusion

(5) Ensure the continuity of sustainable and healthy FSs by building resilience to shocks, stresses and vulnerabilities.

The five action tracks are interlinked and should not be independently considered when positioned in an FS framework to ensure an overarching FS concept (von Braun et al. 2021).

2. Nutrition and Sustainability Shaping the Future of Food Systems

The concept of "optimal nutrition" stemmed from the public health efforts to maintain a healthy life and prevent NCDs. However, nations have now realised that diets also need to embrace the practices to protect the future of our planet. In 2012, the FAO introduced the concept of "healthy and sustainable diets (HSDs)", which is defined as "diets with low environmental impacts that contribute to food and nutrition security and healthy life for present and future generations" (FAO 2012). Such diets need to be available, accessible, affordable, culturally acceptable and should not harm the environment (Global Panel on Agriculture and Food Systems for Nutrition 2020). Achieving healthy diets from sustainable FSs requires the involvement of everyone since significant changes need to be made in the current dietary patterns and eating/consumption habits, including the reduction in food losses and waste, and major improvements in how our food is produced (Willett et al. 2019). Fanzo et al. (2021) foresee that this transformation is achievable, but add that rigorous monitoring is required to keep progress on track.

2.1 Healthy and Sustainable Diets (HSDs)

A healthy diet promotes human health, prevents disease and safeguards planetary health by (1) providing an adequate (but not excess) amount of nutrients from foods that are nutritious and healthy (2) devoid of health-harming substances through all stages of food production (Neufeld et al. 2021). In the EAT-Lancet Commission report, Willett et al. (2019) defined the specifics of healthy dietary patterns for the better of the Earth's various systems; this is called a "healthy reference diet (HRD)", which is also quoted as a "win-win" diet within this context. The report recommends the consumption of the following as part of HRDs:

- The plants as primary protein sources (soybeans, legumes and nuts)
- Fish and other alternative sources of omega-3 fatty acids several times a week
- Optional and modest consumption of poultry and eggs and low intakes of red meat (with very little or no processed meat)
- Plant oils with unsaturated fatty acid content, low amounts of saturated fats and no partially hydrogenated oils
- Carbohydrates primarily from whole grains with a low intake of refined grains and a limited amount of sugar (less than 5% of the energy intake)
- At least five servings of fruits and vegetables, and
- Moderate amounts of dairy products as an optional choice.

Expressed on a daily basis, the HRD consists of around 232 g of whole grains (wheat, corn and others), 50 g of tubers or starchy vegetables, 200–600 g of vegetables, 100–300 g of fruit and 0–500 g of dairy foods. Protein sources can be 0–28 g red meat, 0–58 g poultry, 0–100 g fish and 0–13 g eggs. 0–50 g nuts can be consumed as an alternative to red meat. A total of 50 g (dry weight) of legumes (lentils, beans and peas) and 25 g of soybeans is recommended. Total added fat is listed as 50 g, emphasising the predominant use of unsaturated plant oils. Finally, sugar and other sweeteners consumption is limited to 31 g (Willett et al. 2019).

Dietary patterns with such characteristics are likely to reduce the risk of major chronic diseases and improve health (Orlich et al. 2013, Satija et al. 2017, Schwingshackl et al. 2017, Watling et al. 2022). Eating a variety of foods is not only healthier but also supportive of stronger sustainability as it increases demand for more diverse products (Bene et al. 2019). Although the EAT-Lancet report provided the first evidence-based targets for HSDs, it has received criticism because it did not assess cost or affordability (Schwingshackl et al. 2020, Hirvonen et al. 2020). The study of Hirvonen et al. (2020) showed that HRDs are not affordable for much of the world's low-income populations, sub-Saharan and South Africa in particular. The same study reported that HRDs are on average 60% more expensive than the foods required for nutrient adequacy, partly due to the inclusion of greater quantities of animal-derived foods and fruits and vegetables.

To achieve the required transition to HSDs, everyone has a part to play. Environmental sustainability should be an essential part of dietary guidance while interacting with individuals or groups regarding their dietary choices and setting guidelines for the wider public (Rose et al. 2019). The following are some of the diet-related practices that all consumers can exercise when making food choices:

- Avoiding over-consumption beyond the caloric requirement and opting for a variety of nutrient-dense foods
- Limiting intake of ultra-processed (UP), nutrient-poor and over-packaged food
- Consuming no more than recommended amounts of animal tissue
- Increasing the consumption of plant-based foods (Barbour et al. 2021, Ridoutt et al. 2021).

We have a formidable task in hand since shifting to healthy diets will require significant steps such as a 50% reduction in the global consumption of unhealthy foods (such as red meat and sugar) and a 100% increase in the consumption of healthier ones (such as vegetables, fruits, legumes and nuts). Any slight deviation from those targets (although these rates may differ from region to region) would compromise the timely accomplishment of SDGs (Willett et al. 2019).

2.2 Overconsumption

Adult obesity and overweight continue to rise at a similar rate in rich and developing countries. The prevalence rose from 11.8% in 2012 to 13.1% in 2016 globally, and no country is on track to halt the increase in adult obesity (FAO IFAD UNICEF WFP and WHO 2021, Global Nutrition Report 2021). Similarly, childhood overweight is rising (Global Nutrition Report 2021). The dilemma lies in the fact that the global food production is adequate to meet the energy requirements of everyone but put simply, the distribution is unequal. Some are undernourished due to the inability to access and afford food, and some become overweight or obese due to excess consumption (FAO IFAD UNICEF WFP and WHO 2021). Serafini and Totil (2016) defined obesity as an "unsustainable metabolic condition". They introduced a new term, "metabolic food waste" (MFW), indicating that any food eaten beyond physiological needs should be considered waste due to the excess and undesirable body fat accumulation. The following study provided evidence of the substantial amount of food lost globally through overeating and its impact on the environment (Toti et al. 2019). The study reported that between different food commodities, meat/offals and dairy products/milk/ eggs were the major contributors of MFW. Consuming more calories beyond requirement creates a "lose-lose" situation both for human and planetary health and subsequently undermines the underpinning principles of the HSDs.

Unhealthy (discretionary) foods do not constitute a part of a healthy diet. They are likely to be nutrient-poor but energy-dense and therefore may contribute to excess energy intake. They contribute to poor nutrition and have a harmful environmental impact, as recently established in a study on Australian dietary intake and environmental data (Ridoutt et al. 2021). This study found that discretionary foods increased environmental impact score the most, followed by fresh meat and alternatives and dairy foods and alternatives.

Sugary foods and drinks make up a significant portion of discretionary foods. Excess consumption of added sugar has negative consequences on health since it is a risk factor for cardiovascular disease, type 2 diabetes, both directly and indirectly. A high intake of added sugar can be accompanied by further calorie intake, leading to weight gain and an increased risk of obesity (Stanhope 2016, Mekonnen et al. 2013, Khawaja et al. 2019).

Food processing is necessary to provide safe, digestible, nutritious and palatable food with an extended shelf life to preserve the nutrients. It is also an effective tool to reduce food waste. That said, when foods are processed to the degree that they are stripped of their nutrients and cannot possess any of those benefits, they do more harm than good to health. The consumption of ultra-processed foods (UPFs) has been associated with weight gain and increased risk of NCDs. UPFs are defined as multi-ingredient and industrially-formulated mixtures that contain little (if any) intact foods. They tend to be high in added sugar, sodium and saturated and trans fats and low in valuable nutrients (e.g., protein, fibre and micronutrients) (Monteiro et al. 2019). Moreover, UPFs harm the environment due to carbon emissions and water use associated with their production and being heavily wrapped with materials that are not environmentally friendly. Unfortunately, they make up a significant dietary energy source in high-income countries such as Canada, the USA and UK (Monteiro et al. 2013, Baker et al. 2020). Popkin et al. (2021) emphasised that the food industry has a major role in developing less processed products and called for action to reduce the consumption of UPFs globally.

2.3 Meat Consumption and Alternatives

The World Cancer Research Fund (WCRF) recommends limiting red meat consumption to three portions (350 to 500 g cooked weight) per week, with the optimum for public health outcomes set at 300 g per week and avoiding processed meats. Red meat consumption is associated with an increased risk of colorectal cancer and other NCDs (Boada et al. 2016, Huang et al. 2021). The link between processed meat consumption and cancer is more clearly demonstrated (Handel et al. 2020, Veettil et al. 2021, Watling et al. 2022). Often high in salt, processed meat may increase the risk of high blood pressure and mortality from cardiovascular disease (WCRF 2018).

Western consumers' acknowledgement of the environmental impact of meat consumption is still very low (Sanchez-Sabate and Ruben-Sabate 2019, Moreira et al. 2022). This finding corroborates the need to introduce nationally increased targets to promote the uptake of sustainably and ethically produced meat. The Eating Better alliance calls for an average 50% reduction in meat and dairy consumption in the UK by 2030 and a transformation to "better meat and dairy" as a standard to benefit health, social justice and the environment (Dibb and de Llaguno 2018).

To reduce excessive meat consumption, consumers need to incorporate more plant-based foods and meat alternatives into the diet. Plants are recommended as primary, but not exclusive, sources of proteins in the HRD (Willett et al. 2019). Some diets avoid animal products, but such diets need to be planned carefully and appropriately to prevent micronutrient deficiencies (vitamin B12, zinc, calcium and selenium) (Vesanto et al. 2016, Bakaloudi et al. 2021, WHO/EURO 2021).

Meat alternatives are foods obtained from novel protein sources such as insects, microalgae, cultured meat and plants (Zhang et al. 2022). Novel protein sources have attracted great research interest recently, and their future is promising. Examples of alternative protein sources and their protein content range are presented below (Fasolin et al. 2019). The protein quality of those sources needs to be taken into account when assessing their suitability to substitute animal-derived products. Furthermore, their digestibility, allergenicity and toxicity need to be investigated thoroughly.

- Insects (cricket, grasshopper, honey bee brood and mealworm, 22 to 76% protein)
- Algae (*Aphanothece microscopica, Arthrospira platensis (Spirulina platensis), Chlorella vulgaris*; 12 to 54% protein)
- Fungi (*Aspergillus niger, Fusarium venenatum, Saccharomyces cerevisiae, Torula utilis*; 10 to 61% protein)
- Bacteria (*Rhodopseudomonas* sp., *Rhodopseudomonas faecalis*; 50 to 92% protein)
- Vegetables (amaranth, lupin, navy bean and quinoa; 12–55% protein).

2.4 Availability, Accessibility and Affordability of HSDs

Every human being has the right to adequate, nutritious and safe food. As discussed previously, the FS is failing to supply optimal nutrition to everyone when the following points are considered: (1) Inability of the system to feed the future world population (2) Inability of the system to deliver a healthy diet (3) Inability of the system to produce equal and equitable social and economic benefits (4) Unsustainability of the system and its implications on the environment (Bene et al. 2019).

The food environment refers to the physical, economic, political and socio-cultural context in which consumers engage with the FS to acquire, prepare and consume food (HLPE 2017). Downs et al. (2020) provided an expanding definition of the food environment, encompassing food availability, affordability, convenience, promotion, quality and sustainability properties. The food environment is critical in the FS since it is where complex information-processing procedures for purchasing a food item occur (Blake et al. 2021, Downs et al. 2020). At the same time, the food environment significantly influences what we buy and, in turn, what we eat (WHO/EURO 2015). Healthy and sustainable foods provided by the food environment need to be widely available and affordable for all, but unfortunately, this does not yet seem to be the case.

Many people cannot afford healthy diets due to the high costs of healthy food and income inequalities. Healthy diets cost approximately five times more than energy-sufficient diets (Herforth et al. 2020), and it is the poorest that spend the most significant proportion of income on food (UN World Food Programme 2020). Healthy diets were unaffordable for three billion people in every region of the world in 2019 (FAO IFAD UNICEF WFP and WHO 2021), with sub-Saharan Africa and Southern Asia not being able to afford a nutrient-adequate diet and therefore most prominently affected by the problem (Herforth et al. 2020, Miller et al. 2021).

The food industry can enhance the availability, accessibility and affordability of healthy foods by:

- Expanding food production sustainably to offer more nutrient-rich foods with greater diversity
- Making healthier foods more widely available at a lower cost
- Reducing production costs through technology and innovation
- Implementing methods to ensure the preservation of nutrients along the food chain
- Reducing food loss and waste
- Collaborating with the public sector, non-governmental organisations and academia to develop a partnership strategy to deliver healthy and sustainably obtained foods (IFST 2017, Global Panel 2020, Miller et al. 2021, Reyes et al 2021).

2.5 How Can We Make HSDs More Desirable?

People may decide not to adopt HSDs even when they are accessible and affordable. This is because an array of interlinked factors determine food choice (Contento 2011, Leng et al. 2017, Blake et al. 2021), which inevitably makes the nature of the decision-making process complicated and the outcome difficult to predict. Some of the public health strategies that can be used as efficient strategies to enhance nutrition knowledge and promote healthier choices can be (1) alignment of food-based dietary guidelines (FBDGs) with the latest evidence on nutrition and sustainability, and their more effective use by the general public, (2) wider use of nutrition labels by the food industry and consumers and (3) reinforcement of regulatory measures (Mozaffarian et al. 2018, Popkin et al. 2021, Reyes et al. 2021).

2.5.1 Wider Use of Upgraded FBDGs to Promote Enhanced Knowledge

FBDGs educate consumers on healthy diets and inform programs and policies to ensure healthy diets for all (UNICEF 2020). Policymakers can facilitate HSDs at a national level via FBDGs and food strategies (Barbour et al. 2021, Fanzo et al. 2021). Coleman et al. (2021) reported that this could be achieved by: (1) conveying consistent local and national messages on what comprises an HSD (via FBDGs, food strategies and campaigns) (2) improving the quality of food in public institutions where the local and national governments have the most significant influence, and (3) adopting a "whole food systems" approach, which covers all interconnected elements of the food systems, in schools.

FBDGs need to be upgraded in light of the scientific data and discussions around the broader concept of sustainability and be promoted to enhance knowledge about the implications of dietary choices (Global Panel on Agriculture and Food Systems for Nutrition 2020). Only a few counties have integrated sustainable ways of healthy eating into their FBDGs. Brazil, Germany, Sweden, and Qatar featured sustainability, and some others (the Netherlands, Norway, Estonia, United Kingdom and France) alluded to it in their quasi-official guidelines (Fischer and Garnett 2016). Brazil successfully replaced the nutrient-based approach (commonly known as the "food pyramids") with an overarching food-based approach that incorporates healthy eating patterns and environmental, socio-economic and cultural aspects of sustainability (Coutinho et al. 2021). Brazil's approach is noteworthy since the term "sustainability" is often used to express specific environmental concerns and ignore its equally important social, economic and cultural aspects (Fischer and Garnett 2016).

FBDGs need to convey precise and up-to-date information on the impact of food choices on human and planetary health (Herforth et al. 2019, Global Panel on Agriculture and Food Systems for Nutrition 2020, Springmann et al. 2020). They need to be communicated to and used by the general public, businesses in the FS and policymakers much more effectively. To have a tangible impact on food consumption, they should also be used to inform and guide government policies (e.g., school and hospital meals, industry standards and advertising regulations, etc.) (Fischer and Garnett 2016, Global Panel on Agriculture and Food Systems for Nutrition 2020).

Most DGs have been "not ambitious enough" to bring FSs within planetary boundaries (EAT 2020). One of the specific actions recommended for governments during the transition is investing in the next generation of enhanced FBDGs. These DGs should be science-based, dynamic (responsive to changing conditions), feasible (that can be acted on and not merely giving advice), and constructed with more consideration of the implications of food choices on human and planetary health (Global Panel on Agriculture and Food Systems for Nutrition 2020). Governments are also recommended to explore additional forms of communication/guidance to enhance people's understanding of how food choices influence human and planetary health.

2.5.2 Nutrition Labelling

The impact of nutrition labels in educating the general public on healthy food choices is indisputable. There are different formats for nutrition labels, and the more visual the labels are, the more easily they are understood and hence the greater possibility for them to be used. In this regard, Front of pack (FOP) labels have proven to be successful (Talati et al. 2016, Oswald et al. 2022, Meijer et al. 2021). Most countries mandate the inclusion of a nutrition label on pre-packed foods, but the format and information may vary from country to country (Becker et al. 2015). The importance of nutrition labels as an educative tool for consumers also needs to be recognised by the food industry. A recent study investigated the perceptions of English food manufacturers and retailers (n = 35) on nutrition labels and their use in consumers' purchasing decisions. It revealed that 20% of the respondents did not know the importance of FOPs to the consumers, 48% were unsure whether FOP labelling is mandatory, and 60% claimed that including FOP labels would incur high costs to their businesses (Ogundijo et al. 2022).

2.5.3 Reinforcement of Regulatory Measures

The corporate global supply chains impact the types of foods offered and, hence, food consumption (IPES-Food 2017). This raises a fundamental question: Should food production be driven by profit or its implications on public health and the environment? The food industry continues to produce highly profitable UP foods and drinks, mainly due to the significant incentives they receive. Open trade policies allow transnational food and beverage corporations to easily sell foods associated with unhealthy diets containing high salt, sugar and saturated fat (Willett et al. 2019, Baker et al. 2020).

Governments can implement a spectrum of policies ranging from voluntary to mandatory to support improvements in diets and health (Hawkes et al. 2013, Mozaffarian et al. 2018, Willett et al. 2019, UNSCN 2019). Good examples of this can be seen in many low to middle-income countries, such as Latin America and South Africa, which instituted national policies to change the food environment to facilitate healthier food choices (Popkin et al. 2021, UNSCN 2019).

Fiscal incentives and disincentives can be imposed on consumers, producers and retailers. Disincentives can include levies or sales taxes on unhealthy foods such as sugar-sweetened beverages (the UK, Mexico and South Africa) and UPFs, and also the removal of industry tax benefits (HLPE 2017, Mozaffarian et al. 2018, Popkin et al. 2021).

Comprehensive marketing controls and restrictions for unhealthy foods, which governments introduced proved to be successful—the ban on the marketing of UPFs in Brazilian schools is an excellent example of this. Chile's multi-dimensional obesity prevention programme, which combined the use of FOP labels, marketing restrictions, school-based restrictions and future taxation plans, exemplifies the effective use of a combination of linked strategies rather than single policies implemented in isolation (Popkin et al. 2021).

3. The Food Industry as Part of the Food System

The food industry has an essential role in processing food and making it available for everyone globally (Knorr and Augustin 2021). It has been trying to respond to the worldwide changes and challenges with producing safe and healthy food that consumers are increasingly demanding. "Foodsystems 4.0 for a sustainable world" drives the food industry to adopt advanced technical and scientific knowledge for consumer-oriented food innovation (Rosenthal et al. 2021). The responsibilities of the food industry are vast and diverse, and they can be grouped into two categories: process-targeted and nutrition-targeted.

Process-targeted Responsibilities

- Understanding the importance of food processing technologies for the production of safe, nutritious and organoleptically acceptable foods
- Applying "Industry 4.0" technologies such as digitalisation, data analytics, robotisation and automation
- Use of more sustainable food processing technologies instead of traditional methods (e.g., green technologies such as ultrasound, high-pressure process, green cloud point extraction, pulsed electric fields, green extractors, membrane processes and cold gas plasma)
- Choosing more energy-efficient methods for industrial processes and recommending efficient methods for food preparation at home
- Application of novel fermentation techniques to develop value-added ingredients and food products
- Optimising the parameters of processing techniques (temperature, time, pH, etc.) to reduce the loss of nutrients and bioactive compounds
- Applying processing techniques that consider the food matrix and the function of the constituents
- Carrying out advanced R&D studies to produce nutritionally optimised and sensorily appealing ingredients and food products
- Customisation of food processing to consumer requirements (utilisation of the PAN concept - Preferences, Acceptance and Needs of the consumers)
- Increasing water/resource use efficiency and minimising waste.

Nutrition-targeted Responsibilities

- Use of sustainable and novel alternative protein sources (e.g., algal biomass, insects, and cultured biomass) to reduce the consumption of animal foods
- Modifying fat and simple sugar content in food formulations or replacing them with healthier options
- Enrichment with minerals, proteins and bioactive compounds
- Reducing calorie value of foods by facilitating structural psycho-chemical changes in the food matrix
- Characterisation of bioaccessibility and bioavailability of nutrients and phytochemicals in processed foods with consideration of food matrix effect
- Recovery of valuable components such as bioactive compounds from food waste
- Adopting smaller serving sizes for packaged foods (El and Simsek 2012, Fasolin et al. 2019, Augustin et al. 2020, Knorr and Augustin 2021, Rosenthal et al. 2021, Jagtap et al. 2021).

Industrial-scale food processing enables the preservation and transformation of raw crops into edible, safe, healthy, sustainable and nutritious food products (Fasolin et al. 2019). However, negative perceptions about the food industry and processed foods, accompanied by a biased approach to the role of food processing in ensuring food safety and food diversity, still exist. This could be addressed by sharing more information with consumers on ultra-processed, minimally processed and unprocessed foods, as classified by the NOVA system (Monteiro et al. 2019). Bhawra et al. (2021) developed a tool, Food Processing Knowledge (FoodProK) score, which determined consumers' knowledge about the healthiness of foods processed to different extents. Such tools can help improve processed food categorisation in NOVA systems.

4. Novel Protein Sources

With the limited protein sources that we currently have, it is impossible to meet the protein requirement of the increasing population. The daily amount of protein (of animal or vegetable origin) per capita is estimated at 52 g in 2030 and 54 g in 2050 (FAO 2017, FAO IFAD UNICEF WFP and WHO 2020). Many national DGs recommend meeting protein requirements from various foods that support health and sustainability. For example, in the 2015–2020 Dietary Guidelines for Americans, the "protein foods" group consists of meat, poultry, seafood, beans and peas, eggs, processed soy products, nuts and seeds (USDA 2015–2020).

The EAT-Lancet report reiterates that humanity should no longer consume more food than necessary (Willett et al. 2019). The consumption of animal-based foods is more dominant in developed countries such as the USA than in developing countries. These countries must adopt more plant-based protein foods and novel alternative protein sources to maintain adequate total protein intake (Karmaus and Jones 2021). However, when making recommendations for developing and underdeveloped countries, amino acid (AA) composition and digestibility of proteins in plant-based foods and alternative proteins need to be considered (WHO 2007). This is because protein deficiency develops when the diet consistently provides little protein or lacks essential amino acids (EAAs). Health problems due to protein deficiency, especially in the long-term, are slowed growth, impaired brain and kidney functions, early ageing, poor immunity, loss of muscle mass, and inadequate nutrient absorption (Wu 2016).

Both diet and environment have been identified as important motivating factors in reducing meat consumption in European Union (EU) countries. However, this differs between consumers in northern and southern European countries to varying degrees. In northwest Europe, which consists of the ten richest EU countries with the largest share in economic and social sustainable development indicators, the vast majority of consumers acknowledged their role in reducing meat as part of HSDs to a greater extent than their stakeholders such as food chain or governmental actors (de Boer and Aiking 2022). A study in the UK investigated the consumption of plant-based alternative foods (vegetables, legumes, nuts and seeds) and compared it against meat and dairy reduction targets. The study reported that from 2008 to 2019, the consumption rate of plant-based foods, plant-based meat alternatives and plant-based milk consumption increased. The highest increase (approximately three times) was noted in plant-based milk consumption, while meat intake was significantly reduced. Although plant-based foods are expected to become more popular and be consumed more widely, ultra-processed forms of these products cannot contribute significantly to a healthy diet and may even lead to other health problems (Alae-Carew et al. 2022).

The food industry seeks alternative protein sources for human consumption in response to global concerns about food security and protein malnutrition due to the growing population. New protein sources such as seaweed or algae, peas, rapeseed, duckweed, and insects have received a lot of research interest over the past decade. While numerous R&D projects are ongoing, those sources should be critically evaluated for protein quality. More studies should be conducted on the metabolic behaviour, health effects, and physiological and cognitive functions of proteins (van der Spiegel et al. 2013, Karmaus, and Jones 2021). Further studies should consider safety aspects such as toxic chemicals, allergenicity, impurities and contaminants (EFSA 2020).

In the light of the nutritional recommendations stated in the framework of HRDs by Willett et al. (2019), the rest of this chapter will discuss the scientific advancements about several new protein sources (insects and microalgae) and innovative applications to reduce the sugar content of foods.

4.1 Insects

The most restrictive criterion in fortifying foods with protein is the high energy/protein ratio when high-quality protein sources are used. Insects can contribute to the diet as an economical source with their high protein quality and low energy content (Van Huis 2013). In general, edible insects are accepted as sources of quality protein and EAAs, also essential fatty acids, vitamins (A, D, thiamine, riboflavin, folic acid) and minerals (iron, calcium, copper, selenium and zinc) (Van Huis 2013, Köhler et al. 2019).

In January 2018, EFSA (European Food Safety Agency) approved the consumption and sale of insect-based foods in the EU member states with the "New Food Regulation" (No. 2015/2283) (European Commission 2015). It also issued an opinion on the safety of dried yellow mealworms (Turck et al. 2021a) in January 2021, frozen and dried formulations from migratory locusts (Turck et al. 2021b) in July 2021, and finally frozen and dried formulations from house crickets in August 2021 (Turck et al. 2021c). After the (EU) 2015/2283 (European Commission 2015) Regulations, the consumption of insects as an alternative food source in EU member countries is expected to become increasingly widespread.

Edible insects such as crickets (*Acheta domesticus*), mealworms (*Tenebrio molitor*), Mopani worms (*Gonimbrasia belina*), black soldier fly larvae (BSFL) (*Hermetia illucens*), ants, water bugs and silkworms are considered good alternative protein sources with their high protein and fat content. In addition, they have a high feed conversion ratio, low environmental footprint and higher economic value (Fasolin et al. 2019, Bessa et al. 2020, Imathiu 2020).

There are more than 2,000 edible insect species, and their nutritional values vary widely between them. There is insufficient information about the nutritional composition of insects due to this diversity. The protein content of larvae, beetle, grasshopper, cricket, termites, bee and dragonfly vary between 35 and 77% of dry matter. However, insect protein digestibility differs widely due to chitin, a polysaccharide that is a major constituent of their exoskeleton. Some studies reported 7-98% digestibility without the exoskeleton (Kim et al. 2019). *In vitro* protein digestibility of biscuits enriched with 20% cricket flour was reported as 45% (Bas 2020). More studies are needed on the digestibility of insect proteins and chitin in humans.

Even within the same species, the nutrient profile of the insects can differ depending on the growth conditions (farmed or caught in the wild), stage of development (larva or adult), the environment, season and their diet. Köhler et al. (2019) reported that the Bombay locust, scarab beetle, house cricket, and mulberry silkworm could be used as human food depending on their nutrient contents. Their AA scores ranged between 0.58 to 1.04 when the suggested pattern of FAO (2013) was used as a reference. Insects can be a good source of iron since it exists in a highly biologically available form. Buffalo worms were found to exhibit higher iron bioavailability when referenced to $FeSO_4$ (Latunde-Dada 2016). Chitin and chitosan, which are abundant in the shells of insects, suppress pathogenic microorganisms in the intestines and act as prebiotics (Stull et al. 2018, Montowska et al. 2019, Saadoun et al. 2022). However, those compounds also possess anti-nutritional effects such as binding minerals, dietary fats and fat-soluble vitamins and cause their excretion from the colon.

Baiano (2020) reviewed edible insects' nutritional, microbiological, chemical properties and toxicological safety. They suggested that insects could be considered microbiologically safe but might contain harmful residues such as drugs or heavy metals. Cross-reactions between allergens in some insects also need to be examined. More detailed studies on insects are required on their allergenicity, pesticide residue levels, pathogenic bacterial contaminations and heavy metal contents (such as arsenic, cadmium, mercury and lead).

Using insect flour or protein extract as masked forms in food formulations is a good strategy to increase consumer acceptance. However, comprehensive data on their functional properties before being added to food is required. Their structural and thermal properties that affect their technological and functional behaviours such as solubility, emulsification, oil and water binding capacity and gelling, need to be researched and reviewed (Gonzalez et al. 2019, Bolat et al. 2021). During processing, such properties can cause changes in techno-functionality due to structural modifications whilst affecting their nutritional properties positively or negatively. For example, Maillard glycation between amino groups of proteins and carbonyl groups of reducing sugars is expected to occur during processing and storage. A study reported that glycation of black soldier fly larvae BSFL with glucose caused the surface structure to become looser and more porous. Compared to native BSFL protein, BSFL-glucose conjugates exhibited greater thermal stability at 90°C. Conjugation can offer possibilities to introduce new food structures that are more functional, but this can be negated by the undesirable changes in the nutritional value of proteins (Mshayisa et al. 2021).

Traditional methods (e.g., alkaline extraction, isoelectric precipitation) are simple and easy to use. Still, they can have a low extraction rate and negatively affect the extracted protein's functional properties. Therefore, obtaining efficient results with green technologies can be advantageous. For example, the High Hydrostatic Pressure (HHP) treatment of mealworm and cricket powders significantly affected protein functional properties and antioxidant activity (Bolat et al. 2021). To use edible insects as a sustainable protein source, more studies are needed on insect proteins' (1) extraction utilising green technologies, (2) techno-functional properties and (3) bioactivity and use as ingredients in nutritious food formulations.

4.2 Microalgae

Microalgae represent extensive biodiversity with about 40,000 identified species. Studies evaluating the importance of microalgae on health and human nutrition and their potential use in the food and nutraceutical industries have gained momentum after 2018 (Sathasivam et al. 2019, Ververis et al. 2020, Abu-Ghosh et al. 2021, Kusmayadi et al. 2021). Microalgae species such as *Chlorella vulgaris, Arthrospira platensis, Euglena gracilis, Auxenochlorella prototlecoides, Chlamydomonas reinhardtii, Dunaliella barwil* were granted "Generally Recognized as Safe (GRAS)" status by Food and Drug Administration (FDA) and EFSA (Torres-Tiji et al. 2020). The confirmed species of *Chlorella* are *C. vulgaris* (Canada, EU, and Japan), *C. pyrenoidosa* (EU and China), *C. prototlecoides* (US, Japan), *C. luteoviridis* (EU), *C. sorokiniana* and *C. regularis* (Canada). *Dunaliella salina* (China and Canada), *Euglena gracilis* (Canada, China, and Japan). Recently, dried *Euglena gracilis* biomass was also reported to be safe by EFSA Panel on Nutrition, Novel Foods, and Food Allergens at the recommended level as a new food (Turck et al. 2020).

Novel microalgae-based protein and peptides can be used to substitute animal and plant-based protein sources because algae can produce all EAAs. The profiles of microalgal AAs are similar to high-quality protein sources, including casein and egg. EAA composition in microalgal proteins from various species are comparable to AAs in traditional protein sources such as red meat but higher than plant-based protein sources (Singh et al. 2021). The total amount of protein in the whole cell is species-dependent and can range from 30 to 55% of the dry weight (could be as high as 70% in species such as *Arthrospira platensis*) (Barka and Blecker 2016).

Microalgae-derived hydrolysed protein is not only a protein source for the diet but also a source of peptides that exhibit important techno-functional properties. It has good AA balance and digestibility and affects health positively (Etemadian et al. 2021, Wang et al. 2021). Over the last decade, *in silico*, *in vitro* or *in vivo* studies identified various biological activities performed by the microalgae-derived bioactive peptide sequences: antimicrobial (Guzmán et al. 2019, Tejano et al. 2019), antiproliferative, and anti-amyloidogenic (Olasehinde et al. 2019), antioxidant (Lai et al. 2019), antihypertensive (Lin et al. 2018), hypocholesterolemic, and anticancer (Petruk et al. 2018), and antithrombotic activity among others (Tejano et al. 2019).

The first step in the processing of microalgae is the upstream process (cultivation), which directly impacts the quality, nutrient composition, content and yield throughout the process (Daneshvar et al. 2021). All sequential processes (harvesting, drying, lysis, extraction, purification) from biomass harvesting to the final product are downstream processes (Nitsos et al. 2020). Enzymatic hydrolysis (utilising glycosidases, glucanases, cellulose, pectinase, and lipases) is an environmentally friendly and low-energy alternative degradation technique commonly performed under mild conditions to promote the recovery of intracellular components (Al-Zuhair et al. 2017). Cell lysis disrupts the structure of the cell wall by increasing its porosity and thus enabling the extraction of intracellular components. It is an essential step for extracting proteins and obtaining bioactive peptides from algae. Steam explosion, bead milling, high-pressure homogenisation and physical methods (drying, freeze-drying, ultrasonication, microwave radiation, pulsed electric field) are some of the new lysis methods (Alavijeh et al. 2020).

Powders from different microalgae species such *Spirulina platensis*, *Chlorella vulgaris*, *Tetraselmis* sp. and *Phaeodactylon* sp and/or *Dunaliella* sp. are used in many formulated products such as pasta, cookies and wheat crackers to enrich their protein and AA content (Singh et al. 2021, Terriente-Palacios and Castellari 2022). Terriente-Palacios and Castellari (2022) investigated the AA profiles of commercial pasta and crackers enriched with algae protein from different species. In addition to the high protein and EAA content, the products (particularly those prepared with red algae species) contained sulfonic acid derivative taurine, its precursor hypotaurine and homologous homotaurine. Such AA derivatives can protect against diseases, free radicals and heavy metals. Taurine is a semi-essential amino acid and has become increasingly important in treating the central nervous system and cardiovascular system's disorders, cancer and other metabolic diseases such as type 2 diabetes. Because taurine is derived from cysteine and found naturally in fish and meat, microalgae can be considered a good sustainable protein source for vegetarian and vegan diets (Terriente-Palacios and Castellari 2022).

The use of microalgal proteins in foods has been limited for many reasons. Using whole biomass without extracting the intracellular protein results in low protein digestibility in humans. Moreover, there is insufficient data on the molecular properties and bioactivity of microalgae protein hydrolysates obtained through a combination of lysis methods (Nitsos et al. 2020). Recently, Singh et al. (2021) reviewed the future insights, obstacles, and suggestions for selecting the appropriate extraction technologies of protein and bioactive compounds from microalgae. More studies are required to determine the best strain, processing techniques, cell lysis and the appropriate food matrix to increase the digestibility and AA bioavailability of microalgae proteins (Lafarga 2019).

5. Sugar Reduction

The worldwide prevalence of obesity nearly tripled between 1975 and 2016 (WHO 2020). Dental caries, type 2 diabetes and cardiovascular diseases have become common, especially among children and adolescents, due to excessive sugar intake. Excessive sugar consumption causes health problems associated with an increased risk of chronic diseases such as non-alcoholic fatty liver, obesity, diabetes, cardiovascular diseases, some types of cancer and cognitive decline (Rippe and Angelopoulos 2016). The WHO recommends limiting free sugar consumption to 10% of total daily energy intake or further decreasing it to 8%. Many countries develop and implement initiatives to reduce the consumption of sugar-rich foods and/or beverages (WHO 2018, Caporizzi et al. 2021).

In September 2021, EFSA's Expert Panel on Nutrition announced a draft opinion for the safe intake of dietary sugars from all sources, following a comprehensive scientific review of over 30,000 publications (EFSA 2021). The report recommended that the intake of added free sugars should be as low as possible and be limited in Europe, as corroborated by scientific evidence. From a public health perspective, the FS must adopt a more cautionary approach regarding the production of high-sugar products (Deliza et al. 2021).

Sugar has many techno-functional properties apart from imparting sweetness to foods, such as adding volume, water retention, reducing water activity and providing crispness. The incorporation of sugar prevents the growth of microorganisms and extends the shelf life of products (McCain et al. 2018, Caporizzi et al. 2021, Deliza et al. 2021). Reducing, replacing or eliminating sugar in manufactured foods is challenging for the food industry. Because such processes are likely to affect foods' overall quality and sensory properties adversely. Consumers generally describe sugar-reduced products as "unpalatable" (WHO 2017, Ilyasoglu and El 2021). Multiple approaches such as partial/ complete sugar substitution, direct/gradual reduction of the amount of sugar in the product, and enhancing sweetness via multisensory interactions exist (Dötsch et al. 2009, Di Monaco et al. 2018).

The direct/progressive reduction of sugar depends on legal restrictions, consumer preference and the recipe to be used (Di Monaco et al. 2018). Gradual sugar reduction is a slow-progressing method; consumers become accustomed to lower sugar concentrations without noticing any change (Oliveira et al. 2016). Bulk agents such as polydextrose, inulin, resistant starch and fructooligosaccharides are often used to compensate for the undesirable structural changes caused by reduced sugar content (Auerbach and Dedman 2012).

The use of artificial sweeteners instead of reducing sugar content could cause metabolic diseases; therefore, the sugar perception thresholds of consumers should be lowered by gradually reducing the sugar content and sweetness intensity, especially in processed foods (Deliza et al. 2021). The study emphasised that policymakers should provide responsive regulatory approaches to encourage the food industry to take action and that food scientists play a crucial role in promoting new approaches.

The use of flavouring agents (such as vanilla or fruit extracts) that are complementary to (or enhancing) the flavour profile of the original product brings in taste-aroma interaction (Stieger and van de Velde 2013, Di Monaco et al. 2018, Caporizzi et al. 2021). Model systems containing sucrose (placed between hydrogel structures and layers of mixed agar/gelatin gel) have also been reported as an effective approach for increasing the perception of sweetness. The results showed that it was possible to reduce the sucrose content by 20% without causing any reduction in flavour intensity (Mosca et al. 2010).

These approaches can bring in some technological difficulties; employing more than one approach in a product may increase its acceptability. Some new methods that can overcome these technological problems are promising. For example, modifying the microstructure of the food can compensate for the loss in the perception of sweet taste due to reduced sugar levels. This relies on the fact that food structure and taste perception during oral digestion are closely related. The reduction of the sugar content of foods can be accomplished by the following strategies:

- Ensuring the inhomogeneous distribution of sugar in the food structure
- Modification of the mechanical properties of food
- Increasing the release of liquid from the food matrix
- Modification of bolus rheology (Stieger and van de Velde 2013, Kuo and Lee 2014, Busch et al. 2013, Ilyasoglu and El 2021).

The inhomogeneous spatial distribution of sucrose is a new approach to increase sweetness. For the non-homogeneous distribution of sucrose in liquid and semi-liquid foods, air bubbles, layered gels that differ in sucrose distribution, and oil-in-water (O/W) or water-in-oil (W/O) emulsion systems have been used (Hutchings et al. 2018, Ilyasoglu and El 2021). Using double or multiple emulsion systems can allow sugar to be dispersed non-homogeneously in the food structure. This is achieved by keeping the sugar content in the product constant and using sugar-free ingredients (e.g., air, oil, etc.) to occupy some volume in the product. This enables the perception of higher sweetness levels by the taste receptors compared to the food with the same total sugar content and the same volume. The oil phase in an emulsion is the portion that contributes to the bulk of the food but does not contain sugar, resulting in a higher sugar concentration in the water phase and a more intense perception. The outermost water in the water-in oil-in water (W/O/W) multiple emulsions interacts with the tongue surface and affects taste perception (flavour and aroma). Therefore, an increase in flavour intensity can be achieved by adding sucrose only to the outer water phase (Ilyasoglu and El 2021).

Inhomogeneous sucrose distribution was also studied in cake formulations, where it was demonstrated that inhomogeneous distribution could allow the reduction of sucrose content without changing the perceived sweetness (Dadali and Elmaci 2021). The study found that the perceived sweetness of the inhomogeneous formulations with reduced sucrose content (15% reduction) was similar to the control sample (no reduction) and higher than the homogeneous formulation at the same reduction rate.

Pruksasri et al. (2020) integrated two processes (nanofiltration (NF) and mechanical pre-fractionation) into conventional juice production to reduce the sugar content of cloudy apple juice. The processes resulted in a 30% reduction in sugar without causing a significant loss in bioactive compounds. It was concluded that the process could be potentially applied on an industrial scale when technological and economical aspects are considered.

6. Conclusion

It has been a challenge to achieve the goals set by the UN for 2030 due to the recent pandemic, ongoing conflicts, climate change and economic instability. To remedy the long withstanding issues of global food insecurity and malnutrition, FSs should adopt a more holistic and coordinated approach. The five action tracks defined at the UN Food System Summit should be implemented in all countries considering the legal regulations and nations' nutritional status and deficiencies.

All stakeholders in the FS need to work together to achieve a comprehensive set of actions that comprises: (1) improving food environments and consumers' food choices, (2) emphasising the inherent link between dietary choices and environmental sustainability, (3) raising consumer awareness by providing information/guidance and via nutritional/public health interventions, (4) enhancing sustainable production foods at a sufficient scale and by using methods to build resilience to difficult conditions and finally, (5) developing more inclusive policies and strategies for the disadvantaged groups working in the agri-food sector.

Incorporating sustainable alternative protein sources into food products, implementing sustainable processing technologies to provide safe and sufficient food supply, and enhancing the nutritional and health-related benefits of foods will be important targets for the FSs of the future.

Abbreviations

Amino acids (AAs)

Black soldier fly larvae (BSFL)

Cardiovascular disease (CVD)

Dietary Guidelines (DGs)

EFSA (European Food Safety Agency)

Essential amino acids (EAAs)

Food-based dietary guidelines (FBDGs)

Food systems (FSs)

Front of pack (FOP)

Generally Recognized as Safe (GRAS)

Healthy reference diet (HRD)

Healthy and sustainable diets (HSDs)

High Hydrostatic Pressure (HHP)

Metabolic food waste (MFW)

Nanofiltration (NF)

Non-communicable diseases (NCDs)

Sustainable development goals (SDSs)

The Food and Agricultural Organization (FAO)

The World Cancer Research Fund (WCRF)

Ultra-processed (UP)

Ultra-processed foods (UPFs)

O/W Oil in water

W/O/W Water-in oil-in water

References

Abu-Ghosh, S., Dubinsky, Z., Verdelho, V. and Iluz, D. 2021. Unconventional high-value products from microalgae: A review. Bioresource Technology, 329. https://doi.org/10.1016/j.biortech.2021.124895.

Alae-Carew, C., Green, R., Stewart, C., Cook, B., Dangour, A.D. and Scheelbeek, P.F.D. 2022. The role of plant-based alternative foods in sustainable and healthy food systems: Consumption trends in the UK. Science of the Total Environment, 807. https://doi.org/10.1016/j.scitotenv.2021.151041.

Alavijeh, R.S., Karimi, K., Wijffels, R.H., van den Berg, C. and Eppink, M. 2020. Combined bead milling and enzymatic hydrolysis for efficient fractionation of lipids, proteins, and carbohydrates of Chlorella vulgaris microalgae. Bioresource Technology, 309: 123321. https://doi.org/10.1016/J.BIORTECH.2020.123321.

Al-Zuhair, S., Ashraf, S., Hisaindee, S., Darmaki, N. Al, Battah, S., … Chaudhary, A. 2017. Enzymatic pre-treatment of microalgae cells for enhanced extraction of proteins. Engineering in Life Sciences, 17(2. https://doi.org/10.1002/elsc.201600127.

Auerbach, M. and Dedman, A.-K. 2012. Bulking agents - multifunctional ingredients. pp. 435–471. *In*: O`Donnell, K. and Kearsley, M. W. (Eds.). Sweeteners and Sugar Alternatives in Food Technology (2nd ed.) Wiley-Blackwell.

Augustin, M.A., Sanguansri, L., Fox, E.M., Cobiac, L. and Cole, M.B. 2020. Recovery of wasted fruit and vegetables for improving sustainable diets. In Trends in Food Science and Technology (Vol. 95). https://doi.org/10.1016/j.tifs.2019.11.010.

Baiano, A. 2020. Edible insects: An overview on nutritional characteristics, safety, farming, production technologies, regulatory framework, and socio-economic and ethical implications. Trends in Food Science & Technology, 100: 35–50. https://doi.org/10.1016/J.TIFS.2020.03.040.

Bakaloudi, D.R., Halloran, A., Rippin, H.L., Oikonomidou, A.C., Dardavesis, T.I., … Chourdakis, M. 2021. Intake and adequacy of the vegan diet. A systematic review of the evidence. Clinical Nutrition, 40(5): 3503–3521. https://doi.org/10.1016/J.CLNU.2020.11.035.

Baker, P., Machado, P., Santos, T., Sievert, K., Backholer, K., … Lawrence, M. 2020. Ultra-processed foods and the nutrition transition: Global, regional and national trends, food systems transformations and political economy drivers. Obesity Reviews, 21(12. https://doi.org/10.1111/OBR.13126.

Barbour, L.R., Woods, J.L. and Brimblecombe, J.K. 2021. Translating evidence into policy action: which diet-related practices are essential to achieve healthy and sustainable food system transformation? Australian and New Zealand Journal of Public Health, 45(1): 83–84. https://doi.org/10.1111/1753-6405.13050.

Barka, A. and Blecker, C. 2016. Microalgae as a potential source of single-cell proteins. A review. Biotechnology, Agronomy and Society and Environment, 20(3). https://doi.org/10.25518/1780-4507.13132.

Bas, A. 2021. Production And Nutritional Evaluation of Biscuits Enriched With Cricket Flour (Acheta domesticus) [MSc thesis]. Ege University, Turkiye.

Becker, M.W., Bello, N.M., Sundar, R.P., Peltier, C. and Bix, L. 2015. Front of pack labels enhance attention to nutrition information in novel & commercial brands. Food Policy, 56: 76–86. https://doi.org/10.1016/J.FOODPOL.2015.08.001.

Béné, C., Oosterveer, P., Lamotte, L., Brouwer, I.D., de Haan, S., … Khoury, C.K. 2019. When food systems meet sustainability – Current narratives and implications for actions. World Development, 113: 116–130. https://doi.org/10.1016/J.WORLDDEV.2018.08.011.

Bessa, L.W., Pieterse, E., Sigge, G. and Hoffman, L.C. 2020. Insects as human food; from farm to fork. Journal of the Science of Food and Agriculture, 100(14). https://doi.org/10.1002/jsfa.8860.

Bhawra, J., Kirkpatrick, S.I., Hall, M.G., Vanderlee, L. and Hammond, D. 2021. Initial development and evaluation of the food processing knowledge (FoodProK) Score: A Functional Test of Nutrition Knowledge Based on Level of Processing. Journal of the Academy of Nutrition and Dietetics, 121(8): 1542–1550. https://doi.org/10.1016/j.jand.2021.01.015.

Blake, C.E., Frongillo, E.A., Warren, A.M., Constantinides, S.V., Rampalli, K.K. and Bhandari, S. 2021. Elaborating the science of food choice for rapidly changing food systems in low-and middle-income countries. Global Food Security, 28: 100503. https://doi.org/10.1016/J.GFS.2021.100503.

Boada, L.D., Henríquez-Hernández, L.A. and Luzardo, O.P. 2016. The impact of red and processed meat consumption on cancer and other health outcomes: Epidemiological evidences. Food and Chemical Toxicology, 92: 236–244. https://doi.org/10.1016/J.FCT.2016.04.008.

Bolat, B., Ugur, A.E., Oztop, M.H. and Alpas, H. 2021. Effects of High Hydrostatic Pressure assisted degreasing on the technological properties of insect powders obtained from Acheta domesticus & Tenebrio molitor. Journal of Food Engineering, 292. https://doi.org/10.1016/j.jfoodeng.2020.110359.

Busch, J.L.H.C., Yong, F.Y.S. and Goh, S.M. 2013. Sodium reduction: Optimizing product composition and structure towards increasing saltiness perception. Trends in Food Science & Technology, 29(1): 21–34. https://doi.org/10.1016/J. TIFS.2012.08.005.

Caporizzi, R., Severini, C. and Derossi, A. 2021. Study of different technological strategies for sugar reduction in muffin addressed for children. NFS Journal, 23(April): 44–51. https://doi.org/10.1016/j.nfs.2021.04.001.

Coleman, P.C., Murphy, L., Nyman, M. and Oyebode, O. 2021. Operationalising the EAT– Lancet Commissions' targets to achieve healthy and sustainable diets. The Lancet Planetary Health, 5(7): e398–e399. https://doi.org/10.1016/S2542-5196(21)00144-3/ATTACHMENT/6B196535-C36D-4F4F-B8EA-2FE420071748/MMC1.PDF.

Contento, I. 2011. Overview of determinants of food choice and dietary change:Implications for nutrition education. In Nutrition Education:Linking research, theory and practice (2nd ed., pp. 26–42). Jones and Bartlett Publishers.

Coutinho, J.G., Martins, A.P.B., Preiss, P.V., Longhi, L. and Recine, E. 2021. UN Food System Summit Fails to Address Real Healthy and Sustainable Diets Challenges. Development, 64(3): 220–226. https://doi.org/10.1057/S41301-021-00315-Y.

Dadalı, C. and Elmacı, Y. 2019. Reduction of sucrose by inhomogeneous distribution in cake formulation. Journal of Food Measurement and Characterization, 13(4): 2563–2570. https://doi.org/10.1007/S11694-019-00176-7.

Daneshvar, E., Sik Ok, Y., Tavakoli, S., Sarkar, B., Shaheen, S.M., ... Bhatnagar, A. 2021. Insights into upstream processing of microalgae: A review. In Bioresource Technology (Vol. 329). https://doi.org/10.1016/j.biortech.2021.124870.

de Boer, J. and Aiking, H. 2022. Do EU consumers think about meat reduction when considering to eat a healthy, sustainable diet and to have a role in food system change? Appetite, 170. https://doi.org/10.1016/j.appet.2021.105880.

Deliza, R., Lima, M.F. and Ares, G. 2021. Rethinking sugar reduction in processed foods. Current Opinion in Food Science, 40: 58–66. https://doi.org/10.1016/j.cofs.2021.01.010.

Di Monaco, R., Miele, N.A., Cabisidan, E.K. and Cavella, S. 2018. Strategies to reduce sugars in food. Current Opinion in Food Science, 19: 92–97. https://doi.org/10.1016/J.COFS.2018.03.008.

Dibb, S. and de Llaguno, E.S. 2018. Principles for eating meat and dairy more sustainably: the 'less and better' approach. http://bit.ly/lessandbettermeat (Accessed 9 Jul 2023).

Dietz, W.H. and Pryor, S. 2022. How Can We Act to Mitigate the Global Syndemic of Obesity, Undernutrition, and Climate Change? Current Obesity Reports, 11: 61–69. https://doi.org/10.1007/S13679-021-00464-8.

Dötsch, M., Busch, J., Batenburg, M., Liem, G., Tareilus, E., ... Meijer, G. 2009. Strategies to reduce sodium consumption: A food industry perspective. Critical Reviews in Food Science and Nutrition, 49(10). https://doi.org/10.1080/10408390903044297.

Downs, S.M., Ahmed, S., Fanzo, J. and Herforth, A. 2020. Food Environment Typology: Advancing an Expanded Definition, Framework, and Methodological Approach for Improved Characterization of Wild, Cultivated, and Built Food Environments toward Sustainable Diets. Foods, 9(4): 1–32. https://doi.org/10.3390/FOODS9040532.

EAT. 2020. Diets for a better future. https://eatforum.org/knowledge/diets-for-a-better-future/(Accessed 9 Jul 2023).

EFSA. 2020. Insights on Novel foods Risk Assessment Outline of the Presentation Introduction Other Trends in Novel Foods - Food Supplements Q & A. https://www.efsa.europa.eu/sites/default/files/event/2020/108th-plenary-meeting-nda-panel-open-observers-presentation.pdf (Accessed 9 Jul 2023).

EFSA. 2021. EFSA explains draft scientific opinion on a tolerable upper intake level for dietary sugars. https://www.efsa.europa.eu/sites/default/files/2021-07/sugars-factsheet-en.pdf (Accessed 9 Jul 2023).

El, S.N. and Simsek, S. 2012. Food technological applications for optimal nutrition: An overview of opportunities for the food industry. Comprehensive Reviews in Food Science and Food Safety, 11(1). https://doi.org/10.1111/j.1541-4337.2011.00167.x .

Etemadian, Y., Ghaemi, V., Shaviklo, A.R., Pourashouri, P., Sadeghi Mahoonak, A.R. and Rafipour, F. 2021. Development of animal/ plant-based protein hydrolysate and its application in food, feed and nutraceutical industries: State of the art. Journal of Cleaner Production, 278(123219): 1–23. https://doi.org/10.1016/j.jclepro.2020.123219.

European Commission. 2015. REGULATION (EU) 2015/2283 OF THE EUROPEAN PARLIAMENT AND OF THE COUNCIL of 25 November 2015 on novel foods (Official Journal L 327/1. https://eur-lex.europa.eu/legal-content/EN/TXT/PDF/?uri=CELEX:32015R2283#:~:text=1.,human%20health%20and%20consumers'%20interests. (Accessed 9 Jul 2023).

Fanzo, J., Haddad, L., Schneider, K.R., Béné, C., Covic, N.M., ... Rosero Moncayo, J. 2021. Viewpoint: Rigorous monitoring is necessary to guide food system transformation in the countdown to the 2030 global goals. Food Policy, 104: 102163. https://doi.org/10.1016/J.FOODPOL.2021.102163.

FAO. 2013. Dietary protein quality evaluation in human nutrition. Report of an FAQ Expert Consultation. FAO Food Nutr Pap., 92: 1–66. https://www.fao.org/ag/humannutrition/35978-02317b979a686a57aa4593304ffc17f06.pdf (Accessed 9 Jul 2023).

FAO. 2017. The future of food and agriculture – Trends and challenges. https://www.fao.org/3/i6583e/i6583e.pdf (Accessed 9 Jul 2023).

FAO, IFAD, UNICEF, WFP and WHO. 2021. The State of Food Security and Nutrition in the World 2021. Transforming food systems for food security, improved nutrition and affordable healthy diets for all. https://www.fao.org/documents/card/en/c/cb4474en (Accessed 9 Jul 2023).

Fasolin, L.H., Pereira, R.N., Pinheiro, A.C., Martins, J.T., Andrade, C.C.P., ... Vicente, A.A. 2019. Emergent food proteins – Towards sustainability, health and innovation. Food Research International, 125: 108586. https://doi.org/10.1016/J.FOODRES.2019.108586.

Fischer, C.G. and Garnett, T. 2016. Plates, pyramids, planet Developments in national healthy and sustainable dietary guidelines: a state of play assessment. https://www.fao.org/sustainable-food-value-chains/library/details/en/c/415611/ (Accessed 9 Jul 2023).

Global Nutrition Report. 2021. The State of Global Nutrition. https://globalnutritionreport.org/reports/2021-global-nutrition-report/(Accessed 9 Jul 2023).

Global Panel on Agriculture and Food Systems for Nutrition. 2020. Future Food Systems: For people, our planet, and prosperity. https://www.glopan.org/wp-content/uploads/2020/09/Foresight-2.0_Future-Food-Systems_For-people-our-planet-and-prosperity.pdf (Accessed 9 Jul 2023).

González, C.M., Garzón, R. and Rosell, C.M. 2019. Insects as ingredients for bakery goods. A comparison study of H. illucens, A. domestica and T. molitor flours. Innovative Food Science and Emerging Technologies, 51: 205–210. https://doi.org/10.1016/j.ifset.2018.03.021.

Guzmán, F., Wong, G., Román, T., Cárdenas, C., Alvárez, C., ... Rojas, V. 2019. Identification of Antimicrobial Peptides from the Microalgae Tetraselmis suecica (Kylin) Butcher and Bactericidal Activity Improvement. Marine Drugs, 17(8). https://doi.org/10.3390/md17080453.

Händel, M.N., Rohde, J.F., Jacobsen, R., Nielsen, S.M., Christensen, R., ... Heitmann, B.L. 2020. Processed meat intake and incidence of colorectal cancer: a systematic review and meta-analysis of prospective observational studies. European Journal of Clinical Nutrition, 74(8): 1132–1148. https://doi.org/10.1038/s41430-020-0576-9.

Hawkes, C., Jewell, J. and Allen, K. 2013. A food policy package for healthy diets and the prevention of obesity and diet-related non-communicable diseases: the NOURISHING framework. Obesity Reviews, 14(S2): 159–168. https://doi.org/10.1111/OBR.12098.

Herforth, A., Arimond, M., Álvarez-Sánchez, C., Coates, J., Christianson, K. and Muehlhoff, E. 2019. A Global Review of Food-Based Dietary Guidelines. Advances in Nutrition (Bethesda, Md.): 10(4): 590–605. https://doi.org/10.1093/ADVANCES/NMY130.

Herforth, A., Bai, Y., Venkat, A., Mahrt, K., Ebel, A. and Masters, W.A. 2020. Cost and affordability of healthy diets across and within countries Background paper for The State of Food Security and Nutrition in the World 2020. https://doi.org/10.4060/cb2431en.

Hirvonen, K., Bai, Y., Headey, D. and Masters, W.A. 2020. Affordability of the EAT-Lancet reference diet: a global analysis. The Lancet. Global Health, 8(1): e59–e66. https://doi.org/10.1016/S2214-109X(19)30447-4.

HLPE. 2017. Nutrition and food systems. A report by the High Level Panel of Experts on Food Security and Nutrition of the Committee on World Food Security. https://www.fao.org/3/i7846e/I7846E.pdf (Accessed 9 Jul 2023).

Huang, Y., Cao, D., Chen, Z., Chen, B., Li, J., ... Wei, Q. 2021. Red and processed meat consumption and cancer outcomes: Umbrella review. Food Chemistry, 356: 129697. https://doi.org/10.1016/J.FOODCHEM.2021.129697.

Hutchings, S.C., Low, J.Y. Q. and Keast, R.S.J. 2019. Sugar reduction without compromising sensory perception. An impossible dream? Critical Reviews in Food Science and Nutrition, 59(14): 2287–2307. https://doi.org/10.1080/10408398.2018.1450214.

IFST. 2017. Food System Framework - a Focus on Sustainability. https://www.ifst.org/sites/default/files/IFST Sustainable Food System Framework.pdf (Accessed 9 Jul 2023).

Ilyasoglu Buyukkestelli, H. and El, S.N. 2021. Enhancing sweetness using double emulsion technology to reduce sugar content in food formulations. Innovative Food Science and Emerging Technologies, 74: 102809. https://doi.org/10.1016/j.ifset.2021.102809.

Imathiu, S. 2020. Benefits and food safety concerns associated with consumption of edible insects. NFS Journal, 18. https://doi.org/10.1016/j.nfs.2019.11.002.

IPES-Food. 2017. Unravelling the food-health nexxus. Addressing prectices, political economy and power relations to build healthier food systems. https://www.ipes-food.org/_img/upload/files/Health_FullReport(1).pdf (Accessed 9 Jul 2023).

Jagtap, S., Saxena, P. and Salonitis, K. 2021. Food 4.0: Implementation of the Augmented Reality Systems in the Food Industry. Procedia CIRP, 104, 1137–1142. https://doi.org/10.1016/j.procir.2021.11.191.

Karmaus, A.L. and Jones, W. 2021. Future foods symposium on alternative proteins: Workshop proceedings. Trends in Food Science and Technology, 107: 124–129. https://doi.org/10.1016/j.tifs.2020.06.018.

Khawaja, A.H., Qassim, S., Hassan, N.A. and Arafa, E.S.A. 2019. Added sugar: Nutritional knowledge and consumption pattern of a principal driver of obesity and diabetes among undergraduates in UAE. Diabetes & Metabolic Syndrome, 13(4): 2579–2584. https://doi.org/10.1016/J.DSX.2019.06.031.

Kim, T.-K., Yong, H.I., Kim, Y.-B., Kim, H.-W. and Choi, Y.-S. 2019. Edible Insects as a Protein Source: A Review of Public Perception, Processing Technology, and Research Trends. Food Science of Animal Resources, 39(4): 521–540. https://doi.org/10.5851/kosfa.2019.e53.

Knorr, D. and Augustin, M.A. 2021. From value chains to food webs: The quest for lasting food systems. Trends in Food Science and Technology, 110: 812–821. https://doi.org/10.1016/j.tifs.2021.02.037.

Köhler, R., Kariuki, L., Lambert, C. and Biesalski, H.K. 2019. Protein, amino acid and mineral composition of some edible insects from Thailand. Journal of Asia-Pacific Entomology, 22(1): 372–378. https://doi.org/10.1016/j.aspen.2019.02.002.

Kuo, W.Y. and Lee, Y. 2014. Effect of Food Matrix on Saltiness Perception-Implications for Sodium Reduction. Comprehensive Reviews in Food Science and Food Safety, 13(5): 906–923. https://doi.org/10.1111/1541-4337.12094.

Kusmayadi, A., Leong, Y.K., Yen, H.-W., Huang, C.-Y. and Chang, J.-S. 2021. Microalgae as sustainable food and feed sources for animals and humans - Biotechnological and environmental aspects. Chemosphere, 271: 129800. https://doi.org/10.1016/j.chemosphere.2021.129800.

Lafarga, T., Acién-Fernández, F.G., Castellari, M., Villaró, S., Bobo, G. and Aguiló-Aguayo, I. 2019. Effect of microalgae incorporation on the physicochemical, nutritional, and sensorial properties of an innovative broccoli soup. LWT, 111: 167–174. https://doi.org/10.1016/j.lwt.2019.05.037.

Lai, Y.C., Chang, C.H., Chen, C.Y., Chang, J.S. and Ng, I.S. 2019. Towards protein production and application by using Chlorella species as circular economy. Bioresource Technology, 289: 121625. https://doi.org/10.1016/j.biortech.2019.121625.

Latunde-Dada, G.O., Yang, W. and Vera Aviles, M. 2016. In Vitro Iron Availability from Insects and Sirloin Beef. Journal of Agricultural and Food Chemistry, 64(44): 8420–8424. https://doi.org/10.1021/acs.jafc.6b03286.

Leng, G., Adan, R.A.H., Belot, M., Brunstrom, J.M., De Graaf, K., ... Smeets, P.A.M. 2017. The determinants of food choice. Proceedings of the Nutrition Society, 76(3): 316–327. https://doi.org/10.1017/S002966511600286X.

Lin, Y.H., Chen, G.W., Yeh, C.H., Song, H. and Tsai, J.S. 2018. Purification and identification of angiotensin I-Converting enzyme inhibitory peptides and the antihypertensive effect of chlorella sorokiniana protein hydrolysates. Nutrients, 10(10): 1397. https://doi.org/10.3390/nu10101397.

McCain, H.R., Kaliappan, S. and Drake, M.A. 2018. Invited review: Sugar reduction in dairy products. Journal of Dairy Science, 101(10): 8619–8640. https://doi.org/10.3168/jds.2017-14347.

Meijer, G.W., Detzel, P., Grunert, K.G., Robert, M.C. and Stancu, V. 2021. Towards effective labelling of foods. An international perspective on safety and nutrition. Trends in Food Science and Technology, 118: 45–56. https://doi.org/10.1016/J.TIFS.2021.09.003.

Mekonnen, T.A., Odden, M.C., Coxson, P.G., Guzman, D., Lightwood, J., ... Bibbins-Domingo, K. 2013. Health benefits of reducing sugar-sweetened beverage intake in high risk populations of California: results from the cardiovascular disease (CVD) policy model. PloS One, 8(12): e81723. https://doi.org/10.1371/JOURNAL.PONE.0081723.

Miller, K.B., Eckberg, J.O., Decker, E.A. and Marinangeli, C.P.F. 2021. Role of Food Industry in Promoting Healthy and Sustainable Diets. Nutrients, 13(8): 2740. https://doi.org/10.3390/NU13082740.

Monteiro, C.A., Cannon, G., Lawrence, M., Costa Louzada, M.L. and Machado, P.P. 2019. The NOVA food classification system and its four food groups. Ultra-Processed Foods, Diet Quality, and Health Using the NOVA Classification System. http://www.wipo.int/amc/en/mediation/rules (Accessed 9 Jul 2023).

Monteiro, C.A., Moubarac, J.C., Cannon, G., Ng, S.W. and Popkin, B. 2013. Ultra-processed products are becoming dominant in the global food system. Obesity Reviews, 14(S2): 21–28. https://doi.org/10.1111/OBR.12107.

Montowska, M., Kowalczewski, P.Ł., Rybicka, I. and Fornal, E. 2019. Nutritional value, protein and peptide composition of edible cricket powders. Food Chemistry, 289: 130–138. https://doi.org/10.1016/j.foodchem.2019.03.062.

Moreira, M.N.B., da Veiga, C.P., da Veiga, C.R.P., Reis, G.G. and Pascuci, L.M. 2022. Reducing meat consumption: Insights from a bibliometric analysis and future scopes. Future Foods, 5: 100120. https://doi.org/10.1016/J.FUFO.2022.100120.

Mosca, A.C., Velde, F. van de, Bult, J.H.F., van Boekel, M.A.J.S. and Stieger, M. 2010. Enhancement of sweetness intensity in gels by inhomogeneous distribution of sucrose. Food Quality and Preference, 21(7): 837–842. https://doi.org/10.1016/j.foodqual.2010.04.010.

Mozaffarian, D., Angell, S.Y., Lang, T. and Rivera, J.A. 2018. Science and Politics of Nutrition: Role of government policy in nutrition—barriers to and opportunities for healthier eating. The BMJ, 361: k2426. https://doi.org/10.1136/BMJ.K2426.

Mshayisa, V.V., Van Wyk, J., Zozo, B. and Rodríguez, S.D. 2021. Structural properties of native and conjugated black soldier fly (Hermetia illucens) larvae protein via Maillard reaction and classification by SIMCA. Heliyon, 7(6): e07242. https://doi.org/10.1016/j.heliyon.2021.e07242.

Neufeld, L.M., Hendriks, S. and Hugas, M. 2021. The Scientific Group for the UN Food Systems Summit Healthy diet: A definition for the United Nations Food Systems Summit 2021. https://sc-fss2021.org/ (Accessed 9 Jul 2023).

Nitsos, C., Filali, R., Taidi, B. and Lemaire, J. 2020. Current and novel approaches to downstream processing of microalgae: A review. Biotechnology Advances, 45: 107650. https://doi.org/10.1016/j.biotechadv.2020.107650.

Ogundijo, D.A., Tas, A.A. and Onarinde, B.A. 2022. The perceptions of food manufacturers and retailers in England on nutrition labels and their use in consumers' purchasing decisions. University of Lincoln PGR Showcase Conference.

Olasehinde, T.A., Odjadjare, E.C., Mabinya, L.V., Olaniran, A.O. and Okoh, A.I. 2019. Chlorella sorokiniana and Chlorella minutissima exhibit antioxidant potentials, inhibit cholinesterases and modulate disaggregation of β-amyloid fibrils. Electronic Journal of Biotechnology, 40: 1–9. https://doi.org/10.1016/j.ejbt.2019.03.008.

Oliveira, D., Reis, F., Deliza, R., Rosenthal, A., Giménez, A. and Ares, G. 2016. Difference thresholds for added sugar in chocolate-flavoured milk: Recommendations for gradual sugar reduction. Food Research International, 89: 448–453. https://doi.org/10.1016/j.foodres.2016.08.019.

Orlich, M.J., Singh, P.N., Sabaté, J., Jaceldo-Siegl, K., Fan, J., … Fraser, G.E. 2013. Vegetarian dietary patterns and mortality in Adventist Health Study 2. JAMA Internal Medicine, 173(13): 1230–1238. https://doi.org/10.1001/JAMAINTERNMED.2013.6473.

Oswald, C., Adhikari, K. and Mohan, A. 2022. Effect of front-of-package labels on consumer product evaluation and preferences. Current Research in Food Science, 5: 131–140. https://doi.org/10.1016/J.CRFS.2021.12.016.

Petruk, G., Gifuni, I., Illiano, A., Roxo, M., Pinto, G., … Monti, D.M. 2018. Simultaneous production of antioxidants and starch from the microalga Chlorella sorokiniana. Algal Research, 34: 164–174. https://doi.org/10.1016/j.algal.2018.07.012.

Popkin, B.M., Barquera, S., Corvalan, C., Hofman, K.J., Monteiro, C., … Taillie, L.S. 2021. Towards unified and impactful policies to reduce ultra-processed food consumption and promote healthier eating. The Lancet. Diabetes & Endocrinology, 9(7): 462–470. https://doi.org/10.1016/S2213-8587(21)00078-4.

Pruksasri, S., Lanner, B. and Novalin, S. 2020. Nanofiltration as a potential process for the reduction of sugar in apple juices on an industrial scale. LWT, 133: 110118. https://doi.org/10.1016/j.lwt.2020.110118.

Reyes, L.I., Constantinides, S.V., Bhandari, S., Frongillo, E.A., Schreinemachers, P., … Blake, C.E. 2021. Actions in global nutrition initiatives to promote sustainable healthy diets. Global Food Security, 31: 100585. https://doi.org/10.1016/j.gfs.2021.100585.

Ridoutt, B.G., Baird, D. and Hendrie, G.A. 2021. Diets within planetary boundaries: What is the potential of dietary change alone? Sustainable Production and Consumption, 28: 802–810. https://doi.org/10.1016/J.SPC.2021.07.009.

Rippe, J.M. and Angelopoulos, T.J. 2016. Added sugars and risk factors for obesity, diabetes and heart disease. International Journal of Obesity, 40(S1): S22–7. https://doi.org/10.1038/ijo.2016.10.

Rose, D., Heller, M.C. and Roberto, C.A. 2019. Position of the Society for Nutrition Education and Behavior: The Importance of Including Environmental Sustainability in Dietary Guidance. Journal of Nutrition Education and Behavior, 51(1): 3–15.e1. https://doi.org/10.1016/J.JNEB.2018.07.006.

Rosenthal, A., Maciel Guedes, A.M., dos Santos, K.M.O. and Deliza, R. 2021. Healthy food innovation in sustainable food system 4.0: integration of entrepreneurship, research, and education. Current Opinion in Food Science, 42: 215–223. https://doi.org/10.1016/j.cofs.2021.07.002.

Saadoun, H.J., Sogari, G., Bernini, V., Camorali, C., Rossi, F., … Lazzi, C. 2022. A critical review of intrinsic and extrinsic antimicrobial properties of insects. Trends in Food Science and Technology, 122: 40–48. https://doi.org/10.1016/j.tifs.2022.02.018.

Sanchez-Sabate, R. and Sabaté, J. 2019. Consumer Attitudes Towards Environmental Concerns of Meat Consumption: A Systematic Review. International Journal of Environmental Research and Public Health, 16(7): 1220. https://doi.org/10.3390/IJERPH16071220.

Sathasivam, R., Radhakrishnan, R., Hashem, A. and Abd_Allah, E.F. 2019. Microalgae metabolites: A rich source for food and medicine. Saudi Journal of Biological Sciences, 26(4): 709–722. https://doi.org/10.1016/j.sjbs.2017.11.003.

Satija, A., Bhupathiraju, S.N., Spiegelman, D., Chiuve, S.E., Manson, J.A.E., … Hu, F.B. 2017. Healthful and Unhealthful Plant-Based Diets and the Risk of Coronary Heart Disease in U.S. Adults. Journal of the American College of Cardiology, 70(4): 411–422. https://doi.org/10.1016/J.JACC.2017.05.047.

Schwingshackl, L., Schwedhelm, C., Hoffmann, G., Lampousi, A.M., Knüppel, S., … Boeing, H. 2017. Food groups and risk of all-cause mortality: a systematic review and meta-analysis of prospective studies. The American Journal of Clinical Nutrition, 105(6): 1462–1473. https://doi.org/10.3945/AJCN.117.153148.

Schwingshackl, L., Watzl, B. and Meerpohl, J.J. 2020. The healthiness and sustainability of food based dietary guidelines. BMJ, 370, m2322. https://doi.org/10.1136/BMJ.M2417.

Serafini, M. and Toti, E. 2016. Unsustainability of Obesity: Metabolic Food Waste. Frontiers in Nutrition, 3, 40. https://doi.org/10.3389/FNUT.2016.00040/BIBTEX.

Singh, S., Verma, D.K., Thakur, M., Tripathy, S., Patel, A.R., … Aguilar, C.N. 2021. Supercritical fluid extraction (SCFE) as green extraction technology for high-value metabolites of algae, its potential trends in food and human health. Food Research International, 150: 110746. https://doi.org/10.1016/j.foodres.2021.110746.

Springmann, M., Spajic, L., Clark, M.A., Poore, J., Herforth, A., … Scarborough, P. 2020. The healthiness and sustainability of national and global food based dietary guidelines: Modelling study. The BMJ, 370, m2322. https://doi.org/10.1136/bmj.m2322.

Stanhope, K.L. 2016. Sugar consumption, metabolic disease and obesity: The state of the controversy. Critical Reviews in Clinical Laboratory Sciences, 53(1): 52–67. https://doi.org/10.3109/10408363.2015.1084990.

Stieger, M. and Van de Velde, F. 2013. Microstructure, texture and oral processing: New ways to reduce sugar and salt in foods. Current Opinion in Colloid and Interface Science, 18(4): 334–348. https://doi.org/10.1016/j.cocis.2013.04.007.

Stull, V.J., Finer, E., Bergmans, R.S., Febvre, H.P., Longhurst, C., … Weir, T.L. 2018. Impact of Edible Cricket Consumption on Gut Microbiota in Healthy Adults, a Double-blind, Randomized Crossover Trial. Scientific Reports, 8(1): 10762. https://doi.org/10.1038/s41598-018-29032-2.

Talati, Z., Pettigrew, S., Kelly, B., Ball, K., Dixon, H. and Shilton, T. 2016. Consumers' responses to front-of-pack labels that vary by interpretive content. Appetite, 101: 205–213. https://doi.org/10.1016/J.APPET.2016.03.009.

Tejano, L.A., Peralta, J.P., Yap, E.E.S. and Chang, Y.W. 2019. Bioactivities of enzymatic protein hydrolysates derived from Chlorella sorokiniana. Food Science and Nutrition, 7(7): 2381–2390. https://doi.org/10.1002/fsn3.1097.

Terriente-Palacios, C. and Castellari, M. 2022. Levels of taurine, hypotaurine and homotaurine, and amino acids profiles in selected commercial seaweeds, microalgae, and algae-enriched food products. Food Chemistry, 368: 130770. https://doi.org/10.1016/j.foodchem.2021.130770.

Torres-Tiji, Y., Fields, F.J. and Mayfield, S.P. 2020. Microalgae as a future food source. In Biotechnology Advances (Vol. 41, p. 107536). https://doi.org/10.1016/j.biotechadv.2020.107536.

Toti, E., Di Mattia, C. and Serafini, M. 2019. Metabolic Food Waste and Ecological Impact of Obesity in FAO World's Region. Frontiers in Nutrition, 6: 126. https://doi.org/10.3389/FNUT.2019.00126/BIBTEX.

Turck, D., Bohn, T., Castenmiller, J., De Henauw, S., Hirsch-Ernst, K. I., ... Knutsen, H.K. 2021. Safety of frozen and dried formulations from whole house crickets (Acheta domesticus) as a Novel food pursuant to Regulation (EU) 2015/2283. EFSA Journal, 19(8): 6779. https://doi.org/10.2903/j.efsa.2021.6779.

Turck, D., Castenmiller, J., De Henauw, S., Hirsch-Ernst, K.I., Kearney, J., ... Knutsen, H.K. 2020. Safety of dried whole cell Euglena gracilis as a novel food pursuant to Regulation (EU) 2015/2283. EFSA Journal, 18(5): 6100. https://doi.org/10.2903/j.efsa.2020.6100.

Turck, D., Castenmiller, J., De Henauw, S., Hirsch-Ernst, K.I., Kearney, J., ... Knutsen, H.K. 2021. Safety of dried yellow mealworm (*Tenebrio molitor larva*) as a novel food pursuant to Regulation (EU) 2015/2283. EFSA Journal, 19(1): 6343. https://doi.org/10.2903/j.efsa.2021.6343.

Turck, D., Castenmiller, J., De Henauw, S., Hirsch-Ernst, K.I., Kearney, J., ... Knutsen, H.K. 2021. Safety of frozen and dried formulations from migratory locust (*Locusta migratoria*) as a Novel food pursuant to Regulation (EU) 2015/2283. EFSA Journal, 19(7). https://doi.org/10.2903/j.efsa.2021.6667.

UN Department of Economic and Social Affairs. 2017. World Population Prospects: The 2017 Revision, Key Findings and Advance Tables. Working Paper No. ESA/P/WP/248. https://population.un.org/wpp/publications/files/wpp2017_keyfindings.pdf (Accessed 9 Jul 2023).

UN World Food Programme. 2020. The cost of a plate of food. https://cdn.wfp.org/2020/plate-of-food/ (Accessed 9 Jul 2023).

UNEP. 2016. Food Systems and Natural Resources. A Report of the Working Group on Food Systems of the International Resource Pane. https://www.resourcepanel.org/file/133/download?token=6dSyNtuV (Accessed 9 Jul 2023).

UNICEF. 2020. Review of National Food-based Dietary Guidelines and associated guidance for children, adolescents and women. https://www.unicef.org/media/102761/file/2021-Food-based-Dietary-Guidelines-final.pdf (Accessed 9 Jul 2023).

UNSCN. 2019. Food environments: Where people meet the food system. https://www.unscn.org/uploads/web/news/UNSCN-Nutrition44-WEB.pdf (Accessed 9 Jul 2023).

US Department of Health and Human Services and USDA. 2015. Dietary Guidelines for Americans 2015-2020 (8th Edition. http://health.gov/dietaryguidelines/2015/guidelines/ (Accessed 9 Jul 2023).

van der Spiegel, M., Noordam, M. Y. and van der Fels-Klerx, H. J. 2013. Safety of novel protein sources (insects, microalgae, seaweed, duckweed, and rapeseed) and legislative aspects for their application in food and feed production. Comprehensive Reviews in Food Science and Food Safety, 12(6): 662–678. https://doi.org/10.1111/1541-4337.12032

Van Huis, A. 2013. Potential of insects as food and feed in assuring food security. Annual Review of Entomology, 58, 563–583. https://doi.org/10.1146/annurev-ento-120811-153704

Veettil, S.K., Wong, T.Y., Loo, Y.S., Playdon, M.C., Lai, N.M., ... Chaiyakunapruk, N. 2021. Role of Diet in Colorectal Cancer Incidence: Umbrella Review of Meta-analyses of Prospective Observational Studies. JAMA Network Open, 4(2): e2037341. https://doi.org/10.1001/JAMANETWORKOPEN.2020.37341.

Ververis, E., Ackerl, R., Azzollini, D., Colombo, P.A., de Sesmaisons, A., ... Gelbmann, W. 2020. Novel foods in the European Union: Scientific requirements and challenges of the risk assessment process by the European Food Safety Authority. Food Research International, 137: 109515. https://doi.org/10.1016/j.foodres.2020.109515.

Vesanto, M., Winston, C. and Susan, L. 2016. Position of the Academy of Nutrition and Dietetics: Vegetarian Diets. Journal of the Academy of Nutrition and Dietetics, 116(12): 1970–1980. https://doi.org/10.1016/J.JAND.2016.09.025.

von Braun, J., Afsana, K., Fresco, L., Hassan, M. and Torero, M. 2021. Food Systems – Definition, Concept and Application for the UN Food Systems Summit. https://www.un.org/sites/un2.un.org/files/scgroup_food_systems_paper_march-5-2021.pdf (Accessed 9 Jul 2023).

Wang, K., Luo, Q., Hong, H., Liu, H. and Luo, Y. 2021. Novel antioxidant and ACE inhibitory peptide identified from Arthrospira platensis protein and stability against thermal/pH treatments and simulated gastrointestinal digestion. Food Research International, 139: 109908. https://doi.org/10.1016/j.foodres.2020.109908.

Watling, C.Z., Schmidt, J.A., Dunneram, Y., Tong, T.Y.N., Kelly, R.K., ... Perez-Cornago, A. 2022. Risk of cancer in regular and low meat-eaters, fish-eaters, and vegetarians: a prospective analysis of UK Biobank participants. BMC Medicine 2022 20:1, 20(1): 1–13. https://doi.org/10.1186/S12916-022-02256-W.

WCRF. 2018. Recommendations and public health policy implications. https://www.wcrf.org/wp-content/uploads/2021/01/Recommendations.pdf (Accessed 9 Jul 2023).

WHO. 2007. Protein and Amino Acid Requirements in Human Nutrition: Report of a Joint WHO/FAO/UNU Expert Consultation. https://apps.who.int/iris/bitstream/handle/10665/43411/WHO_TRS_935_eng.pdf?sequence=1&isAllowed=y (Accessed 9 Jul 2023).

WHO. 2017. Incentives and disincentives for reducing sugar in manufactured foods: an exploratory supply chain analysis: a set of insights for Member States in the context of the WHO European Food and Nutrition Action Plan 2015–2020. https://apps.who.int/iris/bitstream/handle/10665/345828/WHO-EURO-2017-3404-43163-60439-eng.pdf?sequence=1&isAllowed=y (Accessed 9 Jul 2023).

WHO. 2018. Guideline: Sugars Intake for Adults and Children. https://www.who.int/publications/i/item/9789241549028 (Accessed 9 Jul 2023).

WHO. 2020. Obesity and Overweight. https://www.who.int/publications/i/item/9789241549028 (Accessed 9 Jul 2023).

WHO-EUROPE. 2015. Using price policies to promote healthier diets. https://www.euro.who.int/__data/assets/pdf_file/0008/273662/Using-price-policies-to-promote-healthier-diets.pdf (Accessed 9 Jul 2023).

WHO-EUROPE. 2021. Plant-Based Diets and Their Impact on Health, Sustainability and the Environment: A Review of the Evidence. https://apps.who.int/iris/bitstream/handle/10665/349086/WHO-EURO-2021-4007-43766-61591-eng.pdf?sequence=1&isAllowed=y (Accessed 9 Jul 2023).

Willett, W., Rockström, J., Loken, B., Springmann, M., Lang, T., … Murray, C.J.L. 2019. Food in the Anthropocene: the EAT–Lancet Commission on healthy diets from sustainable food systems. The Lancet, 393(10170): 447–492. https://doi.org/10.1016/S0140-6736(18)31788-4.

Wu, G. 2016. Dietary protein intake and human health. Food & Function, 7: 1251. https://doi.org/10.1039/c5fo01530h.

Zhang, C., Guan, X., Yu, S., Zhou, J. and Chen, J. 2022. Production of meat alternatives using live cells, cultures and plant proteins. Current Opinion in Food Science, 43: 43–52. https://doi.org/10.1016/J.COFS.2021.11.002.

2

Food Sustainability Index

Rosangela Rodrigues Dias,* Rafaela Basso Sartori and
Adriane Terezinha Schneider

1. Introduction

There are several definitions that describe a sustainable food system. From a general perspective, it can be defined as a food system that provides people with perfectly healthy and nutritious food, in a way that economic, social, and environmental bases are not compromised for future generations. This signifies presenting economic sustainability, social sustainability, and environmental sustainability. It is clear that making food systems sustainable, fair, and inclusive is not an easy task and significant paces need to be taken to prepare them for the future (Burey et al. 2022).

Although some actions to transform these systems may be succeeding, it is noteworthy that the global food system is currently not sustainable with respect to environmental, social, and economic elements. Food production has profound impacts on the environment. It consumes a large amount of natural resources that contribute to biodiversity loss and climate change. The consequences of lack of access to healthy and nutritious food, in turn, are casting a shadow on the future of entire communities and countries; obesity, for example, has become a public health problem with negative effects on economic prosperity. These are some of the substantive issues that make the current food system unsustainable, and although the change is not simple, the benefits are clear (EC 2020).

In this sense, with the aim of promoting knowledge about food sustainability, the Food Sustainability Index (FSI) was developed recently. FSI is a collaboration between the Economist Intelligence Unit and the Barilla Center for Food and Nutrition. This is a global study that collects data from dozens of countries to highlight holistic solutions and best practices for food sustainability. Sustainable agriculture, nutritional challenges, and food loss and waste are the three pillars used to assess the performance of food systems. The performance of these systems is assessed through indicators and sub-indicators that address environmental, social, and economic aspects (BCFN 2022).

Some of these indicators and sub-indicators, in the case of the sustainable agriculture pillar, refer to land use, water management, and biodiversity, and how much food loss and waste refer to strategies, targets, and legislation for reducing food waste. To assess the pillar of nutritional challenges criteria such as dietary patterns, physical activity level, and overweight prevalence were analyzed.

Department of Food Science and Technology, Federal University of Santa Maria (UFSM), Roraima Avenue, 1000, 97105-900, Santa Maria, RS, Brazil.
* Corresponding author: ro.rosangelard@gmail.com

Without a doubt, the FSI provides a tool that can shed light on the path to a sustainable food system. Furthermore, it highlights some of the main challenges that need to be overcome such as access to food, promotion of nutritious diets, and an increasingly responsible food supply chain (Birdwell et al. 2021).

In this chapter, the central objective is to provide a landscape of the food sustainability index and the pillars that govern a food system capable of meeting the demands for nutritious, fair, and ecologically correct food. What emerges from this chapter is the view that both pillars are interconnected and require holistic solutions and the joint collaboration of all sectors and actors in society.

2. Food Sustainability Index (FSI)

The Food Sustainability Index (FSI) is a global study that examines food sustainability through a multidisciplinary approach. The three pillars included in the FSI are discussed in detail below.

2.1 Sustainable Agriculture

Discussions about the environmental and social impacts of conventional agriculture, in the mid-1980s, joined global environmental issues (destruction of forests, acid rain, environmental accidents, greenhouse effect), leaving the agronomic environment and institutions and reaching the consumers. Concerned about the quality of the products they are ingesting and the environmental damage caused by the conventional agricultural model, consumers started to interfere in the production system, through the demand for healthy products, which were produced respecting the environment and the health of workers. So the term "sustainable agriculture" came up. In this context, the Brundtland Report was essential for the concept of sustainability, previously restricted to other branches of the economy, to be extended to agriculture. Also entitled "Our Common Future", it was created in October 1987 by the World Commission on Environment and Development (WCED) and points to the incompatibility between sustainable development and current production and consumption patterns (UN 1987, Purvis et al. 2019).

In general, the concept of sustainable agriculture can be described as a system of agricultural practices, based on scientific innovations through which it is possible to produce healthy food with respect to land, air, and water, as well as the health and rights of farmers. The goal of sustainable agriculture is to satisfy humanity's need for healthy food, improve the quality of the environment, maintain the natural resource base, use non-renewable and agricultural resources more effectively, implement natural biological cycles, and sustainably support the rural economy, development, and quality of life of farmers (FAO 1988). Therefore, sustainability is much more than ensuring the protection of the natural resource base. To be sustainable, agriculture must meet the needs of present and future generations for its products and services, ensuring profitability, the environment, health, and social and economic equity (EOS 2019).

The various approaches to agricultural sustainability must take into account a number of factors, constraints, and opportunities that are determined by the availability of agricultural resources, as well as meeting the needs of individuals in communities. The configuration of agriculture and the means to stimulate sustainability processes will require continuous adjustment, innovation, and improvement in strategies, policies, and technologies in order to maximize productivity and consequently achieve a balance between social, economic, and environmental aspects. All these processes reflect the evolution of society's values and accumulated knowledge, which have a major impact on how sustainability goals are defined in practice. Within this complex system, specific constraints and natural and socioeconomic boundaries will define what fits into the sustainable operational space (Holling 2000, Rockström et al. 2009, Movilla-Pateiro et al. 2019). A growing challenge for sustainability today is identifying and balancing interactions, benefits, and trade-offs that result from different agricultural configurations. Furthermore, the institutions that govern agricultural production (which determine production, type of technologies and practices, and level of returns obtained) are fundamental levers for regulating the type and distribution of products and services that can be derived from agriculture (FAO 2019).

Legal and institutional responses to the scope of agricultural sustainability have so far been quite limited. For example, no instruments legally linked to combating drought and desertification have apparently been developed (UN 2018). Instead, Building Instruments and Soft Law Instruments have contributed to the development of rights and innovative means of conflict resolution. Greenhouse gas emissions, through the 2015 Paris Agreement, the water footprint through the Global Food Security Index (GFSI), and the rights of workers linked to the United Nations (UN), are also widely discussed in international agreements and treaties.

In any case, agriculture and food are at the heart of Agenda 2030 not only for their systemic nature but also because it is a fundamental tool to achieve the Sustainable Development Goals (SDG). According to studies carried out by Caron et al. (2018), the question of why the combination of food and agriculture systems are essential to advancing the 2030 Agenda can only be answered considering several factors in common: (i) food shortages and rural poverty are causes most of the political problems—instability, conflict, violence, and migration; (ii) the agriculture sector is one of the most responsible for the use of existing energy resources; (iii) agricultural practices are closely linked to the environment, health, and climate change; and (iv) the agriculture sector is the main source of income for most of the poor and vulnerable people. According to Food and Agriculture Organization (FAO 2019), agriculture is closely linked to poverty reduction, where more than 30% of the world's workers are employed in the agricultural and rural sectors.

Given the complexity of global food and agriculture supply chains, a collaboration between these two sectors is key to success in promoting sustainable best practices. For example, through joint actions, governments, industry, non-governmental organizations, and research centers can achieve concrete measures to eradicate hunger while preserving the environment (UN 2019). While companies pursue sustainability strategies through external partners (e.g., Carrefour with the World Cocoa Foundation and the International Federation for Human Rights), support for small farmers and traditional agricultural practices, including coverage cutting, different rotation patterns, and no-till, can improve soil quality, retain water, organic matter and achieve food yields that provide balanced nutrition and market increase. In this case, Danone, for example, provides a Livelihood Fund for families in the agricultural sector and invests in low-carbon agriculture projects to produce higher quality products through traditional practices. TechnoServe uses existing no-till approaches to promote better yields and prevent land depletion and deforestation, while many governments are currently providing support to low-income farmers through training programs (Fixing Food 2018).

On the other hand, technology is also enabling significant progress and precision farming and new digital tools can help, increasing the efficiency and sustainability of farming while improving yields (Beluhova-Uzunova and Dunchev 2019). Initiatives such as Climate Smart Agriculture focused on farmer resilience, access to finance, the supply chain, and agriculture-related deforestation, work through a partnership with governments, research centers, universities, and other institutional partners. The Connecterra monitors and Indigo Agriculture's database also provide traceability and are able to predict the best investment times in the industry. Finally, as venture capital and impact investing can support these types of businesses, governments also need to play a role in promoting and facilitating the financing of new technologies and tools to make sustainable agriculture possible in most parts of the world (Fixing Food 2018). The countries most likely to attract investments in this area are shown in Fig. 1.

2.2 Nutritional Challenges

Food security is commonly defined as physical, social, and economic access to sufficient, safe food that meets the dietary needs and food preferences of each individual. However, food security is necessary but not sufficient for nutrition. Nutritional security considers care, health, and hygiene practices, in addition to food safety (FAO 1999, Lam et al. 2012). FAO defines nutritional security

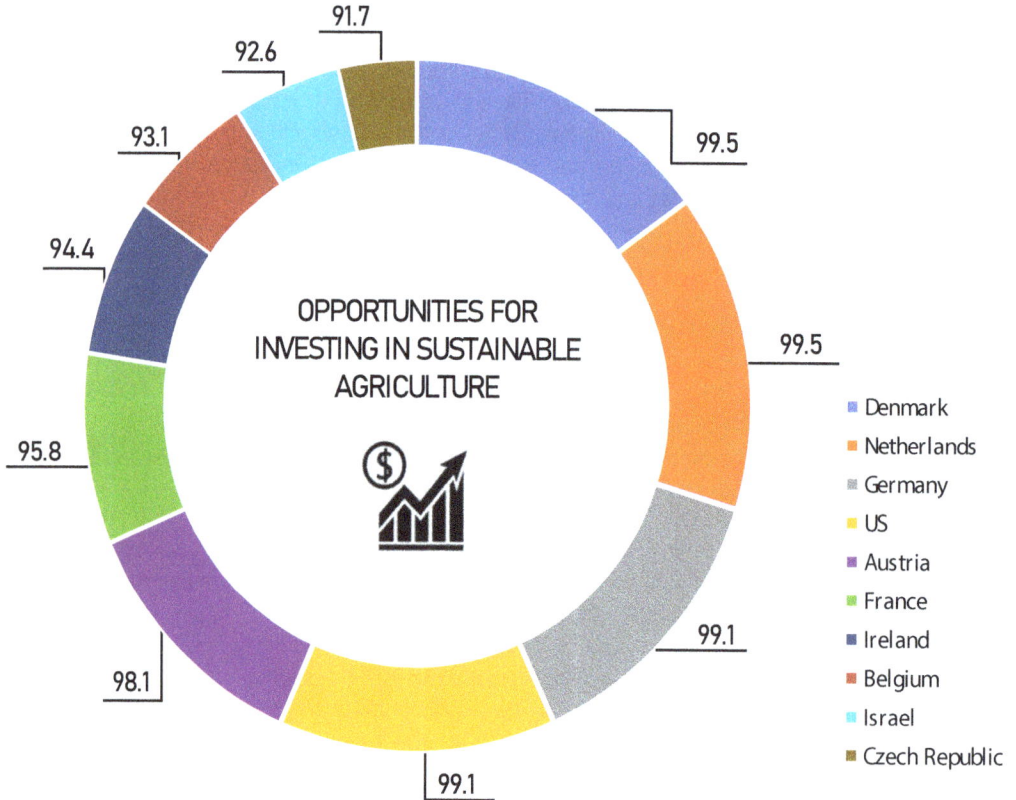

Notes: Scores are scaled from 0 to 100, where 100 = the highest

Figure 1. Top ten countries most likely to attract investment in sustainable agriculture.

as "the situation that exists when secure access to an adequately nutritious diet is associated with a sanitary security, environment, adequate services, and healthcare, in order to ensure a healthy environment and active living for all." It was proposed that the two terms should be brought together as "food and nutrition security" to better reflect the importance of nutrition's role in sustainable food security and make the explicit distinction between quantity (energy) and quality (food diversity) at the individual level, a goal expressed in SDG 2 (Charlton 2016).

Perhaps the most surprising of the dietary paradoxes is the parallel existence of obesity and malnutrition, very common in developing countries. Developing countries currently lack public resources to fight hunger and rapidly spreading diseases due to weight gain. Policy options to combat obesity include public education campaigns, promotion of healthy physical activity, implementation of mandatory labeling on food packaging, tax measures on foods high in fat, sugar, and salt, and restrictions on advertising aimed at children and adolescents (Agovino et al. 2018). On the other hand, micronutrient deficiencies are often underestimated. Micronutrient deficiencies are still common in low and middle-income countries and cause health problems including anemia, atrophy, and infant mortality (Tanumihardjo et al. 2007).

When it comes to a healthy diet, opinions differ widely and many debates about what constitutes a healthy diet will likely continue. World Health Organization (WHO) is updating its guidelines on fat and carbohydrate intake, which include recommendations on dietary fiber as well as fruit and vegetables. WHO (2017) strongly recommends that all individuals should reduce their intake of free sugars to less than 10% and total energy intake to no more than 5%, and even advises to reduce salt intake in foods and switch from trans fat to polyunsaturated fat. However, what is clear is that the increasing amounts of processed foods and junk food consumed in recent years have contributed to an increased incidence of diabetes and cardiovascular disease.

According to the WHO, more than 1.9 billion adults, aged 18 and over, are overweight. Of these, more than 650 million are obese. Approximately 39 million children under 5 years of age were overweight or obese in 2020, and more than 340 million children and adolescents aged between 5 and 19 years were overweight or obese in 2016 (WHO 2021). Maternal and child overweight and obesity are increasingly contributing to obesity, diabetes, and non-communicable diseases in adults. These high prevalence rates of these conditions and the heavy burden of morbidity, disability, and death can pose a threat and create a devastating financial problem, straining a country's health services and harming its economic and social well-being. Therefore, urgent action is needed to resolve this alarming and growing problem (Agovino et al. 2018).

Certainly, access to nutritious food has been compromised by food price volatility, conflict, and natural disasters. Besides that, there is no doubt that poverty and increasing social inequalities are the main factors for food and nutrition insecurity. That's because more than 1 billion people live in extreme poverty and around 870 million are undernourished globally (Movilla-Pateiro et al. 2019).

In this sense, some partnerships have emerged to develop better practices in the formulation and offerings of healthier foods. Some of these include the Global Alliance for Improved Nutrition (GAIN), a non-profit organization that fights malnutrition and increases access to safe, nutritious food, especially for vulnerable communities. GAIN has supported initiatives such as NutriRice, which makes white rice as nutritious as possible, developed by DSM and Bühler Group. Other examples of large companies include Danone's acquisition of a company specializing in organic food and beverages (WhiteWave) and Nestlé, which acquired a vegetable protein manufacturing company (Sweet Earth). In addition, McDonald's also recently announced new nutrition standards on its menu (Fixing Food 2018).

Although challenging, some countries are making progress in accessing and introducing healthy food at the national level. For example, Brazil has combined public policies and programs to combat the root causes of hunger (such as the Zero Hunger Program, Food Purchase Program, and the National School Feeding Program) (Henz and Porpino 2017). In 2016, France and Italy introduced legislation to prevent food waste and actions that prioritize the redistribution and donation of safe and edible food by affiliated supermarkets. These initiatives dramatically reduced food waste that was later implemented by several other countries such as Australia, Canada, the Czech Republic, Slovakia, Slovenia, Sweden, and the United States (EC 2019).

Undoubtedly, the most important and recommended practice today, in addition to fighting hunger, is to apply guidelines on the need for choices, healthy behaviors, and especially clarifying what constitutes a balanced nutrient diet for consumers. The Food Sustainability Index (FSI) features several indicators on which to capture the health effects of food and the greatest progress towards achieving sustainable performance; these estimates are a powerful tool to support decision-making on health policy and resource allocation (WHO 2021).

Finally, based on data from WHO Global Health (2018) (Fig. 2), Japan and Cyprus lead the ranking for the best healthy life expectancy. In contrast, Nigeria, Mozambique, and Sierra Leone are among the worst performers in this category. The conclusion of these results is that, in low-income countries, hunger and micronutrient deficiency remain relentlessly high. Indeed, progress in ending all forms of malnutrition is still limited and insufficient and remains a concern for most regions and nations.

2.3 Food Loss and Waste

There are millions of people one step closer to death from lack of food at the table. Unfortunately, the number of people affected by hunger in the world takes on an even more cruel outline in light of another alarming problem: food loss and wastage. Tons and tons of perfectly edible food are lost or thrown away every day. It is estimated that a third of everything that is produced globally is not

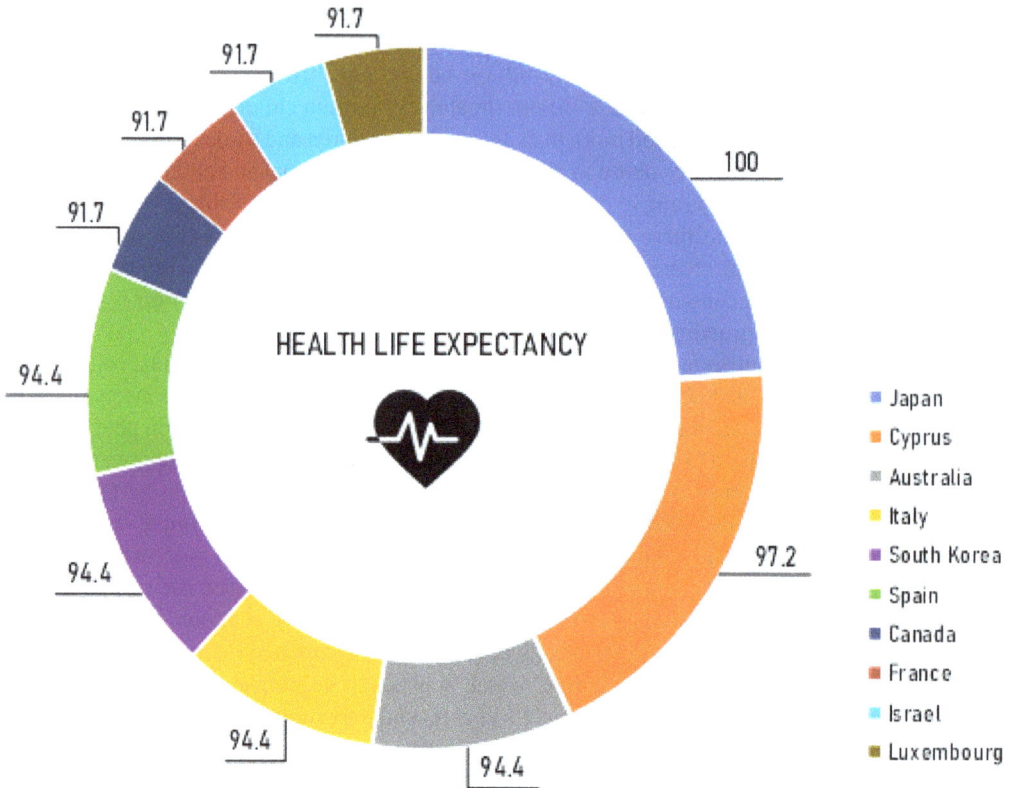

Figure 2. Top ten countries' most healthy life expectancy.

consumed by loss or waste. The good news is that food production is enough to feed everyone on the planet. However, although this is true, it will not be possible to end hunger or meet the global demand for food without addressing the high level of food loss and waste (Ortiz-Gonzalo et al. 2021).

Undoubtedly, leaving aside the issue of food loss and wastage is also missing chances to improve food security and environmental-economic prosperity, which are faces of the same coin. As for the environment, it is important to highlight that the way we produce and consume food today represents one of the greatest threats to the planet. That's because the loss and wastage of food also mean wasting limited natural resources. Food production has a huge impact on climate change and livelihoods. And since a third of everything that is we produce is wasted, a third of that impact is generated without reason (Li et al. 2022).

In North America, for example, around 168 million tons of food are wasted each year. The loss and wastage of food per capita in Canada, the United States, and Mexico are approximately 396, 415, and 249 kg/person/year, respectively. Most of it is wasted at the consumer level (40%), followed by the pre-harvest stage (29%), processing (12%), post-harvest (10%), and distribution, retail, and foodservice (9%). Some of the causes of food loss and wastage at these stages are (i) inaccurate supply and demand forecasts (ii) classification standards for size and quality, (iii) cold chain deficiencies, (iv) inadequate infrastructure and machinery, (v) damage during production and transportation, (vi) remittances rejection, (vii) strict management, (viii) excess inventory, and (ix) inadequate handling and storage (CEC 2017, Chauhan et al. 2021).

Also, according to the report of the Commission for Environmental Cooperation (CEC), the result of all this loss and waste is the emission of about 193 million tons of CO_2/year besides the waste of 22 million hectares/year, 39 million m³/year of landfill space, 18 billion m³/year of water, 3.9 million tons/year of fertilizers and an estimated loss of biodiversity of 319 million/year. Besides that, 274 million homes a year could be supplied with energy and 260 million people a year could be fed with the lost or wasted food across the entire food supply chain (CEC 2017).

The obstacles to mitigating food loss and waste are complex and, to face the challenge, the joint work of government agencies, businesses, industries, institutions, and international organizations is necessary. In this narrative, policymakers can implement legislation to encourage the reduction of food waste and penalize those who are throwing away perfectly edible food. Companies, in turn, can invest in their operations and consumers can assume responsibility for their actions. Effective solutions can indeed be found if multiple stakeholders are involved. Besides that, why and what is being lost needs to be disseminated so that meaningful and lasting solutions can be implemented at every stage of the supply chain and so that impacts are minimized and what is produced reaches its destination—on the plate of all people worldwide (Wang et al. 2021, Bhattacharya and Fayezi 2021).

Actions to minimize surplus production and avoid the generation of avoidable loss and waste is an approach that can be effective and involves reducing portions' sizes in foodservice environments and improving cold chain management. Another very promising approach is the rescue of food for human consumption that would otherwise be discarded or wasted. This approach involves supporting food collection organizations that aim to increase food access for food-insecure people, explore financial incentives for food donation, and develop regulations that protect food donors from liability. These two approaches are indisputably preferable to recycling and disposal (Ishangulyyev et al. 2019, Moraes et al. 2021).

Fortunately, different actors are engaged in solving the problem of food loss and waste through their programs, policies, and guidelines. However, in terms of countries, it can be said that many, despite having some plan or strategy for food waste, do not have laws and regulations. This is something that needs to be changed. By the way, the Food Sustainability Index (FSI) can be consulted on the performance of low-, middle- and high-income countries that have been at the forefront of policies and measures to reduce the problem of food loss and waste (Fixing Food 2018, Birdwell et al. 2021).

Finally, it is worth emphasizing that when dealing with food loss and waste, which is, in fact, a question of public interest, the collaboration between different sectors and actors in society is critical. Solutions need to be holistic and everyone needs to do their part (Filimonau and Ermolaev 2021).

3. Conclusion

The challenges that the current global food system affront is hard to overestimate. However, the moment of climate emergency that we live in concentrated the minds on the need to address food sustainability, which was already a need for a long time. It is becoming clear that we are not providing healthy food and enough for everyone within the limits of natural resources. This is evidenced in this chapter. The chapter addresses the Food Sustainability Index and the pillars that govern the performance of food systems. What is verified here is the view that holistic solutions must be adopted and this involves the collaborative efforts of governments, public and private sectors, as well as civil society.

References

Agovino, M., Cerciello, M. and Gatto, A. 2018. Policy efficiency in the field of food sustainability. The adjusted food agriculture and nutrition index. Journal of Environmental Management 218: 220–233.

BCFN. 2022. Food Sustainability Index. https://www.barillacfn.com/en/food_sustainability_index/(accessed 06 January 2022).

Beluhova-Uzunova, R.P. and Dunchev, D.M. 2019. Precision farming—concepts and perspectives. Problems of Agricultural Economics 3(360): 142–155.

Bhattacharya, A. and Fayezi, S. 2021. Ameliorating food loss and waste in the supply chain through multi-stakeholder collaboration. Industrial Marketing Management 93: 328–343.

Birdwell, J., Koehring, M., Fischer, D.H., Chow, M., Tokas, S., Jain, M., et al. 2021. Fixing Food 2021: an Opportunity for G20 Countries to Lead the Way.

Burey, P.P., Panchal, S.K. and Helwig, A. 2022. Sustainable food systems. In Food Engineering Innovations Across the Food Supply Chain (pp. 15–46). Academic Press.

Caron, P., Ferrero y de Loma-Osorio, G., Nabarro, D., Hainzelin, E., Guillou, M., Andersen, I., et al. 2018. Food systems for sustainable development: Proposals for a profound four-part transformation. Agronomy for Sustainable Development 38(4): 41.

CEC. 2017. Characterization and Management of Food Loss and Waste in North America. Montreal, Canada: Commission for Environmental Cooperation. 289 pp.

Charlton, K.E. 2016. Food security, food systems and food sovereignty in the 21st century: a new paradigm required to meet. Sustainable Development Goals. Nutrition and Dietetics 73(1): 3–12.

Chauhan, C., Dhir, A., Akram, M.U. and Salo, J. 2021. Food loss and waste in food supply chains. A systematic literature review and framework development approach. Journal of Cleaner Production, 126438.

EC. 2019. Redistribution of surplus food: Examples of practices in the Member States. EU Platform on Food Losses and Food Waste. https://ec.europa.eu/food/safety/food-waste/eu-actions-against-food-waste/eu-platform-food-losses-and-food-waste_en (accessed 06 January 2022).

EC. 2020. Towards a Sustainable Food System. https://ec.europa.eu/info/sites/default/files/research_and_innovation/groups/sam/scientific_opinion_-_sustainable_food_system_march_2020.pdf (accessed 06 January 2022).

EOS. 2019. Sustainable Agriculture Changes the Concept of Farming. https://eos.com/blog/sustainable-agriculture-changes-the-concept-of-farming/(accessed 02 January 2022).

FAO. 1988. Report of the FAO Council, 94th Session. Rome.

FAO. 1999. The state of food insecurity in the world. Rome.

FAO. 2019. Tracking progress on food and agriculture-related SDG indicators 2021. http://www.fao.org/sdg-progress-report/en/(accessed 09 January 2022).

Filimonau, V. and Ermolaev, V.A. 2021. Mitigation of food loss and waste in primary production of a transition economy via stakeholder collaboration: a perspective of independent farmers in Russia. Sustainable Production and Consumption.

Fixing Food. 2018. Best practices towards the sustainable development goals. Barilla Center for Food & Nutrition. www.barillacfn.com (accessed 07 January 2022).

Henz, G.P. and Porpino, G. 2017. Food losses and waste: how Brazil is facing this global challenge? Horticultura Brasileira 35: 472–482.

Holling, C.S. 2000. Theories for sustainable futures. Ecology and Society 4(2): 7.

Ishangulyyev, R., Kim, S. and Lee, S.H. 2019. Understanding food loss and waste—Why are we losing and wasting food? Foods 8(8): 297.

Lam, V.W.Y., Cheung, W.W.L., Swartz, W. and Sumaila, U.R. 2012. Climate change impacts on fisheries in West Africa: implications for economic, food and nutritional security. African Journal of Marine Science 34(1): 103–117.

Li, C., Bremer, P., Harder, M.K., Lee, M.S., Parker, K., Gaugler, E.C., et al. 2022. A systematic review of food loss and waste in China: Quantity, impacts and mediators. Journal of Environmental Management 303: 114092.

Moraes, N.V., Lermen, F.H. and Echeveste, M.E.S. 2021. A systematic literature review on food waste/loss prevention and minimization methods. Journal of Environmental Management 286: 112268.

Movilla-Pateiro, L., Mahou-Lago, X.M., Doval, M.I. and Simal-Gandara, J. 2019. Toward a sustainable metric and indicators for the goal of sustainability in agricultural and food production. Critical Reviews in Food Science and Nutrition 61(7): 1108–1129.

Ortiz-Gonzalo, D., Ørtenblad, S.B., Larsen, M.N., Suebpongsang, P. and Bruun, T.B. 2021. Food loss and waste and the modernization of vegetable value chains in Thailand. Resources, Conservation and Recycling 174: 105714.

Purvis, B., Mao, Y. and Robinson, D. 2019. Three pillars of sustainability: in search of conceptual origins. Sustainability Science 14: 681–695.

Rockström, J., Steffen, W., Noone, K., Persson, Å., Chapin, F.S., Lambin, E.F., et al. 2009. A safe operating space for humanity. Nature 461(7263): 472–475.

Tanumihardjo, S.A., Anderson, C., Kaufer-Horwitz, M., Bode, L., Emenaker, N.J., Haqq, A.M., et al. 2007. Poverty, obesity, and malnutrition: an international perspective recognizing the paradox. Journal of the American Dietetic Association 107(11): 1966–1972.

UN. 1987. Report of the world commission on environment and development: our common future. Oxford University Press, Oxford.

UN. 2018. Gaps in international environmental law and environment-related instruments: towards a global pact for the environment Report of the Secretary-General, A/73/419, 30.

UN. 2019. World 'off track' to meet most Sustainable Development Goals on hunger, food security and nutrition. UN News. Global perspective human stories. https://news.un.org/en/story/2019/07/1042781 (accessed 03 January 2022).

Wang, Y., Yuan, Z. and Tang, Y. 2021. Enhancing food security and environmental sustainability: A critical review of food loss and waste management. Resources, Environment and Sustainability 4: 100023.

WHO. 2017. Proposed policy priorities for preventing obesity and diabetes in the Eastern Mediterranean Region. EMRO Technical Publications Series 46: 1–72.

WHO. 2021. Obesity and overweight. https://www.who.int/news-room/fact-sheets/detail/obesity-and-overweight (accessed 01 January 2022).

3

Green and Sustainable Foodomics

Norelhouda Abderrezag[1,2] and *Jose Antonio Mendiola*[2,*]

1. What is Foodomics?

Before answering the question that titles this section, it is necessary to clarify some points and put it in context. Since the 1990s, chemical analysis of biological samples is turning dramatically. It was almost entirely focused on singles aspects of single molecules in each test. However, the evolution of analytical technologies and data science has made it possible to analyze thousands of molecules simultaneously, which has opened up research to study how these molecules interact with each other. Therefore, in order to easily identify this way of work, the suffix –omics is added at the main.

Foodomics is a new and global concept that include all the working areas related to food and nutrition domains. The Foodomics concept was first cited in 2009 (Cifuentes 2009). In this sense, the Foodomics strategy is used to control safety, quality, traceability, and hygiene management of food and to evaluate functional and nutritional factors in the field of food science (Váldes et al. 2021). This approach encompasses the use of multiple omics tools (see Fig. 1), such as metabolomics, proteomics, transcriptomics, genomics and lipidomics, which allow comprehensive and high throughput information of quality, safety, composition and nutritional values of the studied compound (Balkir et al. 2021).

In last few years, omics techniques are increasingly gaining importance for validation and development in food technology, beside corresponding quality control of all the processes. A brief description of these tools is provided below.

Genomics is an interdisciplinary field of biology studying the structure, function, mapping, and editing of genomes (Kaur 2013). In this sense, genome is the organism's complete set of DNA. The first genome was sequenced in 1977 by Frederick Sanger, who developed a sequencing technique for DNA to sequence the first complete genome, called phiX174 virus, which opened the doorway to the possibility in the field of genomics. In fact, his pioneer work in this field was awarded with the Nobel Prize in 1980 with Wally Gilbert and Paul Berg for pioneering DNA sequencing methods. Nowadays, Genomics is segmented in three main areas: comparative, structural and functional genomics. The first one comprises comparing the genomes of organisms, useful to detect evolutionary variations across genomes. Structural genomics implicates the physical nature of genomes, which also includes the mapping and sequencing of genomes. Functional genomics, finally, is related to

[1] Laboratory of Environmental Processes Engineering, University of Salah Boubnider Constantine 3, Constantine, Algeria.
[2] Foodomics Laboratory, Bioactivity and Food Analysis Department, Institute of Food Science Research CIAL (CSIC-UAM), C/Nicolas Cabrera 9, 28049 Madrid, Spain.
* Corresponding author: j.mendiola@csic.es

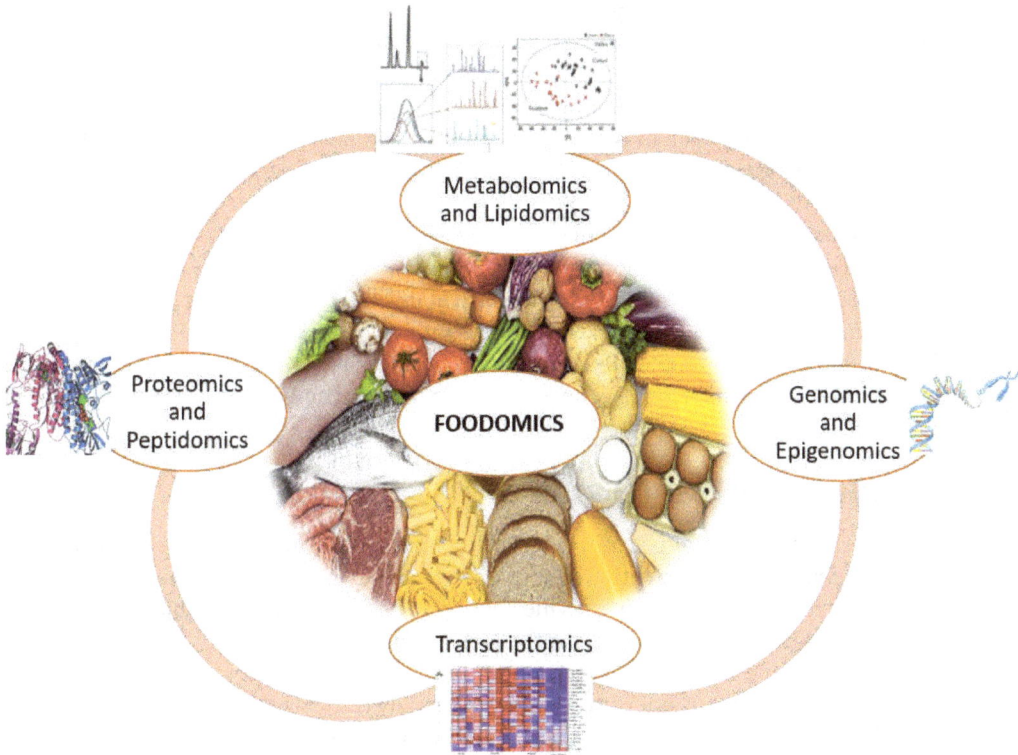

Figure 1. Foodomics connections with other –omic disciplines.

the expression and function of the genome. Genomics can also involve the research of interactions among genes and among genes and the environment.

Transcriptomics is a discipline related to the study of the expression levels of RNAs, in particular cell, tissue, organ, or whole organism (Cifuentes 2013). Being the transcriptome, the complete set of RNA transcripts are generated by the genome, under particular conditions or in a specific cell. It is generally studied using high-throughput methods, such as microarray analysis. Comparison of transcriptomes allows the identification of genes that are differentially expressed in distinct cell populations, or in response to different treatments.

Proteomics is a comprehensive analysis of the whole protein content in a biological system at specific moment, which is called the proteome. The proteome is not steady; it varies from cell to cell and varies over time. To some degree, the proteome replicates the underlying transcriptome. Nevertheless, many aspects besides the expression level of the relevant gene also modulate protein activity (frequently assessed by the reaction rate of the processes in which the protein is implicated). Proteomics also included the study of protein modifications (post-translational modifications include phosphorylation and glycosylation, among many others), interactions and quantification. These variations, together with alternative splicing in cells, render the proteome substantially more complex than the transcriptome. Proteomic studies can identify proteins that enable to control the disease progression (Alvarez-Rivera et al. 2021).

Metabolomics is one of the major omics disciplines in life sciences that explores small molecules (generally less than < 1000 Da) in various biological matrices (Cifuentes 2013). It is considered as chemical profiling of a natural matrix including cells, biofluids such as serum, plasma, excrement, tissue, cerebrospinal fluid, and exhaled gas. There are three fundamental approaches in metabolomics: target analysis, metabolic profiling, and metabolic fingerprinting. Target analysis aims at the quantitative measurement of selected metabolites, the pure standard compounds are usually analyzed, and hence there is no doubt regarding metabolite identification. In contrast, untargeted analysis (metabolic profiling) focuses on the study of a group of related metabolites or a specific metabolic pathway, while metabolic fingerprinting aims to compare the patterns of metabolites that change in response to the cellular environment (Ibáñez and Simó 2013). Other specialized subfields have emerged from metabolomics. Lipidomics represents a comprehensive study of the structure and function of global lipid profile in biological system (Balkir et al. 2021), and glycomics "*is the systematic study of all glycan structures of a given cell type or organism*" (Rudd et al. 2017).

All these omics studies produce great amount and complexity of data in a short time. Consequently, it is mandatory to develop computational strategies to convert the obtained data into functional and useful information such as the bioinformatics, and statistical tools.

Furthermore, the combination between different omics technologies is possible and it can be significantly adequate to understand and to sustain scientifically the health benefits from food ingredients, as well as help in traceability and food security. This can be considered the essence of Foodomics.

2. What's Green Analytical Chemistry?

In last decades, Green Analytical Chemistry arose from analytical chemistry; it is a concept that incorporates sustainable development values to analytical laboratories. All researchers should introduce sustainability values to their activities and fields of expertise. The objective of Green Analytical Chemistry is to implement efficient methods with reduced number of processing steps in sample preparation, and minimize the generation of wastes stream, and to provide an efficient quantification of analytes, as well as the use of solvent generally recognized as safe (GRAS) (Santana and Meireles 2021). Green Analytical Chemistry emerged as a subdiscipline of Green Chemistry, so let's start seeing what Green Chemistry is.

Our world is becoming increasingly concerned about the environment and how this is affected (and how it will be affected in the future) by different chemical and engineering activities at both, industrial and laboratory scale. Since initial 1990s, the Green Chemistry movement has been investigating techniques to reduce the risks of harmful chemical exposure to humans and environment. Simply expressed, Green Chemistry diminishes or eliminates the use or production of hazardous substances from chemical products and processes and improves all types of chemical products and processes by reducing impacts on human health and environment. Anastas and Warner stated in 1998: "*Green Chemistry is the use of chemistry techniques and methodologies that reduce or eliminate the use or generation of feedstocks, products, byproducts, solvents, reagents, etc., that are hazardous to human health or the environment*" (Anastas and Warner 1998). Green Chemistry technologies involve every type of chemical processes, from synthesis, separation, catalysis, monitoring, analysis and, of course, reactions. There are three main characteristics that lead in the 12 principles that govern Green Chemistry: waste, hazard (health, environmental, and safety), and energy.

Analytical methods are chemical processes in which primary information and knowledge, solvents, analytes or samples, reagents, energy, and instrument measurements are used as inputs to solve a specific problem, being qualitative or quantitative. The outputs of those analytical processes can be qualitative and/or quantitative composition of the analytes in the starting sample. Nevertheless, analytical methods may also comprise side effects (e.g., energy utilization, wastes or by-products that could damage the environment or pose risks for operators, etc.). These side effects of analytical methods, along with waste generation and management, are also the responsibility of method developers and users (Garrigues et al. 2010). All these aspects must be considered as the key elements to consider during development of green analytical methods. The analytical community is, day-by-day, getting more environmentally sensitive. The concept of improving analytical methods by reducing the use of reagents and solvents is at the core of many theoretical developments, and its importance is growing by the day.

In this sense, Keith et al. (2007) defined Green Analytical Chemistry as "*the use of analytical chemistry techniques and methodologies that reduce or eliminate solvents, reagents, preservatives and other chemicals that are hazardous to human health or the environment and that may also enable faster and more energy-efficient analysis without compromising performance criteria*". Analytical methodologies can also have side effects (e.g., energy usage, and wastes that could create risks for operators and damage the environment). One of the key points of side effects is the waste generation and waste management, which are also the responsibility of method developers and users (Garrigues et al. 2010). Analytical chemistry offers the data necessary to make conclusions about human and environmental health. The main point of analytical chemists is getting fast, precise, and accurate results; the green challenge is to join the informational requirements of chemists themselves, industry, and society while lowering the human and environmental impact of the analyses. Food analysis, in this sense, has the same requirements as any other analytical challenge. The natures of the matrix, the analytes, and the way signal is generated influence enormously the ease of creating a green analytical method. In this sense, the methods that do not use pretreatment, use less reagents, or work with aqueous or easily biodegradable solvents have a green advantage. Therefore, there is a big path for improving analytical methods in the green way to achieve the ideal Green Analytical Chemistry.

In order to develop greener chemical processes, "12 Principles of Green Chemistry" to guide chemists were formulated by Anastas and Warner (Garrigues et al. 2010). Unfortunately, those principles did not include explicitly some important concepts, greatly applicable to environmental impact, like the need for life cycle assessment, the inherency of a product or process, or the chance of heat recovery exothermic reactions or heat integration. Few years later, Anastas and Zimmerman (2003) formulated the "12 Principles of Green Engineering". In this line, Gałuszka et al. (2013) formulated the "12 Principles of Green Analytical Chemistry". In order to remember easily all these principles that directs the green approach to chemistry, some mnemonic have been proposed to remember them using the words PRODUCTIVELY, IMPROVEMENTS and SIGNIFICANCE for Green Chemistry, Green Engineering and Green Analytical Chemistry, respectively. The meaning of them can be seen in Fig. 2.

Also, it must be kept in mind that Analytical Chemistry helps the society in general, in the sense that many decisions are made on the basis of analytical information. Therefore, there is social responsibility underlying Analytical Chemistry (well, also, in every aspect of Science, but let's focus). Social responsibility in analytical chemistry denotes transferring information to the whole society but in an understandable, complete, and contextualized manner (Marcinkowska et al. 2019). Consequently, Green Analytical Chemistry can be considered the analytical arm of sustainable development, as it requires the incorporation of all aspects of sustainability (economic, social and environmental) to transform it into Sustainable Analytical Chemistry.

The progression of omics disciplines is helping the other analytical areas to develop clean analytical methods. The combination of modern analytical techniques with innovations in microelectronics and miniaturization drives the development of robust and effective analytical devices for successful control of processes and contamination. Combining miniaturization in analytical equipments with progresses in chemometrics is also of interest. The evolution of bioinformatics and chemometrics has reinforced expansion of solvent free methodologies based on mathematical treatment of signals acquired by direct measurements on untreated solid or liquid samples.

However, it must be also considered that those equipments are not cheap nor affordable and the social dimension of Green Analytical Chemistry must link with sustainability to lead to remarkable and new definitions such as "*Democratic Analytical Chemistry*" or "*Equitable Analytical Chemistry*". Highlighting the significance of Green Analytical Chemistry lies in its capacity to make analytical procedures economically viable and accessible to countries, institutions, and individuals worldwide, thereby benefiting society as a whole. These concepts have been detailed and exemplified in the following reviews by Marcinkowska et al. (2019) and Ballesteros-Vivas

GREEN CHEMISTRY	GREEN ENGINEERING	GREEN ANALYTICAL CHEMISTRY
P Prevent wastes	I Inherently non-hazardous and safe	S Select direct analytical technique
R Renewable materials	M Minimize material diversity	I Integrate analytical processes and operations
O Omit derivatization s teps	P Prevention instead of treatment	G Generate as little waste as possible and treat it properly
D Degradable chemical products	R Renewable material and energy inputs	N Never waste energy
U Use safe synthetic methods	O Output-led design	I Implement automation and miniaturization of methods
C Catalytic reagents	V Very simple	F Favor reagents obtained from renewable source
T Temperature, pressure ambient	E Efficient use of mass, energy, space and time	I Increase safety for operator
I In-process monitoring	M Meet the need	C Carry out *in-situ* measurements
V Very few auxiliary substances	E Easy to separate by design	A Avoid derivatization
E E-factor, maximize feed in product	N Networks for exchange of local mass and energy	N Note that the sample number and size should be minimal
L Low toxicity of chemical products	T Test the life cycle of the design	C Choose multi-analyte or multi-parameter method
Y Yes, it is safe	S Sustainability throughout product life cycle	E Eliminate or replace toxic reagents

Figure 2. 12 Principles of Green Chemistry, Green Engineering and Green Analytical Chemistry, respectively, according to mnemonic rules proposed by Tang et al. 2008, Galuszka et al. 2013.

et al. (2021). We recommend that readers of this chapter read it in order to delve into low-cost, high-benefit analytical methods of Green and Sustainable Analytical Chemistry; in fact, some of them could be applied within a "Citizen Science" framework. In this sense, and taking into account that sample preparation is one of the most environmentally problematic issues, López-Lorente et al. (2022) described the 10 principles of Green Sample preparation and they have developed the idea of connecting Sample Preparation with Green and Sustainable Analytical Chemistry. Green Sample Preparation may contribute to achieve the following SDG (Sustainable Development Goals) established by the 2030 Agenda of the United Nations, https://www.un.org/sustainabledevelopment/ (last accessed Feb. 2022) see Fig. 3.

- Reducing amount of hazardous chemicals released to the environment may contribute to good health and well-being (SDG 3)
- The quality of water (SDG 6) may improve after eliminating solvents and toxic solvents. Besides, it will protect sea life (SDG 14) and in a certain way terrestrial life (SDG 15)
- The use of safer solvents and procedures is related to SDG 8 (improving the quality of work)
- Working with natural and/or renewable sorbents for extraction purposes, and reusing and recycling of materials, is connected to responsible consumption and production (SDG 12) as well as reduction (size economy), recycling or elimination of discarded material that will contribute toward sustainable cities and communities (SDG 11)
- Decreasing energy consumption is related with allowing population the access to more affordable and clean energy (SDG 7)
- And, of course, the implementation of all these steps can have a mitigating effect on climate change and the reduction of greenhouse gas emissions (SDG 13).

Figure 3. Green Sample Preparation principles and their connections with Sustainable Development Goals (SDG). Reprinted from Lopez-Lorente et al. (2022) with permission of Elsevier.

3. Green Analytical Chemistry + Foodomics + SDG = Green Sustainable Foodomics

So, after understanding the meaning of Foodomics and Green Analytical Chemistry comes the turn of the combination: Green Foodomics.

Nowadays, food analysis is one of the most challenging areas in analytical chemistry. Among all the plethora of analytical steps, sample preparation can be considered as the main source of environmental toxicants, followed by separation techniques.

One of the main challenges in Foodomics, along with food security, is to help the rational design and the development of novel foods designed for disease prevention and/or health improvement, and to be able to explain the molecular mechanisms involved with reinforcing scientific proof. Extracting bioactive compounds is one of the first phases in the scheme of functional foods' design and, therefore, the design of green environment-friendly methods to concentrate or isolate them from natural sources is a very relevant research field in Foodomics. Obviously, the greenest sample preparation method comes when no sample preparation/concentration is needed, but in real world with real samples related to Food Science, it's very far to be common. Green extraction methods are eco-friendly alternative to classical sample preparation methods; among them, high-pressure techniques (pressurized liquid extraction (PLE), subcritical water extraction (SWE) and supercritical fluid extraction (SFE)), and low-pressure techniques (microwave-assisted extractions (MAE), ultrasound assisted extraction (UAE) and pulsed electric field (PEF)) are feasible for extracting target compounds from complex matrices (Demirhan et al. 2017). These methodologies are based on the use of minimal quantities of solvents for extraction of target compounds, with certain selectivity. The use of pressure, temperature or other factor may enhance the efficiency and speed up the extraction processes. The use of one or other technique with a particular solvent depends, mainly, on the polarity of the target compound and/or the compressed fluid. *Similia similibus solvuntur* principle also applies when dealing with compressed fluids. Changes in temperature and/or pressure of the solvent can modify their physicochemical properties (density, diffusivity, viscosity, and dielectric constant), which also modify their selectivity and solvating power, especially when dealing with supercritical fluid extraction. The selection of the method of sample preparation can be complicated, since it often requires considering several parameters simultaneously. First, it is important to assess the chemical and physical properties of the compound of interest, including volatility, polarity, solubility and stability (thermal, oxidative, hydrolytic).

Latest trends in the design of Green Foodomics processes for concentrating or extracting bioactive compounds emphasize on not only the search for new green, food-grade solvents but also the idea of multiple integrated processes, which can also be called biorefinery or downstream processing. The idea comes from the crude oil (petroleum) refining where different products are obtained from the same raw material in a single process. So, in biorefinery, they use this philosophy but starting from biological matrices. In food science, those matrices can be fruits, seeds, by-products, or algae. One example can be found in the process developed by Gilbert- López et al. to fractionate bioactive compounds from the microalga *Isochrisis galbana* (Gilbert-López et al. 2015). They used supercritical fluid extraction to obtain carotenoids and the un-extracted material was sequentially treated with different solvents in increasing polarity in order to extract as much different things as possible, as can be seen in Fig. 4.

One of the best things of Green Chemistry is the possibility of measuring "how green is a process"; in fact, there is a whole discipline called "green metrics" to quantify it. Nevertheless, despite green metrics being highly useful in analytical chemistry, they are not very used. Green metrics offer qualitative and quantitative data on the real environmental impact of analytical methods and help identifying "black points" of the analytical method under development. Several metrics have been developed of varying comprehensiveness and complexity, namely, analytical Eco-Scale (Gałuszka et al. 2012), HPLC Environmental Assessment Tool (HPLC-EAT) (Gaber et al. 2011), Green Analytical Procedure Index (GAPI) and the advance Complex GAPI software (Płotka-Wasylka and Wojnowski 2021), Analytical Method Volume Intensity (AMVI) (Hartman et al. 2011), among many others. These tools use different criteria from the Green Analytical Chemistry philosophy and the results are generally based on pictograms that picture the degree of agreement with evaluated criteria (López-Lorente et al. 2022). But, probably, the most powerful tool to quantify the environmental performance of chemical processes and, therefore, analytical methods is LCA (Life Cycle Assessment). For LCA, analytical methods are considered as independent processes, where inputs are samples, reagents, solvents, energy and, of course, instrumental measurements. The boundaries of the LCA measurement can be done step by step (i.e., clean up, derivatization,

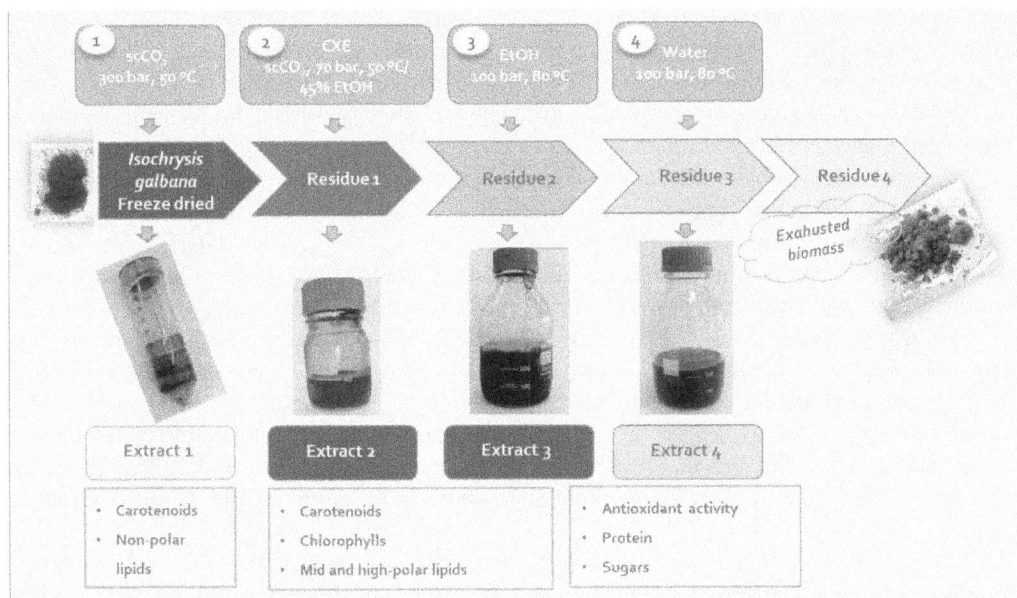

Figure 4. Combination of compressed fluids extraction steps within a biorefinery concept to concentrate bioactive compounds from the microalga Isochrysis galbana. Reprinted with permission of RSC publishing from Gilbert-López et al. 2015.

conservation, measurement, solvent disposal) or considering the method as a whole, the first way has the advantage of identifying "the blackest point" of the method. LCA is a complex methodology, and we encourage readers who want learn about it to read the book of Horne et al. (2009) or Curran et al. (2016).

Sustainable food production requirements are challengingly growing. Besides, the globalization of the food market and the growth of world population are generating pressure to achieve more sustainable processes. Of course, traceability, food security, and authenticity guarantee become of extreme importance to assure consumers' safety (Aguilar et al. 2019). One of the main problems for employing the abovementioned Green Analytical Chemistry principles to food analysis is the complexity of the samples to analyse. For instance, to determine the toxicity and presence of existence contaminants, additives, new chemicals, etc., which due to the low levels at which these toxic substances are found, makes it difficult to avoid sample preparation phases and the development of simpler or, at least, direct procedures. Despite these disadvantages, various approaches have arisen to reach successful procedures following the action lines of Green Chemistry, all the more so involving the participation of the general population as it is shown in initiatives such as "*Citizen science*" (Vohland et al. 2021). Different strategies have been proposed for the analysis of organic and inorganic contaminants, measuring hazardous constituents generated during food processing or for foodstuffs characterization.

Cheng et al. (2020) addressed the limitations of sample preparation by using a direct sampling method in which the surfaces of food samples were brushed with a metallic probe and then analyzed by tandem MS. This approach allowed the screening of over 300 pesticides in less than 1 minute and even allowed the quantification of 15 of them. Another good example of green method was done by de Almeida et al. (2021), who converted an invasive plant in an efficient and low-cost sorbent for the extraction of lead from tea samples; moreover, such extraction was followed by a simple and fast detection method (flame atomic absorption spectrometry, FAAS) providing a sustainable, effective and fast analytical procedure.

Besides Food Safety, the analysis of food bioactives is the main pillar of Foodomics. As a result, there are also Green Foodomics applications in this field. The search and isolation of bioactive compounds for agri-food products novelties (fortified foods, nutraceutical, functional foods, etc.) and those helping us to understand how food can contribute to good health and well-being pose significant challenges. In this sense, food scientists are carrying out valuable efforts to extract and analyze the main bioactive compounds that can be found in foods, by-products and natural products, along with the design of analytical platforms for evaluation of bioactivity using Foodomics strategies. Many classical methodologies have found replacement in novel procedures that can improve their environmental performance. For instance, Assirati et al. (2020) proposed a high throughput Green Analytical Chemistry platform to extract and tentatively identify a wide range of metabolites existent in sugarcane solid by-products, as well as facilitated the possible scaling up towards a biorefinery platform. The extraction method was based on two-liquid-phase ultrasound-assisted and extraction with probe and two-liquid-phase dynamic maceration. The greenness of this method was quantified using two different metrics: Analytical-Eco Scale; the Analytical Greenness Calculator. The replacement of organic solvents by supercritical carbon dioxide done by Bueno et al. (2020) is also interesting; besides, it was the first time that neuroprotective compounds were obtained from the edible microalga *Duanliella salina*. Also, the greenness of this procedure was quantified comparing it with a classical solid-liquid extraction method by means of Life Cycle Assessment (LCA).

Portable devices opened the door to a more democratic and sustainable Analytical Chemistry. The advancement in technology miniaturization combined with increasing sensitivity of detector makes portable detectors more reliable. Their use can be more reachable and widespread globally (including less developed countries), thus contributing to reach SDGs in a more fair and responsible way (Chemat et al. 2019). Even the smartphone that nowadays everybody carry in their pocket can be used as a detector, as Böck et al. depicted in their review (Böck et al. 2020). For example, Gu et al. (2013) developed a highly sustainable method that combines a very low-cost separation method using a simple filter paper, and very cheap detection method using a smartphone camera with open access image processing software and novel green solvents such as Natural Deep Eutectic Solvents (NaDES). This method allowed monitoring antioxidant capacity in NaDES media, which mimicks this important biological role as nature does, because NaDES are naturally formed within the cells by common metabolites, and they play a crucial role in the production, transport, storage and usage of compounds as phenols.

"Green Chemistry" and "Green and Sustainable Analytical Chemistry", it is evident from the discussion above how they can contribute to the improvement of the Sustainable Development Goals (SDGs). In the field of Sustainable Food Science, the SDG2 (zero hunger) is particularly important. It requires food production systems that are efficient, sustainable and able to meet the different dimensions of food security: availability, global food access, utilization, and stability over time. Biorefinery processes increase the potential of food bioeconomy, as powerful food production approach founded in the circular biobased economy, for large-scale production of high-quality foods, since these involve improved utilization of biological resources by generating value from all parts of particular biomass feedstock sources. Green Analytical Chemistry principles and Green Chemistry principles are closely correlated as we have seen. Their implementation of these principles in Food Science not only promotes food security and zero-hunger (SDG 2) accomplishment but also helps to strengthen other SDGs, as can be seen in Fig. 5. Those principles could improve clean water (SDG6), health and welfare (SDG3), clean-energy production and consumption (SDG7), industrial innovation and infrastructure (SDG9), promoting the sustainable cities and communities' development (SDG11), as well as responsible consumption and production (SDG12) (Ballesteros-Vivas et al. 2021).

4. Conclusions and Future Outlook

This work describes the bases of Green Foodomics but also heed its social dimension in order to achieve Green and Sustainable Foodomics. Food analysis is currently facing numerous challenges

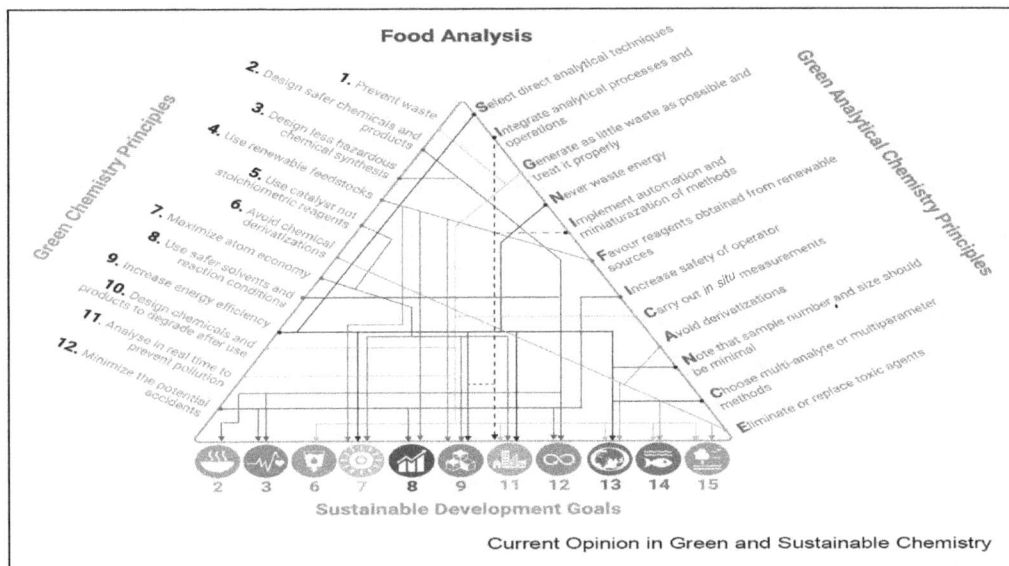

Figure 5. Connections of Food Analysis with Sustainable Development Goals through Green Chemistry Principles and Green Analytical Chemistry Principles. Reprinted from Ballesteros-Vivas et al. (2021) with permission of Elsevier.

in the areas of sustainability, food quality, bioactivity and safety assessment. Overcoming these hurdles has the potential to significantly contribute to the achievement of the SDGs. The relationship among Green Chemistry, Green Analytical Chemistry principles and SDG is proposed as a way of realizing how to reach Green Sustainable Foodomics. Some methodologies and novel procedures coupling this philosophy have been exemplified in this work. Nevertheless, the research in this line continues. There is still a lot of work to do but this goal is getting closer.

Acknowledgments

This work has been funded by the project 202170E057 (Bioprospección de plantas endémicas de Argelia mediante tecnologías de extracción sostenibles) and INCGLO0019 (Bioprospección de recursos agrícolas locales, una vía para alcanzar los Objetivos de Desarrollo Sostenible), both funded by CSIC.

N. Abderrezag is grateful to the Salah Boubnider Constantine 3 University, Constantine, Algeria for the fellowship.

References

Aguilar, C.N., Ruiz, H.A., Rubio Rios, A., Chávez-González, M., Sepúlveda, L., Rodríguez-Jasso, R.M., et al. 2019. Emerging strategies for the development of food industries. Bioengineered 10: 522–537.

Alvarez-Rivera, G., Valdés, A., Leon, C. and Cifuentes, A. 2021. Foodomics – Fundamentals, state of the art and future trends. In Food Chemistry, Function and Analysis; Jorge Barros-Velazquez, Ed.; ISBN 9781839163005.

Anastas, P.T. and Warner, J.C. 1998. Green Chemistry: Theory and Practice; Oxford University Press: Oxford.

Anastas, P.T. and Zimmerman, J.B. 2003. Design of through the 12 principles green engineering. Am. Chem. Soc., 37: 95–101 A, doi:.org/10.1021/es032373g.

Assirati, J., Rinaldo, D., Rabelo, S.C., Bolzani, V. da S., Hilder, E.F. and Funari, C.S. 2020. A green, simplified, and efficient experimental setup for a high-throughput screening of agri-food by-products—from polar to nonpolar metabolites in sugarcane solid residues. J. Chromatogr. A 1634, doi:10.1016/j.chroma.2020.461693.

Balkir, P., Kemahlioglu, K. and Yucel, U. 2021. Foodomics: a new approach in food quality and safety. Trends Food Sci. Technol. 108: 49–57.

Ballesteros-Vivas, D., Socas-Rodríguez, B., Mendiola, J.A., Ibáñez, E. and Cifuentes, A. 2021. Green food analysis: current trends and perspectives. Curr. Opin. Green Sustain. Chem. 31.

Böck, F.C., Helfer, G.A., da Costa, A. Ben, Dessuy, M.B. and Ferrão, M.F. 2020. PhotoMetrix and colorimetric image analysis using smartphones. J. Chemom. 34.

Bueno, M., Vitali, C., Sánchez-Martínez, J.D., Mendiola, J.A., Cifuentes, A., Ibáñez, E., et al. 2020. Compressed CO2 technologies for the recovery of carotenoid-enriched extracts from *Dunaliella salina* with potential neuroprotective activity. ACS Sustain. Chem. Eng. 8: 11413–11423, doi:10.1021/acssuschemeng.0c03991.

Demirhan, B., Kara, H.E., Ş and Demirhan, B.E. 2017. Overview of green sample preparation techniques in food analysis. pp. 137–144. *In*: Mark, T. Stauffer (ed.). Ideas and Applications Toward Sample Preparation for Food and Beverage Analysis; Vol. 32.

Chemat, F., Garrigues, S. and de la Guardia, M. 2019. Portability in analytical chemistry: a green and democratic way for sustainability. Curr. Opin. Green Sustain. Chem. 19: 94–98.

Cheng, S.C., Lee, R.H., Jeng, J.Y., Lee, C.W. and Shiea, J. 2020. Fast screening of trace multiresidue pesticides on fruit and vegetable surfaces using ambient ionization tandem mass spectrometry. Anal. Chim. Acta 1102: 63–71, doi:10.1016/j.aca.2019.12.038.

Cifuentes, A. 2009. Food Analysis and Foodomics. J. Chromatogr. A, 1216, 7109.

Cifuentes, A. 2013. Foodomics: Principles and Applications. In Advanced Mass Spectrometry in Modern Food Science and Nutrition; pp. 1–13.

Curran, M.A. 2016. Life cycle assessment in the agri-food sector: case studies, methodological issues, and best practices. Int. J. Life Cycle Assess. 21: 785–787, doi:10.1007/s11367-015-0977-5.

de Almeida, O.N., Menezes, R.M., Nunes, L.S., Lemos, V.A., Luzardo, F.H.M. and Velasco, F.G. 2021. Conversion of an invasive plant into a new solid phase for lead preconcentration for analytical purpose. Environ. Technol. Innov. 21. doi: 10.1016/j.eti.2020.101336.

Gaber, Y., Törnvall, U., Kumar, M.A., Ali Amin, M. and Hatti-Kaul, R. 2011. HPLC-EAT (Environmental Assessment Tool): a tool for profiling safety, health and environmental impacts of liquid chromatography methods. Green Chem. 13: 2021–2025, doi:10.1039/c0gc00667j.

Gałuszka, A., Migaszewski, Z.M., Konieczka, P. and Namieśnik, J. 2012. Analytical eco-scale for assessing the greenness of analytical procedures. TrAC - Trends Anal. Chem. 37: 61–72.

Gałuszka, A., Migaszewski, Z. and Namieśnik, J. 2013. The 12 principles of green analytical chemistry and the significance mnemonic of green analytical practices. TrAC - Trends Anal. Chem. 50: 78–84.

Garrigues, S., Armenta, S. and Guardia, M. de la. 2010. Green strategies for decontamination of analytical wastes. TrAC - Trends Anal. Chem. 29: 592–601.

Gilbert-López, B., Mendiola, J.A., Fontecha, J., Van Den Broek, L.A.M., Sijtsma, L., Cifuentes, A., et al. 2015. Downstream processing of *Isochrysis galbana*: a step towards microalgal biorefinery. Green Chem. 17: 4599–4609, doi:10.1039/c5gc01256b.

Gu, S.H., Nicolas, V., Lalis, A., Sathirapongsasuti, N. and Yanagihara, R. 2013. Complete genome sequence and molecular phylogeny of a newfound hantavirus harbored by the Doucet's musk shrew (*Crocidura douceti*) in Guinea. Infect. Genet. Evol. 20: 118–123, doi:10.1016/j.meegid.2013.08.016.

Hartman, R., Helmy, R., Al-Sayah, M. and Welch, C.J. 2011. Analytical method volume intensity (AMVI): a green chemistry metric for HPLC methodology in the pharmaceutical industry. Green Chem. 13: 934–939, doi:10.1039/c0gc00524j.

Ibáñez, C. and Simó, C. 2013. Ms-Based Metabolomics in Nutrition and Health Research. Foodomics Adv. Mass Spectrom. Mod. Food Sci. Nutr. 245–270, doi:10.1002/9781118537282.ch9.

Kaur, S. 2013. Genomics. Brenner's Encycl. Genet. Second Ed., 2: 310–312, doi:10.1016/B978-0-12-374984-0.00642-2.

Keith, L.H., Gron, L.U. and Young, J.L. 2007. Green analytical methodologies. Chem. Rev. 107: 2695–2708.

López-Lorente, Á.I., Pena-Pereira, F., Pedersen-Bjergaard, S., Zuin, V.G., Ozkan, S.A. and Psillakis, E. 2022. The ten principles of green sample preparation. TrAC - Trends Anal. Chem. 148.

Marcinkowska, R., Namieśnik, J. and Tobiszewski, M. 2019. Green and equitable analytical chemistry. Curr. Opin. Green Sustain. Chem. 19: 19–23, doi:10.1016/j.cogsc.2019.04.003.

Płotka-Wasylka, J. and Wojnowski, W. 2021. Complementary green analytical procedure index (ComplexGAPI) and software. Green Chem. 23: 8657–8665, doi:10.1039/d1gc02318g.

Ralph Horne, Tim Grant and Verghese, K. 2009. Life cycle assessment: principles, practice and prospects; CSIRO Publishing; CSIRO Publishing.

Rudd, Pauline G. Karlsson, Niclas Khoo, Kay-Hooi H. and Packer, N. 2017. Chapter 51: Glycomics and Glycoproteomics. In: Varki, A., Cummings, R.D. and Esko, J.D. (eds.). Essentials of Glycobiology [Internet]. 3rd edition. The Consortium of Glycobiology: California.

Santana, Á.L. and Meireles, M.A.A. 2021. Green analytical chemistry for food industries; Elsevier Inc.; Elsevier Inc., ISBN 9780128218839.

Tang, S.Y., Bourne, R.A., Smith, R.L. and Poliakoff, M. 2008. The 24 principles of green engineering and green chemistry: "Improvements Productively." Green Chem. 10: 268–26, doi:10.1039/b719469m.

Valdés, A., Álvarez-Rivera, G., Socas-Rodríguez, B., Herrero, M., Ibáñez, E. and Cifuentes, A. 2021. Foodomics: analytical opportunities and challenges. Anal. Chem. acs.analchem.1c04678, doi:10.1021/ACS. ANALCHEM.1C04678.

Vohland, K., Land-Zandstra, A., Ceccaroni, L., Lemmens, R., Perello, J., Ponti, M., et al. 2021. Editorial: the science of citizen science envolves. pp. 1–12. *In*: The Science of Citizen Science; Springer International Publishing.

Part II
Next Generation and Future Technology for Sustainable Foods

4

Pseudocereals as Superfoods
Complementarity with No Competition for Food and Nutrition Security

Nhamo Nhamo* and Abidemi Olutayo Talabi

1. Introduction

Global food insecurity has become the most significant challenge for a growing affluent population in the 21st century (Godfray et al. 2010). Though fast-paced industrialization has increased average incomes and job opportunities for some locations, the access and availability of food that meets the required taste is dwindling. Similarly, there are observable limited quantities of food and relatively low-quality food standards in places such as sub-Saharan Africa. For instance, nearly 811 million people went hungry and were severe to moderately food insecure in 2020 (FAO, IFAD, UNICEF, WFP and WHO 2021), and the figures seem to be increasing (Von Grebmer et al. 2008). In most of the low-income countries, hunger is closely related to poor health and reduced participation in the work-related activities. In agriculture-based rural economies, this situation creates unbreakable poverty cycles transcending generations. The causes of the current mismatch between demand and supply of food have roots in the narrow range of genetic materials that are exploited for food and the recent impacts of climate variability on agriculture production.

Production figures of major food crops have plateaued in the last half a century (Gadgil and Rupa Kumar 2006). Yields of mainstream cereals crop, i.e., maize, rice, and wheat, have stagnated causing glitches in food chain supply and availability of industrial products. The yield levels of legumes, forage crops and fruit tree have not significantly increased either. In sub-Saharan Africa, for instance, the gap between population increases and production levels of starch and protein sources is steeply increasing. There is a looming food insecurity in the form of starch, protein and other crop based nutrient element deficit. Put together with the risk the community faces from environmental disasters and pandemics, there is an urgent need to re-strategize on crop-based food sources including diversifying sources. The long-term investments in breeding for superior cultivars in a handful of crops are threatened with diminishing returns, hence the need to consider supporting superfoods that are currently orphaned and have demonstrated resilience to biophysical stresses.

Population hot spots are closely associated with good agricultural soils and climate conditions suitable for production of mainstream crops. High population densities have led to over-harvesting

International Center for Biosaline Agriculture, P O Box 14660, Dubai, United Arab Emirates.
* Corresponding author: n.nhamo@biosaline.org.ae, nnhamo@gmail.com

of natural resources and intensive use of limited land and other resources. Land degradation driven by extensive agricultural practices in population hotspots has led to reduced agricultural production and unsustainable fragmentation of productive land. In many instances, climate extremes have also weighed in and caused huge losses in agricultural systems and produce due to communities' disaster risk unpreparedness. Food spoilage and waste has also been reported to be high in population hot spots, thereby driving poverty levels higher. Climate change seem to have compounded the population growth rates and food insecurity associated with poverty in hotspots. Climate smart technologies are required to reduce extreme weather impacts and play a role in resolving other layers of challenges associated with population hot spots in developing countries. If left unchecked, climate change driven weather extremes will derail production of major crops given the sharp increase in frequency of droughts, cyclones, and floods. There are opportunities of exploring climate smart technologies on orphaned crops in less populated regions of the world where indigenous communities have relied on diverse crops for food and nutrition.

Since the time of the green revolution of the 1960s, mainstream crops have been promoted to avert widespread food shortages at the expense of the development of minor crop. It has become apparent that major crops on their own cannot support a 10 billion population's demand on starch, protein, vitamins, and minerals (Gilland 2002, Nadathur et al. 2017). High yielding varieties, orange fleshed cultivar and iron fortified crops have had a resounding impact, but more is still required. Minor crops need to be part of the solution in feeding a growing population and making sustainable production systems with low carbon foot prints a possibility in many communities (Jnawali et al. 2016). For communities that have developed their livelihoods on the minor crops, it is probably the right time for the "old indigenous knowledge" to be revealed for the benefit of all. Similarly, there is a need to capture the climate smart drivers from across the board and focus on some super-foods that can save the world from hunger and poor health.

The term "Superfoods" was originally used to describe a collection of crop-derived foods that are rich in compounds and nutrient which support human health and wellbeing (Tacer-Caba 2019). However, the loose use of the word superfoods has a broader set of food crops compared to the original list that included blueberries, broccoli, and salmon. The extended list of superfoods now includes but not limited to seafood (e.g., cod, tuna, or trout), nuts such as almonds or cashew nuts, legumes, green leafy vegetables, and small grained crops (amaranth, buckwheat, chia, quinoa or teff). Other reports have picked black rice for instance as a superfood because of its unique attributes that are clearly different from those of the other forms of rice. Regardless, here we use superfoods to apply to a range of nutritionally superior crops that can contribute to human nutrition and health, of which pseudocereals are an integral part of the set.

Pseudocereals are dicotyledon crops belonging to the Polygonaceae family that are cultivated or exploited from the wild that resemble cereal crops except in their nutraceutical attributes. Pseudocereals are also known as pseudo-grains, they share some characteristic with the positive grains, ancient grains, and some orphaned starch sources for traditional communities. Pseudocereals such as buckwheat have been described as having high nutraceutical, industrial, and medicinal value and can be used for multiple purposes (Pirzadah and Malik 2020). It is in this context that pseudocereals' status and value to human food and nutrition security will be discussed in this chapter.

The chapter describes the status of pseudocereals' production, and utilization. The role played by superfoods and potential of pseudocereals in providing nutrition security for various communities are discussed. This chapter analyses nutritional contribution, crop management practices, value chain development and the use of IoT in pseudocereals crops. Advances in systems agronomy and research for development of pseudocereals were also discussed. The agriculture-health nexus is discussed given the importance of pseudocereals' contribution to nutrition and human wellbeing now and in the future.

2. Approach

The chapter focusses on pseudocereals' importance now and in the future. Discussions in this chapter address the nutritional composition, contribution to food systems and health diets, the crop production systems, components of value chain development, potential threats from abiotic stress factors and the application of the fourth industrial revolution technologies in increasing efficiencies in pseudocereals' supply chain. The three broad areas of application of these arguments and discussion points are sustainable food diversification, food supply chains challenges and use of the internet of things (IoT) in agriculture.

2.1 Importance of Pseudocereals (superfoods) in Food Diversification

Food and nutrition are central to human life and human development. Recommended daily healthy diets are made up of a range of energy, nutrient sources and fluids that supports active living. Though the drive to increase production and intake of few main crops was the main food security rhetoric of the 20th century, there is growing recognition that putting emphasis on the balancing role of minor crops will improve human health and reduce dietary diseases. Pseudocereals are a part of the superfoods that have been neglected by most people and hence need to be popularized and their production increased. In this chapter, pseudocereals will be used as examples of how alternative diets, which include superfoods, could be strengthened through improved production and related value chain components.

2.2 Food Supply Chain Improvement

Crop value chain development is an inclusive process involving activities of players ranging from input providers, producers, processors, packers, marketers, and distributors for consumed products (Fernandez-Stark et al. 2012). There is evidence that food insecurity can emanate from distribution glitches or weak local production and distribution systems as experienced during the different phases of the COVID19 pandemic in recent years. Similarly, advanced distribution systems of the major crops are not inclusive and adequate by design; hence, they do not accommodate the minor crops, the category in which superfoods fall. Both local and inter-regional food supply chains area are important in supporting food and nutrition security across many communities and more work is required to move towards an inclusive system.

Production of pseudocereals has improved since the nutritional and nutraceutical value has been made public knowledge from the human health perspective. Production in the short to medium term will be constrained by both down- and up-steam factors found in the current environment. There are a range of drivers that need to be critically analyzed for the supply chains of superfood to support the increasing demand and to lower the cost of accessing the food products.

2.3 Fourth Industrial Revolution Technologies

The internet of things (IoT) has increased visibility, communication and hence connectivity of production systems and potential markets in recent years. Smallholder farmers produce these crops for a variety of reasons including food security, income generation, leisure, and pass time, and in support of their cultural traditions. The level of investment has remained low in these lowly resource endowed groups of farmers due to the small margins or profits which they derive from their enterprises. They also produce food stuffs which feed into not-so-popular value chains. The current intersection of developments, i.e., increased food requirement, dietary diseases with solutions in superfood production and the need for farm processes data for modeling, predictions and designing improvements, point to the importance of applying the internet of thing on smallholder farms to support the enhancement of food supply chains for a healthy population.

3. Pseudocereals as Superfoods

Pseudocereals are dicotyledonous crops with close similarities with the dominant cereal crops. Their 'superfoods' status is a relatively new development emanating from the recognition of their benefits to health patients with nutrient deficiencies or health disorders. Amaranth, buckwheat, chia, quinoa or teff are sources of quality nutrients, complex carbohydrates, protein, and vitamins.

3.1 Nutrition

Pseudocereals are trusted sources of calories and proteins in traditional communities where the major cereals often succumb to droughts, pests, and diseases. Therefore, the nutrition of adults and children in such indigenous communities is based on the most abundant pseudocereals, or minor grains found locally. These crops that are mostly produced using minimal external inputs, such as fertilizers or pesticide, fill in the gaps in protein supply created by the dependence on animal-based protein sources. In places where per capita consumption of animal-based protein foods is low, crop-based protein sources have been associated with human and livestock health improvement. Similarly, in locations where climate impacts ravage cereals such as maize or rice, pseudocereals and millets form the basis of the calories basket.

3.1.1 Pseudocereals for Calories

Calorie consumption is an important indicator of food sufficiency in many developing countries and globally. Cereal crops are currently leading global sources of calories (35%) ahead of sugarcane (24%) and fruits and vegetables (23%) for human health (Pardey et al. 2014). Of the leading cereals, rice leads the pack contributing 21% of global calories and 27% for developing countries. Wheat comes second and maize is third, the latter contributes about 15% of global per capita calorie consumption. There is a wide variation in the per capita consumption of calorific crops across regions and countries within a region.

The average per capita consumption of cereals has been estimated at about 147 kg per capita per (pc) year. Among the leading cereals, rice tops the pack in terms of consumption with a high of 268 kg pc per year, followed by wheat and maize is last. In southern Africa where maize is staple source of starch, consumption averages of 85 kg pc per year are much higher compared to other sub-regions of Africa (20–30 kg pc per year) (Shiferaw et al. 2011). A popular starch source, cassava that is grown in more than 56% sub-Sahara African countries, is consumed at about 100 kg pc per year (Hahn 1989). In developing countries where total available cereals annually are not adequate to support food sufficiency, small grains and pseudocereals make up the difference during the hunger months of the year. In rural Asia and sub-Saharan Africa, consumption of small-grained crops such as pearl millet averages about 30 kg pc per year and pseudocereals have the potential to substitute about 30% of starch requirements for the communities. Data shown in Table 1 suggest that pseudocereals can be viewed as reliable sources of energy and carbohydrate in many communities where food security threatens livelihoods. Though consumption figures (kg pc per year) of pseudocereals are rarely quoted in literature, the contribution of superfood crops to energy, carbohydrates and nutritional elements is widely reported and acknowledged.

The chemical characterization of some pseudocereals (Table 1) shows that the energy content ranges from 1435 to 2351 kJ per 100 g DM, with a mean of 1646 kJ. The concentration of carbohydrates was observed to range between 42 g and 85 g per 100 g DM (Table 1). The carbohydrate contents of pseudocereals (e.g., buckwheat) have been found to be slightly lower than that of cereal crops on average. However, specific crop data shows that some pseudocereals superfoods contain 3–5% more starch than average cereals, e.g., amaranth, canihua and quinoa (Bender and Schönlechner 2021). Two other notable differences have been observed in the structure and composition of starch. Pseudocereals have much smaller starch granules (less than 3 μm) compared to cereals. The amylose content is lower in amaranth, canihua and quinoa than in cereals while buckwheat has higher waxy

Table 1. Energy (kcal), carbohydrate (g) and fiber (%) composition (per 100 g), of selected pseudocereals (amaranth, broomcorn millet, buckwheat, canary seed, chia, quinoa), and teff important for human nutrition.

Crop	Carbohydrate (g)	Energy (KJ)	Fiber (g)	Fat (g)	Protein (%)	References
Amaranth (*Amaranth* spp.)	68.1	1,554	11.1	10.6	17.5	Caselato-Sousa and Amaya-Farfán (2012), Schoenlechner (2016)
Proso millet (*Panicum miliaceum* L.)	70.4	1,582	14.2	3.1	12.5	Habiyaremye et al. (2017), Das et al. (2019)
Buckwheat (*Fagopyrum esculentum*)	80.8	1,435	18	3.4	11.0	Pirzadah and Malik (2020)
Canary seed (*Phalaris canariensis*)	69.8	1,660	7.3	7.9	23.7	Abdel-Aal et al. (2011)
Chia (*Salvia hispanica* L.)	42.1	2,351	34.4	40.2	24.2	Kulczyński et al. (2019)
Quinoa (*Chenopodium quinoa* Willd.)	66.7	1,539	14.7	6.3	16.5	Schoenlechner (2016)
Teff (*Eragrostis tef*)	85.0	1,418	9.8	5.7	11.1	Nascimento et al. (2018), Gebru et al. (2020)
Fonio (*Digitaria exilis*)	74.1	1,470	1.03	1.9	12.2	Babarinde et al. (2020)
Breadnut (*Artocarpus camansi*)	76.2	1,803	-	29.0	19.9	Ragone (2006)
Mean	70.4	1,646	13.8	12.0	16.5	
Standard deviation	12.2	290	9.8	13.4	5.2	

starch content relative to cereals (Kringel et al. 2020, Repo-Carrasco-Valencia and Arana 2017, Zhu 2016).

As research on superfoods gains pace, several information gaps will need to be filled. With regards to energy and carbohydrate supply, systematic characterization data is urgently required from natural populations and those incorporated in the mainstream farming systems. Such data will be largely influenced by the food processing methods and post harvesting handling procedures applied on the superfoods. More research is required in this area to include how indigenous communities have been utilizing processing methods to obtain energy from such crops, seed storage methods and production systems. There is scope for innovations in the development of products from pseudocereals and their promotion into the food value chain systems.

Under the agriculture for health thematic research area, questions such as the bioavailability of nutrient elements from pseudocereals superfoods and how their use vary across age groups and gender will be of interest. This information could be important in determining the products and dishes which adults and children can consume. Minimum daily recommended consumption rates will also need to be determined for healthy diets.

3.1.2 Superfoods for Readily Available Proteins

Much of the world's population relies on plant-based protein for their nutrition and wellbeing (Nadathur et al. 2016). Though a sharp increase in animal-protein rich diets was closely related to the growth in the global economy, arguments supporting plant-based diets seem to be winning recent debates on production systems that are climate sensitive and environmentally protective (Pardey et al. 2014, Sabate and Soret 2014). In developing countries, the current increase in consumption of animal meat products is being driven by fast-paced urbanization trends and related improvement in per capita income of educated young population actively involved in economic activities (Sharma et al. 2019). However, world-over, the number of undernourished people is on the increase (> 800 million) and malnutrition accounts for more than 30% of deaths in children under the age of 5 years. There is evidence of large-scale shortages in protein-rich food sources in several communities leading to stunted growth or poor child-development.

Research on superfoods that are rich in protein and other minerals was driven by the quest to fill in the huge gap in the access and availability of protein-rich foods that exists globally. To date, protein energy undernutrition is common in many children in the developing world and is exhibited through marasmus or kwashiorkor. Acute deficiency of protein in children is a major cause of death and wasting while in adults the compound effects lead to low productivity and secondary infections. Consumption of plant-based protein food serves more than 60% of the world (De Boer and Aiking 2011). Pseudocereals have gained popularity in recent years and the consumption rates are increasing fast because of their macronutrient content and quality of proteins.

Amaranth and quinoa are leading contenders for fastest popularity rating in recent years. Amaranth is widely grown for its leaf and seed. Amaranth's leaves contain 17–30% protein (Oliveira and De Carvallo 1975) while the grain averages 14% in protein (Pedersen et al. 1987). Similarly, buckwheat, breadnut, canary seed and chia are among pseudocereals with high protein contents that support healthy protein-rich diets in human beings.

Besides pseudocereals, research is widening on protein sources that can be scaled and could be produced at low-cost and low carbon footprints. The potential of microalgae, water lentils and a range of indigenous mushrooms has been discussed. Microalgae, for example *Arthrospira* sp., has shown potential for production at scale and use as protein supplements both as fresh and in the form of dry processed products (Barka and Blecker 2016).

Protein contents of pseudocereals presented in Table 1 demonstrate their potential to complement the major crops in supplying plant-based protein sources to humans. The mean protein content for the selected pseudocereals was 16.5% with a minimum of 11% and a maximum of 24%.

3.2 Healthy Diets

Human health is central to development and the attainment of aspiration of humanity. Food is a major source of body nourishment and energy. Globally, diets are a function of income as it determines ability and willingness to pay, access to and availability of quality food. However, in marginalized locations where production is dependent on the abiotic and biotic factors of production, the diminishing production potential can jeopardize opportunities for healthy diets. For the urban population and population hotspots where land limitation prohibits production and individual purchasing capacity is high, food value chain performance becomes an important determinant of access to food. Both adults and children require a balanced diet for proper body function and wellbeing. Unbalanced diets have led to many health conditions and if unchecked, can cost economies huge health bills.

3.2.1 Human Malnutrition

Malnutrition has been defined as the severe lack of adequate nutrition because of consumption of insufficient food, inappropriate foods, or the inability of the body to utilize available food. More health aligned definitions of malnutrition such as a disordered nutritional state resulting from a combination of inflammation and a negative nutrient balance, leading to changes in body composition, function and outcome have been put forward by several workers (Soeters et al. 2017). There is an overlap between malnutrition and the availability and quality of foods. A growing number of cases of malnutrition reported in developing countries has been explained by heavily skewed eating habit often relying on one starch source and limited accompanying nutrient sources. In parts of sub-Saharan Africa where either maize or cassava dominate daily calorific intake, protein energy malnutrition, vitamin deficiencies and other mineral limitation in the food-baskets are apparent and impacts cut across all age groups. Most elemental deficiencies of micronutrients and vitamins are difficult to diagnose during the early stages and this has led to occurrence of "silent hunger" in many communities. Solutions of averting silent hunger are found in the basic designs of the diets that are used by the majority. While designing and adhering to balanced diets sound a straightforward recommendation from the health professions, the reality especially in resource constrained communities, unlike in developed countries, is a more complex discussion. Accessing balanced diets that reduce malnutrition is an important factor that can contribute to non-communicable diseases' prevention at community level.

The statistics on deaths from non-communicable diseases (NCDs) reported by the World Health Organization (WHO) have shown a clear relationship to country incomes, with the countries accounting for a large share of the malnutrition burden. For example, about 85% of the NCD related deaths occurred in low- and middle-income countries in 2016, and adults in these countries faced a higher risk compared to counterparts in developed countries. Globally, NCDs account for 71% of the 57 million total annual deaths (WHO 2018b). Four leading NCDs responsible for most deaths include cardiovascular (44%), cancers (22%), respiratory diseases (9%) and diabetes (4%) (WHO 2018a). In roads have already been made to match food types and prevention and remediation of several NCDs. The latest strategies developed by the WHO focus on the risk factors driving morbidity and mortality due to NCDs, i.e., tobacco use, unhealthy diet, physical inactivity, and harmful use of alcohol (WHO 2013). Health diets and increased consumption of nutrient-rich superfoods can significantly contribute to the reduction in mortalities from NCDs in developing countries.

Several superfoods have been recommended for the reduction of incidences, severity and in some cases correct disorders or health conditions in humans. Positive reviews on the consumption of some pseudocereals seem to contribute to the reduction of NCDs. Table 2 shows some of the conditions and NCDs known to respond to the change in dietary habits leading to health restoration and the recommended superfoods. Here cardiovascular, celiac, diabetes and sight are used as examples of the effectiveness of pseudocereals and other superfoods in dealing with prevalent NCDs and health conditions.

Cardiovascular diseases: Several risk factors associated with heart diseases ave been studied intensively and these include high blood pressure, and high blood cholesterol. High blood pressure and cholesterol have been controlled using plant-based products as substitutes of animal meat products. Major benefits in this substitution are protein energy and other nutrients. Pseudocereals such as (1) amaranth, (2) quinoa and (3) chia that contain high levels of protein have been used to supply the nutrient in non-meat diets.

(1) Amaranth leaves constitute an important source of protein in many communities of Africa and Asia. The amaranth leaves steamed, fried, or cooked in peanut butter constitute some of the dishes that are commonly consumed together with cereals as main meals. Amaranth seed is a rich source of proteins and other nutrients. Amaranth is widely available in many communities and its use can be improved through awareness and demonstration of the benefits in humans.

(2) Quinoa seed is a rich source of quality protein and other benefits suitable for vegetable-based diets. Besides being rich in protein, quinoa has widely been characterized for its nine amino acids important for human health.

(3) Chia seed is well known for its contribution to edible oils. With a protein content of 15–25%, dietary fiber 18–35% and fat content of 15–35%, chia stands out as a nutritious crop. The bioactive compounds and the nutraceutical capabilities make its products important in the management of human health conditions (Katunzi-Kilewela et al. 2021).

Table 2. Human malnutrition leading to health condition that can be addressed by use of pseudocereals and other plant-based nutrient source of nutrients.

Health condition	Dietary requirements	Superfoods supportive of diets	References
Cardiovascular	Low cholesterol foods	Amaranths, buckwheat, millets and quinoa	Caeiro et al. 2022
Coeliac/Celiac disease	Gluten free foods	Amaranths, buckwheat, millets and quinoa	Caeiro et al. 2022
Diabetes	Low glycemic foods	Amaranth and quinoa	Tang and Tsao (2017)
Eyesight	Vitamin A rich foods	Amaranth, quinoa and squash	Lale (2017), Eshete et al. (2016)

Celiac disease: Celiac disease is also known as gluten-dependent enteropathy. It is an autoimmune condition resulting in intestinal mucosal injury occasioned by the consumption of dietary gluten present in the endosperm of some cereals, namely barley, rye, and wheat (Webster-Gandy et al. 2020). The only confirmed treatment for celiac disease is consumption of gluten-free diets; however, some gluten-free diets are deficient in other essential nutrients. Thus, pseudo cereals such as amaranth, buckwheat, millets and quinoa which are gluten-free with high nutritional profile are considered candidate crops for a one stop dietary option for celiac patients (Caeiro et al. 2022).

Diabetes: This is a chronic metabolic disorder responsible for elevated levels of blood glucose. This condition could occur when there is deficiency of insulin, a regulatory hormone for blood sugar level in the body (Type 1 diabetes). There could also be a situation in which the body is unable to effectively utilize the insulin in the body due to overweight and this is referred to as type 2 diabetes. While type 1 requires intake of insulin, type 2 requires a special diet for the patients to help in weight loss. Amaranth flours have been found healthier than those of wheat, as they are not only nutritious, but very useful in regulation of the blood sugar levels such that sugar is released into the blood stream in a slow manner, thus preventing adding of weight. Quinoa, another pseudocereal with a glycemic index of 53 (lower than the benchmark of 55), is a choice crop for diabetic patients because the protein and fiber in grains helps to slow down the rate of digestion. Thus, amaranth and quinoa are anti-obesity and anti-diabetic healthy super crops (Tang and Tsao 2017).

Sight: Consumption of pseudo cereals have also been found useful in the prevention and treatment of eye defects. For example, quinoa seeds and leaves have been used to cure a range of diseases including eye defects in several regions of the world (Lale 2017). According to Eshete et al. (2016), leaves of amaranths are used for treatment of eye defects in Ethopia.

3.2.2 *Vitamin and Antioxidant in Pseudocereals*

Healthy diets are important for human wellbeing and child development. Unhealthy eating habits have been the major cause of wasting, increased hospitalization, and poor cognitive development in children. In adults, hidden hunger during early year of development cause complex health conditions and exacerbate the impact of curable diseases. Micronutrients, vitamins, antioxidant, and other plant-based compounds are an essential component of healthy eating habits. However, in developing countries where calories and proteins are in short supply, there was little emphasis on the vitamins and antioxidant for a long time. It is only in recent times that community-wide programs on healthy foods for the children during the first 1000 days have been popularized together with water and sanitation interventions. More is still required for adults' diets in order to reduce frequencies and severity of illness. Several pseudocereals crops have been analyzed and can potentially support the daily per capita requirements of vitamin, micronutrients, and antioxidants.

Vitamins, defined here as chemical compounds found in food that are essential for human body building, are required in relatively smaller quantities compared to calories and proteins. Similarly, micronutrients and antioxidants that are important for a balanced diet are absorbed in small quantities by the human body. Table 3 shows a range of vitamins that can be supplied by pseudocereals.

Both plant and animal products are reliable sources of vitamins and micronutrients for daily per capita requirements. However, in many communities in developing countries, diets do not naturally contain enough such nutrients and hence food fortification with essential vitamins and micronutrients is a common practice. Food fortification, defined here as the deliberate inclusion of requisite amount of nutrient element in affordable foods, has assisted in averting community wide deficiencies in vitamins and micronutrients. Common examples of fortified processed foods include mealie meal from maize, salt, and wheat flour. Several beverages are also fortified with nutrients.

Pseudocereals such as quinoa contain bioactive compounds beneficial for human health and wellbeing, like antioxidants, polyphenols, flavonoids, vitamins, and minerals that impart various health benefiting characteristics to this grain.

Table 3. Vitamin profiles (mg per 100 g) of selected pseudocereals important for supporting healthy human diets.

	Folic Acid	Niacin	Riboflavin	Thiamin	Vit B5	Vit B6	Vit E	References
	mg/100 g							
Amaranth (*Amaranth* spp.)	0.09	0.56	0.13	o.13	0.24	0.50	1.92	Ebert et al. (2011), Rodríguez et al. (2020)
Proso millet (*Panicum miliaceum* L.)	0.10	1.32	0.22	0.63		0.38		Devi at al. (2014), Das et al. (2019)
Buckwheat (*Fagopyrum esculentum*)	0.10	1.85	0.10	0.33	0.44	0.58	0.32	Saeed et al. (2021), Rodríguez et al. (2020)
Canary seed (*Phalaris canariensis*)	0.10	1.10	0.12	0.79				
Chia (*Salvia hispanica* L.)								
Quinoa (*Chenopodium quinoa* Willd.)	0.10	0.88	0.57	0.49	0.62	0.21	2.08	Sekhavatizadeh et al. (2021), Rodríguez et al. (2020)
Teff (*Eragrostis tef*)	0.02	0.80	0.10	0.51				Nascimento et al. (2018)
Fonio (*Digitaria exilis*)								
Breadnut (*Artocarpus camansi*)								

4. Pseudocereals: Abiotic and Biotic Stresses Factors

Yields of mainstream crops have plateaued during the past 30 years due to both abiotic and biotic factors of production. This trend casts some doubts over the capacity of doubling production to meet the needs of a growing population by 2050. The situation is alarming in areas where average crop yields are reducing due to complex interaction between the current production practices, soil degradation, biotic stress and climate change and variabilities. Pseudocereals have been reported to be climate resilient albeit limited data for detailed analysis and narrow exposure of the superfoods to high intensity and extreme weather conditions. The production practices of pseudocereals will need to be fine-tuned to tackle impending threats from climate change, soil degradation and biotic stresses.

4.1 Climate Change

In marginal areas, several permutations of climatic plant stress factors are found, and they can include heat and drought stresses. Germplasm diversity found in quinoa is a leading explanation why the crop has high tolerance to drought and other climatic conditions (Ruiz et al. 2014). Amaranth has also exhibited tolerance to a range of extreme weather conditions. It has both salinity and drought tolerance, and this can be exploited to extend the geographic regions where food can be obtained. Relative to other crops, amaranth has been grown in conditions where cereals will crush due to climatic conditions (Olufolaji et al. 2010, Grundya et al. 2020, Ebert et al. 2011, Mlakar et al. 2009).

With regards to climate change, there are wide information gaps which need attention with regards to climate resilience of pseudocereals. A systematic review is required to explore the quality of data available and to explore its use in developing integrated systems which include pseudocereals. Climate resilience information is also important to define production niches and mega environments in which pseudocereals can be extended to. In most locations where cereals are hard hit by climate change, alternative crops which include pseudocereals are needed.

4.2 Soil Degradation

Pseudocereals harvested from natural stands and those from agricultural fields are affected by extensive production practices where limited inputs are added systematically and hence net nutrient mining dominates. Coupled with low inherent soil fertility, nutrient mining is considered one of the major yield limiting factors in the smallholder farming systems. Similarly, an increasing feature limiting productivity of many soil groups is soil salt content because of salty water management or due to salt rich parent material. Regarding salt affected soil, salinity stands out as the major problem, followed by soil acidity and in places soil sodicity is an impediment to production of crops.

Soil salinity, defined here as the condition of a soil with exchangeable salt of more than 2 dS/m and a pH lower than 8.5, is the leading cause of reduced production in salt affected soils globally. Natural causes of soil salinity are (a) geochemical (b) saltwater intrusion and (c) low soil fertility. Globally human induced salinity is increasing at a fast pace and to date 76.3 million ha are affected by salinity while secondary salinization accounts for about 45.4 million ha. Crop yield losses are estimated to be between 10 and 40% and a cumulative financial loss due to salinity has been estimated to be US$11.4 billion/year (Pitman and Lauchli 2002).

Majority of mainstream cereals, e.g., maize, rice and wheat are not tolerant to soil and water salinity. Several pseudocereals have shown tolerance to water and soil salinity and therefore the latter could be used in developing solutions to an increasing soil salt problem. More work is required in this area to determine the benefits from such interventions.

4.3 Biotic Stresses

Biotic stress on major crops has reduced production by as much as 60% of the genetic potential. All three factors, i.e., diseases, pests and weeds have been found to cause major losses. The management of biotic stress factors is important for both cereals and pseudocereals. Data on improved production practices on pseudocereals is scarce owing to limited research efforts and this situation hampers the discussions on the development of the crops. The similarities in the species range that infect the two groups of crops have not been widely published. More work is required in this area to characterize tolerance levels and remedial practices' effectiveness.

5. Value Chain Development

Agricultural systems' improvement in recent years has targeted increased efficiencies across all components of the value chain. Across many pseudocereals, a huge amount of effort will be required to improve value chain's functionality. Several factors have been identified as deal breakers in improving amaranth, buckwheat, and quinoa value chains. These include, but are not limited to, first, input market weaknesses, i.e., limited improved high yielding seeds, fertilizer formulations suitable for improved productivity, and access to agrochemicals. Second, production constraint which include improved agronomic practices, low mechanization, and poor biotic stress management. Third, the consumer demand for pseudocereal products is narrow and found in small pockets, market linkages are weak, and the unit price of produce is not supported by major food distributors. Fourth, supportive policies on pseudocereals' utilization limit the human contribution from farmers, extension personnel, and researchers in both private and public sectors (Aderibigbe et al. 2022, Hall et al. 2014). Pseudocereals' value chains have a huge capacity to contribute to the food systems' delivery and complement a handful mainstream crops which the world currently relies on. Current underutilization is related to value chain bottlenecks, especially seed systems, distribution networks and the output market structures.

5.1 Seed Systems and Input Markets

Improved genetic performance through breeding has contributed to more than 60% of production capacity developed during the first green revolution era. Beyond the success of the 1960–1970s, several breeding objectives were followed resulting in varietal development targeting a range of mega environments. Improved high yielding varieties introduced into the input market systems increased the production potential for crops-based starch, protein, vitamins, and minerals' sources for both humans and livestock. However, other supportive services that shape diverse and functional seed systems were not attended to until more recently towards the end of the 20th century. For instance, linking farmers to input through private sector agro-dealer networks was only promoted recently in limited areas. Similarly, other inputs that ensure sustainable production such as fertilizers and agrochemicals either remained unavailable or were too costly for most farmers, thereby limiting access and utilization. With increased awareness of the importance of seed systems and related inputs, there is need for making such provisions for pseudocereals in the 21st century if dietary diversity, food, and nutrition security are to be achieved. Increasing access to improved high yielding and climate smart pseudocereals seeds is central to their contribution to current agriculture value chains.

Breeding for improved high yielding pseudocereals has not been supported widely because of deemed limited importance attached to pseudocereals and many other species outside the mainstream crops currently utilized for food, feed, and fiber. To advance the cause of pseudocereals, breeding programs can take advantage of species diversity found in both domesticated cultivars and those harvested from natural (*in situ*) stands found in traditional communities. Modern and advanced breeding techniques, developed and tested using mainstream crops, can be unleashed on prominent pseudocereals and the results could be phenomenal (e.g., Talabi et al. 2022). However, investing time and resources for successful breeding programs need some practical initiatives, policy support and the buy-in of value chain players with a view of utilizing pseudocereals to supplement the mainstream crops for improved food and nutrition security across the various socio-economic classes of society.

Current support for seed services for pseudocereals can be described as skewed towards a few interest-groups, dedicated foundations, institutes in advanced research centers and universities and those promoting the welfare of marginalized communities with access to some of the prominent species. The number of requests for pseudo cereals' accessions from gene banks and other formal sources show minimal traffic when compared to mainstream cereals and legumes. Data on breeding effort and progress over time is rare and scattered for the whole range of pseudocereals, making systematic temporal analysis difficult at present. However, information on genetic variability in domesticated cultivar such as amaranth, buckwheat and quinoa seem to be widely available in literature. Similarly, there is evidence of whole or partial genome sequencing for a few crop species. Aderibigbe et al. (2022) reported the importance and participation of farmers' groups in community seed bank activities and limited breeding. Several seed systems' players have been organized to support seeds of the main crops in the recent past and once convinced of the impending need for supporting pseudocereals and superfoods, the collaborative effort of farming communities, advanced research institutions, national breeding programs and private sector seed houses will be required.

A range of pseudocereals' inputs are important in developing improved agronomic practices for the crops. Most important is the provision of the suitable yield enhancing primary inputs for specific sites as guided by the interactions with biophysical factors of production. To improve efficiencies, there is need for supporting mechanization across pseudocereals' value chains. Mechanization has been identified as the missing link in the mainstream crop value chains and its improvement was closely linked to both food and nutrition, and income security of producers.

5.2 Distribution Systems

Postharvest management of pseudocereals include processing, storage and more importantly, the distribution networks for onward value addition and delivery to end users. At production centers and in communities where pseudocereals are currently produced, documented distribution systems are underdeveloped. This scenario limits adoption rates and explains the high rates of switching from pseudocereals' production by farmers. Smallholder farmers often struggle to participate in lucrative market due to transportation costs, low market prices and mostly limited information on the distribution network supporting their produce. As popularity of the importance of pseudocereals as superfoods increases, there is need for researchers and extensionists to work together and promote an easy-to-understand produce distribution network for farmers who adopt production of the crops.

Mainstream cereals such as maize, rice and wheat have developed produce distribution networks supported by suitable infrastructure either built by national governments or by key private sector players. Similarly, in countries and regions of leading producers of cereals, strong networks of grain traders' boards and associations have been formed to support the local and international distribution systems. More lobbying is required to build similar association around prominent pseudocereals that can contribute to food and nutrition security.

5.3 Output Markets

Information on commodity markets for several crops is now widely available for use by local farmers in making the important decision of when and where to participate in trade. Daily and hourly updates on the performance of some high value commodities is now available on mobile phones, a major achievement of the promotion of e-marketing systems that started some 30 years ago. Without proper value chain development, pseudocereals may not have been a part of these global market platforms for a long time.

The major and most important reported indices in output markets are the commodity price and the delivered quantities. Pseudocereals' producers need these data for their long-term planning of production and market participation. Similarly, the pricing information is often used by farmers in determining participation in the output markets locally and globally. To date, there is limited information online on quantities delivered at the large output markets and the prices obtainable at different times of the year.

While e-commerce and e-markets have served the value chain stakeholders of the mainstream crops, the fourth industrial revolution (4IR) with advanced and improved utilization of both information and technology is available for improving crop value chains. Opportunities now exist to serve several farming communities and enable their participation in markets far-afield. Pseudocereals' awareness and policy development will benefit from the data produced by multiple sensing devices and satellite imagery and processed using machine learning techniques.

6. Pseudocereals' Value Chains and Machine Learning

The capabilities brought about by the "Internet of Things" (IoT), a key feature of the fourth industrial revolution (4IR), present numerous opportunities for increasing information availability on pseudocereals. Modern computing capabilities can be applied in the processing of nutrient contents' data in relation to wild species and indigenous production practices of several communities and genetic variability. There is need to analyze the nutrient quality data from pseudocereals in relation to processing techniques, time of storage and forms of products developed for consumption. The relationship between production practices and the biophysical factors is important in determining the suitable germplasm for growing under a range of climate change scenarios. The potential value chain efficiencies of key pseudocereals' crops need to be explored. Estimation of plant stands of species that are approaching extinction in the wild and area under cultivation for the improved cultivars can also be studied further as more data is generated.

In population hot spots where food demands are diverse and relatively large, data storage and downloading for processing is extremely important. Machine learning and artificial intelligence applications have been lauded for the unique multidimensional data processing capabilities. In developing sustainable agronomic practices, both numerical and categorical data frames are important sources of new insights and improved technological interventions into improved practices from agro-industrial systems, which provide useful information for farmers about the soil.

As production practices of superfoods evolve, there is need for keeping track of the new interventions and innovations. Smallholder agricultural production has suffered the fate of uncharacterized production unit with no record on history and scientific explanation of the combinations of practices which explain the success stories. There are opportunities of creating data records, storing, and processing the same for improved production practices for the smallholder farmers. Often using data from different environments and extrapolation has been a major source of error, e.g., in fertilizer recommendations and other cropping practices. However, the initial investment in data logger, sensors and satellite data collection tools comes at a huge cost to smallholder farmers. More sensitization is required in investing on data collection tools. Similarly policy makers also need to support the investing in data collection from smallholder farms.

Pseudocereals are a source of healthy foods and it's becoming a global trend to know the details of the communities where the food source is coming from and answering questions on how climate smart and environmentally friendly are the production practices. Multiple sensing agricultural devises can bridge this gap and answer to the clarion calls from marginalized communities who need support from consumers of produce located across the globe. With the IoT technologies, product tracking and feedback mechanisms are very important inputs and outputs of an improved modern relationship between producers and consumers. Similarly, the dynamics of practices associated with pseudocereals harvested from the wild can easily be reported in a transparent manner. The community of practice in the production, processing, and value addition as part of the developed value chain can be improved by using the combination of multiple sensing gadgets and machine learning for improved efficiencies in adopted crops. Connecting producing communities and the consumers will also be possible if more investment and supportive policies are put in place to support pseudocereals' proliferation and utilization as supplementary sources of food and nutrition for human and livestock.

The marginalization of pseudocereals and the communities currently producing the superfoods is rampant; more work will be required to remove the prejudices by engaging stakeholders and lobbying for policies that support the development of pseudocereals' value chains.

7. General Discussion

7.1 Germplasm Diversity and Breeding Effort

Wild population and their distribution are important sources of variability required in breeding for improved cultivars of pseudocereals and other superfoods. Successful breading programs rely on wide genetic variability; there is a need for more effort in characterizing the gene pools associated with pseudocereals and superfoods, especially those found in marginal area and those that are used as the principal food sources by traditional communities.

7.2 Sustainable Agronomic Practices

The overarching goals of food and nutrition security for all and deployment of carbon neutral technologies in agriculture will only be attained when the food and material needs of population hot spots are satisfied while natural resources are used sustainably in marginal areas. The world must grapple with the needs of a growing affluent society with fast changing food tastes and, at the same time, address the development gaps that exist in marginal communities. In all the high potential areas, the food value chain systems are well developed and functional, while in low potential marginal areas more innovative natural resource management practices will need to be developed and applied without delays. Improved agronomic practices on pseudocereals and other minor crops are required in both high and low potential areas to keep pace with the population increase. Similarly, there is need to quicken the development of climate smart agricultural practices applicable to pseudocereals. It seems that climate variability will challenge the current agronomic practices in way that the majority of farmers and researchers are yet to understand. Regardless of the trajectory, climate smart practices leading to yield gains from management of nutrient and water cycles, pests and disease management and deployment of improved germplasm will dominate the discourse on improved agronomic practices.

Fertilizer management is important in pseudocereals such as quinoa as there is evidence of yield quality and quantity increase following appropriate N-fertilizer application (Wang et al. 2020). Studies on the interaction between fertilizer and water showed an increased drought tolerance following application of N fertilizer under controlled environment (Alandia et al. 2016). Judicious application of mineral and organic fertilizer on amaranth has yielded better results (Alam et al. 2007), a phenomenon also observed in irrigation (water).

The use of organic fertilizer as nutrient sources and as building block of soil-C is associated with successful methods of farming. However, in marginal areas the amount of organic matter available for incorporation into the soil quickly diminishes. Therefore, the need to make the best use of the available crop residues and other organic materials of different quality is extremely important for marginal areas. In areas where pseudocereals are grown, there is need for developing the scientific basis of balancing the use of organic materials for plant growth and nutrition, organic amendments as soil-C storage and organic management methods to keep emissions from soil-C low from such systems. Equally important is reducing the amount of forest lost to produce pseudocereals in locations where improved soil management techniques are least adopted.

7.3 Complementarity without Competition

Pseudocereals play an important complementary role in the performance of cropping systems, for instance, when they are grown together in intercrops or in rotation with cereal food crops. It is now evident that the current cropping systems need to evolve to include in one field several crop species which are beneficial to human health and are supportive of ecosystem services that ensure sustainable production. If cropping systems could be designed to increase diversification of food products and hence nutrient and nutraceutical sources, then healthier diets will be easier to develop at local and international level. Such cropping systems could contribute to the challenges faced in communities where sharp test of food products are driving markets. Developing value addition products which include cereals, legumes and pseudocereals is an area with great potential for scientific discovery.

Unlike previous crop production models, where a few crops were promoted at the expense of the rest, there is now an urgent need for consensus on a global food mix constituted by major and minor crop to promote healthy diets. The examples of dietary diseases and solutions emanating from superfoods, the food quality improvements as demonstrated by mixtures of major and minor crops, e.g., use of chia seeds in wheat bread, all point to the need of a shift in mindset and promotion of a range of crop types. Complementarity between pseudocereals and other crop types leads to practical solutions to dietary human health issues and hence the need to reduce negative competition in production of nutritious crops.

8. Conclusions

The majority of prominent pseudocereals qualify as superfoods as they provide high nutrient content and heathy foods for human bodily needs. Pseudocereals are good sources of energy and starch, proteins, vitamins and minerals and anti-inflammatory organic compounds. Though currently underutilized, pseudocereals can complement mainstream crops such as maize, rice and wheat in reducing the hunger burden in many communities. Superior quality gluten free and low glycemic foods can be provided to persons with healthy conditions if barriers to production of pseudocereals are removed. Similarly, food-based vitamins' supplements and minerals can be delivered to the needy using superfoods. World blindness will be lower with increased availability of heathy crop-based foods.

Production practices of pseudocereals are rarely documented. In many occasion, management practices are extrapolated from major food crop. There are also wrong signals from proponents of minimal investment on pseudocereals. More research on systems' agronomy is required, especially on improved varieties. There is evidence that pseudocereals are resilient to both land degradation and climate change; however, more work is required in this area so that the relative superiority of these crops compared to mainstream crops can be empirically determined. The potential of utilizing pseudocereals as solutions to the currently increasing challenges on salt affected soil, such as salinity, soil acidity and sodicity, need further research. Similarly, the heat and drought tolerance across the whole range of crops need to be reviewed.

The value chains of the majority of pseudocereals are underdeveloped because of the history of relying on a handful of cereals and legumes for food and nutrition over centuries. Developing value chains, in particular the seeds and input, distribution network and the output markets, is of paramount importance in the 21st century if hunger and malnutrition is to be averted. Investing in big data equipment on small and marginalized farms is an important step in the use of both multiple sensing devices and machine computing and learning capabilities. Digital connection between producers and consumers of pseudo cereals is highly feasible using modern-day internet of thing technologies. The interest of policy makers in strengthening generation and use of big data from small farms requires serious lobbying.

Complementarity between mainstream crops and pseudocereals superfood crop is evident and strategies used in reducing the population threatened by food shortages need to take this into account. Promotion of healthy diets, diverse foods and improved distribution networks locally and globally is central in avoiding widespread hunger and malnutrition.

References

Abdel-Aal, E.S.M., Hucl, P., Miller, S.S., Patterson, C.A. and Gray, D. 2011. Microstructure and nutrient composition of hairless canary seed and its potential as a blending flour for food use. Food Chemistry 125(2): 410–416.

Aderibigbe, O.R., Ezekiel, O.O., Owolade, S.O., Korese, J.K., Sturm, B. and Hensel, O. 2022. Exploring the potentials of underutilized grain amaranth (*Amaranthus* spp.) along the value chain for food and nutrition security: a review. Critical Reviews in Food Science and Nutrition 62(3): 656–669.

Akin-Idowu, P.E., Odunola, O.A., Gbadegesin, M.A., Ademoyegun, O.T., Aduloju, A.O. and Olagunju, Y.O. 2017. Nutritional evaluation of five species of grain amaranth–an underutilized crop. International Journal of Sciences 6(1): 18–27.

Alam, M.N., Jahan, M.S., Ali, M.K., Islam, M.S. and Khandaker, S.M.A.T. 2007. Effect of vermicompost and NPKS fertilizers on growth, yield and yield components of red amaranth. Australian Journal of Basic and Applied Sciences 1(4): 706–716.

Alandia, G., Jacobsen, S.E., Kyvsgaard, N.C., Condori, B. and Liu, F. 2016. Nitrogen sustains seed yield of quinoa under intermediate drought. Journal of Agronomy and Crop Science 202(4): 281–291.

Babarinde, G.O., Adeyanju, J.A., Ogunleye, K.Y., Adegbola, G.M., Ebun, A.A. and Wadele, D. 2020. Nutritional composition of gluten-free flour from blend of fonio (*Digitaria iburua*) and pigeon pea (*Cajanus cajan*) and its suitability for breakfast food. Journal of Food Science and Technology 57(10): 3611–3620.

Barka, A. and Blecker, C. 2016. Microalgae as a potential source of single-cell proteins. A review. Base.

Bender, D. and Schönlechner, R. 2021. Recent developments and knowledge in pseudocereals including technological aspects. Acta Alimentaria 50(4): 583–609.

Caciro, C., Pragosa, C., Cruz, M.C., Pereira, C.D. and Pereira, S.G. 2022. The role of pseudocereals in celiac disease: reducing nutritional deficiencies to improve well-being and health. Journal of Nutrition and Metabolism, 2022.

Caselato-Sousa, V.M. and Amaya-Farfán, J. 2012. State of knowledge on amaranth grain: a comprehensive review. Journal of Food Science 77(4): R93–R104.

Das, S., Khound, R., Santra, M. and Santra, D.K. 2019. Beyond bird feed: proso millet for human health and environment. Agriculture 9(3): 64.

De Boer, J. and Aiking, H. 2011. On the merits of plant-based proteins for global food security: marrying macro and micro perspectives. Ecological Economics 70(7): 1259–1265.

Devi, P.B., Vijayabharathi, R., Sathyabama, S., Malleshi, N.G. and Priyadarisini, V.B. 2014. Health benefits of finger millet (*Eleusine coracana* L.) polyphenols and dietary fiber: a review. Journal of Food Science and Technology 51(6): 1021–1040.

Ebert, A.W., Wu, T.H. and Wang, S.T. 2011. Vegetable amaranth (*Amaranthus* L.). AVRDC Publication, (11–754), p. 9.

Eshete, M.A., Asfaw, Z. and Kelbessa, E. 2016. A review on taxonomic and use diversity of the family Amaranthaceae in Ethiopia. Journal of Medicinal Plants Sciences 4(2): 185–194.

FAO, IFAD, UNICEF, WFP and WHO. 2021. The State of Food Security and Nutrition in the World 2021.Transforming food systems for food security, improved nutrition and affordable healthy diets for all. Rome, FAO, pp. 240. https://doi.org/10.4060/cb4474en.

Fernandez-Stark, K., Bamber, P. and Gereffi, G. 2012. Inclusion of small-and mediumsized producers in high-value agro-food value chains. Inter-American Development BankMultilateral Investment Fund (IDB-MIF), May.

Gadgil, S. and Rupa Kumar, K. 2006. The Asian monsoon—agriculture and economy. pp. 651–683. *In*: The Asian Monsoon. Springer, Berlin, Heidelberg.

Gebru, Y.A., Sbhatu, D.B. and Kim, K.P. 2020. Nutritional composition and health benefits of teff (*Eragrostis tef* (Zucc.) Trotter). Journal of Food Quality, 2020. doi.org/10.1155/2020/9595086.

Gilland, B. 2002. World population and food supply: can food production keep pace with population growth in the next half-century? Food Policy 27(1): 47–63.

Godfray, H.C.J., Beddington, J.R., Crute, I.R., Haddad, L., Lawrence, D., Muir, J.F., et al. 2010. Food security: the challenge of feeding 9 billion people. Science 327(5967): 812–818.

Grundy, M.M., Momanyi, D.K., Holland, C., Kawaka, F., Tan, S., Salim, M., et al. 2020. Effects of grain source and processing methods on the nutritional profile and digestibility of grain amaranth. Journal of Functional Foods 72: 104065.

Habiyaremye, C., Matanguihan, J.B., D'Alpoim Guedes, J., Ganjyal, G.M., Whiteman, M.R., Kidwell, K.K., et al. 2017. Proso millet (*Panicum miliaceum* L.) and its potential for cultivation in the Pacific Northwest, US: a review. Frontiers in Plant Science 7: 1961.

Hahn, S.K. 1989. An overview of African traditional cassava processing and utilization. Outlook on Agriculture, 18(3): 110–118.

Hall, J., Matos, S.V. and Martin, M.J. 2014. Innovation pathways at the base of the pyramid: Establishing technological legitimacy through social attributes. Technovation 34(5-6): 284–294.

Jnawali, P., Kumar, V. and Tanwar, B. 2016. Celiac disease: overview and considerations for development of gluten-free foods. Food Science and Human Wellness 5(4): 169–176.

Katunzi-Kilewela, A., Kaale, L.D., Kibazohi, O. and Rweyemamu, L.M. 2021. Nutritional, health benefits and usage of chia seeds (*Salvia hispanica*): A review. African Journal of Food Science 15(2): 48–59.

Kringel, D.H., El Halal, S.L.M., Zavareze, E.D.R. and Dias, A.R.G. 2020. Methods for the extraction of roots, tubers, pulses, pseudocereals, and other unconventional starches sources: a review. Starch-Stärke 72(11-12): 1900234.

Kulczyński, B., Kobus-Cisowska, J., Taczanowski, M., Kmiecik, D. and Gramza-Michałowska, A. 2019. The chemical composition and nutritional value of chia seeds—current state of knowledge. Nutrients 11(6): 1242.

Lale, E.F.E. 2017. An alternative food and medicinal crop: Quinoa (*Chenopodium quinoa* L.). In International Congress on Medicine and Aromatic Plants, May 10–12, 2017. Pg 288–292.

Mlakar, S.G., Bavec, M., Turinek, M. and Bavec, F. 2009. Rheological properties of dough made from grain amaranth-cereal composite flours based on wheat and spelt. Czech Journal of Food Sciences 27(5): 309–319.

Nadathur, S., Wanasundara, J.P. and Scanlin, L. (eds.). 2016. Sustainable protein sources. Academic Press.

Nadathur, S.R., Wanasundara, J.P.D. and Scanlin, L. 2017. Proteins in the diet: challenges in feeding the global population. In Sustainable Protein Sources (pp. 1–19). Academic Press.

Olaniyi, J.O. 2007. Evaluation of yield and quality performance of grain amaranth varieties in the Southwestern Nigeria. Research Journal of Agronomy 1(2): 42–45.

Oliveira J.S. and DE Carvallo, M.F. 1975. Nutritional value of some edible leaves used in Mozambique. Econ. Bot. 29: 255.

Olufolaji, A.O., Odeleye, F.O. and Ojo, O.D. 2010. Effect of soil moisture stress on the emergence, establishment, and productivity of Amaranthus (*Amaranthus cruentus* L.). Agriculture and Biology Journal of North America 1(6): 1169–1181.

Pardey, P.G., Beddow, J.M., Hurley, T.M., Beatty, T.K. and Eidman, V.R. 2014. A bounds analysis of world food futures: Global agriculture through to 2050. Australian Journal of Agricultural and Resource Economics 58(4): 571–589.

Pirzadah, T.B. and Malik, B. 2020. Pseudocereals as super foods of 21st century: recent technological interventions. Journal of Agriculture and Food Research 2: 100052.

Pitman, M.G. and Läuchli, A. 2002. Global impact of salinity and agricultural ecosystems. In Salinity: Environment-plants-molecules (pp. 3–20). Springer, Dordrecht.

Ragone, D. 2006. *Artocarpus camansi* (breadnut). The Breadfruit Institute, National Tropical Botanical Garden, Hawaii.

Repo-Carrasco-Valencia, R. and Arana, J.V. 2017. Carbohydrates of kernels. Pseudocereals: chemistry and technology, pp. 49–69.

Rodríguez, J.P., Rahman, H., Thushar, S. and Singh, R.K. 2020. Healthy and resilient cereals and pseudo-cereals for marginal agriculture: molecular advances for improving nutrient bioavailability. Frontiers in Genetics, p. 49.

Ruiz, K.B., Biondi, S., Oses, R., Acuña-Rodríguez, I.S., Antognoni, F., Martinez-Mosqueira, E.A., et al. 2014. Quinoa biodiversity and sustainability for food security under climate change. A review. Agronomy for Sustainable Development 34(2): 349–359.

Sabate, J. and Soret, S. 2014. Sustainability of plant-based diets: back to the future. The American Journal of Clinical Nutrition 100(suppl_1): 476S–482S.

Schoenlechner, R. 2016. Properties of pseudocereals, selected specialty cereals and legumes for food processing with special attention to gluten-free products. Journal of Land Management, Food and Environment 67(4): 239–248.

Sekhavatizadeh, S.S., Hosseinzadeh, S. and Mohebbi, G. 2021. Nutritional, antioxidant properties and polyphenol content of quinoa (*chenopodium quinoa* willd) cultivated in Iran. Future of Food: Journal on Food, Agriculture and Society, 9(2).

Sharma, V.P., De, S. and Jain, D. 2019. Managing agricultural commercialisation for inclusive growth in South Asia. Gates Open Res. 3(507): 507.

Shiferaw, B., Prasanna, B.M., Hellin, J. and Bänziger, M. 2011. Crops that feed the world 6. Past successes and future challenges to the role played by maize in global food security. Food Security 3(3): 307–327.

Soeters, P., Bozzetti, F., Cynober, L., Forbes, A., Shenkin, A. and Sobotka, L. 2017. Defining malnutrition: a plea to rethink. Clinical Nutrition 36(3): 896–901.

Tacer-Caba, Z. 2019. The concept of superfoods in diet. In The Role of Alternative and Innovative Food Ingredients and Products in Consumer Wellness (pp. 73–101). Academic Press.

Talabi, A., Vikram, P., Thushar, S., Rahman, H., Ahmadzai, H., Nhamo, N., et al. 2022. Orphan crops: a best fit for dietary enrichment and diversification in highly deteriorated marginal environments. Frontiers in Plant Science, 13.

Tang, Y. and Tsao, R. 2017. Phytochemicals in quinoa and amaranth grains and their antioxidant, anti-inflammatory, and potential health beneficial effects: a review. Molecular Nutrition & Food Research 61(7): 1600767.

Von Grebmer, K., Fritschel, H., Nestorova, B., Olofinbiyi, T., Pandya-Lorch, R. and Yohannes, Y. 2008. Global hunger index: the challenge of hunger 2008 (No. 594-2016-39943).

Wang, N., Wang, F., Shock, C.C., Meng, C. and Qiao, L. 2020. Effects of management practices on quinoa growth, seed yield, and quality. Agronomy 10(3): 445.

Webster-Gandy, J., Maddenand, A. and Holdsworth, M. (eds.). 2020. Oxford Handbook of Nutrition and Dietetics 3e. Oxford University Press.

WHO. 2013. Global action plan for the prevention and control of noncommunicable diseases 2013–2020. World Health Organization, Geneva.

WHO. 2018a. Global Health Estimates 2016. Deaths by Cause, Age, Sex, by Country and by region, 2000–2016. World Health Organization, Geneva.

WHO. 2018b. Noncommunicable diseases country profiles 2018. World Health Organization, Geneva: License: CC BY-NC-SA 3.0 IGO.

Zhu, F., 2016. Buckwheat starch: Structures, properties, and applications. Trends in Food Science & Technology, 49, pp.121-135.

Zhu, F. 2020. Dietary fiber polysaccharides of amaranth, buckwheat and quinoa grains: a review of chemical structure, biological functions and food uses. Carbohydrate Polymers 248: 116819.

5

Plant-Based Foods

Karthik Pandalaneni,[1] *Ozan Kahraman,*[2] *Junzhou Ding,*[3] *Ragya Kapoor*[4]
and *Hao Feng*[4,*]

1. Introduction

1.1 Why is "Plant-based" Important?

A plant-based diet is defined as an eating pattern dominated by fresh or minimally processed plant foods and decreased consumption of meat, eggs, and dairy products. According to a definition by Ostfeld (2017), a plant-based diet "consists of minimally processed fruits, vegetables, whole grain, legumes, nuts, seeds, herbs, and spices and excludes all animal products, including red meat, poultry, fish, eggs, and dairy products". Over the last few years, there has been a tremendous consumer interest in reducing meat consumption and choosing plant-based foods and plant-based proteins. According to a report by the Good Food Institute, the retail market for plant-based foods was worth $7 Billion in 2020, up from $5.5 billion in 2019 (Good Food Institute 2021). Furthermore, plant-based food (i.e., plant-based meat, cheese, milk, yogurt, etc.) sales in 2020 grew 2 times faster than overall food sales and 3 times the rate of their animal-based counterparts. The number one and the biggest reason motivating the consumers to move toward plant-based diet is an increased awareness about the benefits of these foods on human heath (Aschemann-Witzel and Peschel 2019). There are numerous studies showing the positive correlation between high consumption of plant based foods and decreased risk of cardiovascular diseases and certain cancers (Massera et al. 2016). The second biggest driver is the moral and ethical concerns related to the climate and resource impact of animal-based products (Circus and Robison 2019). Plant-based diets are known to be more sustainable as compared to diets rich in animal-based products. This is mainly because they use lesser natural resources and therefore, are less taxing on the environment (Aschemann-Witzel et al. 2020). According to some studies, the land required to produce soy proteins is 6–17 times lesser than land required to raise feed to produce animal protein. Similarly, the ratio of water used in production of soy protein compared with the same quantity of animal protein is 4 to 26 (Sabaté and Soret 2014). There are certain other factors driving the shift towards plant-based diets which include taste as well as desire to try new flavors. With this heightened consumer interest in health focused and sustainable plant-based products, several plant-based and alternative protein

[1] Plant Protein Innovation Center, University of Minnesota, Minneapolis, MN 55455.
[2] Applied Food Sciences, Coralville, IA 52241.
[3] Food Science and Processing Research Center, Shenzhen University, Shenzhen, China.
[4] Department of Food Science and Human Nutrition, University of Illinois, Urbana, IL 61801.
* Corresponding author: hfeng@ncat.edu, haofeng@illinois.edu

sources are being studied and developed in order to provide sustenance for people around the world (Aschemann-Witzel et al. 2020). The following section will discuss a few of these protein sources along with their characteristics.

1.2 Sources of Plant-based Food Ingredients

Fruits, vegetables, whole grains, and plant-based protein sources are the building blocks of any plant-based diet. Plant-based proteins are mainly derived from sources such as legumes which include beans, peas, and lentils. Among all other plant-based proteins, soybean is the largest source of plant-protein and is considered to be a complete protein as it contains all the essential amino acids with a nutritional and biological value roughly equivalent to that of animal protein (Watanabe et al. 2018). Additionally, pea protein is a relatively new type of plant protein majorly obtained from leguminous crop of field pea. Compared to soybean or other proteins, pea protein is characterized by its high digestibility and relatively less allergenic response (Zhao et al. 2020). Moreover, it has a well-balanced amino acid profile, with high amount of lysin and is widely available at a low cost. Recently, proteins derived from industrial hemp plant (*Cannabis sativa* L.) are gaining popularity in the food industry. Hemp proteins have an adequate level of essential amino acids and are characterized by good digestibility. However, hemp plant is difficult to grow and due to various governmental regulations on it, it can be costly for both suppliers and manufacturers.

Some other sources of plant-based proteins include grains such as rice and wheat. Even though rice is not as rich a protein source as soybean or pulses, some of the byproducts of rice processing such as rice bran, endosperm and broken rice kernels can serve as a cheap source of plant-protein. Similarly, wheat is an important cereal crop that can be utilized for valuable protein.

1.3 Plant-based Products and Brands in the Market

As a fast growing food category, plant-based foods and ingredients have penetrated into many food sectors, including dairy, meat, seafood, and egg alternatives. According to the data report from Markets and Markets, the dairy alternatives market was valued at 22.6 billion in 2020 and is projected to reach 40.6 billion by 2026. Plant-based dairy categories in aggregate grew by 24% in the U.S.; plant protein-based milk accounts for 15% of all dollar sales of U.S. retail milk (Spins and GFI). Plant-based meat grew 45% in dollar sales from 2019. Over 290 million units of plant-based meat were sold in the past year, a unit increase of 36%. Plant-based meat sales now account for 2.7% of all dollar sales for retail packaged meat and approximately 1.4% of all dollar sales for total retail meat. In 2020, plant-based seafood sales account for less than 1% of dollar sales, compared to the conventional U.S. seafood market worth tens of billions of dollars. The sales ($27 million) of plant-based egg grew by 168% in the past year, which experienced the greatest dollar sales growth. Table 1 lists these four plant-based categories' major products and key brands in the global market.

2. Dairy Alternatives

Among the plant-based foods' categories, a notable and consistent growth in sales in the dairy alternatives category is reported to be 3.6 billion dollars just from the U.S. in 2020 (Anon. 2021). This growth in dairy alternatives sector is majorly contributed by the plant-based milk category; however, consumer interest and significant innovations brought in by other dairy alternative categories such as plant-based yogurt, cheese, ice cream, and butter are contributing factors as well. This section discusses the properties and application challenges of some key categories in dairy alternatives.

Table 1. Plant-based products and brands in the market.

Categories	Major Products	Key Brands (Partial List)
Dairy alternatives	Plant protein milk (Soy, Almond, Coconut, Peanuts, Cashew, Oats, Rice, Quinoa, Chia, Flax, Hemp, etc.); Cheese Yogurt, Butter, Ice cream, Creamers	Blue Diamond Growers (USA), Hain Celestial (USA), Panos Brands (USA), Califia Farms (USA), Good Karma Foods (USA), Yoconut Dairy Free (USA), Yumbutter (USA), Magnum by Unilever (Global), Danone "Silk, So Delicious" (France), Triballat Noyal (France), Sun Opta (Canada), Earth's Own Food Company (Canada), Daiya Foods (Canada), Sanitarium Health Food Company (Australia), Freedom Foods Group (Australia), Pureharvest (Australia), Valsoia (Italy), Döhler (Germany), Oatly (Sweden), Kikkoman (Japan), Lo Lo (China), Yin Lu (China), Dali Foods (China), Vitasoy (China), Coconut Palm (China)
Meat alternatives	Vegan Burger Patty, Steak, Nugget, Bacon, Ground Beef/Pork, Meat Balls, Hotdogs/Links	Impossible (USA), Beyond (USA), PlantEver by Cargill (USA & China), The Vegetarian Butcher by Unilever (Global), Harvest Gourmet by Nestle China, Sweet Earth by Nestle USA, Be Leaf (USA), Lightlife (USA), Pure Farmland (USA), Abbot's Butcher (USA), Gardein (USA), All Vegetarian (USA), Quorn (USA), Loving Hut (USA & Asia), Worthington (USA), Maple Leaf Foods (Canada), Shuangta Food (China), Whole Perfect Food (China), Hong Chang Vege Farm (China), Sulian Food (China)
Seafood alternatives	Plant-based sashimi, fish fillets, fishcakes, fish burgers, tuna spread, crab cakes, coconut shrimp, abalone, scallops	New Wave Foods (USA), Ocean Hugger Foods (USA), The plant based seafood co (USA), Good Catch (USA), Gardein (USA), All Vegetarian (USA), Jinka (USA), Quorn (USA), Fuji Boeki (Japan)
Egg alternatives	Ready-to-pour liquid egg, powder mix, cookie dough, mayo, and sauces	JUST Egg (mung bean protein-based) (USA), Vegan Egg by Follow Your Heart (USA), Vegg power scramble and vegan egg yolk by The Vegg New York (USA); Eggcitables (chickpea-based) (Canada)

2.1 Plant-based Milk

Non-dairy or plant-based milk alternatives have seen a tremendous growth because of consumer interest in such products. This increased interest is not only because of absence of allergens (with few exceptions) and lactose but also because consumers now are seeking those products which have health-promoting effects and are sustainable as well. However, to compete with bovine milk in terms of flavor, functionality, and nutritional properties, plant-based milk manufacturers are on a continuous expedition to evaluate new plant-based sources. Although the protein digestibility corrected amino acid score (PDCAAS) and functional properties of soy-based milk are at par with bovine milk, its allergenicity, and genetically modified (GM) food status initiated the search for more "label-friendly" plant-based sources.

Some of the common plant sources that are being used for dairy alternative milk production include cereals (oat, rice, corn, and spelt), legumes (soy, peanut, lupin, and cowpea), nuts (almond, coconut, hazelnut, pistachio, and walnut), seeds (sesame, flax, hemp, and sunflower), and pseudo- cereals (quinoa, teff, and amaranth). The production process of plant-based milk on an industrial scale includes several steps such as pretreatment, grinding, thermal treatment, extraction, emulsification, homogenization, and sterilization/pasteurization. In the pretreatment process, methods such as dehulling, soaking, roasting, and defatting are employed. This process improves the overall acceptability of the final product. The pretreatment step is followed by grinding and thermal processes. In these steps, the anti-nutritional compounds and enzyme activity present in the raw materials are reduced. These steps produce a mixture of dissolved and suspended components containing proteins, fiber, oil, and sugar. This mixture undergoes high shear mixing to break any larger particles and forms an emulsion. The emulsion is usually stabilized by the addition of oils, lipids, and emulsifiers. The obtained product is homogenized and heat-treated for food safety assurance and storage stability before packaging. In order to improve consumer acceptability and nutritional composition of the plant-based milk, various compounds that enhance aroma, flavor, and color and vitamins and minerals are added during formulation (Vogelsang- O'Dwyer et al. 2021, McClements et al. 2019, Sethi et al. 2016, Kyriakopoulou et al. 2021).

Although the protein content of plant-based milk is low (0.6–2.8%) when compared to bovine milk (3.4–3.6%), the presence of bioactive compounds makes them suitable as dairy alternatives. The bioactive compounds differ based on the plant source and include compounds such as β-glucan, phytosterols, isoflavones, lignans, and omega-3-fatty acids (Sethi et al. 2016, Munekata et al. 2020). However, the production of plant-based milk is still posed with challenges such as poor solubility, poor nutritional composition, low emulsion stability, and sensory issues such as presence of off-flavors. Therefore, the acceptability of plant-based milk can be further increased by improving the sensory and nutritional profiles. Sensory issues caused by the poor functional properties of the plant-based protein, emulsion stability, and flavor can be addressed during pretreatment and enzyme hydrolysis. The nutritional properties of plant-based milk can be enhanced by fortification and blending different plant-based proteins. Blending two or more plant-based proteins based on their amino acid profile and functional properties can result in a plant-based milk with a balanced nutritional profile and improved PDCAAS score. Table 2 presents the key characteristics of plant-based milk, challenges in their formulation and processing, and potential technical solutions.

2.2 Plant-based Yogurt

Plant-based yogurt is another key dairy alternative category which has seen increased sales in the past few years. In the U.S., plant-based yogurt alternatives are expected to be valued at 1.3 billion US dollars by 2027. Plant-based yogurt has proven to be popular, especially among millennials, due to its nutritional benefits from bioactive compounds obtained during the fermentation process (Craig and Brothers 2021). In addition to that, their ability to satisfy the consumers' organoleptic needs with an appealing taste and other attributes such as sweet, moist, soft, and smooth have contributed to their increased market value.

Plant-based yogurts are generally prepared by fermenting hydrated, homogenized, and heat-treated plant-based products derived from different sources such as cereals, legumes, nuts, seeds, and pseudo-cereals (Brückner-Gühmann et al. 2019, Mäkinen et al. 2016, Pandey et al. 2021). Some of the major challenges faced by the plant-based yogurt manufacturers are often associated with the stability, appearance, texture, flavor, and nutritional properties of the product. While fermentation of plant-based sources imparts desirable flavors and health benefits from having prebiotics and probiotics, acidification during the fermentation often leads to poor gel formation and syneresis. In order to avoid phase separation during storage, hydrocolloids are used in the product formulation. The addition of oil and gelling agents such as natural gums, starches, and proteins also improves the mouthfeel of the yogurts and stabilizes the particles in suspension to avoid syneresis (Greis et al. 2020, Craig and Brothers 2021). Recent literature also suggested the use of processing methods like high-pressure processing (HPP), heat treatment, and pH adjustment during the protein pretreatment process to improve the texture and stability of plant-based yogurt (Klost et al. 2020, Kyriakopoulou et al. 2021, Sim et al. 2020). The addition of dietary fibers such as pectin, locust bean gum, and inulin further reduces the wateriness and thinness and improves the texture and stability of the plant-based yogurts.

The functional properties of plant-based sources must be evaluated and considered when selecting them to be used in the formulation as they significantly contribute to the stability, texture, and nutritional properties of plant-based yogurts. Plant-based yogurts prepared using soy, coconut, cashew, and almond sources received consumers' interest because of their similarity to dairy yogurts. Plant-based source blends are also proven to enhance the nutritious and organoleptic properties of plant-based yogurts. Yogurt developed using coconut and soy milk resulted in a product with a good source of proteins and fatty acids (Tangyu et al. 2019, Grasso et al. 2020). The protein content

Table 2. Some major plant-based sources for dairy-alternative applications, challenges in processing, and technical solutions (Adapted from (Munekata et al. 2020, Sethi et al. 2016, Paul et al. 2020)).

Category	Plant-based sources	Protein (g)	Lipids (g)	Total Carbohydrates (g)	Functional compounds	Advantages	Challenges	Technical solutions
Soy	80	7	4	4	Isoflavones, Phytosterols	Decrease blood pressure level Hypolipidemic effects Effective against chronic disease Recommended against osteoporosis Higher bone density and lower rates of fracture α-Galactosidase activity	Beany flavor due to action of lipoxygenase on unsaturated fatty acids Presence of inhibitors	Vacuum treatment at high temperature, hot grinding, blanching in boiling water, alkaline soaking, use of soy protein isolates, addition of flavoring compounds Denaturation and inactivation by heat
Almond	40	1	3	2	Beta-sitosterol, campesterol and stigmasterol folate, vitamin E (mainly α-tocopherol), niacin (B3), Arabinose	Powerful antioxidant Low-calorie High in vitamin E Prebiotic properties	Presence of allergenic protein amandin Limited cariogenic properties in presence of sucrose Very low-protein Supports growth of pathogenic microorganisms	
Oat	80	2.5	4	-	β-Glucan	Hypocholesterolaemic reduce blood glucose level Increases solution viscosity and can delay gastric emptying time Anti-pathogenic effect	Poor emulsion stability due to high starch content Presence of inhibitors such as phytates	Enzymatic hydrolysis of starch by alpha and beta-amylase Treatment with phytase in order to liberate inorganic phosphate from phytic acid
Rice	130	1	2	27	Phytosterols (_-sitosterol and -oryzanol)	Lowers cholesterol and hypertension Best choice for people with multiple allergies Anti-inflammatory	Poor emulsion stability due to high starch content	Enzymatic hydrolysis of starch by alpha and beta amylase or glucosidase

Sesame	140	1.5	6	16.5	Lignans (sesamin, sesamolin, sesaminol)	Neutraceutical properties such as antioxidative, hypocholesterolemic, anticarcinogenic, antitumor, and antiviral activities	Anti-nutritional factors such as oxalates Low solubility of sesame proteins in water as they are salt soluble and are susceptible to heat denaturation Bitterness and chalkiness	Decortication to remove oxalates as they are confined to outer hull Alkali soaking, roasting, defatting, germination, microwave heating to improve functional properties of protein Roasting and alkali soaking have been observed to improve the overall acceptability and flavor
Hemp	70	2	6	1	Omega 3-fatty acids	Reduce both motion and toxin induced vomiting Anti-thrombotic -anti vasoconstrictive Anti-inflammatory -anti-neuroinflammatory activity	High dosage can induce toxicity and act against inhibiting inflammatory cytokines production	

of the plant-based yogurts can be improved using plant sources such as legumes, pea, and soy (Yang et al. 2021, Craig and Brothers 2021). The unpleasant sensory characteristics related to the plant-based sources and produced in the fermentation process often result in beany off-flavors in the final product. These flavor characteristics can be improved by carefully selecting the strain or mixture of strains used in the fermentation process, pretreating the plant-based source to extract any off-flavor causing compounds, and adding different flavors to the final product (Yang et al. 2021, Kyriakopoulou et al. 2021).

2.3 Plant-based Cheese

Plant-based cheese alternatives (PBCAs) are one of the most emerging plant-based foods in the market due to their cost-effectiveness and ability to meet the needs of consumers with specific dietary requirements (Bachmann 2001). In addition, PBCAs serve as consumers' functional, healthy, and diversified plant-based food options. The global market for alternative cheese products is expected to reach 4 billion US dollars by 2024 (Grasso et al. 2020).

Bachmann (2001) outlined two types of production processes for cheese alternatives. The first process involves conventional cheese-making methods using plant-based milk, whereas the second production process uses blending as a primary unit operation to manufacture PBCAs. Various compounds such as thickeners, coagulants, plant-based fats, flavors, colors, preservatives, and water are added during blending to mimic the texture of dairy cheese. Thickeners such as carrageenan, xanthan, and guar gum are commonly used to improve the firmness and delay syneresis in PBCAs. Various starches such as tapioca, rice, maize, pea fiber, and potato are being explored to be used in this application, while modified starch from corn and potato are often used as thickeners and moisturizers. Several compounds such as acids (lactic acid, tartaric acid, malic acid, citric acid, and glucono-delta-lactone), salts (calcium sulfate, calcium chloride, calcium acetate, calcium lactate, and magnesium sulfate), and enzymes are used as coagulants in the processing of PBCAs. The addition of fats contributes to the emulsion stability and mouthfeel of PBCAs. Coconut, palm, sunflower, rapeseed, soybean, and safflower oils are often used as fat sources in the formulation of PBCAs. Among these oils, coconut oil is deemed desirable because of its higher saturated fatty acids and high melting point. The presence of higher saturated fatty acids also improves the emulsion property in the matrix of PBCAs (Ferawati et al. 2021, Mefleh et al. 2022, Jeske et al. 2018). These non-protein ingredients not only play an important role in replacing the physicochemical characteristics of animal proteins (Mattice and Marangoni 2020) but also contribute to key desirable properties of PBCAs such as shreddability, spreadability, meltability, stretchability, and flavor that drive consumers' acceptance. These key desirable properties are also influenced by the processing conditions such as time-temperature combinations and blending speed.

Modifying inclusion levels of stabilizers, salts, acids, and fat along with processing conditions helps to engineer PBCAs with desired functional properties. As PBCAs are oil-in-water emulsions with fat droplets acting as emulsifiers, functional attributes can be improved by modifying the distribution of fat, protein, and carbohydrates. Partial hydrogenation of oil improves texture, melting profile, mouthfeel, and texture, whereas starches contribute to a decrease in melting profile (Lobato-Calleros et al. 1997, Bachmann 2001, Grasso et al. 2020).

While there is a lot of research done on improving functional and flavor attributes, the nutrition profile of PBCAs needs significant improvement. PBCAs contain low protein and calcium content, which makes them nutritionally inferior when compared to dairy cheese. Although vitamins and minerals can be added to the PBCAs during the last step of the process, the presence of high amounts of carbohydrates, fats, and salts negatively influences the consumers' acceptability. Plant-based protein sources must be identified based on their structural and functional properties, antinutrients, and flavor profile to be used in the formulation of PBCAs. Legumes are identified as a potential source to produce PBCAs; however, pretreatments such as boiling, and roasting are suggested to reduce the levels of anti-nutritional factors and off-flavors (Ferawati et al. 2021). Blends of various plant protein sources, specifically legumes, exhibited great potential to be used in the formulation of PBCAs to improve their nutritional attributes (Mefleh et al. 2022).

2.4 Plant-based Ice Cream

Recently, there has been a significant shift in consumers' interest towards "better-for-you", functional, and non-dairy ice creams. A negative perception of dairy-based desserts like ice cream has motivated manufacturers to launch plant-based versions of their dairy ice creams to get consumers' attention. The brands of ice creams with lower fat and sugar contents and labeled as a good source of protein were observed to have strong growth in the market (Sipple et al. 2022).

The first step in making plant-based ice cream is to prepare an ice cream mix by blending solid and liquid ingredients to form a hydrocolloidal dispersion. This ice cream mix is further whipped to incorporate air into an aqueous matrix. Some of the common ingredients used in the ice cream formulation are fats, solids-not-fat, sweeteners, stabilizers, emulsifiers, water, and flavors. The ingredients used, such as the solids-not-fat, are the main source of protein. The continuous phase of ice cream is comprised of concentrated and unfrozen sugar solution, whereas the dispersed phase contains tiny, flocculated fat globules with some of them surrounding air bubbles. These fat globules are in turn surrounded by proteins and emulsifiers forming a complex multiphase system (Kot et al. 2021, Beegum et al. 2021, Aboulfazli et al. 2015, Romulo et al. 2021).

Macromolecules such as fats, proteins, and complex carbohydrates play an important role in the texture and flavor of plant-based ice cream. Fats improve the melting resistance and ice recrystallization by stabilizing the air phase and increasing the level of fat aggregation. Fats are also a key contributor to improving mouthfeel which is a desirable attribute in ice cream (Hyvönen et al. 2003). Dairy solids are replaced by plant protein isolates in plant-based ice cream formulation; however, plant-based proteins that exhibit good emulsifying and gelation properties are desirable for this application. Plant proteins can form a thicker interfacial layer at oil/water interfaces when compared to dairy proteins because of their larger molecular size and structure restriction through disulfide crosslinks (Le Roux et al. 2020). Plant sources such as fruits, tubers, legumes, and cereals are widely utilized in plant-based ice cream production, of which pea, soy, and coconut are very familiar. The addition of sweeteners not only gives desirable taste but also influences the solids' content to contribute to improving the viscosity and texture of the ice cream. The addition of sweeteners also reduces the freezing point and avoids large crystal formation during storage. Stabilizers aid in improving the viscosity and delay the rate of meltdown thus increasing the stability of ice cream during temperature fluctuations, whereas emulsifiers help water and fat to hold together to delay recrystallization of ice during storage (Kot et al. 2021, Romulo et al. 2021).

2.5 Plant-based Butter

Consumers' interest in plant-based butter and spreads has increased considerably in the past few years. Some of the common plant-based butter and spreads that are widely available in the market are obtained from nuts or seeds such as peanuts, almonds, cashews, pumpkin seeds, pistachios, cashews, soy, sunflower seeds, and sesame seeds. In addition to the mentioned sources, manufacturers' attention was drawn to legumes such as chickpeas due to their rich protein content. These plant-based products are referred to as butter if it contains at least 90% of nut/seed/legume as ingredients, whereas they are referred to as spreads if the product has at least 40% of nut/seed/legume as ingredients (Gorrepati et al. 2015).

The industrial production of plant-based butter involves basic steps such as roasting the nuts followed by blending ingredients. Various ingredients such as sugars, salts, stabilizers, and emulsifiers are added to improve the flavor profile and stability of the product during storage. The addition of stabilizers prevents the phase separation in plant-based butter during storage, while the addition of anti-oxidation agents in the product formulation is recommended when higher amounts of unsaturated oil are present in the plant-based source. The texture and mouthfeel of plant-based butter are determined by the particle size distribution of the fats and time-temperature conditions used during blending.

Further research to develop products that can deliver functional, nutritional, flavor attributes comparable to dairy counterparts is required to improve consumer acceptability of dairy alternatives.

3. Meat Alternatives

Meat alternatives, sometimes also called meat substitutes, mock meat, or faux meat, are food products that contain similar physical and chemical properties of meat products. Plant-based meat alternatives are made from plants with an intention to imitate the appearance, flavor, and the fibrous texture of animal meat (Boukid 2021). In the past decade, there has been an increase in the consumer demands towards plant-based meat products and a growing interest in their development and production. The increased interest in healthy diets and concerns about ethical issues, safety, processing, and high prices of meat products are the main drivers of this shift in consumer preferences towards the consumption of plant-based meat alternatives (Kumar et al. 2017, Xazela et al. 2017). The key factors affecting the consumers' acceptance towards consumption of plant-based meat alternatives were summarized in a previous study (Boukid 2021).

According to the Good Food Institute 2020 State of the Industry Report, the number of plant-based meat products manufacturing companies has increased, and many large animal-based meat companies entered the plant-based market, including plant-based meat alternative products into their portfolios. These not only helped plant-based meat products to become more available and accessible, but also enabled them to reach a price parity with animal-based meat. Therefore, with their high nutritional value, cost-effectiveness and improved physical attributes, plant-based meat alternatives carry a great potential to meet the protein demand of growing world population by reducing the concerns about environmental impact and sustainability problems associated with animal source proteins.

In this section, texturized vegetable proteins (TVP) and high moisture meat analogues (HMMA) will be introduced which are defined as "food products made from edible protein sources and characterized by having a structural integrity and identifiable structure such that each unit will withstand hydration and cooking, and other procedures used in preparing the food for consumption" by the USDA (1971). Although there are various texturizing techniques to obtain textured protein such as high-temperature induced shearing, wet-spinning, freeze structuring and 3D printing, a majority of the textured vegetable protein products are produced using extrusion technology which can resemble the texture and the appearance of meat while maintaining or improving the nutritional quality of products (Boukid 2021, Riaz 2011). Singh et al. (2021) have brought together a comprehensive list showing the major companies/brands involved in this business with their product range.

Extrusion cooking has been defined as a process in which a material is forced to flow under a variety of conditions through a barrel using a screw system and pushed through a mold at a predetermined rate to obtain products in a variety of forms (Alam et al. 2016). Proteins in raw food materials lose their native structures during an extrusion process due to the exposure of thermal and mechanical energies which results in the formation of a continuous and viscoelastic mass. As illustrated in Fig. 1., while the product moves through the extrusion system in the presence of heat, moisture and shear, its molecules are realigned, crosslinked and reformed causing it to transform into an expandable structure in the final product.

Through the adjustments of extrusion process parameters, various TVP products can be obtained and depending on their moisture contents they can be categorized into two groups, namely, low moisture TVP (LM-TVP) and high moisture TVP (HM-TVP), which contain about 10–40% and 40–70% moisture, respectively (Zhang et al. 2019). The structural changes happening during the production of these two types of meat analogues are illustrated in Fig. 2.

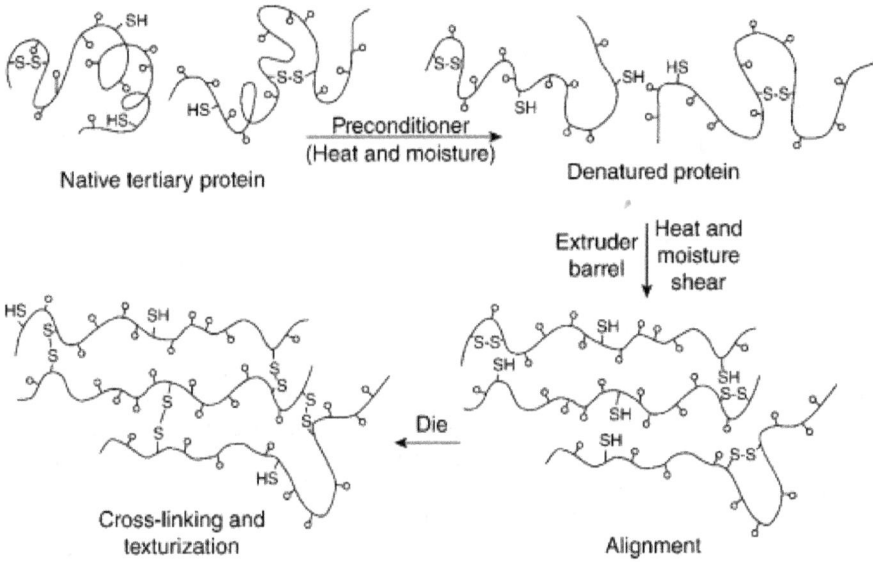

Figure 1. Protein denaturing with extrusion (Riaz 2011).

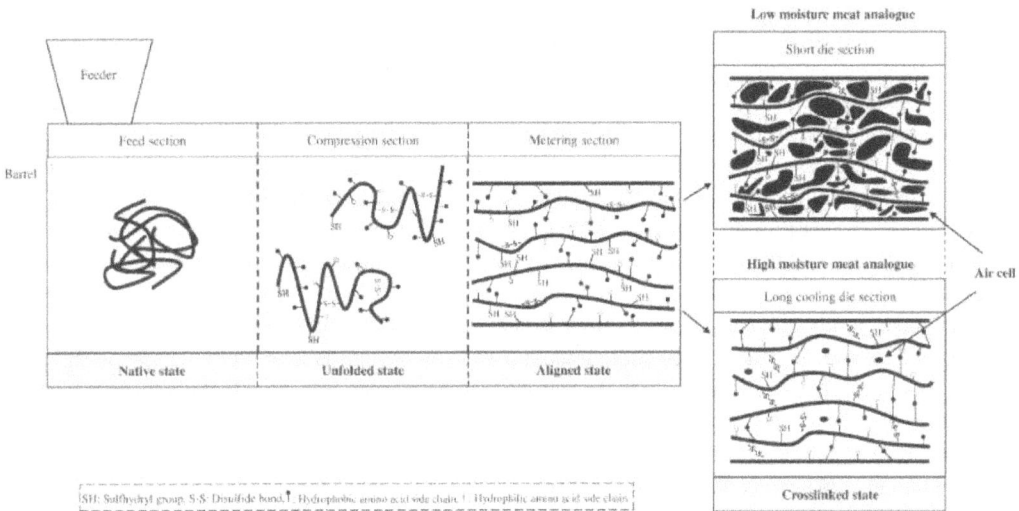

Figure 2. Mechanism of protein denaturation and texturization of low- and high-moisture meat analogue (Adapted from Samard et al. 2019).

3.1 Texturized Vegetable Protein (TVP)

Low moisture texturized vegetable proteins are mostly produced by a dry extrusion process carried out at low to intermediate moisture level, generally below 40%. As illustrated in Fig. 3, during this process, as the viscoelastic mass passes through the mold, the superheated vapor inside it confronts with a sudden pressure drop resulting in expanding the protein matrix. The expanded material is then cut into predetermined sizes, further dried until its moisture content drops below 2% and finally cooled down and packed for storage. In low moisture extrusion, the final product resembles a sponge, and it needs to be rehydrated prior to consumption. Because of its spongy structure, it absorbs the water quickly once dipped in a cream, sauces, or other liquids (Singh et al. 2021). Therefore, the most common applications of low moisture texturized vegetable proteins are meat extenders, such as nuggets, burgers, sausages and hams, and ground meat substitutes (Ryu 2020).

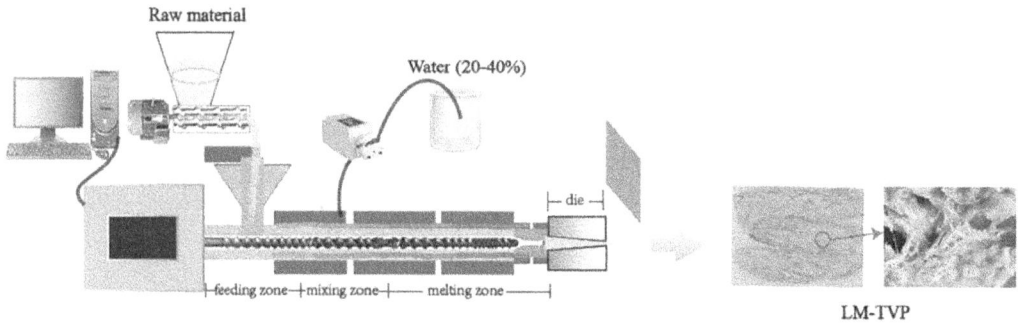

Figure 3. Illustration of a typical low moisture extrusion process (Zhang et al. 2019).

Generally, in low moisture extrusion process, the feed moisture content is between 10–30% and it usually occurs with a short retention time (5–15 seconds) and high temperature (100– 200°C). Depending on extrusion process parameters, feed moisture content, physical characteristics of the viscoelastic mass, and the properties of extruded material can vary. TVP products have been made by using many different plant protein sources such as soybeans, wheat, rapeseed, and peas (Riaz 2011). It has been shown that process parameters have a significant impact on extruder responses and/or final product characteristics, including but not limited to specific mechanical energy, textural characteristics, sensorial attributes, protein solubility, quality, and structure in low-moisture extrusion cooking (Riaz 2004, Yu et al. 2013).

In a recent study, fortification of rice starch with pea protein and pea fiber was achieved by using low moisture extrusion process and effects of blending ratios, different screw speed levels, die temperatures and feed moisture contents on expansion behaviors, and microstructures of the extrudates were investigated (Beck et al. 2018). According to their results, processing parameters have a significant impact on the physical characteristics of the final extrudates. It was shown that, while up to 25% pea protein isolate and 16% pea fiber can enhance the expansion relative to the control group, the expansion is significantly lowered above these inclusion levels of pea protein isolate and fiber.

In another study, the use of rice protein isolate in producing low moisture meat analogue was investigated (Lee et al. 2022). Rice protein isolate and soy protein isolate were blended at different levels (25:75, 50:50, 75:25 and 100:0 (w/w)) and extruded at a low moisture level using a twin screw extruder for manufacturing a meat analogue. The physicochemical analysis results showed that while replacing soy protein isolate with rice protein isolate resulted in a decrease in some physical parameters such as specific mechanical energy, porosity and water absorption capacity, the nutritional quality of meat analogues increased significantly with the addition of rice protein isolate. It was also shown that the total soluble matter of meat analogues made from the combination of soy protein and rice protein isolates were lower than that of the control group which was a commercial texturized vegetable protein product made from defatted soybean flour. This was explained by the hydrophobicity, formation of aggregated bodies and disulfide cross-linking of glutelin. Because low moisture meat analogues mostly require addition of a liquid before consumption, it was suggested that TVPs with less total soluble materials could be better in maintaining the nutritional quality after the rehydration. The morphological properties of the produced meat analogues were also investigated. Figure 4 represents surface, transversal, and longitudinal section morphology of meat analogues before and after the hydration process.

3.2 High Moisture Meat Analogues (HMMA)

High moisture extrusion (HME) process can be used for the production of high moisture meat analogues which have a meat-like texture. As shown in Fig. 5, different from low moisture extrusion process, this technique is generally employed at high water content levels (> 40%) mostly using

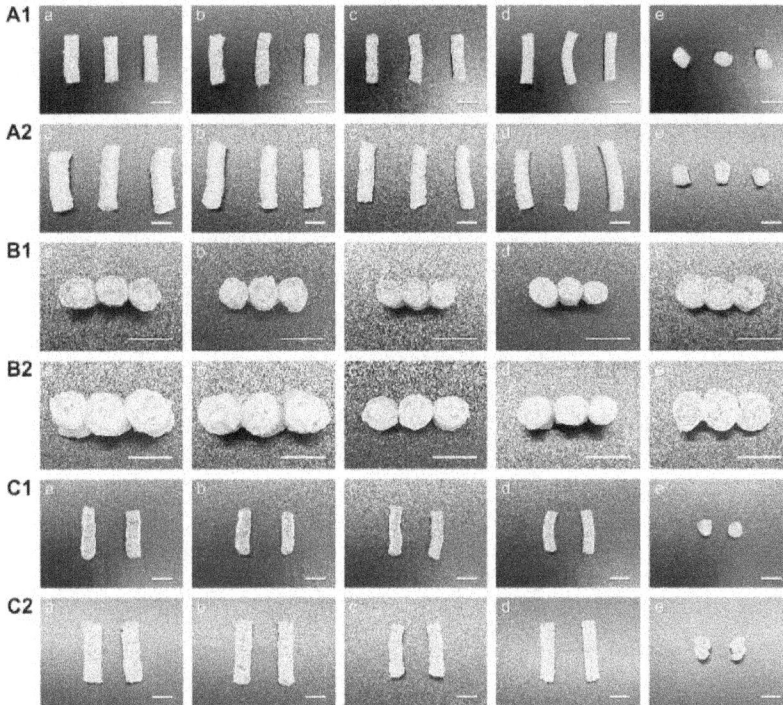

Figure 4. Multi-phase morphology (A) surface, (B) transversal section, and (C) longitudinal section of textured rice proteins (TRPs) [(a) TRP25, (b) TRP50, (c) TRP75, and (d) TRP100] and (e) commercial textured soy protein (C-TSP), (1) before and (2) after rehydration. (TRP samples with rice protein isolate and soy protein isolate at approximate ratios of 25:75, 50:50, 75:25, and 100:0 (w/w) were referred to as TRP25, TRP50, TRP75, and TRP100, respectively (collectively referred to as TRPs)) (Lee et al. 2022).

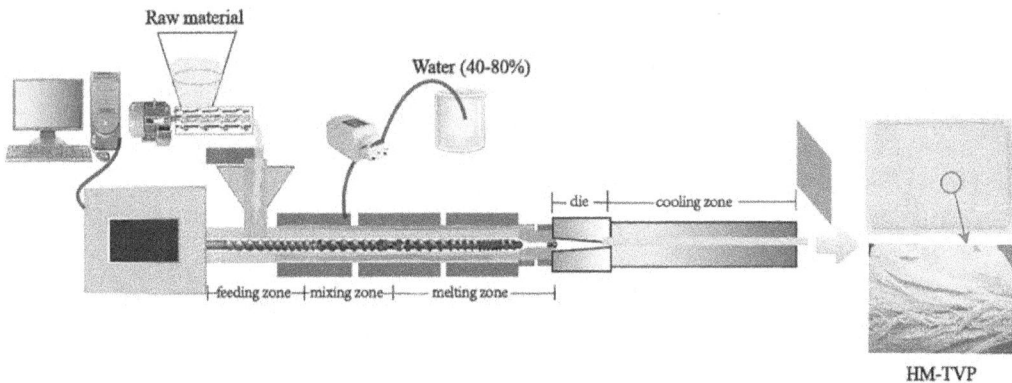

Figure 5. Illustration of a typical high moisture extrusion process (Zhang et al. 2019).

a twin-screw extruder with a cooling die which provides more flexibility in terms of processing parameters and producing a variety of products. During the HME process, after the unfolding, aggregating and realignment of plant proteins with the impact of high temperature and shear in the extruder barrel, the fibrous meat-like structure is formed with the crosslinking of protein molecules in which covalent disulphide and non-covalent bonds have been suggested to play a crucial role (Zhang et al. 2019).

One of the most important factors affecting the formation of fibrous texture of high moisture meat analogue products are the amount and type of proteins in the raw material (Ferawati et al. 2021). It has been recommended that to produce a desirable fibrous texture in the extrudate, the protein content of raw material should exceed a certain threshold which is reported to be around 50%. Therefore, protein concentrate, or isolates are critical in creating meat-like structure in HMMA products (Dekkers et al. 2018). Liu and Hsieh (2008) suggested the protein–protein interactions happened when two commercial soy protein isolates were turned into fibrous meat analogues by high moisture extrusion or gelation process. Their findings showed that both soy protein gels and extrudates contain covalent disulfide bonds and non-covalent interactions but in different proportions leading to differences in their structural properties. While non-covalent bonds play a fundamental role in making protein gels, both non-covalent and covalent disulfide bonds are equally important in forming a fibrous structure of protein extrudates.

One of the challenges with manufacturing high moisture meat analogue products is optimization of the extrusion process to enhance the textural characteristics of extrudates. Cornet et al. (2021) created an overview of plant-based ingredients and combinations thereof, and process conditions used to produce fibrous structures through extrusion processing which is shown in Table 3. The raw material used for a HME process can either be a single component such as one type of protein isolate or concentrate or a multicomponent where two or more materials are blended. It has been suggested that although both types of raw materials are capable of creating an anisotropic meat-like structure, multicomponent systems are more successful in forming a meat-like texture than single component systems due to several reasons such as the formation of a dispersed phase from the added component, the influence of the component on protein aggregation/crosslinking reactions, or changes of the rheological properties (Wittek et al. 2021). In a recent study, the influence of inclusion of whey protein concentrate in soy protein isolate-based meat analogue at varying ratios on the HME process conditions, and anisotropic structure and morphology development of extrudates, and phase behavior of whey protein in the matrix were evaluated. It was observed that with the increase in whey protein concentrate in the blending formula, the anisotropic structure became more distinct and the die pressure in the process decreased significantly presumably because of the decrease in blending viscosity. The multiphase morphology of extrudates was affected by the addition of whey protein concentrate depending on the inclusion ratio. As shown in Fig. 6, although addition of 15% whey protein concentrate did not significantly alter the morphology, at the 30% inclusion rate, it was detected that the structures became more disordered, shorter, and thinner compared to extrudates made with only soy protein isolate and with 15% whey protein concentrate. Their findings suggested that blending of protein- rich sources carries a great potential to improve the structural characteristics of meat analogues produced with HME process.

Another study investigated the potential use of hemp protein concentrate (HPC) in substituting soy protein isolate in the production of high moisture meat analogue using a twin screw co-rotating extruder (Zahari et al. 2020). Four formulations (0% HPC, 20% HPC, 40% HPC and 60% HPC)

Table 3. Twin screw (TSE) and single screw extrusion (SSE) process conditions applied in plant- based ingredients (Obtained from Cornet et al. 2021).

Technique	Material	Moisture Content (%)	Max. Temperature (°C)	References
TSE	SPC	60	100, 140, 160	Pietsch et al. (2019)
TSE	SPC + WG	60	170	Chiang et al. (2019)
TSE	PPI	60	140	Osen et al. (2015)
TSE	SPI + WG	50	148	Zhang et al. (2015)
TSE	SPI	50	150	Fang et al. (2014)
SSE	PPC + DPF	50 – 55	165	Rehrah et al. (2009)

SPC, soy protein concentrate; PPI, pea protein isolate; WG, wheat gluten; SPI, soy protein isolate; PPC, peanut protein concentrate; DPF, defatted peanut flour

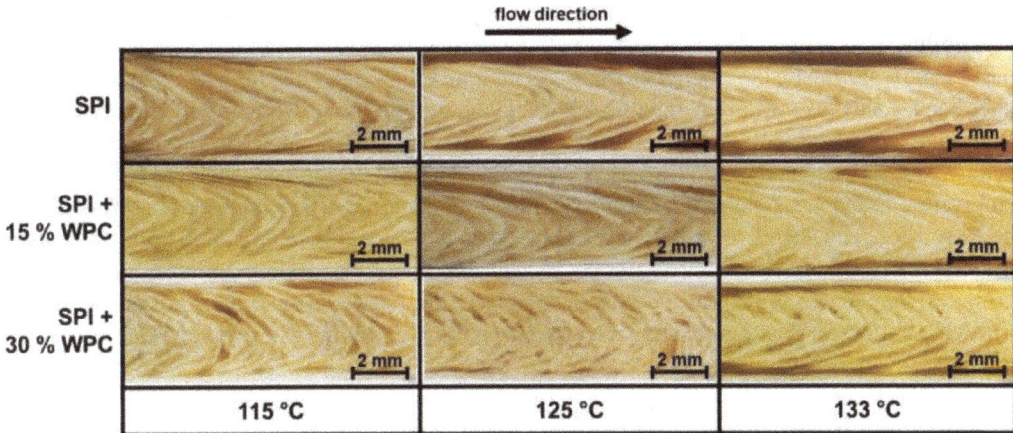

Figure 6. Multi-phase morphology, obtained through cryo-imaging of extrudates for different protein blends (0–30% whey protein concentrate) and material temperatures (115–133°C) (Wittek et al. 2021).

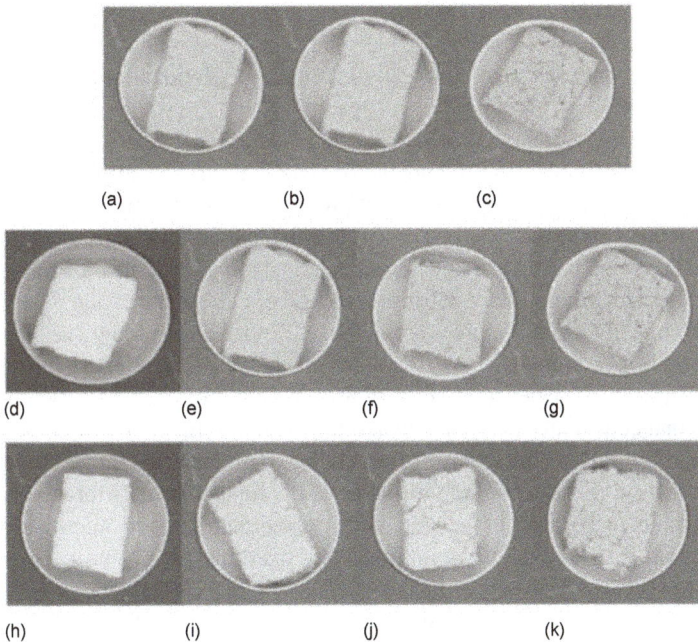

Figure 7. Images of extruded meat analogues with different formulations: (a–c) 20%, 40%, 60% HPC at 65% target moisture; (d–g) 0%, 20%, 40%, 60% HPC at 70% target moisture; (h–k) 0%, 20%, 40%, 60% HPC at 75% target moisture (Zahari et al. 2020).

were extruded at target moisture (65–80%), temperature (40–120°C) and screw speed (500 or 800 rpm) levels. Figure 7 illustrates the meat analogue samples produced with different formulas. Based on the results, it was suggested that blending formula and process conditions have a significant impact on texture profiles and color of extrudates, and soy protein isolate could be substituted by hemp protein by up to 60% in manufacturing high moisture meat analogue.

Similarly, in the study of Palanisamy et al. (2018), lupin protein concentrates and isolate mixture (1:1 w/w) was used in developing a meat analogue with high moisture extrusion, and the impact of various process parameters on the properties of product, including physical attributes and nutritional quality, were assessed. According to the scanning electron microscopy analysis results, increasing temperature and screw speed levels and reducing the water feeding rate can facilitate forming a final product with a compact microstructure and multilayered texture. Additionally, *in-vitro* protein digestibility of lupin extrudates decreased with increasing barrel temperature which was attributed to rise in non-enzymatic browning reactions and thermal cross- linking at high temperatures. On the other hand, Saldanha do Carmo et al. (2021) used one protein- rich raw material in manufacturing meat analogue with HME process and examined the influence of three processing parameters, namely, raw material feed rate, ratio between water feed rate (WFR) and raw material feed rate (FR), temperature in the last zone of the extruder, on the functional, sensorial, and textural properties of the extrudates. They achieved producing a meat analogue solely based on a faba bean concentrate that had a desired elasticity and firmness, as well as a good bite-feeling and satisfying sensory attributes comparable to similar products available in the market at optimum process conditions of a temperature ranging between 130 and 140°C, WFR/FR ratio of 4 and feed rate at 11 rpm (1.10 Kg/h).

4. Seafood Alternatives

Plant-based seafood alternatives are an emerging category in the plant-based foods section due to rising ethical concerns on overfishing. Some of the common plant sources used to manufacture plant-based seafood alternatives are soy, wheat gluten, algae, seaweed, microalgae, mushrooms, and vegetables (Grossmann and McClements 2021). These plant sources not only provide the protein content to the final product but also contribute to the key functional attributes such as gelation, emulsification, water holding, and oil binding capacity as well.

Other non-protein ingredients such as salts, fat, and dietary fibers are used during product formulation to obtain desirable texture and flavor. The addition of salt influences the solubility, denaturation, and gelation properties of the plant proteins during thermal or shear stress processes. The amount of salt that can be added is dependent on the plant protein being used in the formulation (Tahergorabi and Jaczynski 2012). Seafood, especially fish, is rich in omega-3 polyunsaturated fatty acids (PUFAs) which hold many nutritional benefits. One of the challenges faced by product developers is that none of the plant sources are rich in PUFAs, hence leading to a plant-based seafood product with poor nutritional content. For functional attributes and to improve the texture, vegetable fat is added, so their droplets can act as fillers in the gel matrix of plant-based seafood alternatives. Another additive that contributes to the health and the texture of the final product is dietary fiber. To maximize the benefit of dietary fiber, they are chosen based on their functional properties to improve water holding, emulsifying, gel-forming properties of the final product (Moreno et al. 2016).

One of the main challenges associated with developing seafood alternatives is its fibrous structure. Most companies are instead aiming to develop products with similar texture, smell, taste, and appearance to processed seafood products (Kazir and Livney 2021). Emerging technologies such as hydro spinning, electrospinning, extrusion, and 3D printing are opted as structuring techniques to give a fibrous structure to plant-based proteins. However, these techniques might pose some limitations when scaling up the production.

To increase sales and improve consumers' acceptability of plant-based seafood, manufacturers should aim at meeting the nutritional properties of plant-based seafood along with its sensory properties.

5. Egg Alternatives

Plant-based egg is one of the alternative products that was introduced in past few years that grew up to have sales of 27 million US dollars in 2020. However, the growth in sales from 2019–2020 was 168% of total plant-based sales, making this one of the fastest-growing products in the plant-based category (Anon. 2021). The interest in plant-based egg alternatives is driven due to sustainability and ethical benefits associated with the ill-treatment of chickens in slaughterhouses and increased greenhouse gas emissions because of current chicken farming practices (Grossmann and McClements 2021). While the other plant-based alternative products are designed to provide the sensory, flavor, and nutritional attributes of the original product, egg alternatives are developed for dual purposes. Egg alternatives are developed to meet consumers' expectations to be used as an ingredient in the food applications such as bakery, ice creams, mayonnaise, and pasta in addition to being able to make omelets, scrambled and folded eggs. For this reason, identifying plant proteins that have similar functional properties as egg protein is crucial.

Plant proteins with emulsifying and foaming properties to stabilize oil-in-water emulsions are desirable for the applications such as salad dressings, mayonnaise, sauces, etc. Soy protein concentrate and soy milk are some of the suitable sources that can be used for these product applications. In addition to soy, a few other plant-based protein sources such as sunflower, pea, tomato seed, wheat, white lupin, and faba bean were also successful in these product applications. Stabilizers such as xanthan gum and guar gum are commonly used in these product applications to stabilize the emulsions (Nikzade et al. 2012).

Similarly, plant proteins with gelling properties similar to eggs are desirable for liquid plant-based egg applications. However, denaturation temperatures of plant proteins are often higher when compared with egg proteins (63–93°C). Therefore, to obtain the desirable final texture, plant proteins often have to be heated at higher temperatures or for a longer period (Grossmann and McClements 2021). Soy, pea, and mung bean are commonly used plant protein sources for this product application. However, the literature suggests that off-flavors like beany or green are often observed when soy or pea are used as plant protein sources because of the presence of saponin, aldehyde, and ketone compounds (Söderberg 2013). Other ingredients that are used to formulate the liquid plant-based eggs are stabilizers, emulsified oil, flavorings, pH modulators, preservatives, and colorant. The addition of emulsified oil simulates functional attributes of lipoproteins in eggs giving plant-based egg products their desirable mouthfeel and texture. Stabilizers in the formulation increase the viscosity of the plant-based fluid egg delaying phase separation during storage. Curcumins and carotenoids are often preferred as natural coloring pigments to give plant-based egg liquid its characteristic yellowish color (McClements and Grossmann 2021, Alcorta et al. 2021).

Although competing with the nutritional profile of egg can be challenging when formulating plant-based egg alternatives, choosing plant protein or plant protein blends with desirable functionality and amino acid profile will result in a successful product.

6. Challenges and Future Directions

6.1 Cut down the Cost of Production and Retail Price

On the production side, to realize the price of plant-based food parity with the price of conventional real meat and milk still face challenges. According to the market research of Global Information Inc., the labeled price of plant-based meat and milk on the retail store shelves is generally expensive to some degree. Figure 8 shows the average retail prices at U.S. multi-outlet grocery stores in the 52 weeks ending July 11, 2021.

Plant-based meat alternative manufacturers have stated that the price parity is especially important for their products while the current production cost of plant-based burger is higher than regular meat burger. An important impact factor is the scale of manufacturing gross. The meat alternatives are made on a relatively small scale while animal-sourced products remain affordable to consumers. On the consumption side, these novel plant-based foods are new to consumers, which have used lots of undistributed ingredients and processing technologies. Therefore, they are often labeled at higher prices to gain higher margins than their traditional counterparts. There is a trend that the rise of key players and investment in plant-based foods will reduce the production cost and profit margin for finished products' sales.

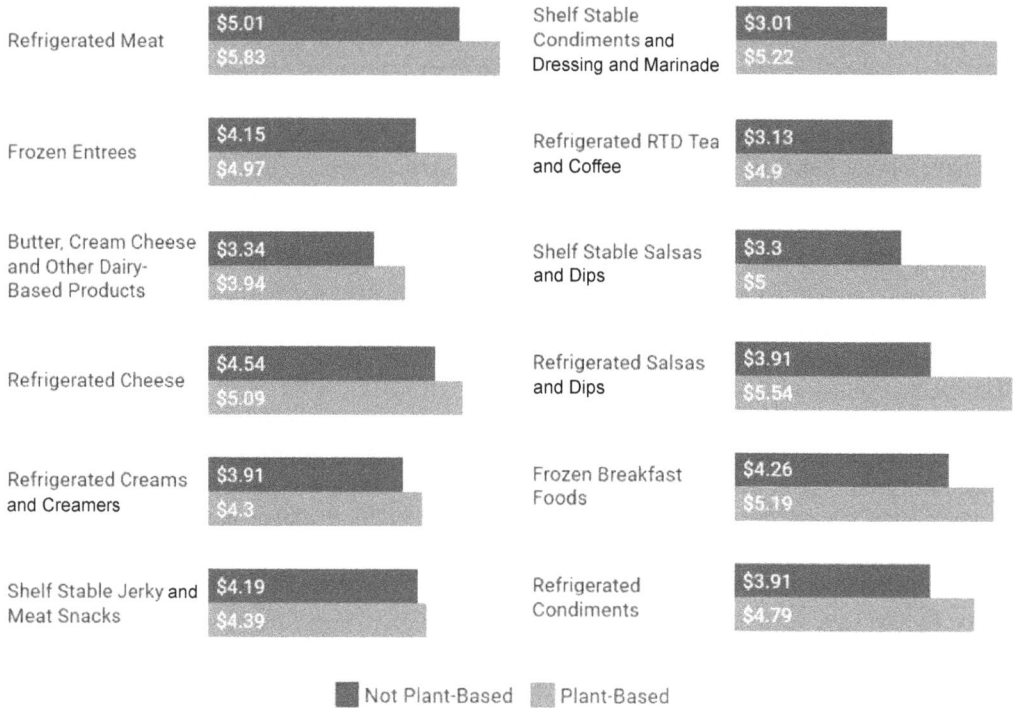

	Not Plant-Based	Plant-Based
Refrigerated Meat	$5.01	$5.83
Frozen Entrees	$4.15	$4.97
Butter, Cream Cheese and Other Dairy-Based Products	$3.34	$3.94
Refrigerated Cheese	$4.54	$5.09
Refrigerated Creams and Creamers	$3.91	$4.3
Shelf Stable Jerky and Meat Snacks	$4.19	$4.39
Shelf Stable Condiments and Dressing and Marinade	$3.01	$5.22
Refrigerated RTD Tea and Coffee	$3.13	$4.9
Shelf Stable Salsas and Dips	$3.3	$5
Refrigerated Salsas and Dips	$3.91	$5.54
Frozen Breakfast Foods	$4.26	$5.19
Refrigerated Condiments	$3.91	$4.79

Figure 8. Average retail price comparison at U.S. grocery stores.

6.2 Challenges with Current Processing Equipment

Except plant-based dairy alternatives, the substance of processing of plant-based novel foods, such as meat alternatives, seafood alternatives, and egg alternatives, is "restructure". Kinney et al. (2019) from GFI summarized a general process flow diagram of whole muscle meat products, which includes six major steps:

Step 1: TVP production;
Step 2: Hydration/Marination;
Step 3: Coating;
Step 4: Cooking;
Step 5: Extended Shelf-Life Processes;
Step 6: Packaging and Storage (refrigerate or freeze).

In current market, whole muscle plant-based meat attracts tons of attention, which displays the fibrous texture analogous aiming to match the mouthfeel of real whole muscle foods, such as beef steak, chicken breast, and pork chop. Both dry and wet Textured Vegetable Protein (TVP) produced by extrusion has been used. Few world-wide major manufacturers for TVP extrusion have been shown in Figure 9. This list includes, but is not limited to, the global industrial machineries, such as Coperion, Buhler, Clextral, Wenger, etc., as well as some emerging extrusion machinery in China, such as Jinan DG machinery, Nanjing KY solution Group, Changsha Fumach, etc. There also are well-known lab scale extrusion equipment brands and mode, such as Thermo Scientific HAAK Pheomex and Brabender DSE.

Currently, the food extrusion machinery industry is going through a fierce competition stage under the pressure from international giant enterprises and manufacturers, which lead the application of extrusion technology and machine in chemical, plastic, and pharmaceutical industry. These "aircraft carriers" are turning their course now and exploring their new models for food extrusion. On the other hand, hundreds of small extrusion machinery company are booming in

Figure 9. Manufacturer brands of Food Extruder (Partial list).

China, especially in Nanjing City and Jinan City. The positive point is that the general cost to build an extrude line has been cutting down, which makes it much more affordable than the big brands. These small manufacturers are able to modify and polish the design for each buyer, who wants to use the extrusion equipment to produce unique restructured food. Furthermore, some manufacturers are offering OEM business for whole extrusion line design.

Upgrading advanced technology needs stable talent team; however, the excellent engineers are always hunted by these small private brand owners. Nowadays, those extrusion masters are spraying everywhere, which result in the difficulty to the human resource management. On the market side, sometimes low prices are exchanged for low quality and techno-services. Formulation and implementation of international regulations for quality supervision of plant-based food extrusion equipment still faces challenges.

6.3 Improve the Production and usage of Plant Protein Ingredients

Meeting the food demands of global population within limited planetary resource is a challenge. Henchion et al. (2021) have informed the development of pathways to the sustainable animal production by analyzing the trends in animal-sourced foods, including the significance of "animal-sourced protein relative to plant-protein sources". The ratio of animal-sourced to plant protein varies across regions, ranging from 0.29 in Africa to 1.08 in Europe in 2017. Therefore, to deal with the challenge of sustainable livestock production and consumption, the importance of planning regional specific strategies should be highlighted. The challenge is not merely one for science and technology but one very much based on wider aspects of the food system and its diverse stakeholders.

Plant proteins from different crops region show quality difference globally, such as soybean, pea, wheat, rice, lentils, chicken pea, peanuts, sunflower, pumpkin seed, and others. Volatile prices and quality of raw materials often cause troubles. That is why breeding and crop development are challenges for plant proteins' supplier. Figure 10 shows those major supplier brands worldwide.

A single type of plant protein is difficult to create the satisfactory functional performance. "Blends" is a key word in the production, such as plant proteins mix. It is crucial to seek low-cost plant protein sources and create novel combinations with other food ingredients for the production of plant-based meat, egg, milk, and seafood analogues. Researchers are mapping the rheological properties (strain, stress, and elasticity) of dense plant protein blends. An interesting phenomenon on the market is that meat—plant protein blends just meet the flexitarian's need. The traditional meat brands have begun to blend meat with plant-based analogues while promoting these "hybrid meat" products as more sustainable and healthier.

Figure 10. Worldwide suppliers of plant protein ingredients (Partial list).

Are those plant-based ingredients and products really sustainable? I would say: "Not all of them, at least not ideal enough". There are two main reasons. Firstly, those plant proteins are not equally sustainable, some type of tree nuts consume more water than cereal grains and beans (SPINS and C.A). Fortune has stated similar opinion in their recent report, 'The State of Plant-Based Innovation' (SPINS Natural Enhanced Channel and SPINS Conventional Multi Outlet 2021). Secondly, the current produce methods of extracting plant proteins by acid-base method may not be "green". The wastewater treatment and the energy consumption are facing challenges. Drying processing is the final key step to produce plant protein powders. Does the spray drying bring out the best quality of protein powder? Some novel drying technologies, such as ultrasonic drying, deserve more investigation.

6.4 Blockchains in different Supply Chains bring Challenges

Through the "Distributed Ledger Technology", IBM blockchain supply chain solutions provide participants greater visibility across all supply chain activities. IBM Food Trust™ is the first pioneer network to connect participants including ingredient suppliers, distributors, food retailers, and consumers, through a secured, permanent, and shared record of food system data, which can increase food safety and supply chain efficiencies. Carrefour, Walmart, Nestlé, and Tyson Foods are co-operating with IBM to create a more transparent and trustworthy global food supply chain. In the food traceability system, the blockchain technology helps keep a record of the supply chain with time chops and the process info-flow of single unit food from the producing to retailing. Although blockchain tracking already has been proofed to benefit sales of meat, milk, eggs, fresh vegetables and fruits, the application of blockchain technology to meet the new demands for a more efficient novel plant-based food ecosystem is still facing challenges. Specific challenges include, firstly, large-scale application of these new food supply chains being very complicated; furthermore, the number of these novel plant-based food is growing every day. Secondly, a food traceability system involves multiple sources of supplies, and plant-based food suppliers are totally different from conventional animal-derived foods industry (milk, meat, seafood, egg). Thirdly, there is lack of international standards for plant-based food quality and safety control.

On the other side, there are some blockchains' platforms gathering consumer information to understand costumer shopping preferences and pushing recommendations based on the big data analysis. Blockchain technology is also doing co-operative work with the tech-driven loyalty reward industry, such as Qiibee, to retain their customers better. We expect that more companies, associations, and government agencies will apply the blockchain traceability in their plant-based food practices.

6.5 Sensorial and Mouthfeel Challenges—Taste REAL?

One of the major challenges faced by plant-based foods manufacturers is associated with its sensorial properties which affect the consumer acceptance. Animal proteins confer a unique flavor and texture to products that is not easily attained if substituted with plant proteins in dairy or meat analogs. Plant-based proteins are also observed to give a beany, bitter or chalky taste as well as off-odor notes in plant protein products (Rackis et al. 1979). Moreover, plant cells have a rigid structure and therefore plant-based products normally do not have the bite and chewiness of meat and can often feel crumbly (Sun et al. 2021).

In order to overcome these challenges associated with plant proteins and mimic the key quality attributes of dairy and meat products, food scientists need to fully understand the major compounds and drivers behind these undesirable odor, flavor and texture perception in dairy or meat analogues. Furthermore, utilization of processing techniques such as extrusion cooking or freeze structuring to modify the fibrous structure of proteins may be useful to bridge the texture gap of plant-based products (Dekkers et al. 2018).

6.6 Allergenicity Problems

A food allergy can generally be defined as an adverse reaction of the immune system to an otherwise harmless food or food component, called allergens. Overall, approximately 90% of all food allergies are associated with a small number of specific proteins represented by eight major allergenic foods: peanuts, tree nuts, cow's milk, hen's eggs, fish, crustacea (e.g., shrimp), wheat, and soybeans. Among these, soy, peanuts, tree nuts and wheat are four of the six most commonly occurring food allergies caused by proteins that are of plant origin (Protudjer and Mikkelsen 2020). Given the prevalence, plant protein allergy is still a major concern for the food industry and consumers. Consequently, it's important to carefully plan and monitor the vegan diets so that children and adults who are allergic to such proteins have their nutritional needs met. Recently, pea protein is gaining a lot of interest in the manufacture of foods and is reported to have low allergenicity as compared to other legumes such as soyabean or peanut (Zhao et al. 2020). Similarly, buckwheat protein has also been shown to have great potential as a gluten-free alternative plant-based protein to meet the global demand for protein while keeping allergies in view (Jin et al. 2020). Available studies also show the utilization of various traditional and emerging processing technologies to increase the digestibility and reduce the allergenicity of proteins (Huang et al. 2014, Sathe and Sharma 2009). Similarly, other techniques like genetic engineering are also being investigated to enhance the safety of plant-based food products through slicing of allergen-encoding genes (Herman et al. 2003, Singh and Bhalla 2008).

References

Aboulfazli, F., Baba, A.S. and Misran, M. 2015. Effects of fermentation by Bifidobacterium bifidum on the rheology and physical properties of ice cream mixes made with cow and vegetable milks. International Journal of Food Science & Technology 50(4): 942–949.

Alam, M.S., Kaur, J., Khaira, H. and Gupta, K. 2016. Extrusion and extruded products: changes in quality attributes as affected by extrusion process parameters: a review. Critical reviews in food science and nutrition 56(3): 445–473.

Alcorta, A., Porta, A., Tárrega, A., Alvarez, M.D. and Vaquero, M.P. 2021. Foods for plant-based diets: Challenges and innovations. Foods 10(2): 293.

Aschemann-Witzel, J. and Peschel, A.O. 2019. How circular will you eat? The sustainability challenge in food and consumer reaction to either waste-to-value or yet underused novel ingredients in food. Food Quality and Preference 77: 15–20.

Aschemann-Witzel, J., Giménez, A., Grønhøj, A. and Ares, G. (2020). Avoiding household food waste, one step at a time: The role of self-efficacy, convenience orientation, and the good provider identity in distinct situational contexts. Journal of Consumer Affairs, 54(2): 581–606.

Bachmann, H.P. 2001. Cheese analogues: a review. International Dairy Journal, 11(4-7): 505–515.

Beck, S.M., Knoerzer, K., Foerster, M., Mayo, S., Philipp, C. and Arcot, J. 2018. Low moisture extrusion of pea protein and pea fibre fortified rice starch blends. Journal of Food Engineering, 231: 61–71.

Beegum, P.S., Nair, J.P., Manikantan, M.R., Pandiselvam, R., Shill, S., Neenu, S., et al. 2022. Effect of coconut milk, tender coconut and coconut sugar on the physico-chemical and sensory attributes in ice cream. Journal of Food Science and Technology, 1–12.

Boukid, F. 2021. Plant-based meat analogues: From niche to mainstream. European Food Research and Technology, 247(2): 297–308.

Brückner-Gühmann, M., Benthin, A. and Drusch, S. 2019. Enrichment of yoghurt with oat protein fractions: Structure formation, textural properties and sensory evaluation. Food Hydrocolloids, 86: 146–153.

Chiang, J.H., Loveday, S.M., Hardacre, A.K. and Parker, M.E. 2019. Effects of soy protein to wheat gluten ratio on the physicochemical properties of extruded meat analogues. Food Structure, 19: 100102.

Circus, V.E. and Robison, R. 2019. Exploring perceptions of sustainable proteins and meat attachment. British Food Journal, 121(2): 533–545.

Cornet, S.H., Bühler, J.M., Gonçalves, R., Bruins, M.E., van der Sman, R.G. and van der Goot, A.J. 2021. Apparent universality of leguminous proteins in swelling and fibre formation when mixed with gluten. Food Hydrocolloids, 120: 106788.

Craig, W.J. and Brothers, C.J. 2021. Nutritional content and health profile of non-dairy plant-based yogurt alternatives. Nutrients, 13(11): 4069.

Dekkers, B.L., Emin, M.A., Boom, R.M. and van der Goot, A.J. 2018. The phase properties of soy protein and wheat gluten in a blend for fibrous structure formation. Food Hydrocolloids, 79: 273–281.

do Carmo, C.S., Knutsen, S.H., Malizia, G., Dessev, T., Geny, A., Zobel, H., et al. 2021. Meat analogues from a faba bean concentrate can be generated by high moisture extrusion. Future Foods, 3: 100014.

Fang, Y., Zhang, B. and Wei, Y. 2014. Effects of the specific mechanical energy on the physicochemical properties of texturized soy protein during high-moisture extrusion cooking. Journal of Food Engineering, 121: 32–38.

Ferawati, F., Hefni, M., Östbring, K. and Witthöft, C. 2021. The application of pulse flours in the development of plant-based cheese analogues: Proximate composition, color, and texture properties. Foods, 10(9): 2208.

Gorrepati, K., Balasubramanian, S. and Chandra, P. 2015. Plant based butters. Journal of Food Science and Technology, 52: 3965–3976.

Grasso, N., Alonso-Miravalles, L. and O'Mahony, J.A. 2020. Composition, physicochemical and sensorial properties of commercial plant-based yogurts. Foods, 9(3): 252.

Grasso, N., Roos, Y.H., Crowley, S.V., Arendt, E.K. and O'Mahony, J.A. 2021. Composition and physicochemical properties of commercial plant-based block-style products as alternatives to cheese. Future Foods, 4: 100048.

Greis, M., Sainio, T., Katina, K., Kinchla, A.J., Nolden, A., Partanen, R., et al. 2020. Dynamic texture perception in plant-based yogurt alternatives: Identifying temporal drivers of liking by TDS. Food Quality and Preference, 86: 104019.

Henchion, M., Moloney, A.P., Hyland, J., Zimmermann, J. and McCarthy, S. 2021. Trends for meat, milk and egg consumption for the next decades and the role played by livestock systems in the global production of proteins. Animal, 15: 100287.

Herman, E.M., Helm, R.M., Jung, R. and Kinney, A.J. 2003. Genetic modification removes an immunodominant allergen from soybean. Plant Physiology, 132(1): 36–43.

Hyvönen, L., Linna, M., Tuorila, H. and Dijksterhuis, G. 2003. Perception of melting and flavor release of ice cream containing different types and contents of fat. Journal of Dairy Science, 86(4): 1130–1138.

Jeske, S., Zannini, E. and Arendt, E.K. 2018. Past, present and future: The strength of plant-based dairy substitutes based on gluten-free raw materials. Food Research International, 110: 42–51.

Kazir, M. and Livney, Y.D. 2021. Plant-based seafood analogs. Molecules, 26(6): 1559.

Kinney, M., Weston, Z. and Bauman, J. 2019. Plant-based meat manufacturing by extrusion. Washington, DC: The Good Food Institute.

Klost, M., Giménez-Ribes, G. and Drusch, S. 2020. Enzymatic hydrolysis of pea protein: Interactions and protein fractions involved in fermentation induced gels and their influence on rheological properties. Food Hydrocolloids, 105: 105793.

Kot, A., Kamińska-Dwórznicka, A., Galus, S. and Jakubczyk, E. 2021. Effects of different ingredients and stabilisers on properties of mixes based on almond drink for vegan ice cream production. Sustainability, 13(21): 12113.

Kumar, P., Chatli, M.K., Mehta, N., Singh, P., Malav, O.P. and Verma, A.K. 2017. Meat analogues: Health promising sustainable meat substitutes. Critical Reviews in Food Science and Nutrition, 57(5): 923–932.

Kyriakopoulou, K., Keppler, J.K. and van der Goot, A.J. 2021. Functionality of ingredients and additives in plant-based meat analogues. Foods, 10(3): 600.

Le Roux, L., Mejean, S., Chacon, R., Lopez, C., Dupont, D., Deglaire, A., et al. 2020. Plant proteins partially replacing dairy proteins greatly influence infant formula functionalities. LWT, 120: 108891.

Lee, J.S., Choi, I. and Han, J. 2022. Construction of rice protein-based meat analogues by extruding process: Effect of substitution of soy protein with rice protein on dynamic energy, appearance, physicochemical, and textural properties of meat analogues. Food Research International, 161: 111840.

Liu, K. and Hsieh, F.H. 2008. Protein–protein interactions during high-moisture extrusion for fibrous meat analogues and comparison of protein solubility methods using different solvent systems. Journal of Agricultural and Food Chemistry, 56(8): 2681–2687.

Lobato-Calleros, C., Vernon-Carter, E.J., Guerrero-Legarreta, I., Soriano-Santos, J. and Escalona-Beundia, H. 1997. Use of fat blends in cheese analogs: Influence on sensory and instrumental textural characteristics. Journal of Texture Studies, 28(6): 619–632.

Mäkinen, O.E., Wanhalinna, V., Zannini, E. and Arendt, E.K. 2016. Foods for special dietary needs: Non-dairy plant-based milk substitutes and fermented dairy-type products. Critical Reviews in Food Science and Nutrition, 56(3): 339–349.

Massera, D., Graf, L., Barba, S. and Ostfeld, R. 2016. Angina rapidly improved with a plant-based diet and returned after resuming a Western diet. Journal of Geriatric Cardiology: JGC, 13(4): 364.

Mattice, K.D. and Marangoni, A.G. 2020. Physical properties of plant-based cheese products produced with zein. Food Hydrocolloids, 105: 105746.

McClements, D.J., Newman, E. and McClements, I.F. 2019. Plant-based milks: A review of the science underpinning their design, fabrication, and performance. Comprehensive Reviews in Food Science and Food Safety, 18(6): 2047–2067.

McClements, D.J. and Grossmann, L. 2021. A brief review of the science behind the design of healthy and sustainable plant-based foods. NPJ Science of Food, 5(1): 17.

Mefleh, M., Pasqualone, A., Caponio, F. and Faccia, M. 2022. Legumes as basic ingredients in the production of dairy-free cheese alternatives: a review. Journal of the Science of Food and Agriculture, 102(1): 8–18.

Moreno, H.M., Herranz, B., Pérez-Mateos, M., Sánchez-Alonso, I. and Borderías, J.A. 2016. New alternatives in seafood restructured products. Critical Reviews in Food Science and Nutrition, 56(2): 237–248.

Munekata, P.E.S., Rocchetti, G., Pateiro, M., Lucini, L., Domínguez, R. and Lorenzo, J.M. 2020. Addition of plant extracts to meat and meat products to extend shelf-life and health-promoting attributes: An overview. Current Opinion in Food Science, 31: 81–87.

Nikzade, V., Tehrani, M.M. and Saadatmand-Tarzjan, M. 2012. Optimization of low-cholesterol–low-fat mayonnaise formulation: Effect of using soy milk and some stabilizer by a mixture design approach. Food Hydrocolloids, 28(2): 344–352.

Osen, R., Toelstede, S., Eisner, P. and Schweiggert-Weisz, U. 2015. Effect of high moisture extrusion cooking on protein–protein interactions of pea (*Pisum sativum* L.) protein isolates. International Journal of Food Science & Technology, 50(6): 1390–1396.

Ostfeld, R.J. 2017. Definition of a plant-based diet and overview of this special issue. Journal of Geriatric Cardiology: JGC, 14(5): 315.

Palanisamy, M., Töpfl, S., Aganovic, K. and Berger, R.G. 2018. Influence of iota carrageenan addition on the properties of soya protein meat analogues. Lwt, 87: 546–552.

Pandey, S., Ritz, C. and Perez-Cueto, F.J.A. 2021. An application of the theory of planned behaviour to predict intention to consume plant-based yogurt alternatives. Foods, 10(1): 148.

Paul, A.A., Kumar, S., Kumar, V. and Sharma, R. 2020. Milk Analog: Plant based alternatives to conventional milk, production, potential and health concerns. Critical Reviews in Food Science And Nutrition, 60(18): 3005–3023.

Pietsch, V.L., Bühler, J.M., Karbstein, H.P. and Emin, M.A. 2019. High moisture extrusion of soy protein concentrate: Influence of thermomechanical treatment on protein-protein interactions and rheological properties. Journal of Food Engineering, 251: 11–18.

Protudjer, J.L. and Mikkelsen, A. 2020. Veganism and paediatric food allergy: two increasingly prevalent dietary issues that are challenging when co-occurring. BMC Pediatrics, 20(1): 341.

Rehrah, D., Ahmedna, M., Goktepe, I. and Yu, J. 2009. Extrusion parameters and consumer acceptability of a peanut-based meat analogue. International Journal of Food Science & Technology, 44(10): 2075–2084.

Riaz, M.N. 2004. Texturized soy protein as an ingredient. Proteins in Food Processing, 517–558.

Riaz, M.N. 2011. Texturized vegetable proteins. In Handbook of Food Proteins (pp. 395–418). Woodhead Publishing.

Romulo, A. and Meindrawan, B. 2021, July. Effect of Dairy and Non-Dairy Ingredients on the Physical Characteristic of Ice Cream. In IOP Conference Series: Earth and Environmental Science, 794(1): 012145. IOP Publishing.

Ryu, G.H. 2020. Extrusion cooking of high-moisture meat analogues. In Extrusion cooking (pp. 205–224). Woodhead Publishing.

Sabaté, J. and Soret, S. 2014. Sustainability of plant-based diets: back to the future. The American Journal of Clinical Nutrition, 100(suppl_1): 476S–482S.

Samard, S., Gu, B.Y. and Ryu, G.H. 2019. Effects of extrusion types, screw speed and addition of wheat gluten on physicochemical characteristics and cooking stability of meat analogues. Journal of the Science of Food and Agriculture, 99(11): 4922–4931.

Sathe, S.K. and Sharma, G.M. 2009. Effects of food processing on food allergens. Molecular Nutrition & Food Research, 53(8): 970–978.

Sethi, S., Tyagi, S.K. and Anurag, R.K. 2016. Plant-based milk alternatives an emerging segment of functional beverages: a review. Journal of Food Science and Technology, 53: 3408–3423.

Sim, S.Y., Hua, X.Y. and Henry, C.J. 2020. A novel approach to structure plant-based yogurts using high pressure processing. Foods, 9(8): 1126.

Singh, M.B. and Bhalla, P.L. 2008. Genetic engineering for removing food allergens from plants. Trends in Plant Science, 13(6): 257–260.

Singh, M., Trivedi, N., Enamala, M.K., Kuppam, C., Parikh, P., Nikolova, M.P., et al. 2021. Plant-based meat analogue (PBMA) as a sustainable food: A concise review. European Food Research and Technology, 247: 2499–2526.

Sipple, L.R., Racette, C.M., Schiano, A.N. and Drake, M.A. 2022. Consumer perception of ice cream and frozen desserts in the "better-for-you" category. Journal of Dairy Science, 105(1): 154–169.

Söderberg, J. 2013. Functional properties of legume proteins compared to egg proteins and their potential as egg replacers in vegan food.

Tahergorabi, R. and Jaczynski, J. 2012. Physicochemical changes in surimi with salt substitute. Food Chemistry, 132(3): 1281–1286.

Tangyu, M., Muller, J., Bolten, C.J. and Wittmann, C. 2019. Fermentation of plant-based milk alternatives for improved flavour and nutritional value. Applied Microbiology and Biotechnology, 103: 9263–9275.

Vogelsang-O'Dwyer, M., Zannini, E. and Arendt, E.K. 2021. Production of pulse protein ingredients and their application in plant-based milk alternatives. Trends in Food Science & Technology, 110: 364–374.

Watanabe, K., Igarashi, M., Li, X., Nakatani, A., Miyamoto, J., Inaba, Y., et al. 2018. Dietary soybean protein ameliorates high-fat diet-induced obesity by modifying the gut microbiota-dependent biotransformation of bile acids. PLoS One, 13(8): e0202083.

Wittek, P., Zeiler, N., Karbstein, H.P. and Emin, M.A. 2021. High moisture extrusion of soy protein: Investigations on the formation of anisotropic product structure. Foods, 10(1): 102.

Xazela, N.M., Hugo, A., Marume, U. and Muchenje, V. 2017. Perceptions of rural consumers on the aspects of meat quality and health implications associated with meat consumption. Sustainability, 9(5): 830.

Yang, M., Li, N., Tong, L., Fan, B., Wang, L., Wang, F., et al. 2021. Comparison of physicochemical properties and volatile flavor compounds of pea protein and mung bean protein-based yogurt. Lwt, 152: 112390.

Yu, H., Park, S.H., You, B.S., Kim, Y.M., Yu, H.S. and Park, S.S. 2013. Effects of extrusion speed on the microstructure and mechanical properties of ZK60 alloys with and without 1 wt% cerium addition. Materials Science and Engineering: A, 583: 25–35.

Zahari, I., Ferawati, F., Helstad, A., Ahlström, C., Östbring, K., Rayner, M., et al. 2020. Development of high-moisture meat analogues with hemp and soy protein using extrusion cooking. Foods, 9(6): 772.

Zhang, J., Liu, L., Liu, H., Yoon, A., Rizvi, S.S. and Wang, Q. 2019. Changes in conformation and quality of vegetable protein during texturization process by extrusion. Critical Reviews in Food Science and Nutrition, 59(20): 3267–3280.

Zhao, H., Shen, C., Wu, Z., Zhang, Z. and Xu, C. 2020. Comparison of wheat, soybean, rice, and pea protein properties for effective applications in food products. Journal of Food Biochemistry, 44(4): e13157.

6

Veggie Meat Alternatives
A Virtue-Ethical View

Carlo Alvaro

1. Introduction

In this chapter, I discuss some practical and ethical aspects concerning those processed plant-based, meat-like food products referred to as veggie meat alternatives or plant-based meats or mock meats or fake meats or meat analogues (and other such terms). I would like to suggest that the production, sale, and consumption of veggie meat alternatives should raise important ethical questions. For example, in what sense, and to what extent, do veggie meat products prevent or exacerbate environmental degradation and global health problems? Assuming that such products can lead to positive health and environmental results, would it be ethical to produce, sell, buy, and consume them?

Based on the very nature of the subject, my discussion tangentially encompasses ethical veganism for at least two reasons: veggie meats were originally created for vegans, and the ideal goal of veggie meats seems to be the reduction and, eventually, the substitution of traditional meat with plant-based meat. Veganism, however, is not a diet, but rather a life choice to act in such a way as to avoid as much animal exploitation as possible. As a dietary practice, vegans refrain from eating the flesh of animals or insects or animal byproducts, such as dairies, eggs, honey, squid ink, castoreum, and so on. For a number of reasons varying from fashion to fitness to morality to religion, vegans consume exclusively plant-based food. Their diets may vary; yet all vegans consume plant-based substances. Veganism as a philosophy and as a diet goes back to the ancient world (Spencer 1996). For millennia, veganism meant consuming all sorts of vegetables, until human ingenuity created veggie meat alternatives. Plant-based products such as tofu (Shurtleff and Aoyagi 2013), tempeh (Babu et al. 2009), and seitan (Day 2011) have been around for centuries.

In addition to vegetables, fruit, grains, legumes, tempeh, seitan, and tofu, especially during the past twenty years, many vegans and vegetarians have been enjoying ultra-processed plant-based products that are often crafted to smell, taste, and behave like animal flesh or like animal byproducts. Nowadays, such products have become quite popular. Just look in most grocery stores and supermarkets to find a special isle displaying a vast array of veggie meats as well as a variety of plant-based analogues of eggs, cheese, honey, yogurt, and more. Also, in recent years, many meat-

Social Science Department, New York City College of Technology, 300 Jay Street, Brooklyn, NY 11201, USA, 917-658-1901.

based food joints, such as McDonald's, Wendy's, Burger King, and White Castle, have been offering veggie meat alternatives in addition to their meat-based products.

Although veggie meat products were originally created for vegans, today they are manufactured for and enjoyed by many non-vegans, too. Some people consume them for ethical reasons, others consume them simply because they taste good, and yet others consume them in the hope of losing weight or due to other touted health benefits. One thing is certain, that since their inception, the popularity of and the demand for veggie meat alternatives has been growing stronger and stronger and will likely keep on growing. Some may even argue that veggie meat alternatives are a possible solution to many ethical and environmental problems, such as the limited sustainability of the livestock sector, animal suffering, food insecurity and shortage, the problem of human overpopulation, environmental degradation, the negative health-related aspects of animal-based diets, and climate change. In this chapter, I discuss such questions and argue that, while on a practical level the production and consumption of veggie meat alternatives may seem to contribute to a solution to such problems, we should not stop at these considerations. In fact, when we reflect on the morality of veggie meat alternatives from a virtue-ethical perspective concerning human excellence and flourishing, we may discover previously-unexplored ethical reasons for not promoting the production, sale, and consumption of veggie meat alternatives.

2. The Origin and Purpose of Veggie Meats

Let me begin by saying something about veggie meat alternatives and their evolution. Believe it or not, they have a long history. A popular example is a wheat gluten product that nowadays goes by the name of "seitan". Seitan dates back to sixth century China. Buddhist monks accidentally discovered seitan by letting wheat dough soak in water, which causes the starch to dissolve thereby obtaining pure wheat gluten (Shurtleff et al. 2014). The mass of gluten obtained is a stretchy and chewy edible substance that is then seasoned and cooked just as if it were a piece of traditional meat. Seitan is possibly the simplest of all veggie meats. Today, most grocery stores even sell wheat gluten flour to make home-made seitan.

Also, veggie meats can be made from soybeans, legumes, oil seeds, and fungi (Bonny et al. 2015, Asgar et al. 2010). Another popular product is tofu, which was consumed as early as 965 CE. Other products include yuba and tempeh, also used for decades by some people (Bohrer 2019, Shurtleff and Aoyagi 2014). These products are still used to this day in many vegan dishes that often emulate meat-based dishes. In the West, the Seventh-day Adventists company Loma Linda Foods started in the late 1800s making the earliest examples of soy- and wheat-based meats (Land 2005). Also, in the sixties, British scientists discovered the microfungus *Fusarium venenatum*, a high-protein substance that was used to make meat alternatives (Wiebe 2002).

Thus, veggie meat alternatives have a long history. Since their inception, the market for veggie meat alternatives has been growing exponentially as a result of increasing social demands. The growth is due to many factors, the most important of which is that veggie meat products have become more and more appetizing. In fact, veggie meats are not consumed only by vegans and vegetarians. Today, even traditional meat eaters enjoy veggie meat alternatives. Not only do people enjoy veggie meats, but also all sorts of plant-based analogues, such as plant-based eggs, milk, cheese, and more. Veggie meat alternatives are so popular that some predict that the market will increase to over $21.23 billion US dollars by 2025 (Bohrer 2019, Shurtleff and Aoyagi 2014). It is important to note that such a market increase does not imply that more and more people are becoming vegans. Rather, it shows that meat eaters, too, have been consuming veggie meats.

Speaking of their sensory qualities, when I first became a vegan in the 1980s, I recall that veggie meat alternatives were not widely available or popular. Vegans were very few and they would consume plant-based whole foods, with the occasional, insipid block of tofu. Veggie meats were unknown by many and rather unappetizing. It was not until the 1990s that veggie meats started making a considerable impact upon the market owing to the fact that companies like Tofurkey,

Quorn, Gardenburger, and Boca Burger started offering a variety of products that had better taste, texture, and appearance than the previous generation of veggie meat alternatives. Still, at that stage, veggie meat alternatives never appealed to anyone other than some adventurous vegans or vegetarians. It would seem safe to assume that meat alternative companies, initially, did not intend to create products that would appeal to meat eaters. In my experience, these early veggie meats looked like some synthetic food out of a sci-fi B-movie, and while their taste was not revolting, it wasn't delicious, either. Meat alternatives in those days neither tasted like meat nor like vegetables. They were a hybrid food item. I still remember that they had a peculiar, unidentifiable, artificial taste to them that many described as tasting like medicine; they had a strange color and an odd texture. Therefore, I always avoided them and preferred to consume fresh fruit and vegetables.

By the 2000s, the taste, texture, and physical aspect of these products underwent yet another considerable improvement. Yet, they were still made specifically for vegans and vegetarians. They were not marketed as meat alternatives that would please anyone, including meat eaters. Rather, they were proposed as a sort of special plant-based food that offered something new and spicy to the typical vegetarian or vegan diet of tofu and lettuce.[1] Not only meat eaters, but even the very vegetarians and vegans did not find these products to be exciting. As plant-based company Sweet Earth's CEO Kelly Swette remarks, "When we launched in 2012, most vegetarian products looked almost like hospital food" (Vox Creative 2019). But this would soon change.

Since the early 2000s, veggie meats changed yet again. Companies like Sweet Earth, Beyond Meat, Gardein, and Impossible Burger (to name just a few) started offering plant-based meat alternatives that boasted uncanny taste, smell, and behavior almost indistinguishable from those of traditional meat. Not to mention the vast variety of plant-based cheeses that melt and stretch and do all those things that dairy cheese does, egg replacers, vegan honey, and more. Veggie meat producers nowadays are manufacturing products that can entice and satisfy a wider audience, especially meat-eating customers. Obviously, the goal is to bring meat eaters to buying and integrating veggie meats into their lifestyles and diets. As already observed, vegans hope that meat eaters will eventually give up meat altogether and consume only veggie meats. The question is whether meat eaters will eventually do so. As good as such products might be, and as even better they could be in the future, I cannot shake off the feeling that there will always be traditional meat eaters and that meat will never be substituted by veggie meats. Thus far, it does not look that way. According to National Purchase Diary Panel (NPD), "U.S. consumers have not given up on beef burgers but are willing to mix things up every now and then" (Seifer 2019). The figures speak for themselves. Surprisingly, in 2019, approximately 95% of people who purchased meatless burgers were meat eaters, and 228 million servings of plant-based burgers and sandwiches were purchased at traditional fast food restaurants. But the consumption of meat is not decreasing. In fact, it is growing exponentially (Saiidi 2019, Linnane 2019, Cooper 2021).

3. Animal Agriculture and The Environment

One of the main reasons veggie meat alternatives are so popular is that their taste pleases both vegans and meat eaters. Moreover, it may be argued that veggie meats have several benefits, two of which are their obvious lucrative business aspect, and their purported environmental friendliness. The production of veggie meats seems to involve practices and substances that do not pollute as much as the livestock sector's producing traditional meat. Furthermore, veggie meat production and consumption do not involve (direct) animal exploitation and suffering.

Regarding sustainability, the current world population is approximately eight billion and counting. It is estimated that it will reach nine billion by 2050 (Bonny et al. 2015). While affluent countries enjoy an abundance of food, world hunger and food insecurity still exist. Since the number of people in the world is steadily increasing, in order to satisfy their appetite for meat, the amount of meat that is produced needs to be increased, too. However, this will not be possible due to many factors, limited arable land and pastureland, scarce food and water resources, animal welfare issues, and the negative global impacts on the environment and climate change (Poore and Nemecek 2018, Dent 2020).

Let's consider the current state of intensive animal agriculture for a moment, which is known to be a leading cause of environmental degradation due to the excessive production of greenhouse gases, factory farm waste that contaminates waters and soil, and other environmental issues (Cameron et al. 2017, Koneswaran and Nierenberg 2008). Vegan diets are said to be environmentally friendly—at least friendlier than animal-based diets (Tuomisto 2019, Premack 2016). Even according to the United Nations, it is imperative that human beings consume less meat and more plant-based food because animal-based diets are no longer sustainable. As the UN states, giving up meat-based and adopting plant-based diet is necessary to "save the world from the worst impacts of climate change" (Carus 2010). For one thing, animal-based diets require considerably more natural resources than vegetarian and vegan diets.

Also, farm animals are known for carrying diseases. Vector-borne diseases are caused by infections transmitted to people by insects. Such diseases are caused, and become more severe, because of the environmental changes that result from the practices of animal agriculture, such as deforestation and reduction of biodiversity. For example, forest clearance in the early 1960s in Bolivia for the purpose of creating arable land caused the migration of mice that concentrated and overpopulated other areas; the mice then carried along a viral fever known as Machupo that killed one seventh of the population. A similar epidemic also happened in Argentina. Most diseases, like HIV, influenza, the Nipah virus, are vector-borne diseases (LeDuc 1989, Charrel and Lamballerie 2003). While the COVID-19 is not spread by eating animal-based food, the COVID-19 is a zoonotic virus that, like other zoonotic viruses, come from wet-markets, commercial factories, dairy, and meat farms. The COVID-19, SARS, the bird flu, the swine flu, cholera, and Ebola are all zoonotic plagues that start with animals in these types of situations and spread to humans. Those who consume animal products, therefore, contribute more to the emergence of zoonotic and vector-borne diseases that cause illnesses and kill people and animals than those who consume vegan diets.

Another problem is that intensive animal agriculture uses too much agricultural land. To be precise, too much means that "livestock takes up nearly 80% of global agricultural land, yet produces less than 20% of the world's supply of calories…seventy percent of Earth's arable land is used to grow food to feed animals, while many people in the world are malnourished or starving" (Ritchie and Roser 2017). These facts show that animal-based diets are very inefficient. Vegetarian diets, generally, require less arable land than meat-based diets. Meat-based diets contribute more to land degradation than plant-based diets (Pimentel and Pimentel 2003, Ritchie 2021). Animal agriculture's use of land is not efficient because feeding animals requires considerably large areas of arable land, plant protein, and water. Therefore, plant-based diets make it possible to feed more people than animal-based diets, and more adequately, by using the same amount of plant protein that is now required to feed farm animals. If animal farming were discontinued, it could be possible to use less arable land (as well as other resources) to grow human food.

Another issue is the depletion of minerals from the soil, which results in a chain of negative effects on the environment. The soil is depleted of minerals due to intensive farming; as a result, farmers use phosphorus fertilizers to supplement the low quantities available in the soil. Such a process has led to the buildup of phosphorus in the soil, and in turn the potential for phosphorus to become soluble. Dissolved phosphorus is transported from farms to lakes, rivers, and streams causing excessive aquatic plant growth, such as eutrophication (rapid growth of algae). Decomposition of algae leads to hypoxia (suffocation of aquatic ecosystems) in rivers and seas. Eutrophication also can generate Pfiesteria Piscicida, literally a group of fish-killer eukaryotes (Webster 2013).

Regarding water usage, the livestock sector uses more fresh water than any other sector. First, animals consume more water than humans. This sector, as previously mentioned, pollutes more water than any other sectors. Furthermore, farmers use fertilizers and pesticides that cause the formation of nitrates that leak into the groundwater resulting in negative health effects for animals and humans. Farmers in the USA use recombinant bovine somatotropin (rBTS), hormones that pollute waters.

Its use is prohibited in Europe and in other countries. Due to these pollutants, aquacultures are negatively affected. Considering that half of the fish that humans consume are produced in those aquaculture systems, it is not surprising that people develop many related diseases (Hotchkiss et al. 2008, Duffy and Moriarty 2003, Burkholder et al. 2007).

Also, meat-based diets require more fossil fuels than plant-based diets. This makes the farm animals' sector a leading cause of climate change because of greenhouse gases emissions. Although studies vary (some studies suggest that animal agriculture causes 51% of all emission of CO_2: Goodland and Anhang 2009), animal agriculture contributes to the depletion of the ozone layer because of animal digestion's emission of methane, and a loss of Earth's photosynthetic capacity through deforestation, which reduces Earth's capacity to absorb carbon from the atmosphere. Thus, there is no question that eating meat has a tremendously negative environmental impact compared to plant-based diets, and therefore, as many argue, animal agriculture should be discontinued, at least in affluent societies. As Deckers (2016: *passim*) argues, considering the ecological footprint that animal agriculture leaves, governments and individuals who care about the right of all humans to health care have a moral obligation to make a move toward diets that do not have deleterious impacts upon human health and the environment. Since plant-based diets can reduce or in some cases eliminate these environmental impacts, people and governments should encourage plant-based diets and inform people of the negative effects of meat-based diets. But which sort of plant-based diets should be recommended—diets that include veggie meat alternatives, for example? Let us go back to the question of the viability of veggie meat alternatives.

4. Health and Animal Suffering

In the previous section, the discussion focused on how plant-based diets can be environmentally friendlier that animal-based diets. Now let me say something about health-related concerns. It must be noted at the outset that there is a considerable body of research that speaks against the consumption of animal products in general and red and processed meats in particular. Also, it must be noted that some commentators have pointed out that the research on the subject of meat consumption and health is far from being conclusive (Klurfeld 2018, Micha et al. 2010, Carroll and Doherty 2019).

Such studies appear to be either neutral about the health effects of meat consumption or even promote meat consumption. I shall, however, discuss the research that is clearly against meat consumption. But the reader should not regard my choice as fallacious or as an example of confirmation bias; the reason is that I am not using the following data to argue against meat consumption (although I do so elsewhere: Alvaro 2017, 2019, 2020). Nevertheless, the research showing that meat-based products are unhealthful exists, and such a concurrence of opinion cannot be denied being highly significant.

For example, in 2015 the Academy of Nutrition and Dietetics, one of the world's most important organizations in this field, published a study concluding that vegan diets are healthful and can prevent and cure certain health issues and reduce the risk of chronic diseases (Cullum-Dugan and Pawlak 2015, Orlich et al. 2013, Turner-McGrievy et al. 2017). On the other hand, the scientific opinion on animal-based diets is not so encouraging. Granted, these studies are still disputed and, often, overzealous vegans may feel entitled to hasty conclusions. Nevertheless, to mention a few studies, processed meat can cause colorectal cancer, that one serving a day of red meat in adolescence or early adulthood can cause a 22% higher risk of perimenopausal breast cancer (Farvid et al. 2015), and that both processed and unprocessed meat were found to cause an increase in death rates from heart disease, stroke, diabetes, Alzheimer's, and lung disease (Etemadi et al. 2017, Daniel et al. 2011).

In many cases, diseases seem to be linked to the harmful effects of animal protein and to saturated fats (Etemadi et al. 2017). Also, according to the Physicians Committee for Responsible Medicine, even milk and other dairy products can increase the risk of breast, ovarian, and prostate cancers (Lanou 2009, Physicians Committee for Responsible Medicine n.d., Goodman 2020). To be specific, "research shows that three servings of milk a day can increase breast cancer risk by up to 80% and risk of death from prostate cancer by 141%" (Levin 2020). Also consider a study published in *The Journal of Nutrition* that concluded that fish is associated with higher incidence rates of breast cancer (Stripp et al. 2003), and egg consumption is associated with higher risk of cardiovascular disease (Zhong et al. 2019).

5. Animal Suffering

Finally, animals can feel pain and experience suffering. Some writers argue that animals should be regarded as persons and, thus, as moral agents (cf. Rowlands 2012, 2019, Tague 2020). As Marian Stamp Dawkins argues, "it is possible to study the negative emotions we refer to as suffering by the same methods we use in ourselves" (Dawkins 2008). Ethologists and philosophers have questioned the morality of using animals in various ways especially based on the notion of animal suffering. Most (sensible) people realize that animals have the capacity to feel pain and suffering. Such realization, however, fails to be a deterrent factor to the consumption of animal products. Veggie meat alternatives, then, would seem to be a solution. They taste like meat, they can satisfy meat eaters, without the harmful implications that traditional meat causes. Why would meat eaters bother eating veggie meat alternatives? Well, presumably, one reason is curiosity. Another reason is that they taste good as evinced by their popularity among meat eaters. Another is that meat eaters may wish to add more variation and excitement to their diet. Perhaps, the most important reason may be health concerns. Granted, nowadays there is a great deal of confusion concerning the subject of health and nutrition. At any rate, the assumption is that veggie meat alternatives are more healthful than traditional meat; moreover, the production, sales, and consumption of veggie meats does not cause (not directly) animal pain and suffering.

6. Weighing the Health and Environmental Benefits and Harms

I now turn to some purported health benefits and harms of veggie meats. Starting with the most obvious benefit, no one would dispute that fresh, whole vegetables are healthful. Therefore, some people might be under the impression that veggie meat alternatives, since they are made from vegetables, are healthful. However, it is quite obvious that veggie meats are neither fresh nor whole. It may be argued that veggie meat alternatives are more healthful than animal flesh. But it is not yet clear whether that is the case considering the ultra-processed nature and ingredients necessary to produce veggie meats. On closer inspection, meat alternatives are not necessarily healthful. It is not clear whether, and to what degree, animal-based food is healthful (Hu et al. 2019). However, at the very least, all nutrition scientists seem to agree that whole foods very healthful, but ultra-processed foods are not (Ha 2019, Tuso et al. 2015, Storz 2018). Ultra-processed foods are defined as

[F]ormulations of several ingredients which, besides salt, sugar, oils, fats, include food substances not used in culinary preparations, in particular flavors, colours sweeteners, emulsifiers, and other additives used to imitate sensorial qualities of unprocessed or minimally processed foods and their culinary preparations or to disguise undesirable qualities of the final product (Martinez Steele et al. 2016).

Veggie meat alternatives are ultra-processed foods. The problem is that ultra-processed foods can cause people to overconsume calories and are unhealthful (Hall et al. 2019, Elizabeth et al. 2020, Rauber et al. 2020). The current trend of the meat alternative business is to produce products that resemble and emulate traditional meat as much as possible, including the macronutrient profile and calorie content.[2] This is important especially for meat eaters who may consider consuming veggie meat products. The idea is that if they will continue consuming veggie meat products, meat eaters expect to eat products that satisfy them as much as eating animal flesh. This means that veggie meat manufacturers must offer products that contain the same number of calories, amount of protein, fats, flavor, and nutrients capable of giving meat eaters a very similar psychological, culinary, and gustatory experience and level of satiety they normally achieve by eating traditional meat.

Here I would like to draw particular attention to the fact that the principal aim of veggie meats companies is not necessarily a concern for the highest level of health; rather, veggie meat companies' goal is to offer very enticing and palatable products that satisfy both vegetarians/vegans' and meat eaters' palate. Those burger patties must be just as greasy and heavy and dense and filling as beef burgers; and those veggie chicken nuggets must have the bite, the characteristic texture, the saltiness, the flavor, and the smell that traditional chicken nuggets have. That is how the meat alternative business can sell big and grow. After all, the primary goal of the veggie meat alternative business, like all businesses, is to sell as much as possible to as many people as possible. Arguably, there are at least two important benefits: one is that companies like Boca Burgers, Beyond Meat, Impossible Burger, and others, are more environmentally responsible than conventional meat as they do not exploit (at least directly) animals and, consequently, the veggie meat industry would seem to avoid the animal suffering that the livestock industry causes.

The second is that the production of veggie meats does not involve the same waste materials as the production of animal meat, which have environmentally negative repercussions. However, as of today, there are very few studies on the health effects of veggie meat alternatives and, thus, such effects remain uncertain at the moment (which I presume that it will not be long before such research will be available). At any rate, veggie meats are not exactly healthful. Consider, for example, that some of the nutrients used to make veggie meats are meant to be similar to the nutrients that traditional meat contains. Some producers use genetically engineered yeasts, the addition of heme iron, and high amounts of salt and plant fats. As noted, veggie meats must have comparable caloric values to traditional meat. For one thing, such ingredients are designed to be addictive and highly caloric leading to the overconsumption of calories, which in its turn may lead to weight gain, and other possible health issues (Hall et al. 2019).

Another consideration is that veggie meat alternatives may contain several infamous substances known as anti-nutrients, i.e., phytoestrogens, amylase inhibitors, lectin, phytic acids, oxalates, and phytate. Research suggests that such substances can have deleterious effects on our health (Petroski and Minich 2020). To mention a few examples, phytic acid can cause mineral depletion and micronutrient deficiency by reducing the bioavailability of iron, zinc, potassium, copper, magnesium, and calcium. Polyphenols can have a negative impact upon the activities of digestive enzymes, and they can negatively impact the bioavailability of proteins and amino acids (Asgar et al. 2010).

Consider another potentially negative ecological aspect concerning seitan. Since seitan is produced from wheat, there is a concern about its sustainability. Seitan depends on the way that wheat is grown, the amount of land that is used to grow it, and how it is treated, e.g., pesticide, fertilizers, and so on. In order to grow crops of wheat, many farmers use different agrochemicals and pesticides. Such chemicals eventually contaminate waters, thereby harming other forms of life possibly damaging biodiversity (Döring and Neuhoff 2021).

Another substance that can cause similar environmental issues is tofu, which is obtained from soy, and thus all those veggie meats made from soy. One would think that soy is such an unproblematic product. However, according to the World Wildlife Foundation, the "surging demand" for soy is causing deforestation, a great deal of greenhouse gas emissions, and a significant use of water and pesticides (worldwildlife.org n.d.).[3] Soy is a fundamental ingredient of many veggie meat products, such as Tofurkey, Gardein, Lightlife, Boca, Impossible Burger, and others. Therefore, their negative impacts on the environment must be taken seriously. The company Beyond Meat, on the other hand, does not use soy for the production of their products; rather, they use pea protein, which seems to be environmentally friendlier than soy.

Let me say what is positive about veggie meat alternatives. In the first place, regarding a concern with protein consumption, most veggie meat alternatives on the market today provide adequate amounts of proteins. Bohrer (2019: *passim*) examined the nutritional contents in several meat products, such as beef burger, beef meatballs, pork ham, chicken nuggets, and veggie meat analogues. The study found that the protein content in animal-based products and in veggie meat products are approximately the same. However, the advantage of veggie meat products is that they have much less cholesterol and more dietary fiber than traditional meat products. Furthermore, veggie meat products are suitable to those consumers who cannot eat traditional meat products, to those who, due to their religious beliefs cannot eat certain meats (e.g., halal and kosher), and, obviously, veggie meats are convenient to vegans and others who refuse to eat animals for ethical reasons—although it is quite peculiar that vegans would refuse to eat animal flesh but then seek substances that taste like meat.

Regarding the environment, once again, whether veggie meat alternatives appear to have a lower environmental impact than meat products is not entirely clear—although virtually all researchers are positive that, even though veggie meat products do have some negative environmental implications, meat is much, much worse for the environment. For example, a study conducted by the University of Michigan determined that veggie burgers have a significantly lower environmental impact than traditional beef burgers. Specifically, Beyond Burger uses 99 percent less water, 93 percent less land, 43 percent less energy. Also, veggie Beyond Burger emits 90 percent fewer greenhouse gases than animal flesh burgers (Heller et al. 2018). However, it is important to mention that the study just cited is not completely unbiased as it was sponsored by the company Beyond Meat. Moreover, consider that the production of veggie meats, unlike traditional meats, does not involve manure pollution (Santo et al. 2020, Ernstoff et al. 2019).

A significant aspect of the environmental implications of veggie meats is that, in the end, any food production has negative global environment impacts, and this includes veggie meat alternatives. Producing veggie meat alternatives involves farming, and farming involves tilling, which destroys soil carbon. Preparing the soil for farming destroys the soil. Such a process releases carbon in the atmosphere. Thus, the growing of industrial crops contributes too much carbon in the atmosphere (Haddaway et al. 2017). Now, of course I am not suggesting that farming plant food be stopped. But the question is whether if the world substituted all meat with veggie meat, global agricultural land use will be reduced. It would be interesting to know whether producing veggie meats would require more land use than if the world adopted whole food, unprocessed, vegan diets (devoid of veggie meats).

In a study published in *Nature* the researchers, with the Oxford Martin Programme on the Future of Food, evince a certain degree of skepticism regarding the environmental effectiveness of totally plant-based diets. The researchers conclude the following:

We analyse several options for reducing the environmental effects of the food system, including dietary changes towards healthier, more plant-based diets, improvements in technologies and management, and reductions in food loss and waste. We find that no single measure is enough to keep these effects within all planetary boundaries simultaneously, and that a synergistic combination of measures will be needed to sufficiently mitigate the projected increase in environmental pressures (Springmann et al. 2018).

Surprisingly, the researchers do not conclude that the most environmentally efficient diet is veganism or vegetarianism. Rather, they argue that in order to resolve environmental problems, humans should follow a flexitarian diet that, while predominantly focused on vegetables and legumes, still incorporates some meat (521-22). In other words, such a study represents a counterintuitive example of the commonsense opinion that totally vegan diets are our best bet to save the planet from environmental degradation. True, as many researchers point out, humans must work very hard at reducing meat consumption. Yet, if this is correct, what is the point of producing veggie meat alternative, that is, the point besides the fact that many people may enjoy their taste? If the ultimate goal of producing veggie meats is to eventually substitute meat with plant-based meat, but such a switch won't guarantee significant environmental benefits, then the production of veggie meats would provide only hedonistic benefit. Consequently, it would be hard to convince meat eaters to quit meat and start consuming only veggie meats. Granted, one may argue as follows. What's the problem? If meat eaters enjoy veggie meats as well as meat, why can't they have both? But again, the goal is not to have both, but rather to eliminate or significantly reduce meat production. The trouble is that, as I indicated, meat production is increasing in spite of veggie meats. Later, I suggest other issues involving veggie meat alternatives. But at this juncture, having considered the foregoing, I would like to suggest one important reason I am opposed to veggie meat alternatives.

A major positive aspect to veggie meat alternatives is that they are vegan.[4] This means that their production, sale, and consumption do not involve (at least not directly) the exploitation and suffering of animals. Foods that involve no direct animal exploitation and suffering are *prima facie* morally superior to foods that do involve direct animal exploitation and suffering. Consequently, the production, sale, and consumption of veggie meats would seem to be morally superior to the production of animal-based meat. Therefore, veggie meat alternatives would seem to be a morally recommended approach to global food and nutrition. This argument sits well with most ethical approaches to animal ethics because most of such approaches rely on the notion of suffering. Vegan ethicists have argued, one way or the other, that eating animals is immoral or unvirtuous based on deontological, teleological, or alternative principles, such as feminist and care ethics (Ryder 1974, 1989, 1999, 2003, 2010, Singer 1975, 1989, Regan 1983, Rachels 1990, Francione 1996, Nobis 2002, Deckers 2016, Alvaro 2017, 2019, 2020, Hursthouse 2006, Adams 1990, Donovan 2006, Luke 1992). Such arguments are well known to scholars and even to laypeople, and thus I do not explain them here.

Although each of the theories just mentioned offers its unique approach to ethical veganism or vegetarianism, all of them have relied on the notion of suffering. A supporter of veggie meat alternatives may suggest the following argument: veggie meats are environmentally friendlier than traditional meat, and their production is free from animal exploitation and suffering. Their taste, texture, and physical appearance are very similar to their animal-based correspondents. This is important because veggie meats can satisfy (perhaps not just yet, but very soon) all meat eaters to the extent that eventually all meat eaters won't miss traditional meat. Given their nature, they can offer viable culinary alternatives to meat produced by the livestock sector such that they can convince meat eaters to become vegans, or at least convince them to significantly reduce their meat consumption. Since we ought to promote food that has such characteristics, it follows that we ought to promote veggie meat alternatives.[5]

7. Veganism Against Veganism

An important question that we should consider is the goal of veggie meat alternatives—not the immediate, but the ultimate goal—and that has to be to replace actual meat consumption, and, ideally, to eventually eliminate it completely. Another way to put it is that the ultimate goal of veggie meat alternatives is to achieve the vegan goal, which is a vegan world. Thus, consuming veggie meats can ease the transition to a vegan world. First, people would get used to the idea that something can taste good and can be nutritious even though it doesn't involve animal flesh. Meat substitutes are meant to overcome that initial sense of lack and dissatisfaction, that something is missing from a meal when you take the meat away. However, a potential problem to veganism may be, surprisingly, veganism itself. The abundance of veggie meat alternatives that have recently become available on the market has not significantly helped move closer to the vegan goal. Rather, such food items have contributed to veganism becoming a commercial food trend among many others. Thus, it seems that the goal that veggie meat alternatives has set out to accomplish has backfired, for veganism has shifted its focus from saving the animals to promoting consumerism and people's self-indulgence.

In fact, more and more people are consuming vegan products—but this does not necessarily imply that more and more people are becoming vegans.[6] In fact, it looks like meat consumption is increasing, which indicates that meat eaters don't have any intention to become vegans. Rather, they are happy to incorporate veggie meat products. Should vegans be happy about it? Perhaps not so.

As the number of vegans seems to be on the rise, many non-vegan corporations are jumping on the vegan bandwagon by creating new foods for the vegan community that (ironically and sadly) taste, look, and behave more and more like real animal products. Recently, MacDonald's has developed a vegetarian "Happy Meal", TGI Friday's has announced a "bleeding" vegan burger, and Gregg released its controversial vegan sausage roll. Also, the companies Impossible and Beyond Meat have been producing vegan burger patties, meatballs, and vegan ground beef, that quite uncannily emulate the smell, taste, and texture of real meat. The veggie meat company Impossible's products are becoming widely available for sale in grocery stores but have been served for a while at many vegan joints and at many non-vegan fast-food chains, such as White Castle, Applebee's, Bareburger, 5 Napkin Burger, Burger King, to name a few.

The company Beyond Meat's products are available for sale in virtually all supermarkets here in the US. One of the unsettling aspects of Beyond Meat's products is that their burger patties are packaged in the same way as meat patties. Also, they "bleed", a feat that is achieved by the addition of beet juice and brown as they cook as a result of the Maillard Reaction just like real meat patties. A further peculiarity regarding the sale of Beyond Meat products is that they are often placed in the same refrigerators where meat products are kept. Perhaps, their uncanny resemblance to meat products and the marketing strategies that the veggie meat companies adopt are counterproductive for veganism.

In the first place, the existence of veggie meats sends a specific message to the public—that meat is the real deal, the standard against which vegan products have to measure. This only reinforces the notion that eating animals is the norm, while veganism is an option or alternative. Vegan products are viewed not as the real deal, but a second-tier product or a second best, a weird brother, or even a basket case that gives meat eaters a wink. In point of fact, most vegan foods emulate animal products—fake burgers that "bleed", vegan cheese that melts, and other vegan products that taste so similar to animal-based products that it is scary. This is a trend that renders some vegans a rather odd group of individuals who refuse to eat animal products, but at the same time wish to eat animal products and in order to do so they eat foods that taste, look, and smell like animal products. Regardless of whether these products are tasty or popular or are helpful as "transitioning" foods, the main problem is that nowadays veganism is no longer the moral lifestyle choice of eco-militants, but rather just another trend that has carved itself a small niche in the animal-oriented market. Instead of representing the opposition, veganism seems to have become an ally to the meat industry. Instead of launching the message, "Eating animals is wrong, GO VEGAN!", veganism has become a movement that, rather than fighting, tries to mitigate the problem and to find a common ground and understanding and even friendship between animal-based food and the vegan lifestyle by launching the morally relativistic message, "We can all be friends. Who are we vegans to judge? There is nothing wrong with eating meat. Here, meat eaters, try my vegan patty. You won't be able to tell the difference".

Also, the production of veggie meat alternatives has caused veganism to revolve so much around food that it has promoted a great deal of self-indulgence. It has associated veganism with pleasure in such a way that it has overshadowed veganism's nominal goal to end animal suffering and exploitation. My fear is that, by reducing veganism to a question of taste (as research seem to show), veganism is leading us away from the concern of morals and that of human flourishing. In other words, the production of veggie meats, far from moving in the direction of a vegan world, has promoted human intemperance. The fear is that if people value pleasure and taste, instead of morals and human character and flourishing, and consuming traditional meat leads to pleasure and enjoinment, then veggie meat alternatives may never replace traditional meat and veggie meats will represent just another item on the menu.

Furthermore, the aesthetic and gustatory characteristics of veggie meat alternatives are, after all, patterned after real animal flesh products and meat dishes. The very concept, veggie *meat*, should raise ethical concerns, that is, in principle veggie meats cannot be completely cruelty-free. On the one hand, if meat alternatives require constant comparison to traditional meat products, then the success of such products, their taste, texture, and so on, will indirectly rely on animal exploitation. On the other hand, suppose for the sake of argument that all humans went vegan and thereby discontinued all meat production. In such a case, the production and consumption of veggie meats will forever remind people of cooked animal flesh, which would be quite peculiar for all sorts of reasons. For one thing, it might stimulate people's curiosity and encourage them to try the "real deal", which means killing animals again to eat their flesh. In the next section, I will put forth a virtue-based argument. The main idea is that some acts are wrong because they stem from the lack of certain fundamental virtuous character traits.

Take for example the virtue of truthfulness. A truthful individual is not one who just tells the truth most of the time or feels compelled to tell the truth or one who thinks or knows that telling the truth leads to the most favorable and desirable states of affairs. Rather, a virtuously truthful person is one whose moral character is well-developed and reliable. His actions (of truth telling in this case) spring from the individual's love for and understanding of the importance of being truthful, his practical knowledge of whether and when to tell the truth depending on the circumstances. Thus, a person who possesses the virtue of truthfulness will, for example, teach his children to be truthful, choose friends who are truthful, and be dismayed by untruthful behavior or even tactless jokes about being untruthful. Now, I argue that one who chooses to be an ethical vegan as an expression of virtue will refrain from eating animals or promote the consumption of animals and animal products. It is not only the act of eating animals that a virtuous individual would find wrong, but also the very idea of eating animals, the very business of killing animals for food, and the way the body parts and the flesh of animals are used and transformed into substances for human consumption. But consider that, in order to design veggie meat alternatives, it would require constantly comparing veggie meats to actual dead animal meats in terms of consistency, taste, texture, and aesthetic values. So, even if one can claim that the end product, veggie meat, is vegan, the process of developing it is not. In an important sense, it is not cruelty-free. Therefore, from the foregoing analysis based on a virtue-based approach (that I will develop in detail in the following sections), the virtuous individual will not promote or consume veggie meat alternatives—not even in moderation.

8. The Influence of Ethical Theory

In the previous sections, we have considered some positive and the negative aspects of veggie meat alternatives against some positive and negative aspects of traditional meat. We have seen that the production and consumption of veggie meats appears to be environmentally friendlier than traditional meat, though not totally friendly. Also, we have seen that some studies suggest that a flexible diet, predominantly vegetarian but with the inclusion of some meat, is possibly the most environmentally optimal. Furthermore, it would seem that the production and consumption of veggie meat alternatives avoid (direct) animal suffering and exploitation. One possible conclusion from all this is the one that a consequentialist ethics may propose, that is, producing, selling, and consuming veggie meat alternatives (and perhaps a limited quantity of meats from animals that had great lives) is the right action because the right action is that which promotes as much good as possible, and producing, selling, and consuming veggie meats promote as much good as possible.

An alternative ethical view is deontology according to which the right action is that which expresses respect for all rational and autonomous beings and does not treat other rational and autonomous beings as mere means but as ends in themselves. Since producing, selling, and consuming veggie meats do not seem to violate the rights of rational and autonomous beings, then they are right actions. Therefore, if we consider the issue of production and consumption of veggie meat alternatives in terms of either deontic or consequential standards, veggie meats respect the rights of animals and humans and can be conducive to the greatest good for the greatest number. In this section, I would like to question such assumptions. The reason that such assumptions seem plausible, I argue, is due to the persuasive influence of certain ethical attitudes, such as the rights-view, neo-Kantianism, and utilitarianism. Such ethical approaches may be referred to as *act-centered* theories because they are mainly concerned about right action.

As mentioned earlier, virtually all ethical defenses of veganism are ultimately based, or rely heavily, on the concept of suffering. For example, the typical approach that utilitarian ethicists take is to, first, assume that the goal of morality is to act in such ways as to maximize utility and minimize disutility. More specifically, we ought to act in such a way as to produce the greatest overall good for the greatest possible number of sentient beings that can benefit from the good.[7] The second step is to show that all suffering, whether human or non-human, is bad. After all, non-human animals represent another species. So, if we value only our desires and wellbeing, we will evince what some refer to as speciesism, unjustified discrimination among species. Thus, all suffering must be minimized, and all preferences or happiness must be maximized.

Therefore, the conclusion, according to utilitarianism, would be that we ought to avoid and shun food produced in ways that involves suffering, but not because the individual animals or people are special, or because the individual animals or people are directly wronged, but rather because suffering itself is bad, and pleasure/happiness itself is good. Since animal-based food involves suffering, we ought to avoid and avoid animal-based food (this is the sort of argument that Singer (1975) and other utilitarians would propose). According to such a view, then, the production and consumption of veggie meats passes the test with flying colors, and thus it is morally permissible (or at least not wrong or bad) because producing, selling, and consuming veggie meats can produce a lot more satisfaction of preference or happiness and a lot less suffering and dissatisfaction of preferences than animal based food.

If the issue were framed in terms of a non-consequentialist argument, such as rights view or neo-Kantian defenses, the result would not change. Such approaches, unlike consequentialism, rely on concepts of right or of duty. For example, Tom Regan (1983) argues that all animals have a right to life, and consequently all mammals such as cows, pigs, goats over a year of age possess the same basic moral rights as humans. The point of such an argument is that, as Regan argues, many animals are what Regan calls "subjects-of-a-life", and they are "subjects-of-a-life" precisely because they are sentient. So, animals have inherent value and, thus, it is immoral to use them for food. A neo-Kantian example has been championed by Christine Korsgaard. In a lecture, Korsgaard (2004) argues that our autonomous nature is not the only source of moral duty, as Kant argued. In fact, we humans can derive normative value from our animal nature. The moral duties to ourselves arise from the fact that we experience pain and suffering, and while animals are not self-legislative, rational creatures, like us, they are creatures that can experience pain and suffering. As Korsgaard puts it, "…human incentives are simply the same as those of the other animals" (2004). In other words, Korsgaard's neo-Kantian approach, too, relies upon the notion that animals are capable of suffering.

Now the first point I wish to make is that the above mentioned arguments are, by no means, the only defenses of veganism. Second, it is not germane to my argument to give an in-depth critique of such arguments. Third, these theories are the most influential both among academics and laypeople. The reason for my mentioning them is to draw attention to those arguments' lack of ethical completeness. Consequentialist and non-consequentialist moral systems address important factors. However, they offer a narrow approach to morality: according to consequentialist ethics, the main focus is the end result of our actions, the consequences, and the total (or average, depending on the particular theory) of happiness. Thus, so long as our actions can lead to the maximization of utility, we ought to perform such actions. Since the production, sale, and consumption of veggie meat alternatives lead to the greatest good for the greatest number (let's assume so), then we ought to promote the production, sale, and consumption of veggie meat alternatives.

Similarly, non-consequentialist theories may conclude that veggie meat alternatives should be promoted or, at least, that the production, sale, and consumption of veggie meat alternatives are not immoral practices. They are not immoral either because they do not violate anyone's basic rights, especially animals' rights, or because we do not undermine in any way our moral duties. In fact, one may argue, the production and consumption of veggie meat alternatives would favor the fulfillment of our duties to non-human animals.

The aspect to which I want to draw attention is that by framing the question of the morality of veggie meats in terms of modern ethical attitude, one may conclude that producing, selling, and consuming veggie meats are not immoral practices. According to the ethical theories just outlined, an act is *immoral* just if it undermines aggregate utility or it violates one's rights or it fails to fulfill one's duty. If the production, sale, and consumption of veggie meat alternatives do not undermine aggregate utility or violate rights or fail to fulfill one's duty, then it would follow that producing, selling, and consuming veggie meat alternatives are morally permissible. In fact, according to utilitarianism, if it turned out that such practices brought about the greatest good for the greatest number, they would be morally required practices.

Thus, to answer the question I posed at the outset, "Is it ethical to produce, sell, buy, and consume veggie meat alternatives?" from the point of view of modern ethical approaches, the answer is in the affirmative. An important point is that not only consequentialist and deontological ethics (action-based or rule-based ethics) have dominated, but they also have shaped and continue to shape our moral sense and direct the animal/environmental moral discourse. Alternative theories exist (care, eco-feminism, virtue-oriented), but they have had little, if any, influence upon veganism, vegetarianism, food ethics, and environmental concerns among academics and laypeople. Consequentialist and deontological theories have shaped our moral intuition and imagination. But the problem, as I argue, is that such ethical approaches are too narrow or incomplete. What I mean is that such theories leave out important considerations—the most important of which is a consideration of individual moral character. Such theories address questions such as what we ought to do or whether animals have rights or what actions produce the greatest good overall. But curiously, they seem to overlook questions concerning human motivation, character, and flourishing. Thus, when the morality of producing, selling, and consuming veggie meat alternatives is addressed, it seems that most people don't see anything wrong with them. After all, most people seem to assess the moral implications of veggie meats (or moral issues in general for that matter) based on either the utility of it, or the rights of the individuals involved, or based on one's duty. Moreover, such theories focus on the importance of animal suffering. But what if, let us suppose, it turned out that animals do not suffer?

What I would like to suggest is that although utility, rights, and duties, and animal suffering, are important moral criteria, focusing on them limit our understanding of the importance of the concept of human virtue and human flourishing. In my view, once we look at the issue of meat alternatives from the perspective of human excellence and flourishing, such a view may reveal that the practices of producing, selling, buying and consuming veggie meats are inconsistent with several human virtues and unfavorable to flourishing; thus, they should not be promoted but avoided.

9. A Virtue-oriented Approach

The view that I propose has its roots in Aristotelian virtue ethics. On such a view, when we try to understand what morality is about, we must understand the nature of humanity and human goodness. In particular, we must first understand what allows humans to grow, to develop, to thrive, and ultimately to flourish (to perfect).[8] As Philippa Foot (2003) would say, the point of morality is understanding what is the human good, and the human good is understood in terms of what human beings need and benefit from in order to fulfill their nature. In other words, by understanding our nature, we can determine the sorts of virtues necessary for human flourishing.

This view is so different from other ethical approaches because it is concerned with much more than the right action. On the virtue approach, what is right is to be or become the best human being one can by acting virtuously. Acting virtuously means acting justly, compassionately, mercifully, benevolently, mercifully, and more. But to be a virtuous person, one must not only exhibit a certain behavior, say, telling the truth most of the times. Rather, her telling the truth is an expression of her character and her character is a complex psychological disposition consisting of right emotions and attitudes guided by sound reasoning. That is, for every particular circumstance, the agent expresses her virtue in appropriate manners, at the right moment, with the right individuals, in the right amount that is called for. For example, a truthful person won't divulge the truth if the truth might hurt others or be used to hurt others.

Thus, a virtuous person, one who is magnanimous, compassionate, temperate, just, courageous (and more), has a much more expansive ethical outlook than one who espouses an action-centered view. A virtuous person, too, values aggregate good and respect for others, but he values such things not for their own sake but because, in some cases, maximizing happiness and respecting others stem from excellent (virtuous) character traits. The important difference between virtue ethics and other ethical systems, however, is that a virtuous person also appreciates that acting in certain ways—even though in so acting we don't violate others' rights, or our actions do bring about the greatest amount of good—can be intemperate, avaricious, petty, cowardly, callous, and unjust.

My contention is that rights-based, deontological, and consequentialist, utilitarian theories are too ethically narrow, and thus fail to track our moral intuitions regarding certain issues. In an important article, Thomas Hill (1983) describes what happened when a neighbor cut down an old tree and paved his yard with asphalt. Hill notes that most people would be outraged by the destruction of nature. But on what grounds may one condemn the act of a person's cutting down a tree in that person's property and covering his yard with asphalt? Or on what grounds may one condemn bulldozing an entire park to build a petrol station? After all, it might be a case involving the use of eminent domain to, say, clearing out an entire park to build a petrol station that benefits the community and leads to the actual maximization of utility. Moreover, since plants have no rights, and the people who build the petrol station do have the right to cut down trees and build whatever they want there as they see fit, we have no way of denouncing such acts. As Hill suggests, in order to see what is wrong in these examples, we should not seek the problem in the act itself, but rather in the individual's character—in those character traits that motivate that person to so act. Thus, the important question is, "What kind of person (character) would destroy or deprive others from nature's beauty to build something ugly in its place?"

Why, then, is what I have said so far relevant in evaluating the production, sales, and consumption of veggie meat alternatives? In accordance with the view I propose, such practices should be avoided because they stem from a defective moral character, such as the lack of temperance and a misunderstanding or distortion of the role of food in human flourishing. Therefore, by using the concept of temperance and that of human flourishing, we can say that such practices should not be promoted but avoided. A central question here is whether we humans are supposed to give so much importance to food as to be willing to create and consume ultra-processed and lab-manufactured food. In my view, most people place too much importance on food. A virtue-oriented approach, thus, looks at such a question unlike other theories because it does not stop at the considerations of what the best outcome is or what the rights involved are or whether we have certain duty with respect to others or whether, in acting, we respect the moral law. On my proposed approach, we consider, first, the role, nature, and importance of human food and, second, the individual's moral character and motivation for his or her choices and relation to food. While I believe that my proposed approach is superior to others, I would like to note that the goal of the present discussion is not so ambitious. Rather, here I would like to offer an alternative perspective on the moral issue of producing, selling, and consuming veggie meats—a view often overlooked or forgotten. The goal, therefore, is to provide an alternative ethical framework to the dominant ethical views—a framework that would make us pause and consider that what might seem all too obvious at first (that veggie meats are morally unproblematic because they don't violate any rights or because they possibly maximize utilities) can and ought to be questioned.

The first step to my preferred ethical view is to note that all living organisms have a specific and optimal diet, and since human beings are living organisms, they have a specific and optimal (natural) diet, too. By "specific" or "natural" or "optimal", I mean a diet to which an organism has adapted, a diet that contributes to a species's flourishing. Straightaway, I anticipate the possible charge that such an empirical claim may be guilty of the appeal to nature fallacy. The appeal to nature fallacy occurs when one argues that since something—let's call it Φ (it can be food, behavior, etc.)—is found in nature, Φ is good, moral, or healthful (or since Φ is *not* natural, it follows that Φ isn't good, moral, healthful, etc.). Why is such a style of argumentation fallacious? Because there are a number of things that are natural and yet are not good, moral, or healthful and some things that aren't natural (in the sense that they are manufactured or man-made, e.g. chemical disinfectants) and yet can be good or moral. For example, cruelty, hurricanes, and diseases are natural but very bad for us. Conversely, there are many artificial things that are very good for us, such as disinfectants, toothbrushes, the Hubble telescope, or defibrillators.

Now it would be a fallacious strategy to argue as follows: eating whole, unprocessed food is healthful for us because it is natural, and therefore we *ought to* eat whole food but avoid ultra-processed foods (such as veggie meats, in this case). However, this is certainly not my argument here. Rather, based on the physiology and natural history of a species, it is possible to determine which diet or food is optimal, which is suboptimal, and which is deleterious for that species. For instance, the natural and optimal food for squirrels is fruit, nuts, and vegetables, which are found in their natural environment. When these food items are not available, then squirrels would eat whatever they can find—including legumes, dairies, chocolate, and even cooked food discarded by humans, all of which are suboptimal or unhealthful for squirrels. Every species has a specific food that enables the members of that species not just to survive but to thrive and flourish. Similarly, while humans can tolerate many substances they eat, based on our physiology and natural history, there is an optimal diet for humans that enables humans to thrive—not just to survive. My argument is that, if one desires to thrive and reach the highest level of health, one should consider following that diet.

Moreover, based on what makes us thrive and flourish, it is not an unwarranted jump to the conclusion that an excellent human being is one who acknowledges his or her nature and act in ways that benefit his nature. The idea is that consuming substances to which our species adapted enables us to flourish. Thus, consuming a diet that conforms and promotes one's nature and flourishing is rational, and therefore, the virtuous individual consumes in moderation only food conducive to his flourishing. So, the argument is not simply that since certain substances are optimal for us we have a moral *obligation* to consume them. Rather, the argument is that the human species has the function of reasoning well in exercising virtuous action, the fulfillment of which will lead to the flourishing of the human species. Consequently, consuming certain substances that are not meant for human consumption, besides being a form of self-deception and disrespect, will not lead to our flourishing. A member of our species might say, "Who cares, why must I flourish?", but this would be equivalent to a sea water fish that, despite understanding that living in the ocean and eating algae is essential to her flourishing, would rather live in a lake and eat potato chips. Such a fish, in the view that I propose, is considered a defective individual.

Regarding our species, our optimal diet is a diet of raw fruit, leafy greens, and some nuts and seeds. Evidently, many people survive eating sub-optimal substances. But the point is that it is clear that the most healthful food for an organism is unprocessed substances that are conducive to that organism's flourishing. In the case of humans, such substances are undoubtedly fruit and vegetables. Therefore, one who desires the highest level of health should follow such a diet. If this is correct, then it seems plausible to argue that a temperate individual doesn't consume processed foods, not even in moderation, but rather avoids them altogether, when one has a choice.

Another question is whether a diet of raw fruit and greens devoid of processed food is specific and optimal for humans. Humans may have adapted to tolerate certain substances. For example, some people have developed certain enzymes that make it more *tolerable* to consume cereal grains and dairy. But our digestive systems and our bodies have not completely mutated. In any case, the point is that such enzymes had to be developed in order to tolerate grains and dairies, which are not substances conducive to the highest level of human health.

In fact, grains (among many other substances) are not ideal for our digestive system. Prior to tolerating such substances, our ancestors ate fruit and tender leafy greens, which abounded in the tropical areas where they originally lived. Humans have not changed in any significant way such that processed food would be conducive to good health and flourishing. To mention another example, after cattle domestication, some cattle herders must have developed some sort of lactose tolerance. Groups that were not dependent on cattle, such as the Chinese, the Thais, or the American Southwest remained lactose-intolerant. Nevertheless, it is mere tolerance that some people acquired—tolerance is not the same as beneficial. Grains and dairies are by no means optimal for our species. In fact, millions of people—70 percent of the global population—are lactose-intolerant, which implies that many people who regularly consume dairy products do not even know that they are lactose-intolerant (Physicians Committee for Responsible Medicine n.d). They merely tolerate such products (Savaiano and Levitt 1987). The same can be said about grain consumption. Many argue that since the development of agriculture, humans have adapted to digesting grains. But it is quite evident that grains and their byproducts are not optimal for our digestive system. First, they would be virtually inedible if not processed by cooking. Second, grain products are typically consumed with the addition of refined sugar, fats, and other condiments. In fact, it is no mere accident that medical professionals warn people to either consume in moderation or, better yet, not consume breads, pasta, pizza, and other such refine grain products, because they are unhealthful.

Another point is that all living organisms consume non-processed food.[9] No living organisms benefit from ultra-processed food, and there is no reason to think that processed food benefits any organism. Granted, by processing food, humans can derive calories from substances that would otherwise be indigestible (grains and legumes for example). So, in a situation of food insecurity or scarcity, humans would benefit from processing organic substances in the sense that they would be able to survive and not die of starvation. However, the benefits to which I am referring when I consider human food are optimal health and, ultimately, flourishing. Furthermore, it just seems implausible, and indeed suspicious, that only one out of thousands of species—the human species—would benefit from processed food when all other species benefit from whole, unprocessed substances. It would seem incorrect to argue that humans are supposed to or designed to consume processed food. Many individuals get by with consuming sub-optimal or non-optimal substances. Obviously, this does not show that humans *ought not to* consume processed food. In my approach, I do not argue for moral obligations in the Kantian sense. Rather, I propose that consuming substances that are not conducive to flourishing evinces an unvirtuous character; moreover, it displays self-deception. Namely, since humans have a specific diet that is conducive to the flourishing of our species, consuming substances that are deleterious or not optimal and not conducive to our flourishing is tantamount to ignoring and deceiving (whether deliberately or not) our nature.

Consequently, since humans are the product of nature and natural selection—we are part of nature—just like all other species, we benefit from natural and unprocessed substances that grow in nature. Moreover, the human body is not endowed with the same features as carnivores, exceptional speed, sight, bodily strength, strong digestive system, etc., all of which enable carnivores unaided by weapons and fire to kill and consume animal flesh. Humans must use weapons and various technologies to kill, process, cook, and consume animals. It follows that leafy greens and fruit constitute the optimal human diet. As Dr. William C. Roberts aptly puts it,

Although most of us conduct our lives as omnivores, in that we eat flesh as well as vegetables and fruits, human beings have characteristics of herbivores, not carnivores. The appendages of carnivores are claws; those of herbivores are hands or hooves. The teeth of carnivores are sharp; those of herbivores are mainly flat (for grinding). The intestinal tract of carnivores is short (3 times body length); that of herbivores, long (12 times body length). Body cooling of carnivores is done by panting; herbivores, by sweating. Carnivores drink fluids by lapping; herbivores, by sipping. Carnivores produce their own vitamin C, whereas herbivores obtain it from their diet. Thus, humans have characteristics of herbivores, not carnivores (Roberts 2000).

The fact that our ancestors at some point started developing various skills and techniques to hunt, kill, process, and eat animals does not tell us that eating animals is conducive to optimal human health and flourishing. If anything, it tells us that our ancestors underwent food scarcity to the degree that they had to start killing other animals and feeding on their flesh in order to survive. Neither our ancestors, nor contemporary humans, process plants and animal flesh in various ways in order to improve their health or to improve the quality of a substance, but rather in order to increase calorie content and thereby survive through scarce conditions. Obviously, processing food enables the consumption of substances that we cannot consume in their raw state. But it certainly doesn't follow from these facts that processing food is beneficial. In fact, processing food became detrimental to human health. Consider the tremendous decline of human health (Wang et al. 2011). In the U.S. alone, the National Health and Nutrition Examination Survey (NHANES) reports that the obesity rate among adults in 2015–2016 was 39.8 percent. Since 1999, obesity rates have been increasing. It is estimated "65 million more obese adults in the USA and 11 million more obese adults in the UK by 2030" (Wang et al. 2011). According to the American Heart Association's Heart and Stroke Statistics, almost half of U.S. adults suffer from cardiovascular disease (AHA 2019). Taking the United States as an example, Americans are eating more calories than they require (Gould 2017). This steady increase in calories is due to an increased availability of calories. The point is that the cause of such a declined status of human health is due to processing food, which enables high calorie concentration. Eating fruit and greens practically could never cause obesity, heart disease, and other such chronic conditions.

No one would argue, for example, that if we start feeding processed substances to giraffes, we are going to improve their health (Milton 1999). On the contrary, virtually every expert would agree that while giraffes might be able to survive by eating processed substances, such substances are not optimal for the health of giraffes. By the same token, no reason exists to believe that processed substances, whether plant- or animal-based, could be optimal for humans when it is not optimal for other animals. The few thousands of years during which our species foraged had no significant effects on our digestive anatomy (Milton 2000). People nowadays eat cooked substances, but their digestive systems are still the same system that nature designed for eating fruit and leafy greens millions of years ago. As primatologist Katharine Milton writes, "the widespread prevalence of diet-related health problems, particularly in highly industrialized nations, suggests that many humans are not eating in a manner compatible with their biology" (Milton 1999). It is not surprising, then, that the health of modern humans is declining because they are not consuming diets "…on which the primate line has flourished for many tens of millions of years and *produced [us]*" (Dehmelt 2005). As Donald S. Coffey observes,

[Human beings] were not biologically selected by the evolution process to eat the way we do today, and the damage is manifested in prostate and breast cancer. Indeed, all of the present suggestions of the National Cancer Institute and the American Cancer Society as to how Americans might reduce their chances of getting prostate and breast cancer revolve around adapting dietary changes in our lifestyle back toward the early human diet of more fruits; a variety of fresh vegetables and fiber; less burning, cooking, and processing; diminished intake of dairy products, red meat and animal fats... (Coffey 2001).

Thus, the first step was to show that humans were not biologically selected to eat ultra-processed substances, nor were they selected to consume animal flesh. All living organisms can be flexible and eat substances that are not optimal to their health. But it is also true that all living organisms do have a specific diet according to their nature. Also note that virtually all wild living organisms consume as much food as they care for without damaging their functions and health. However, research shows that companion animals and other liminal animals suffer from obesity (German et al. 2018). The interesting question is why this is the case. The most cogent answer is that when animals consume a diet that is not proper and specific to their nature, they experience a decline in their health. Granted, humans can cook grains or animal flesh and somehow manage to survive. But clearly this does not mean that cooked substances or ultra-processed foods such as veggie meats are healthful for us. Rather, we must eat food that agrees with our biological nature—food that is conducive to the highest level of health.

The second step is to bring attention to the fact that human diets have been shaped by many different cultural factors and, moreover, by consumer capitalism. Free market capitalism is an economic system whose principle is to sell as much as possible, irrespective of whether the goods sold are beneficial or detrimental. Consider the tobacco industry as an example, which produces and sells carcinogens to people. Under capitalism, consumer demand is deliberately manipulated through mass-marketing techniques, to the advantage of sellers. In order to sell, people are disciplined into believing that they need to surround their lives with many things when in reality, they don't. I think that from time to time, we ought to remind ourselves of the sheer waste of food and resources that those of us who live in affluent countries cause. We have to acknowledge that while people in affluent countries waste food and live in opulence, more than a billion people live on less than five dollars a day. With regard to diet and food, many (though not all) food companies are not necessarily concerned about people's health; rather, they are concerned about profit. Consequently, it should come as no surprise that in a capitalist society, people are told to consume big and, of course, to eat big and as frequently as possible, which is exactly what has caused the current decline of human health.

Capitalism affects us in more negative ways. In particular, for many people, their jobs are not rewarding, not enjoyable, and not worth doing. As a result, people try to overcome their unfulfilling lifestyles by going out and feasting on luscious food and libation. After all, eating plentifully and luxuriously is one of the few things over which they have control. Unfortunately, hunger crises and famine are rising in Africa and the Middle East where twenty million people are at risk of starvation. Let's not forget that hunger is typically used as a weapon against civilian populations. Under capitalism, people are told what to eat and how much to eat. They are told to consume what increases profit—namely, grains, along with meats, dairy, and other animal products—in other words, processed food. So, it is not surprising that the morality of processed food is seldom questioned.

Have you ever paused to reflect on the fact that in affluent countries, food advertisement is ubiquitous? Ads about food are everywhere—in schools, trains, and buses. And people are constantly reminded to eat and eat some more. The food industry has created the very patterns governing the way we eat. In other words, food corporations have taught us, through the media, not only what food is supposed to be but also how to eat it, when to eat it, how much and how often to eat it, and even with whom to eat it. We are told that we need breakfast, lunch, plenty of snacks, dinner, and possibly another snack prior to going to bed. We are told that we need milk, cereal, pasta, spices, meat, chocolate, coffee, beer, liquor, soda, energy drinks, tobacco, and a host of other useless items.

But what is the point of all this? The point is that the first step that one needs to take in order to understand the morality of producing, selling, and consuming veggie meats (or food in general, at any rate) is to comprehend the moral patterns of food consumption dictated by the food industry. Realize that in our society, we are constantly pressured to overconsume everything—clothing, natural resources, and especially food. Consequently, it is difficult for most people to exercise their temperance when it comes to food. I would like to suggest that living a more minimal life is another important element conducive to human flourishing. The great Roman emperor and Stoic philosopher Marcus Aurelius aptly describes the best approach to life when he writes: "If you seek tranquility, do less. Most of what we say and do is not essential. If you can eliminate it, you'll have more time, and more tranquility. Ask yourself at every moment, 'Is this necessary'?" (Marcus Aurelius 4.24) Following Marcus Aurelius' advice, I would add that if we seek flourishing, we should eat less, because most of the food we eat is not essential. The sort of life to which we are advised to aspire, the life of overconsuming and overeating all sorts of highly sensual and processed foods, is a hedonistic existence not conducive to human flourishing.[10]

As Marcus Aurelius and other stoic philosophers observed, eradicating the unessential from one's life is conducive to flourishing. Simplicity and temperance in life are conducive to human flourishing. Simplicity would seem to be consistent with the virtue of temperance. Therefore, if we assess the issue of producing and consuming veggie meats from the perspective of temperance, we may realize that the practices of producing and consuming veggie meat are inconsistent with a temperate character. This is because the purpose of creating and consuming such products is not the pursuit of human flourishing, but rather the hedonistic desire to achieve pleasure through eating certain substances, which are not conducive to optimal health. Again, as noted earlier, the consumption of such substances can be considered as a form of self-deception.

I am particularly interested in the virtue of temperance because, as Aristotle discusses it, temperance is the relevant virtue in the context of sensual and physical appetites. Temperance encompasses a rational and harmonious approach to physical pleasure, which of course includes food pleasure. Based on my analysis of temperance, a virtuously temperate individual is not indiscriminately attracted by food just because it might emanate a pleasant aroma or it might have a certain texture or flavor. That is because such characteristics can and often do mislead us into consuming substances that are not conducive to good health or positively deleterious. As the saying goes, "I do not live to eat but I eat to live". In other words, the primary purpose of eating for a heterotrophic organism is to acquire energy for various activities and growth from food. To be sure, this does not mean that a virtuous person should derive no pleasure at all from eating. The pleasure of eating has physiological and psychological importance, which is evident as early as infants' feeding at their mother's breast.

The problem in affluent societies is the excess amount and the variety of food available, and hyperpalatable food (food that creates psychological addiction), all of which are leading reasons for the decline of human health. Cooked and ultra-processed substances cause the body to release chemicals that trigger a sense of reward, which often leads to overeating. This is one of the reasons why it is virtually impossible to overeat fruit and salad. Therefore, the temperate individual will consume food for the purpose of nourishment. Moreover, she will choose food that is psychologically and physically rewarding but not addicting. Veggie meat alternatives are hyperpalatable substances. There is an obvious danger regarding hyperpalatable substances because they are specifically designed to override one's ability to control the amount consumed. Some of the most egregious characteristics of such substances are a very high content of calories in rather small portions, high content of fats, sodium, and refined sugar. As Gearhardt et al. suggest, "Foods, particularly hyperpalatable ones, demonstrate similarities with addictive drugs". These researchers show that consuming hyperpalatable foods can be capable of "triggering an addictive process" (Gearhardt et al. 2011).

The obvious danger of ultra-processed, hyperpalatable substances is food addiction, which leads to a number of metabolic problems (Blumenthal and Gold 2010, Merlo et al. 2009). There's an important point to distinguish habits from addictions. Here I use the term addiction not merely to indicate a habit, but rather a much stronger psychophysical condition suggesting a disease for which one is not responsible.

A temperate individual consumes only food that is vitally essential and that is conducive to good health and ultimately to her flourishing. This analysis seems to perfectly fit Aristotle's description of the temperate individual who eats substances in moderation, as Aristotle observes, "as long as they are not incompatible with health or vigor, contrary with what is noble, or beyond his means" (Aristotle 2002). It is clear from the foregoing analysis that veggie meat alternatives are not the type of food substances that are compatible with temperance. The temperate individual is emotionally and physically trained so as to have the right emotions toward food. These enable the virtuously temperate to make the right choices and thereby enjoy eating foods that are healthful and avoid food that is potentially unhealthful and addictive.

This argument benefits from remarkable scientific evidence. First, consider that plant-based diets completely devoid of animal-based food and ultra-processed food are optimal at any stage of life (Craig and Mangels 2009). As the American Dietetic Association (AMA) states, "appropriately planned vegetarian diets, including total vegetarian or vegan diets, are healthful, nutritionally adequate, and may provide health benefits in the prevention and treatment of certain diseases" (Tuso et al. 2013). It must be noted that in the AMA study, by "appropriately planned vegetarian diets", the researchers refer to that plant-based diets devoid of ultra-processed substances, i.e., whole plant-based.

The obvious objection is that one may consume veggie meat in moderation. But in the first place, consider the nutritional profile of veggie meats (discussed earlier) and their hyperpalatable, addictive nature. One may consume them sporadically and perhaps not become addicted to them and maintain good health. But from the point of view of a virtuous agent, the morally temperate attitude toward food in general is to avoid altogether those substances that are not inherently conducive to human physical wellbeing or food that is inherently unhealthful. Why eat in moderation a substance that has the potential to cause health problems in the first place? If a substance is potentially deleterious to our health, or it is in no way beneficial to our flourishing, it follows that such substance is not essential, and consequently, the question of moderation does not apply in this context. For example, taken in small doses, cyanide is not immediately lethal. Drinking moderately may not be immediately lethal. Nevertheless, it would be unnecessary—and not very wise—to take small doses of cyanide or small servings of alcohol.

Thus, the conclusion that a temperate person avoids veggie meat alternatives follows from the following premises: a temperate person avoids—even in moderation—a nonessential or harmful substance. For example, smoking one cigarette per year might not be deleterious or lethal. In fact, many people live long lives despite their smoking cigarettes. But smoking cigarettes, even if one cigarette once per year, is unnecessary and not conducive to good health. Moreover, consider the tobacco business in the first place. Consider that the industry produces for profit a deadly product. Consider that such a product is not only deadly for the users, the smokers, but also can cause harm to nonsmokers who breathe second-hand smoke. Consequently, a virtuous person could not possibly contemplate smoking cigarettes in moderation, even if smoking in moderation didn't cause death. The reason is that the very principle and practices of producing and selling cigarettes are inconsistent with virtue. Since good health is an important element that enables one to flourish and achieve human happiness, it follows that it is sensible not to smoke at all or support in any way the tobacco industry.

Similarly, consuming veggie meat alternatives in moderation might not be deadly or at any rate deleterious to us (hopefully not as deadly as cigarette smoke). However, such meat alternatives can be deleterious for our health. Also, the goal of those who produce them is not human flourishing, but rather profit. Therefore, a virtuous individual would not consume veggie meat alternatives even in moderation because the notion and practices involved in the production, sale, and consumption of veggie meat alternatives are conducive to human flourishing—and they are inconsistent with virtue. Fresh fruit and vegetables are, on the other hand, never dangerous for our health, even when consumed on a daily basis and in abundance. They are not wrong in principle, as they are natural and beneficial to our species. A well-planned diet of whole fruit and vegetables completely devoid of veggie meats can be optimal. Moreover, consuming veggie meats, as previously mentioned, is a form of self-deception because it amounts to consuming food for its taste rather than for its being conducive to human flourishing and happiness. Furthermore, giving up veggie meats does not constitute a sacrifice. It does not represent a significant loss of taste or nutrition. Taste can be easily adjusted, and the taste of veggie meat alternatives is not superior to the taste of whole, plant-based food.

Even if one contemplated consuming infrequently certain non-optimal substances, there are at least two problems. One is that the temperate individual would have no idea, except for an arbitrary idea, of what constitutes moderation. Second, once an individual incorporates a non-essential substance in her diet, there is always a risk that such a substance may eventually be consumed on a regular basis or that the frequency of consumption of such a substance be increased—especially if such substance (as in the case of veggie meats) is hyperpalatable and thus addictive. Consequently, the temperate individual will consume only substances that are specific to his nature, substances that are essential to flourishing, and not primarily for its taste. Consequently, a virtuous individual will not endorse the production or consume veggie meats. In short, the production and consumption of veggie meat alternatives stem from the lack of temperance.

Another problem with the production and consumption of veggie meat alternatives is that it can alienate humans from nature and from our own nature. Such ultra-processed substances that we use for food require an increased dependence on technology, leading humans closer and closer to an estrangement from nature. I think that a virtuous individual must deflect such an estrangement. The eagerness that certain individuals exhibit in embracing and consuming veggie meats is fundamentally due to self-indulgence as well as short-sightedness. For a temperate person, the taste of food and food pleasure are not so important as to relinquish or deviate from our moral integrity and nature by producing, selling, and consuming ultra-processed substances when it is possible and desirable to adjust our taste to whole plant food.

Humans, as Aristotle called them, are rational animals, animals endowed with reason. As animals, humans experience carnal pleasure. But animality is not our unique aspect or function. Our specific nature is a rational nature, and thus our specific function is to reason well. As such, physical pleasures should not be repressed, but should be disciplined and regulated by reason in a way that is conducive to our flourishing. Temperate individuals, through practice and reasoning, control their physical pleasures. They do not merely exercise discipline to curb their appetites and abstain from certain pleasures. Rather, they are individuals who choose correctly only what is conducive to flourishing. Think for example of a person who refrains from smoking cigarettes because she understands that smoking is a deleterious habit, but otherwise enjoys smoking and would do it if it were not harmful. This would be an example of a continent person. But a virtuous person is not merely continent. A virtuous person does not enjoy anything about smoking cigarettes in the first place.

In this view, then, consuming veggie meats is an apt example of what a virtue ethical outlook might regard as self-indulgence—even if consumed infrequently. One may object that this seems a very extreme restriction. I have two points in response to such an objection. First, the idea is that the temperate individual consumes only those substances that are beneficial and conducing to flourishing. Consequently, in the first place there would be no reason for such an individual to ever consume sub-optimal or unhealthful substances. In the second place, the temperate individual does not abstain from or force himself to avoid consuming certain substances. Rather, the temperate individual will adjust his or her taste accordingly and so will not desire unhealthful substances or consume certain substances or engage in sexual intercourse only for taste's or pleasure's sake. In other words, for the temperate individual not consuming certain substances will not be a restriction or a sacrifice. Rather, temperate individuals will not be psychologically dependent on any food in particular.

Second, since certain substances are unhealthful and not optimal for humans, it follows that the temperate individual will not merely avoid consuming substances that are potentially unhealthful for humans, but also shun the production and sales of such substances. Again, the expression goes, "We are supposed to eat to live, not live to eat". Owing to modern technology, people in affluent societies are constantly encouraged to consume new substances. This is a form of alienation from their nature and misplaces their desire for the wrong food. Note that I do not mean to underestimate the role of taste and food pleasure in a flourishing human life. I think that enjoying the taste and olfaction of the foods has its importance, but it should be food that comes at no cost to our health—and, ultimately, is flourishing—or to the environment.

It has to be reemphasized that the distinguishing factor between a temperate individual and an intemperate or a continent one is that the temperate individual has a complete and rational control over his appetites. Consequently, it will never be the case that he will desire unhealthful and unfit substances just because they might look, smell, or taste good. Moral virtues are not mere habits; they are solid and reliable dispositions of one's character to act in certain ways and consistent with reason that are necessary for human flourishing. Moreover, to possess a virtue implies a complex psychological framework concerning right feelings, attitudes, understanding, insight, and experience. Thus, in acting, the virtuous person will act with a certain attitude and feelings and insight in such a way that she will express her virtue in the right amount, the right way, to the right person and circumstances, and (though not for every action) enjoy acting in such a way.

For example, consider the virtue of truthfulness. A truly virtuous person does not just tell the truth most of the times and does it in order to score points or reluctantly fulfill his duty. Rather, he tells the truth, but not indiscreetly. He brings up his children according to such values and even encourages others to do the same. Also, he would not even enjoy jokes about dishonesty and would be saddened by dishonesty in friends. Moreover, he cares about truth for its own sake and feels good about the truth. Therefore, having determined that the consumption of veggie meat alternatives is a form of self-deception, that are hyperpalatable and thus addictive, that they are not the most environmentally friendly food, and are not substances conducive to maximum health, and thus human flourishing, he will reject veggie meat alternatives even in principle. Therefore, the idea of consuming them in moderation and even the ideas of producing them and selling them are inconsistent with a virtuous character. This much, in my view, is implied by a thorough understanding of the virtue of temperance and a virtuous approach to food in general.

As a concluding remark, first, I would like to point out that my discussion about the morality of producing, selling, and consuming veggie meat alternatives is supposed to illustrate what a virtue-oriented approach can enable us to see that other ethical theories fail to track. I have argued that our moral understanding has been shaped by certain theories that favor duty, utility, or rights. One peculiar aspect of such theories is that they advise that we look at aspects of the world outside ourselves or specific characteristics that animals or people have that purportedly should motivate us or show us what we ought to do. I am sure that this sort of approach has some advantages and constructive qualities. However, I believe that their approaches often fail to track our moral intuition and in some cases prevent us to see important aspects of a certain issue, as I hope to have shown that it is so in the case of veggie meat alternatives.

The contribution of a virtue-oriented approach, in particular the analysis of what it is to be temperate, the notion of human function, excellence of character, and that of flourishing, can show us an unexamined side of the issue concerning food ethics. I have attempted to show that most people take it for granted that taste and other forms of psychological and physical pleasure are a moral given. As I have tried to show, we must question such assumptions through an analysis of the role of food and pleasure in human flourishing. When we do so, as I have argued, we realize that from such a point of view, the point of view of virtue, we should not promote, but rather avoid, the production, sale, and consumption of veggie meat alternatives.

References

Adams, C.J. 1990. The sexual politics of meat: A feminist-vegetarian critical theory. New York: Continuum.

Alvaro, C. 2017. Ethical veganism, virtue, and greatness of the soul. Journal of Agricultural and Environmental Ethics 30(6): 765–781.

Alvaro, C. 2019. Ethical veganism, virtue ethics, and the great soul. Lanham, MD: Lexington Books.

Alvaro, C. 2020. Raw veganism: The philosophy of the human diet. Routledge.

American Heart Association. 2019. Nearly half of all adult Americans have cardiovascular disease. ScienceDaily. Retrieved December 20, 2021 from www.sciencedaily.com/releases/2019/01/190131084238.htm.

Anscombe, G.E.M. 1958. Modern moral philosophy. Philosophy 33(124): 1–19.

Aristotle, Ross, W.D. and Urmson, J.O. 1980. The Nicomachean ethics. Oxford (Oxfordshire: Oxford University Press).

Aristotle and Sachs, J. 2002. Nicomachean ethics. Focus Pub./R. Pullins.

Asgar, M.A., Fazilah, A., Huda, N., Bhat, R. and Karim, A.A. 2010. Nonmeat protein alternatives as meat extenders and meat analogs. Compr. Rev. Food Sci. Food Saf. 9: 513–29. doi: 10.1111/j.1541-4337.2010.00124.x.

Aurelius, M. 2002. The Meditations. Random House.

Babu, P.D., Bhakyaraj, R. and Vidhyalakshmi, R. 2009. A low cost nutritious food "tempeh"—A review. World Journal of Dairy & Food Sciences 4(1): 22–27.

Blumenthal, D.M. and Gold, M.S. 2010. Neurobiology of food addiction. Curr. Opin. Clin. Nutr. Metab. Care. 13(4): 359–65. doi: 10.1097/MCO.0b013e32833ad4d4. PMID: 20495452.

Bohrer, B.M. 2019. An investigation of the formulation and nutritional composition of modern meat analogue products. Food Sci. Hum. Wellness. 8: 320–9. doi: 10.1016/j.fshw.2019.11.006.

Bonny, S.P.F., Gardner, G.E., Pethick, D.W. and Hocquette, J.F. 2015. What is artificial meat and what does it mean for the future of the meat industry? J. Integr. Agric. 14: 255–63. doi: 10.1016/S2095-3119(14)60888-1.

Bourassa, L. 2021. Vegan and Plant-Based Diet Statistics for 2021. https://www.plantproteins.co/vegan-plant-based-diet-statistics/. Accessed December 13, 2021.

Burkholder, J., Libra, B., Weyer, P., Heathcote, S., Kolpin, D., Thorne, P.S., et al. 2007. Impacts of waste from concentrated animal feeding operations on water quality. Environmental Health Perspectives 115(2): 308–312. https://doi.org/10.1289/ehp.8839.

Cameron, D.R., Marvin, D.C., Remucal, J.M. and Passero, M.C. 2017. Ecosystem management and land conservation can substantially contribute to California's climate mitigation goals. Proc. Natl. Acad. Sci. U S A 114(48): 12833–12838. doi: 10.1073/pnas.1707811114.

Carroll, A.E. and Doherty, T.S. 2019. Meat consumption and health: Food for thought. Annals of Internal Medicine 171(10): 767–8. doi: 10.7326/M19-2620.

Carus, F. 2010. UN urges global move to meat and dairy-free diet. The Guardian. https://www.theguardian.com/environment/2010/jun/02/un-report-meat-free-diet. Accessed December 19, 2021.

Charrel, R.N. and de Lamballerie, X. 2003. Arenaviruses other than Lassa virus. Antiviral Research 57(1-2): 89–100.

Coffey, S. Donald. 2001. Similarities of Prostate and Breast Cancer: Evolution, Diet, and Estrogens. Urology 57: 31–38.

Cooper, C.C. 2021. Plant-Based Diets: A Primer for School Nurses. NASN Sch Nurse. Jan;36(1): 25–28. doi: 10.1177/1942602X20933233.

Craig, W.J. and Mangels, A.R. 2009. American Dietetic Association. Position of the American Dietetic Association: vegetarian diets. J. Am. Diet Assoc. 109(7): 1266–82. doi: 10.1016/j.jada.2009.05.027. PMID: 19562864.

Cullum-Dugan, D. and Pawlak, R. 2019. Position of the academy of nutrition and dietetics: vegetarian diets. J. Acad. Nutr. Diet. 115(5): 801–810. doi: 10.1016/j.jand.2015.02.033. Retraction in: J. Acad. Nutr. Diet. 2015 Aug; 115(8): 1347. PMID: 25911342.

Daniel, C.R., Cross, A.J., Koebnick, C. and Sinha, R. 2011. Trends in meat consumption in the USA. Public Health Nutrition 14(4): 575–583. https://doi.org/10.1017/S136 8980010002077.

Day, L. 2011. Wheat gluten: Production, properties and application. pp. 267–288. *In*: Phillips, G.O. and Williams, P.A. (eds.). Handbook of Food Proteins. Oxford, UK: Woodhead Publishing.

Dawkins, M.S. 2008. The Science of Animal Suffering. Ethology 114: 937–945.

Deckers, J. 2016. Animal (De)liberation: Should the Consumption of Animal Products Be Banned? London: Ubiquity.

Dent, M. 2020. The Meat Industry is Unsustainable. https://www.idtechex.com/en/research-article/the-meat-industry-is-unsustainable/20231.

Donovan, J. 2006. Feminism and the treatment of animals: From care to dialogue. Signs 31(2): 305–329.

Döring, T.F. and Neuhoff, D. 2021. Upper limits to sustainable organic wheat yields. Scientific Reports 11(1): 12729. https://doi.org/10.1038/s41598-021-91940-7.

Duffy, G. and Moriarty, E.M. 2003. Cryptosporidium and its potential as a food-borne pathogen. Anim. Health Res. Rev. 2003 Dec; 4(2): 95–107. doi: 10.1079/ahr200357. PMID: 15134293.

Elizabeth, L., Machado, P., Zinöcker, M., Baker, P. and Lawrence, M. 2020. Ultra-processed foods and health outcomes: a narrative review. Nutrients, 12(7): 1955. https://doi.org/10.3390/nu12071955.

Ernstoff, A., Tu, Q., Faist, M., Del Duce, A., Mandlebaum, S. and Dettling, J. 2019. Comparing the environmental impacts of meatless and meat-containing meals in the United States. Sustainability 11(22): 6235. MDPI AG. Retrieved from http://dx.doi.org/10.3390/su11226235.

Etemadi, A., Sinha, R., Ward, M.H., Graubard, B.I., Inoue-Choi, M., Dawsey, S.M., et al. 2017. Mortality from different causes associated with meat, heme iron, nitrates, and nitrites in the NIH-AARP Diet and Health Study: population based cohort study. BMJ (Clinical research ed.), 357, j1957. https://doi.org/10.1136/bmj.j1957.

Farvid, M.S., Cho, E., Chen, W.Y., Eliassen, A.H. and Willett, W.C. 2015. Adolescent meat intake and breast cancer risk. International Journal of Cancer 136(8): 1909–1920. https://doi.org/10.1002/ijc.29218.

Foot, P. 2001. Natural goodness. Oxford: Clarendon.

Francione, G.L. 1996. Rain without thunder: ideology of the animal rights movement. Philadelphia: Temple University Press.

Gearhardt, A.N., Grilo, C.M., DiLeone, R.J., Brownell, K.D. and Potenza, M.N. 2011. Can food be addictive? Public health and policy implications. Addiction (Abingdon, England) 106(7): 1208–1212. https://doi.org/10.1111/j.1360-0443.2010.03301.x.

German, A.J., Woods, G.R.T., Holden, S.L. et al. 2018. Dangerous trends in pet obesity. Vet. Rec. 182: 25.

Goodland, R. and Anhang, J. 2009. Livestock and Climate Change: What if the Key Actors in Climate Change are ... Cows, Pigs, and Chicken? https://www.semanticscholar.org/paper/Livestock-and-climate-change%3A-what-if-the-key-in-Goodland-Anhang/6704c7a0777c82357704d82b9ae8007c1197cb07.

Goodman, B. 2020. Rethinking Milk: Science Takes On the Dairy Dilemma. WebMD. https://www.webmd.com/diet/news/20200214/rethinking-mik-science-takes-on-the-dairy-dilemma. Accessed December 21, 2021.

Gould, S. 2017. 6 charts that show how much more Americans eat than they used to. Insider. https://www.businessinsider.com/daily-calories-americans-eat-increase-2016-07. Accessed December 21, 2021.

Ha B. 2019. The Power of Plants: Is a Whole-Foods, Plant-Based Diet the Answer to Health, Health Care, and Physician Wellness? The Permanente journal, 23, 19-003. https://doi.org/10.7812/TPP/19.003.

Haddaway, N.R., Hedlund, K., Jackson, L.E., Thomas Kätterer, Emanuele Lugato, Ingrid K. Thomsen, et al. 2017. How does tillage intensity affect soil organic carbon? A systematic review. Environ. Evid. 6: 30. https://doi.org/10.1186/s13750-017-0108-9.

Hall, K.D., Ayuketah, A., Brychta, R., Cai, H., Cassimatis, T., Chen, K.Y., et al. 2019. Ultra-processed diets cause excess calorie intake and weight gain: an inpatient randomized controlled trial of Ad Libitum food intake. Cell Metabolism 30(1): 67–77.e3. https://doi.org/10.1016/j.cmet.2019.05.008.

Heller, M.C. and Keoleian, A.G. 2018. Beyond meat's beyond burger life cycle assessment: a detailed comparison between a plant-based and an animal-based protein source. CSS Report, University of Michigan: Ann. Arbor. 1–38.

Hill, T. 1983. Ideals of human excellences and preserving natural environments. Environmental Ethics 5(3): 211–224.

Hotchkiss, A.K., Rider, C.V., Blystone, C.R., Wilson, V.S., Hartig, P.C., Ankley, G.T., et al. 2008. Fifteen years after "Wingspread"—environmental endocrine disrupters and human and wildlife health: where we are today and where we need to go. Toxicological sciences: An Official Journal of the Society of Toxicology 105(2): 235–259. https://doi.org/10.1093/toxsci/kfn030.

Hu, F.B., Otis, B.O. and McCarthy, G. 2019. Can plant-based meat alternatives be part of a healthy and sustainable diet? JAMA. 322(16): 1547–1548. doi: 10.1001/jama.2019.13187. PMID: 31449288.

Hursthouse, R. 2006. Applying virtue ethics to our treatment of other animals. *In*: Jennifer, J.W. (ed.). The Practice of Virtue: Classic and Contemporary Readings in Virtue Ethics. Indianapolis, IN: Hackett Publishing.

Klurfeld, D.M. 2018. What is the role of meat in a healthy diet? Animal Frontiers: The Review Magazine of Animal Agriculture 8(3): 5–10. https://doi.org/10.1093/af/vfy009.

Koneswaran, G. and Nierenberg, D. 2008. Beef Production: Koneswaran and Nierenberg Respond. Environmental Health Perspectives 116(9): A375–A376. https://doi.org/10.1289/ehp.11716R.

Korsgaard, C. 2004. Fellow creatures: Kantian ethics and our duties to animals. Tanner Lectures on Human Values 24: 77–110.

Land, G. 2005. Historical dictionary of Seventh-Day Adventists. Lanham, MD. Scarecrow Press.

Lanou, A.J. 2009. Should dairy be recommended as part of a healthy vegetarian diet? Counterpoint, The American Journal of Clinical Nutrition, May 2009 89(5): 1638S–1642S, https://doi.org/10.3945/ajcn.2009.26736P.

LeDuc, J.W. 1989. Epidemiology of hemorrhagic fever viruses. Reviews of Infectious Diseases, 11Suppl4, S730-5.

Leonard, A. 2010. The Story of Stuff: How Our Obsession with Stuff Is Trashing the Planet, Our Communities, and Our Health – and a Vision for Change. Constable & Robinson UK.

Levin, S. 2020. Dietary Guidelines Recommendation for 3 Servings of Dairy Daily Increases Breast, Prostate Cancer Risk. Physicians Committee for Responsible Medicine. https://www.pcrm.org/news/blog/dietary-guidelines-recommendation-3-servings-dairy-daily-increases-breast-prostate-cancer. Accessed December 10, 2021.

Linnane, C. 2019. Alternative meat market could be worth $140 billion in 10 years, Barclays says. Market Watch. https://www.marketwatch.com/story/alternative-meat-market-could-be-worth-140-billion-in-ten-years-barclays-says-2019-05-22. Accessed November 22, 2021.

Luke, B. 1992. Justice, caring, and animal liberation. Between the Species 8(2), Article 13.

Martin, C. Heller et al. 2018 Environ. Res. Lett. 13 044004. DOI 10.1088/1748-9326/aab0ac.

Martinez Steele, E., Baraldi, L.G., Louzada, M.L., Moubarac, J.C., Mozaffarian, D. and Monteiro, C.A. 2016. Ultra-processed foods and added sugars in the US diet: evidence from a nationally represented cross-sectional study. BMJ Open 6(3): e009892.

Merlo, L.J., Stone, A.M. and Gold, M.S. 2009. Co-occurring addiction and eating disorders. pp. 1263–1274. *In*: Riess, R.K., Fiellin, D., Miller, S. and Saitz, R. (eds.). Principles of Addiction Medicine. 4th Edition Lippincott Williams & Wilkins; Kulwer (NY).

Micha, R., Wallace, S.K. and Mozaffarian, D. 2010. Red and processed meat consumption and risk of incident coronary heart disease, stroke, and diabetes mellitus: a systematic review and meta-analysis. Circulation 121(21): 2271–2283. https://doi.org/10.1161/CIRCULATIONAHA.109.924977.

Milton, K. 1999. Nutritional characteristics of wild primate foods: do the diets of our closest living relatives have lessons for us? Nutrition 15(6): 488–98. doi: 10.1016/s0899-9007(99)00078-7. PMID: 10378206.

Milton, K. 2000. Back to basics: why foods of wild primates have relevance for modern human health. Nutrition. 2000 Jul-Aug;16(7-8): 480-3. doi: 10.1016/s0899-9007(00)00293-8. PMID: 10906529.

National Obesity Rates & Trends, NHANES, https://www.stateofobesity.org/obesity-rates-trends-overview/. Accessed November 22, 2021.

Nobis, N. 2002. Vegetarianism and Virtue: Does consequentialism Demand Too Little? Social Theory & Practice, 28(1): 135–156.

Orlich, M.J., Singh, P.N., Sabaté, J., Jaceldo-Siegl, K., Fan, J., Knutsen, S., et al. 2013. Vegetarian dietary patterns and mortality in Adventist Health Study 2. JAMA Internal Medicine 173(13): 1230–1238. https://doi.org/10.1001/jamainternmed.2013.6473.

Petroski, W. and Minich, D.M. 2020. Is there such a thing as "Anti-Nutrients"? A Narrative Review of Perceived Problematic Plant Compounds. Nutrients 12(10): 2929. https://doi.org/10.3390/nu12102929.

Physicians Committee for Responsible Medicine (n.d.). Health Concerns About Dairy: Avoid the Dangers of Dairy With a Plant-Based Diet. https://www.pcrm.org/good-nutrition/nutrition-information/health-concerns-about-dairy. Accessed December 1, 2021.

Pimentel, D. and Pimentel, M. 2003. Sustainability of meat-based and plant-based diets and the environment. The American Journal of Clinical Nutrition 78(3): 660S–663S, https://doi.org/10.1093/ajcn/78.3.660S.

Poore, J. and Nemecek, T. 2018. Reducing food's environmental impacts through producers and consumers. Science. 2018 Jun 1; 360(6392): 987–992. doi: 10.1126/science.aaq0216. Erratum in: Science. 2019 Feb 22;363(6429): PMID: 29853680.

Premack, R. 2016. Meat is Horrible. The Washington Post. https://www.washingtonpost.com/news/wonk/wp/2016/06/30/how-meat-is-destroying-the-planet-in-seven-charts/. Accessed December 12, 2021.

Rachels, J. 1990. Created from animals: The moral implications of Darwinism. Oxford: Oxford University Press.

Rauber, F., Steele, E.M., Louzada, M., Millett, C., Monteiro, C.A. and Levy, R.B. 2020. Ultra-processed food consumption and indicators of obesity in the United Kingdom population (2008–2016). PloS one 15(5): e0232676. https://doi.org/10.1371/journal.pone.0232676.

Regan, T. 1983. The case for animal rights. Berkeley, CA: University of California Press.

Ritchie, H. and Roser, M. 2017. Meat and dairy production. Our World in Data. Published online at OurWorldInData.org. https://ourworldindata.org/agricultural-land-by-global-diets [Online Resource]. Accessed December 12, 2021.

Ritchie, H. 2021. If the world adopted a plant-based diet we would reduce global agricultural land use from 4 to 1 billion hectares. Our World in Data. Published online at OurWorldInData.org. https://ourworldindata.org/land-use-diets [Online Resource]. Accessed December 22, 2021.

Roberts, Robert C. 2015. How virtue contributes to flourishing. pp. 36–49. *In*: Mark Alfano (ed.). Current Controversies in Virtue Theory. New York and London: Routledge.

Roberts, W.C. 2000. Twenty questions on atherosclerosis. Proceedings (Baylor University. Medical Center), 13(2): 139–143. https://doi.org/10.1080/08998280.2000.11927657.

Rowlands Mark. 2012. Can animals be moral? New York: Oxford University Press.

Rowlands, M. 2015. Can animals be moral? Oxford: Oxford University Press.

Rowlands, M. 2019. Can animals be persons? Kettering: Oxford University Press. Bottom of Form Top of Form.

Ryder, R. 1974. Speciesism: The ethics of vivisection. Edinburgh: Scottish Society for the Prevention of Vivisection.

Ryder, R. 1989. Animal revolution: changing attitudes towards speciesism. Cambridge: Cambridge University Press.

Ryder, R. 1999. Painism: Some moral rules for the civilized experimenter. Cambridge Quarterly of Healthcare Ethics, 8: 34–42.

Ryder, R. 2003. Painism: A modern morality. London: Open Gate Press.

Ryder, R. 2010. Painism. pp. 402–403. *In*: Bekoff, M. (ed.). Encyclopedia of animal rights and animal welfare. Santa Barbara, Cal. [etc.]: Greenwood Press, imprint of ABC-CLIO, LLC. Bottom of Form.

Saiidi, U. 2019. Meatless alternatives are on the rise—but so is global meat consumption. CNBC. https://www.cnbc.com/2019/06/18/meatless-alternatives-are-on-the-rise-so-is-global-meat-consumption.html. Accessed December 2, 2021.

Santo, R.E., Kim, B.F., Goldman, S.E., Dutkiewicz, J., Biehl, E.M.B., Bloem, M.W., et al. 2020. Considering plant-based meat substitutes and cell-based meats: a public health and food systems perspective. Front. Sustain. Food Syst. 4: 134. doi: 10.3389/fsufs.2020.00134.

Savaiano, D.A. and Levitt, M.D. 1987. Milk intolerance and microbe-containing dairy foods. J. Dairy Sci. 70(2): 397–406. doi: 10.3168/jds.S0022-0302(87)80023-1. PMID: 3553256.

Seifer, D. 2019. Quick Service Burger Buyers Mix It Up Between Plant-Based and Beef. National Purchase Diary Panel. https://www.npd.com/news/press-releases/2019/quick-service-burger-buyers-mix-it-up-between-plant-based-and-beef/. Accessed December 20, 2021.

Shurtleff, W. and Aoyagi, A. 2013. History of meat alternatives (965 CE to 2014) 1st ed. Lafayette CA, USA: Soyinfo Center.

Singer, P. 1975. Animal liberation: A new ethics for our treatment of animals. New York: Avon Books.

Singer, P. 1989. All animals are equal. pp. 215–226. *In*: Regan, T. and Singer, P. (eds.). Animal rights and human obligations. Oxford University Press. Top of Form.

Spencer, C. 1996. The Heretic's Feast: A History of Vegetarianism. London. Fourth Estate Classic House.

Springmann, M., Clark, M., Mason-D'Croz, D., Wiebe, K., Bodirsky, B.L., Lassaletta, L., et al. 2018. Options for keeping the food system within environmental limits. Nature. Oct; 562(7728): 519–525. doi: 10.1038/s41586-018-0594-0.

Storz, M.A. 2018. Is there a lack of support for whole-food, plant-based diets in the medical community? The Permanente Journal 23: 18–068. https://doi.org/10.7812/TPP/18-068.

Stripp, C., Overvad, K., Christensen, J., Thomsen, B.L., Olsen, A., Møller, S., et al. 2003. Fish Intake is Positively Associated with Breast Cancer Incidence Rate. The Journal of Nutrition 133(11): 366–3669, https://doi.org/10.1093/jn/133.11.3664.

Tague, G. 2020. An ape ethic and the question of personhood. Lanham : Lexington Books.

Tuomisto, H.L. 2019. The eco-friendly burger: could cultured meat improve the environmental sustainability of meat products? EMBO Reports 20(1): e47395. https://doi.org/10.15252/embr.201847395.

Turner-McGrievy, G., Mandes, T. and Crimarco, A. 2017. A plant-based diet for overweight and obesity prevention and treatment. Journal of Geriatric Cardiology: JGC 14(5): 369–374. https://doi.org/10.11909/j.issn.1671-5411.2017.05.002.

Tuso, P., Stoll, S.R. and Li, W.W. 2015. A plant-based diet, atherogenesis, and coronary artery disease prevention. The Permanente Journal 19(1): 62–67. https://doi.org/10.7812/TPP/14-036.

Tuso, P.J., Ismail, M.H., Ha, B.P. and Bartolotto, C. 2013. Nutritional update for physicians: plant-based diets. The Permanente Journal 17(2): 61–66. https://doi.org/10.7812/TPP/12-085.

Vegan News Aggregator. 2020. Morningstar Farms Goes 100% Vegan. https://vegannews.press/2020/02/21/morningstar-farms-vegan/. Accessed December 20, 2021.

Vox Creative. 2019. Behind the Rise of Plant-Based Burgers: Awesome Burger is poised to leap into an ever-expanding market. https://next.voxcreative.com/ad/20853610/plant-based-burgers-history-sweet-earth. Accessed December 20, 2021.

Wang, Y.C., McPherson, K., Marsh, T., Gortmaker, S.L. and Brown, M. 2011. Health and economic burden of the projected obesity trends in the USA and the UK. Lancet. 27; 378(9793): 815–25. doi: 10.1016/S0140-6736(11)60814-3.

Webster, J. 2013. Animal husbandry regained: The place of farm animals in sustainable agriculture. Routledge.

Wiebe, M.G. 2002. Myco-protein from Fusarium venenatum: a well-established product for human consumption. Appl. Microbiol. Biotechnol. 58(4): 421–7. doi: 10.1007/s00253-002-0931-x.Epub 2002 Feb 8. PMID: 11954786.

Worldwildlife.org (n.d). Overview. https://www.worldwildlife.org/industries/soy. Accessed November 28, 2021.

Zhong, V.W., Van Horn, L., Cornelis, M.C., Wilkins, J.T., Ning, H., Carnethon, M.R., et al. 2019. Associations of dietary cholesterol or egg consumption with incident cardiovascular disease and mortality. JAMA, 321(11): 1081–1095. https://doi.org/10.1001/jama.2019.1572.

Endnotes

[1] I don't actually think that vegan or vegetarian diets are boring. Rather, I am voicing the opinion of many meat eaters about vegan and vegetarian diets.

[2] Some vegan products, e.g., ground Beyond Meat and Impossible, are even placed by some supermarkets in the same refrigerators as the real meat.

[3] It must be noted, however, that most of the world's soy is fed to livestoick.

[4] As of today, virtually all veggie meats are 100% plant-based. Some alternative meat products produced by Morningstar Farm may still contain eggs. However, according to the company, by this year Morningstar Farm products will all be vegan. Vegan News Aggregator. (2020). https://vegannews.press/2020/02/21/morningstar-farms-vegan/.

[5] But what if animals didn't suffer? Would there be anything else to be said about it?

[6] Although some vegan population statistics show a small increase in the number of vegans in 2021 compared to the number of vegans in 2014, it is not clear exactly how they can establish such figures. At any rate, it seems that nowadays in the US the figure is 2–6% of the population. See Lacey Bourassa (2021) Vegan and Plant-Based Diet Statistics for 2021. https://www.plantproteins.co/vegan-plant-based-diet-statistics/.

[7] The "good" varies in accordance with the different type of utilitarianism; classical utilitarianism argues that we ought to maximize pleasure (or happiness) and or minimize suffering (or unhappiness). Other forms argue that we ought to promote the greatest satisfaction of preference for the greatest number of sentient beings.

[8] See Roberts, Robert C. 2015. How virtue contributes to flourishing. pp. 36–49. *In*: Mark Alfano (eds.). Current Controversies in Virtue Theory. New York and London: Routledge, 2015 for a clear discussion of the concept of human flourishing and virtue.

[9] By processed I don't mean masticating food. Rather, I mean taking whole food, such as lettuce, or mangoes, or bananas, and process them by using a blender or by modifying such foods and transforming them into something else, e.g., smoothies, ice-cream, and, of course, veggie meat.

[10] For an in-depth discussion on this topic, see Annie Leonard, The Story of Stuff: How Our Obsession with Stuff is Trashing the Planet, Our Communities, and Our Health – and a Vision for Change (Constable and Robinson, 2010).

7

'Lab-grown' Meat and Considerations for its Production at Scale

Paul Cameron,[1] Panagiota Moutsatsou,[2] Matt Wasmuth,[1] Farhaneen Mazlan,[1]
Darren Nesbeth,[1] Qasim Rafiq,[1] Alvin William Nienow[1,4] and
Mariana Petronela Hanga[1,3],*

1. Context of Animal Agriculture Today

The world's population is continuously increasing and is predicted to reach approximately 10 billion by 2050 (Arshad et al. 2017, United Nations 2019, Chriki and Hocquette 2020). The current agricultural practices are struggling to keep up with the increase in population. The Food and Agriculture Organisation (FAO) has forecasted that food production would need to increase by more than 60% by 2050 (FAO 2012, Chriki and Hocquette 2020). However, the resources needed such as fresh water and arable land are already limited (Gerbens-Leenes et al. 2013, Grossi et al. 2019) which would make it challenging to achieve those levels of food production.

A particular concern is the animal agriculture for meat production where the demand is increasing as a direct correlation to the increase in per capita income in developing countries (FAO 2016, Stephens et al. 2018). To keep up with the meat demand, animal agriculture has undergone industrialization which is mainly focused on efficiency as in quantity of meat produced rather than animal welfare, interactions with the environment (e.g., waste treatment) or sustainability (use of resources) (Oliver et al. 2011, Gerber et al. 2015, Stephens et al. 2018, Chriki and Hocquette 2020). Moreover, the livestock sector is also responsible for the production of ~ 18% of greenhouse gas emissions (FAO 2006) and exacerbates public health issues such as antibiotic resistance (Landers et al. 2012) and zoonotic diseases (Machalaba et al. 2015) alongside the ethical concerns on animal welfare (Bryant 2019, Szejda et al. 2021).

[1] Department of Biochemical Engineering, Advanced Centre for Biochemical Engineering, University College London, London, United Kingdom.
[2] Mosa Meat BV, Maastricht, Netherlands.
[3] School of Biosciences, Life and Health Sciences College, Aston University, Birmingham, United Kingdom.
[4] Department of Chemical Engineering, University of Birmingham, Birmingham, United Kingdom.
* Corresponding author: m.hanga@ucl.ac.uk

2. Introduction to Cellular Agriculture

To address the shortcomings of the current animal agriculture practices, efficient and sustainable ways to produce proteins are desperately needed. Cellular agriculture has emerged as a novel field for alternative foods providing a possible solution. Unlike animal farming that uses processes operating at the whole-organism level, cellular agriculture produces animal-derived products through processes operating at the cellular level (Stephens and Ellis 2020).

The term "cellular agriculture" was first introduced in 2015 by Isha Datar, the Executive Director of the US-based charity New Harvest, and it referred to the production of animal-sourced foods using cell culture techniques (Datar et al. 2016, Stephens et al. 2020). Cellular agriculture provides the opportunity to produce animal-sourced products without the animal and in a more sustainable manner, thus addressing not only the current environmental issues associated with traditional animal agriculture, but also the animal welfare aspects and public health (Arshad et al. 2017, Mattick 2018, Milburn 2018, Specht et al. 2018, Stephens et al. 2020). Moreover, cellular agriculture is targeting two of the main UN Sustainable Development Goals of "Zero Hunger" and "Responsible Consumption and Production" (Stephens et al. 2020) as it will allow the production of animal-sourced products by using very few or no animals at all. This concept, once considered futuristic, is not new. It was envisaged by visionaries, one of them being Winston Churchill who stated that "we shall escape the absurdity of growing a whole chicken in order to eat the breast or wing, by growing these parts separately under a suitable medium" (Churchill 1931). Ideally, cellular agriculture will replace entirely the traditional animal agriculture practices which would maximise its benefits. However, realistically with the currently existing technology and industrial production capacity, cellular agriculture will only partially replace it in the near future, while working towards meeting the global rising demand for animal-sourced products.

There are two categories of products that can be produced through Cellular Agriculture and those are: acellular and cellular products. Acellular products are made of organic molecules like proteins or fats and don't contain live or whole cells, while cellular products are made of actual cells. Examples of acellular agricultural products include milk, egg whites, fats and enzymes for cheese making to name a few (Rischer et al. 2020, Stephens et al. 2020). These types of products are produced by using fermentation and recombinant DNA technology where host cells like bacteria, yeast, fungus or algae are genetically modified to gain the capability to produce these organic molecules from other species of interest (e.g., cow, chicken, etc.) (Datar et al. 2016, Stephens et al. 2020). In this chapter, we will only focus on the production of cellular products.

2.1 Cellular Agricultural Products

Unlike acellular products, cellular products are highly complex and comprise multiple cell types in defined ratios and arranged in complex architectures. Lab-grown meat is one such product that will be discussed in detail in the next sections.

Lab-grown meat has been previously referred to by various names including 'cultivated meat', 'cultured meat', 'cell-based meat', 'clean meat', 'tissue-engineered meat', 'synthetic meat' and others (Stephens et al. 2018, 2020, Szejda et al. 2021). Several studies have investigated the link between the nomenclature used for these types of products and consumer acceptance. A more positive consumer reaction was found when terms that invoked neutrality, appeal, and descriptiveness were used, while still differentiating it from animal meat. Such terms included 'cultivated meat' and 'cultured meat' (Szejda et al. 2019, Dillard et al. 2019, Szejda et al. 2021) which are now generally accepted. Other terms such as 'lab-grown meat', 'clean meat' or 'synthetic meat' were less appealing to consumers as they raised concerns about the naturalness of these products (Bryant 2019a, Szejda et al. 2021). The term 'cultivated meat' will also be used here.

Cultivated meat is genuine animal meat produced from animal cells in a controlled environment (Stephens et al. 2018, 2020, Reiss et al. 2021). It is not yet available to consumers in supermarkets, and it is still significantly more expensive than animal meat. At the time of writing this chapter, only one cultivated meat product has received regulatory approval for commercialisation in Singapore only. The cultivated chicken nuggets from Eat Just are available only on an order-by-order basis and in a high-end restaurant at the price of $50 per chicken nugget (BBC 2020).

Despite not yet being readily available, cultivated meat products are expected to have significant advantages over animal meat from multiple perspectives including resources needed, animal welfare, environmental impact and even population health (Stephens et al. 2018, Szejda et al. 2021). Initial projections showed that cultivated meat products would require 45% less energy, 99% less land and will produce 78–96% less greenhouse gas emissions (Tuomisto and de Mattos 2011, Stephens et al. 2018). However, more recent life cycle analyses have shown that the benefits will vary across the different types of cultivated meat (e.g., beef, pork, poultry) (Mattick et al. 2015) and will be influenced by the method of energy production chosen in the manufacturing of these products (Smetana et al. 2015). As it is very likely that these products would need to be produced in large scale bioreactors, there are concerns regarding the amounts of energy necessary for their production and the potential environmental impact in the long term due to the different way that CO_2 and CH_4 would be produced. However, this potential issue could be minimised in the future by innovations in the energy field to achieve decarbonization of energy production (Bodiou et al. 2020).

This chapter will focus specifically on the manufacturing of cultivated meat products and will capture the different elements required for its production, from both an engineering and biological point of view.

3. State of the Cultivated Meat industry

The interest and the investment in the cultivated meat sector has increased significantly over the past couple of years. This growth profile of the industry is expected to continue, with the projected market size reaching 593 million USD by 2032. The North American market is expected to account for over half of this due to their large market size and their affinity for processed meats such as nuggets, meaning poultry is expected to be the most common cultured meat product, followed by cultured pork and beef (Cultured Meat Market Report 2019).

While the concept of cultivated meat was envisaged nearly a century ago (Churchill 1931, Arshad et al. 2017), the first proof-of-concept cultivated beef burger was only showcased to the public in 2013 by Prof. Mark Post, the founder of start-up Mosa Meat (Netherlands). The burger weighted approximately 85 g and required culturing 10,000 individual muscle fibres. The production of this proof-of-concept burger patty took 24 months to achieve the necessary cell numbers and it cost approximately €250,000 (Post 2014). It was developed by shaping cell mass harvested from 2D monolayer cultures, a rudimentary technology by today's standards but a valid proof of concept that worked to generate interest and over 96 million USD in investment to the company over the years. Through industrial scale up and the development of new processes and technologies, Mosa Meat now prices a cultured meat burger at around €9 (Clean Technica 2019). Since then, multiple other start-ups around the world have showcased their proof-of-concept products. For example, in 2016, the US-based start-up, Memphis Meats showcased their cell-based meatball. In 2019, Shiok Meats, a start-up based in Singapore has showcased its cultivated shrimp dumplings, while in 2020, Higher Steaks (UK) showcased their cultivated bacon. Another start-up that received significant funding, over 200 million USD since being founded in 2015, is UPSIDE Foods (US). They primarily focus on the cultivation of poultry and have recently announced the development of a "cell feed" that is entirely animal component free (ACF) addressing one of the current challenges that this industry is facing (Upside Foods 2021).

Currently, JUST Inc., a US-based start-up is the first company to have achieved a major milestone in the field in December 2020 which was to obtain the first regulatory approval for their cultivated chicken nuggets product to be commercialised in Singapore (BBC 2020). Singapore was chosen because of its high dependency on meat imports due to their small landmass not accommodating for any agricultural land, and its commitment to self-sufficiency and cutting carbon emissions. The local regulatory body, the Singapore Food Agency (SFA), also had a relatively simple pathway to commercialisation for novel technologies which has facilitated the approval. This milestone has increased the confidence in this newly developed field and expectations are that more cultivated meat products will follow in the next couple of years following on the success of the plant-based meat alternatives.

4. Considerations for Cultivated Meat Production

Animal meat, or skeletal muscle, has a complex structure comprising several different types of cells with approximately 90% being muscle fibres, while the remaining 10% being connective and fat cells (Listrat et al. 2016) with less than 1% blood (Warriss and Rhodes 1977, Reiss et al. 2021). This composition can differ with species and cut of meat. For cultivated meat products to be successful and accepted by consumers, the sensorial characteristics (texture, colour, taste, flavour) are of utmost importance. These characteristics are derived from its structure including the different cell types which determine the molecular characteristics such as content and type of proteins, presence of myoglobin and volatile compounds that contribute to flavour (Fraeye et al. 2020). The technological challenges in producing a cultivated meat product that closely mimics animal-derived meat are very much dependent on the type of cultivated meat product that is targeted.

Cultivated meat products can be split in two distinct categories—structured or non-structured. Structured meat products include steaks, fillets or chops which have a clearly defined structural organisation. Their production would require more complex processing to achieve that structure and to mimic the *in vivo* microenvironment with the signalling pathways necessary to form mature muscle tissue. Non-structured products refer to ground or finely minced meat products and involve a largely heterogeneous cell mass composed of myocytes and adipocytes at varying differentiation states and various ratios depending on the product. Non-structured cultivated meat products can be commercialised under various forms from nuggets to meatballs, sausages and burger patties. The first proof-of-concept cultivated meat product was non-structured in the form of a beef burger patty (Post 2014), while the first approved for commercialisation cultivated product was also non-structured in the form of chicken nuggets (BBC 2020). It is generally acknowledged that the production of non-structured cultivated meat is more feasible to achieve with the currently available technology (Hocquette 2016, Post and Hocquette 2017). The main challenge in production of thick whole cuts is the absence of a complex blood vessel network throughout the construct to deliver nutrients and gases to cells (Bhat et al. 2010, Fraeye et al. 2020). In practice, this would translate to a perfusion system within the whole cut construct; however, achieving a vascular-like network integrated in the construct to overcome diffusion limitations would be the challenge.

A general bioprocess flow for the production of cultivated meat products is shown in Fig. 1. Regardless of the type of cultivated meat product targeted, scalable production requires considerations on starting cell types, culture methods to be employed and their potential for scalability, bioreactor types with advantages and limitations and culture media formulations, challenges and strategies for achieving cost-efficiency. In this section, all of these aspects will be discussed. Additionally, co-culture strategies and challenges will also be covered.

4.1 Starting Cell Types

Several cell types can potentially serve as starting material for bioprocessing to form the cell types found in meat and these are: muscle satellite cells (MyoSCs), embryonic stem cells (ESCs), induced pluripotent stem cells (iPSCs), mesenchymal stem cells (MSCs) and fibro-adipogenic progenitor cells (FAPs) (Post et al. 2020, Dohmen et al. 2022). In this section, we will discuss each possible cell source with its advantages and limitations (Table 1). For cultivated meat production, the starting cell types must have the ability to self-renew and proliferate in order to achieve the high cell numbers

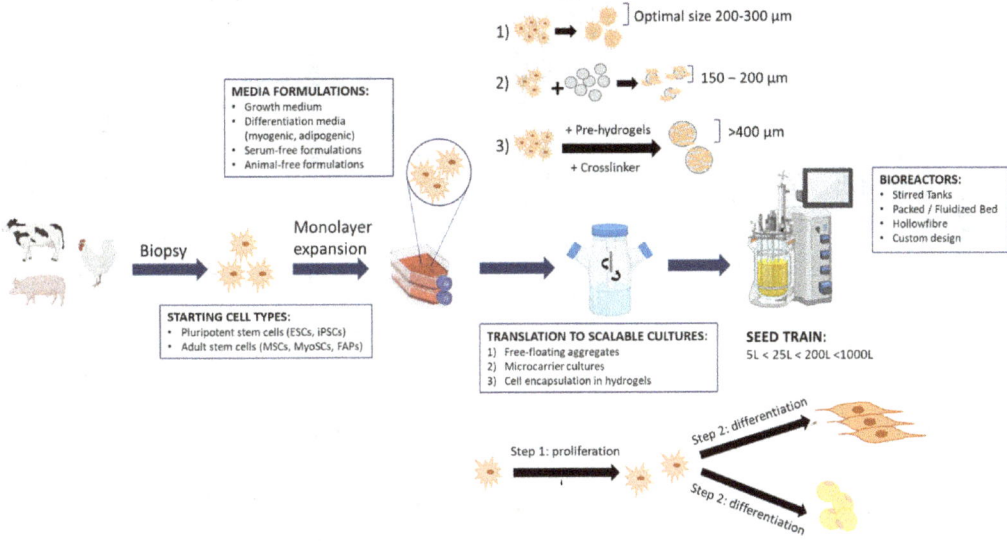

Figure 1. Bioprocess schematic for production of cultivated meat products.

required, but also to possess the capacity to differentiate into the mature cell types that constitute meat (e.g., myocytes, adipocytes, etc.).

Muscle satellite cells (MyoSCs) and their amplifying progeny, myoblasts, are the most reported source and the one used by Mosa Meat to produce the first cultivated beef burger in 2013 (Post 2014, Ding et al. 2018). They reside under the basement membrane of muscle fibres and are capable of differentiating to myocytes which form the multinucleated myotubes that then pack into myofibres (Reiss et al. 2021). MyoSCs are the most-abundant tissue-resident adult stem cells (Bentzinger et al. 2012, Reiss et al. 2021) and they constitute an easily accessible source of cells which can be isolated from animal muscle skeletal tissues obtained through biopsies, using straight forward protocols. They are muscle lineage committed cells, rendering the induction of the differentiation process fairly simple (Ding et al. 2018, Messmer et al. 2022). However, these cells can undergo a finite number of doublings before losing their differentiation potential, rendering them unsuitable for meat production at scale, unless they can be functionally immortalized, similarly to myosatellite model cell lines typically used for the research of skeletal muscle tissues (e.g., L6 and C2C12 cell lines) (Post 2014, Ding et al. 2018, Post et al. 2020).

Another possible cell source with an attractive potential for cultivated meat production is pluripotent stem cells due to their ability to theoretically propagate indefinitely, as well as to differentiate to any cell type present in the three germ layers. ESCs are derived from the inner mass of the blastocyst during the early stages of embryonic development, while iPSCs are generated from somatic cells through cellular reprogramming using specific transcription factors to induce pluripotency (Takahashi and Yamanaka 2006). Whilst isolation of ESCs (Thomson et al. 1998, Reubinoff et al. 2000, Guhr et al. 2006, Chen et al. 2009) and cell reprogramming to obtain iPSCs (Takahashi and Yamanaka 2006, 2016) have been well developed for human cells, these methods are less explored or established for livestock species. ESCs are often difficult to isolate due to the short lifespan of the blastocyst and require highly specialised skills (Reiss et al. 2021). On the other hand, several studies have been published to date attempting to reprogram somatic cells of livestock species to iPSCs, including bovine (Pillai et al. 2019, Bressan et al. 2020, Su et al. 2021), porcine (Cheng et al. 2012, Xu et al. 2019) and avian (Lu et al. 2012, 2015). However, contradictory results have been reported regarding the translatability of protocols optimised for human and mouse pluripotent stem cells to livestock species. ESCs or iPSCs have an increased proliferation capacity compared to MyoSCs as theoretically they can propagate indefinitely. However, whilst the proliferation barrier of 30–40 population doublings of MyoSCs can be overcome with the use

Table 1. Possible cell sources for cultivated meat production with their advantages and limitations.

TYPES OF CELLS	ADULT STEM CELLS			PLURIPOTENT STEM CELLS	
	Myosatellite cells (MyoSCs)	Mesenchymal stem/stromal cells (MSCs)	Fibro-adipogenic progenitors (FAPs)	Embryonic stem cells (ESCs)	Induced pluripotent stem cells (iPSCs)
ADVANTAGES	• Can easily differentiate into muscle (Ding et al. 2018, Messmer et al. 2022) • Obtained through a tissue biopsy	• Can be isolated from a variety of sources: bone marrow, adipose, peripheral blood, umbilical cord tissue (Hill et al. 2019) • Can easily differentiate to fat and have also been reported to be induced towards myogenic differentiation (Gang et al. 2001, Bosnakovski et al. 2005, Stern-Straeter et al. 2014, Okamura et al. 2018)	• Can easily differentiate to fat (Dohmen et al. 2022) • Obtained through a tissue biopsy	• Can differentiate to all cell types required (muscle, fat, connective cells) • Have unlimited proliferative potential	• Can differentiate to all cell types required (muscle, fat, connective cells) • Have unlimited proliferative potential • No ethical issues surrounding their sourcing
LIMITATIONS	• Can only be isolated from muscle tissue • They are primary cells, so can undergo limited divisions before reaching senescence (Post 2014, Ding et al. 2018, Post et al. 2020) • Require protein coatings or activation to enable their culture (Post 2014, Ding et al. 2018) • Require fat from other sources	• Have limited ability to propagate (Dohmen et al. 2022) • Do not require protein coatings or activation (Bosnakovski et al. 2005, Okamura et al. 2018) • Differentiation to muscle is difficult and with limited efficiency (Gang et al. 2001, Bosnakovski et al. 2005, Stern-Straeter et al. 2014, Okamura et al. 2018)	• Have limited ability to propagate (Dohmen et al. 2022) • Can only differentiate to fat requiring addition of muscle from other sources	• Isolated from early-stage embryos with limited availability and they require their destruction (Reiss et al. 2021) • No established cell lines available • Expensive media and components with non-optimal commercially available formulations optimised for human PSCs • Require protein coatings for their maintenance • Not enough research done on ESCs from livestock species due to limited availability to have a good understanding of their limitations or any potential risks	• Require induction of somatic cells through cellular reprogramming using specific transcription factors to induce pluripotency (Lu et al. 2015, Pillai et al. 2019, Bressan et al. 2020, Xu et al. 2019) which is a complex and difficult procedure • No established cell lines available • Expensive media and components with non-optimal commercially available formulations optimised for human PSCs • Require protein coatings for their maintenance • Prone to genetic instability due to extended culture and genetic manipulation (Yoshihara et al. 2017) • Not enough research done on iPSCs from livestock species to have a good understanding of their limitations or any potential risks

of pluripotent stem cells (Post et al. 2020), other limitations relating to their differentiation are posed downstream. The culture of ESCs and iPSCs usually requires the use of expensive induction factors in order to guide the differentiation process into muscle tissue. Additionally, PSCs are prone to genetic instability due to the combined effect of extended cell culture and genetic manipulation (Yoshihara et al. 2017) which raises safety concerns and can complicate regulatory approval of cultivated meat products.

Other adult stem cells such as mesenchymal stem/stromal cells (MSCs) or fibro adipogenic-progenitors (FAPs) are potential alternative cell sources for cultivated meat. MSCs are versatile as they can be easily isolated from multiple tissues including bone marrow, adipose, umbilical cord and blood and peripheral blood, etc. (Hill et al. 2019). Unlike pluripotent stem cells (PSCs) and similarly to MyoSCs and FAPs, MSCs have limited ability to propagate (Dohmen et al. 2022). Moreover, the culture requirements for MSCs including the medium formulation and cost are significantly simplified and cheaper than for PSCs. Unlike MyoSCs and PSCs, MSCs do not require additional protein coatings or activation to enable their culture (Bosnakovski et al. 2005, Okamura et al. 2018). While MyoSCs can only differentiate to muscle cells, requiring addition of fat cells from other sources and FAPs only to adipocytes requiring muscle cell from other sources, MSCs have the ability to differentiate towards both fat and muscle cells, although for the latter, only limited reports are available (Gang et al. 2001, Bosnakovski et al. 2005, Stern-Straeter et al. 2014, Okamura et al. 2018).

4.2 Scalable Culture Methods

To achieve economic success, the cultivated meat products need to be available and affordable. The availability of these products is directly linked to the scalability of the method used to produce them. Scalable cell culture methods include free-floating aggregates, microcarrier culture and encapsulation in hydrogels, and they can be implemented in a range of bioreactor vessels. Whilst microcarrier based culture is typically used during the cell expansion phase for the generation of biomass, encapsulation in hydrogels is mostly used during the differentiation phase. Aggregate based cultures have been demonstrated as suitable for both the expansion and differentiation phases; however, microcarrier use can be extended to be used for differentiation too and hydrogel encapsulation can also be conceivable for cell expansion as well (Torgan et al. 2000, Thakur et al. 2010, Park et al. 2014, Kim et al. 2019). All culture methods attempt to meet the environmental demands of adherent cell types in a scalable culture system.

4.2.1 Free-floating Aggregates

The first culture method that can be applied involves growing cells in "clumps" or aggregates (Mueller-Klieser 1987). This type of culture does not rely on cell adhesion to attachment substrates, and it is only possible for specific cell types. When adherent-dependent cells are not provided with an attachment substrate, the cells attach to one another and survive and proliferate in suspension through the activation of cadherin binding and self-production of extracellular matrix *in situ* (Laperle et al. 2015, Manibog et al. 2016). This approach of cell expansion, particularly when cultured in scalable systems such as bioreactors, is simplified and somewhat advantageous particularly from a downstream processing point of view. Cell-to-cell binding implies 100% purity of the end product, allowing for higher cell densities and making for simpler downstream processing (harvesting and formulation) and cutting out reliance on external suppliers for microcarrier materials, both of which can lower the overall production costs and increase profit margins. However, there are certain concerns when growing cells as aggregates. For example, nutrient and gas diffusion become more difficult as the size of the aggregates increases, setting a defined upper size limit for proliferation. If the size exceeds that limit, necrosis will occur at the centre of the aggregate resulting in cell death, exposure to gradients that further lead to spontaneous differentiation if the cells are potent, as well as heterogeneity in the cell population (van Winkle et al. 2012, Xie et al. 2017, Sart et al. 2017).

The successful culture as aggregates should maintain a tight control over the size of the aggregates in order to avoid the possible issues. Different approaches have been reported to control the aggregate size including using dextran sulphate (Lipsitz et al. 2018), adding heparin to the culture (Li et al. 2011), employing agitation to breakdown the aggregates (Takahashi et al. 2017) or encapsulating cells in hydrogels (Kim et al. 2019). Cell encapsulation in hydrogels will be further discussed in more detail in one of the following sections of this chapter.

Linking back to the possible cell sources for production of cultivated meat, aggregate cultures are more common for pluripotent stem cells (ESCs and iPSCs) as they have an inherent tendency to adhere to each other rather than growing as single cells (van Winkle et al. 2012, Laperle et al. 2015, Sart et al. 2017, Lipsitz et al. 2018, Nogueira et al. 2019, Polanco et al. 2020). A few studies were reported in literature looking at the possibility of growing myogenic cells as aggregates (Wei et al. 2011, Hosoyama et al. 2013, Bodiou et al. 2020). However, doubling times of over 150 h were recorded which suggested that there was very limited cell proliferation. Another study performed on the C2C12 mouse myoblast cell line demonstrated that their culture as aggregates again didn't result in proliferation and the cells were found to express markers of quiescent satellite cells instead (Aguanno et al. 2019) which is not suitable for the production of cultivated meat where cell proliferation is a must in order to achieve the high cell numbers required. Similarly, a few studies have demonstrated the ability to culture mesenchymal stem/stromal cells as aggregates in what are referred to as 'mesenspheroids'. However, it again resulted in limited proliferation, but with an increased differentiation potential and maintenance of multipotency (Baraniak and McDevitt 2012, Isern et al. 2013, Tietze et al. 2019). This approach to their culture can be advantageous to clinical applications but might not be suitable for production of cultivated meat where extensive and fast proliferation is necessary.

4.2.2 Microcarriers

The second possible method of scalable cell culture involves the use of scalable, attachment substrates that would be used in conjunction with bioreactors to provide a homogeneous and dynamic environment for the cells. These attachment substrates are known as microcarriers and are micrometer sized particles with different surface chemistries tuned to allow for cell attachment and proliferation.

The microcarrier concept is not new. It was first introduced 55 years ago when modified dextran particles were used for the first-time to culture cells (van Wezel 1967). Since then, a variety of microcarrier types with different sizes ranges, different surface chemistries and porosities, with or without protein coatings and with or without surface charges (Li et al. 2015, Tavassoli et al. 2018, Chen et al. 2020) have become commercially available. Table 2 shows the list of currently commercially available microcarriers and their characteristics. Several other previously reported microcarriers (Rodrigues et al. 2018, Kalra et al. 2019) have been discontinued from production in the past couple of years. This variety of microcarrier types and configurations were necessary to cover the wide range of applications and cell types. The core material and surface chemistry, as dictated by surface charges or incorporated coatings, are important factors that dictate cell attachment and proliferation. Their physico-chemical properties such as density also play a role in the design of the bioprocess and the cell harvesting step. Typically, the density of these microcarriers is slightly higher than water to allow for an easy suspension and fast sedimentation when required. Density influences the minimal agitation speed (N_{JS}) required for the suspension of microcarriers (Zwietering 1958, Hewitt et al. 2011). For example, for denser microcarriers such as Hillex II (SoloHill/Sartorius), the N_{JS} required would be higher to account for the increased density. On the opposite, the time required for sedimentation would be shorter for these microcarriers.

Microcarriers are now an established technology that offers a large surface area/volume ratio which makes them advantageous for scale-up. They also offer flexibility in terms of the bioreactor types that can be used as they can be compatible with stirred tanks, fluidised or packed beds or even

Table 2. Current list of commercially available microcarriers and their properties.

Microcarrier	Manufacturer	Animal-free	Core material	Surface coating	Surface charge	Porosity	Size range (µm)	Relative density	Nominal surface area (cm²/g)
Plastic	SoloHill/Sartorius	Yes	Crosslinked polystyrene	No	None	Non-porous	125–212	1.022–1.030	360
Plastic Plus	SoloHill/Sartorius	Yes	Crosslinked polystyrene	No	+	Non-porous	125–212	1.022–1.030	360
Hillex II	SoloHill/Sartorius	Yes	Modified polystyrene	Cationic trimethyl ammonium	+	Non-porous	160–200	1.080–1.150	515
Star Plus	SoloHill/Sartorius	Yes	Crosslinked polystyrene	No	+	Non-porous	125–212	1.022–1.030	360
Collagen	SoloHill/Sartorius	No	Crosslinked polystyrene	Type 1 porcine collagen (gelatine)	None	Non-porous	125–212	1.022–1.030	360
ProNectin F	SoloHill/Sartorius	Yes	Crosslinked polystyrene	Recombinant RGD peptides from fibronectin	None	Non-porous	125–212	1.02	360
FACT III	SoloHill/Sartorius	No	Crosslinked polystyrene	Type 1 porcine collagen (gelatine)	+	Non-porous	125–212	1.022–1.030	360
Cytodex 1	Cytiva	Yes	Crosslinked dextran	DEAE	+	Non-porous	140–200	1.03	4,400
Cytodex 3	Cytiva	No	Crosslinked dextran	Denatured collagen	None	Non-porous	120–180	1.04	2,700
Cytopore 1	Cytiva	Yes	Cellulose	DEAE	+, 1.1 meq/g	Porous; 30 µm pores	200–280	1.03	11,000
Cytopore 2	Cytiva	Yes	Cellulose	DEAE	+, 1.8 meq/g	Porous; 30 µm pores	200–280	1.03	11,000
Cultispher G	Merck	No	Crosslinked porcine gelatine	No	None	Porous; 20 µm pores	130–380	1.04	
Cultispher S	Merck	No	Crosslinked porcine gelatine	No	None	Porous; 20 µm pores	130–380	1.04	
Untreated	Corning	Yes	Polystyrene	No	None	Non-porous	125–212	1.026	360

Table 2 contd. ...

...Table 2 contd.

Microcarrier	Manufacturer	Animal-free	Core material	Surface coating	Surface charge	Porosity	Size range (μm)	Relative density	Nominal surface area (cm²/g)
Collagen-coated	Corning	No	Polystyrene	Collagen	None	Non-porous	125–212	1.026	360
Synthemax II	Corning	Yes	Polystyrene	Synthemax II (synthetic coating)	None	Non-porous	125–212	1.026	360
CellBIND	Corning	Yes	Polystyrene	Animal-free attachment enhancer	None	Non-porous	125–212	1.026	360
Enhanced attachment	Corning	Yes	Polystyrene	Oxygen surface treatment	None	Non-porous	125–212	1.026	360

aerated bioreactors (Bodiou et al. 2020). The majority of studies, however, have been carried out in stirred tank bioreactors for a variety of adherent cell types (Hewitt et al. 2011, Chen et al. 2011, Badenes et al. 2015, Gupta et al. 2016, Rafiq et al. 2017, 2018, Hanga et al. 2020, 2021). It is also worth mentioning that the microcarrier cultures of mesenchymal stem/stromal cells and satellite cells in stirred tanks have been highly successful (Molnar et al. 1997, Torgan et al. 2000, Hewitt et al. 2011, Rafiq et al. 2013, Hervy et al. 2014, Heathman et al. 2015, Nienow et al. 2016, Rafiq et al. 2016, Lawson et al. 2017, Rafiq et al. 2017, Verbruggen et al. 2018, Rafiq et al. 2018, Hanga et al. 2020, 2021).

Advantages of microcarrier based systems include the ability to hold adherent cell populations in suspension, promoting even distribution of medium and gases through the mixing action of the impeller and a very straightforward scale up process. However, they are a difficult culture strategy to perfect, with many uncontrolled variables. Common challenges include uneven cell distribution on the carrier surface due to preferential adhesion during cell seeding, limited capacity of some types of cells to populate newly added carriers (bead to bead transfer phenomenon) and hydrodynamic forces acting on the carrier/cell surface during agitation. All the above can stunt cell proliferation and negatively impact differentiation (Wu 1999) if not minimised through controlled approaches. For example, uneven cell distribution and aggregation can be minimised through the control of agitation in the bioreactor vessel. Hanga et al. (2020) used an agitation strategy involving a step increase in agitation every couple of days to minimise the level of aggregation. The bead-to-bead transfer has to date been successfully demonstrated with some cell types including human MSCs (deSoure et al. 2016, Takahashi et al. 2017, Rafiq et al. 2018), Vero (Wang and Ouyang 1999, Yang et al. 2019), HEK293T (Yang et al. 2019) and more recently bovine adipose-derived stem cells (Hanga et al. 2021). Bead-to-bead is a manufacturing approach that facilitates process intensification, therefore resulting in increased cell yields and reduced process times. However, the choice of culture time point for the addition of fresh microcarriers is critical to initiate successful bead-to-bead transfer as certain levels of aggregation can impede the cells capacity to transfer to fresh microcarriers (Hanga et al. 2021). The other limitation of hydrodynamic forces could be addressed by using porous microcarriers as they can aid with damping the hydrodynamic forces on the cells during agitation, since the cells would grow mostly on the inside of the carrier rather than on the surface; however, harvesting of the cells from porous carriers is even more challenging (Wen and Yang 2011).

For the production of cultivated meat products, any of the commercially available microcarriers could be used. However, there are several concerns, particularly regarding the polystyrene-based microcarriers, as the validation of their removal during downstream processing steps is very important to ensure the safety of these products. Ideally, the microcarriers to be used should be edible which would simplify the bioprocess particularly if they are dissolvable/degradable meaning they could be easily broken down to release the cells, thus eliminating additional steps of dissociation and separation. Ideally the building blocks of the degraded microcarrier itself would also be food-grade substances with no need to eliminate them from the product. Tuneable degradation rates have been seen to enhance MSC spread and differentiation capacity due to the more dynamic microenvironment (Toh et al. 2012). It may be possible to use tuneable degradation as a micropatterning technique, targeting and selectively tuning degradation rates of certain regions on the microcarrier to form a topography that can guide cell alignment, fusion and eventually muscle fibre formation (Zhao et al. 2009).

Another option would be that the edible microcarrier itself could serve as a bulking material that could potentially enhance the nutritional and textural profile of the cultivated meat product. A similar type of microcarrier made of pectin (a compound found in fruits) has been previously commercialised, but now discontinued. The Corning dissolvable microcarrier has been used successfully to culture and harvest iPSCs (Rodrigues et al. 2018) and MSCs (Kalra et al. 2019) which are both possible starting sources for cultivated meat production.

4.2.3 Cell Encapsulation in Hydrogels

The third suggested culture method is encapsulation in hydrogels. Hydrogels have their own intrinsic material properties to be exploited and can be processed into different geometries, making this the most versatile technique. Hydrogels can be categorised as either natural or synthetic; natural hydrogels, usually being cheaper, food grade, more biocompatible, are likely to have necessary RGD sequencing helpful for accommodating cell adhesion and capable of being digested or remodelled by matrix metalloproteinases (MMPs), while synthetics have limited batch variability, are often hydrophobic, possess strong covalent bonds within their matrix improving their mechanical strength, service life and absorbability, and are more easily modified for attachment, degradation, and stiffness requirements (Gyles et al. 2017). However, natural hydrogel candidates will have increased batch variability and less room for optimisation due to their lower plasticity in chemical and mechanical tuning, while synthetics likely won't be edible or approved for use in food products without rigorous purity checks. Additionally, they may require external chemical degradation to allow cellular migration and prevent stunting of proliferation. Chemical conjugation and surface functionalisation of synthetics can be expensive but may prove to be a valuable opportunity to generate versatile IP and an optimised process.

The encapsulation can be in a continuous, solid-state hydrogel acting as a scaffold to offer structure, which then degrades as the cells grow or differentiate, or it may be in smaller spheroids, mimicking porous microcarrier systems. Large hydrogel scaffolds have proved advantageous for production of well-differentiated muscle tissue due to their ability to uniformly distribute static tension in conjunction with cell attachment points. These systems have been termed BAM (bio-artificial muscle) constructs (Gholobova et al. 2018).

Novel strategies under development include the immobilisation of growth factors via embedding in the hydrogel scaffold (Silva et al. 2009) where micropatterned deposition could even encourage differentiation down different lineages in different regions or the addition of acetylcholine to encourage multinucleated myotubes to contract, thus accelerating differentiation (Krause et al. 1995). Fibro-adipogenic progenitor and adipose derived stem cells have also been successfully encapsulated in hydrogels such as alginate or collagen and fibrinogen which were then shaped into microfibres, to enhance differentiation into fat tissues (Hsiao et al. 2015, Dohmen et al. 2022).

4.2.4 Culture on Scaffolds

Culture on scaffolds is another method applicable to cultured meat products. Scaffolds can be used both during the expansion and differentiation phases. Cells can be expanded in fibrous carriers which are then either suspended in a stirred-tank reactor (STR) or packed in a packed-bed reactor (PBR). However, the expansion fold is typically low and limited to a few population doublings before the cells cover the surface area provided. They then have to be dissociated and re-seeded into a larger scaffold/surface area, unlike the microcarrier based systems, where cells can migrate on freshly added carriers and several population doublings can be achieved without the need to passage (Hanga et al. 2021). Scaffolds are hence mainly useful for the differentiation stage and are indispensable for the production of structured cultivated meat products (e.g., steaks, fillets, etc.).

Biomaterial-made scaffolds have been used for many years in the field of regenerative medicine and tissue engineering for transplantation purposes. Scaffolds provide internal structure and fall into two distinct categories: permanent or resorbable. Permanent scaffolds tend to be bioinert and remain in place, providing structure to the growing cell mass. The material properties (stiffness and elasticity) can be tuned to direct differentiation of the cells towards a specific lineage. Resorbable scaffolds, on the other hand, typically degrade as the cells differentiate, possibly even releasing bioactive molecules as it breaks down (Nikolova et al. 2019). The tissue formed at the endpoint would be in that case comprised only by the cells with no additional biomaterial and therefore, more likely to be approved for human consumption by the regulatory bodies as long as it can be demonstrated that the degradation components pose no risk.

Scaffolds investigated in the realm of cultivated meat include: decellularised plant tissue (Gershlak et al. 2017, Campuzano and Pelling 2019), fungal mycelium or chitin/chitosan, and recombinant collagen or laminin (Campuzano and Pelling 2019) formed in a variety of topographies. Decellularised plant tissue maintains the animal free product selling point, would add insoluble fibre to meat products to aid digestion and enhance public health, and, being plentiful and easy to grow, can be incorporated into circular economies by taking plant-based food waste, carrying out mass decellularisation and inoculating with animal cells. This would make acquisition cheap and sustainable by cutting down energy for extra plant growth and land usage. Global food wastage is a large contributor to greenhouse gas emissions, responsible for 4.4 billion tonnes of CO_2 equivalent annually (making up 8% of the total) (FAO 2015, Jeswani et al. 2021), meaning utilisation of this waste is key to achieving a sustainable practice. However, decellularisation is a time and energy intensive process, requiring strong detergents and surfactants to strip the native cellular material (White et al. 2017). This would add layers of processing complexity and cost, it would force products to undergo toxicity analysis to ensure no traces of detergents remain, and it would produce significant amounts of chemical waste which then needs to be treated while consuming additional water.

Fungal mycelium or chitin/chitosan are very similar in all these respects but would require less land and less energy than growing green vegetation for decellularisation due to the lower light/nutrient requirements. Recombinant collagen or laminin would be the best option for directing differentiation and kickstarting myogenesis due to their mimicry of physical signalling pathways *in vivo* (Ahmad et al. 2020). However, it would relate to a much higher product price point at commercial sale due to how expensive recombinant biologics are as typically they are produced for medical applications meaning they need to be of a high purity which reflects in their high costs.

4.3 Culture Medium

Cell culture medium is designed to support cell survival when they are removed from their natural *in vivo* environments and placed *in vitro* for analysis or expansion. Mammalian cells cultivated *in vitro* generally need a carbon source, usually glucose, nitrogen sources, i.e., aminoacids, as well as vitamins, minerals and inorganic minerals to maintain physiological osmolality. Growth factors are also essential as they activate signalling pathways and regulate cell behaviour. Animal-derived serum has been traditionally used as a medium supplement in mammalian cell culture as it contains a mixture of proteins and smaller molecules critical for cell maintenance and viability. However, animal-derived serum is subject to ethical concerns because of the way it is sourced. Additionally, it is an undefined component subject to batch-to-batch variation and carrying a risk of viral, bacterial and endotoxin contamination (Gstraunthaler and van der Valk 2013, Fang et al. 2017). For cultured meat production, ideally animal-derived serum should be replaced by non-animal derived and chemically defined alternatives because of these aspects, but also as it is unsustainable for large scale production due to the price volatility which has become apparent in the past few years (Kolkmann et al. 2020). However, given its undefined composition, replacing it with defined and most likely recombinantly derived components is challenging and can only be achieved at high cost. Albumin is also commonly used in serum-free medium formulations because of its versatile functions in *in vitro* cell culture including binding and transport of lipids, metal ions and other factors into the cell, its antioxidant function, and its potential role in cell survival, proliferation and metabolic activity (Francis 2010).

pH buffers such as sodium bicarbonate or HEPES are often added to the medium formulation to negate the effects of acidic or basic products of cellular respiration (Arora 2013). Lastly, as an option, culture medium can also contain 1% antibiotic and/or antimycotic in order to prevent bacterial or fungal contaminations. This approach is useful in the early-stage research where sterility is more difficult to maintain, for example, post-cell isolation from tissue. However, using antibiotics in meat products presents more regulatory hurdles, with large markets such as the EU attempting to scale down their antibiotic use at present. From 2016 onwards, antimicrobial consumption in the EU in livestock populations has been lower than use for humans (JIACRA III 2021). Avoiding antibiotics in medium also prevents further spread of antibiotic resistance. There is also concern about the effect of antibiotics on gene expression, with a recent study identifying 209 PenStrep-responsive genes, including transcription factors likely to alter regulation of other genes (Ryu et al. 2017). Moreover, antibiotics were also found to have a detrimental effect on the growth of myosatellite cells with a reduced cell proliferation by 26% after 6 days of culture (Kolkmann et al. 2020). The same effect was found when using adipose-derived MSCs (Skubis et al. 2017).

Several serum-free media studies for expansion of myosatellite stem cells for cultivated meat production have been published to date (Kolkmann et al. 2020, Messmer et al. 2022). Kolkmann et al. (2020) tested a range of commercially available serum-free media and also two serum replacements for their ability to sustain culture of primary bovine myoblasts. Serum-free media (SFM) tested included Essential 8 (Life Technologies), StemPro MSC SFM (ThermoFisher Scientific), STEMmacs HSC expansion media (Miltenyi Biotec), MesenCult (Stem Cell Technologies) and TeSR E8 (Stem Cell Technologies), while the serum-replacements tested included LipoGro (RMBIO) and XerumFree (TNC Bio). When used with bovine myoblasts, they all performed poorly probably because they were optimised for a different species (i.e., human) and different cell types such as MSCs, HSCs or PSCs. On the opposite, the LipoGro supplement was found to be successful as an FBS replacement for bovine myoblasts. However, it resulted in an adipocyte-like cell phenotype with fat droplets accumulation (Kolkmann et al. 2020).

Myogenic differentiation in serum-based cultures is typically induced by lowering the concentration of serum (serum starvation) (Messmer et al. 2022), while in a serum-free culture, it would be induced by introducing ligands to receptors that are up-regulated during the initial stages of myogenic differentiation such as transferrin, insulin, glucagon and lysophosphatidic acid (LPA) (Messmer et al. 2022). For adipogenic differentiation, typical cocktails include insulin, rosiglitazone, IMBX, hydrocortisone or dexamethasone (Fink and Zachar 2011, Dufau et al. 2021, Dohmen et al. 2022) used in combination or without serum.

Mammalian cells grown *in vitro* are inefficient in the way they metabolize nutrients such as carbon and nitrogen sources, producing a lot of metabolic waste such as lactate and ammonia, which in turn can inhibit cell growth and differentiation at certain concentrations (Schop et al. 2009, 2010, Post et al. 2020). Regular medium changes need to be applied, to remove the waste products and replenish the consumed nutrients. To mitigate this, fed-batch or perfusion processes can be used to perform regular medium exchanges to remove the waste products and to replenish the consumed nutrients resulting in increased cell yields (Europa et al. 2000) and a more effective cell metabolism, perhaps due to lesser substrate concentration fluctuations. Additionally, medium composition can be optimized to drive metabolic pathways towards a more efficient consumption of nutrients. Such strategies have been successfully used to optimize medium use for cell lines in the production of biopharmaceutics (Bell et al. 1995).

Cultured meat is a product that needs to have a price tag several orders of magnitude lower than biosimilars. Therefore, efficient ways of reducing the cost, while increasing cell yields need to be investigated, stretching the current boundaries of cell culture methods applied to the production of biopharmaceuticals. Food grade or even feed grade raw materials of low purity need to be explored for their performance in media formulations. Crude raw materials, such as protein hydrolysates from plant, algal or microbial sources are also potential candidates of nutrient sources to the cells. Hydrolysates can be produced with enzymatic or acidic treatment of the algal, plant or microbial biomass and have been proved effective into promoting cell growth (Ng et al. 2020, Ho et al. 2021). However, hydrolysates are also prone to batch variation and unless a high level or robustness in the production process is achieved, their use can result in significant differences in performance.

Recycling the cell culture medium is another strategy that can be applied to reduce costs and most efficiently use the raw materials. Recycling could be achieved through chemical or biological process.

In the first case, it consists of treating the waste medium in such a way that detrimental metabolic waste products such as lactate and ammonium are extracted from the medium, and the remaining bulk consisting of semi-depleted nutrients (aminoacids, vitamins, minerals, growth factors, etc.) as well as substances produced by the cells. This is, they grow or differentiate through as growth factors and cytokines, and would be re-supplemented and fed back to the bioreactor. Water would also be a valuable ingredient to recycle, as in cell culture, typically MilliQ water is being used and given the volumes needed, it can add significantly to the manufacturing costs. The extracted waste products could even be valorised. For example, ammonium could be further used for the production of fertilizers, while lactate could be purified and used as a supplement in the muscle differentiation phase (Tsukamoto et al. 2018). A biological recycling process would consist of either an at-line or off-line algae culture. Waste medium supplemented with glucose could serve as a feed medium to a microalgae or cyanobacteria culture. CO_2 could in this way be converted to oxygen, ammonia could be metabolized into amino acids and plant/microbial biomass, which can in turn be used to supplement the cell culture medium with nutrients (Okamoto et al. 2020, Haraguchi and Shimizu 2021).

Cost efficiency can also be achieved through metabolic engineering tactics that would enable a more efficient use of media components. If nutrient depletion or metabolite over-saturation leads to the entire volume of media being replaced, water turnover and proportion of metabolites unutilised will be high. Metabolite over-saturation leading to toxicity and cell death is primarily caused by ammonium and lactate in the media. Innovation is required to remove harmful metabolites while continually supplementing required nutrients. Methods discussed in literature include the removal of ammonium ions through zeolite catalysis (Huang et al. 2014), whereby a multi chambered bioreactor system comprised of cell culture reactors and peripheral zeolite packed electrolysis reactors can drop ammonium concentrations from 27.8 mg N/L to 0.3 mg N/L within 1 hour. Copper supplementation in media also appears to increase lactate consumption and so decrease endpoint lactate levels in culture media for CHO cells (Luo et al. 2012). Observed in CHO cells, there was a phenotypic shift from net lactate production (LP) to lactate consumption (LC). Another method to remove lactate is through Lactate Supplementation and Adaptation (LSA) technology (Freund and Croughan 2018). Based on the laws of mass action, growing cells in media supplemented with lactate results in no net lactate production through exploitation of the equilibrium balance and the actual free energy of the oxidation of glucose to lactate, meaning there is a tipping point at the pyruvate oxidation step causing the reaction to flux toward the tricarboxylic acid (TCA) cycle for the continued production of ATP. LSA is a technique which exploits the principles of metabolic engineering, also useful in establishing efficient feeding regimes for cells (Konakovsky et al. 2016).

Left unregulated, cells will naturally consume more glucose and amino acids than required to proliferate. The "feed by need" strategy, where cells are held in a low-glucose environment via continuous batch-feeding in order to lower their consumption of glucose and production of lactate/ ammonium, is often referred to as "metabolic engineering" (Konakovski et al. 2016). Their natural metabolism is lowered via a training feeding regime, making the final production cost (in £/kg) lower, as the cell biomass isn't affected proportional to their consumption.

4.4 Bioreactors

To achieve scale-up of production, the use of bioreactors is a must (Stephens et al. 2018, Allan et al. 2019). The bioreactor's role is to ensure a controlled environment suitable for cell culture. There are multiple categories of bioreactors depending on the method used to achieve mixing and environmental homogeneity, and thus to aid nutrient and gas diffusion to the cells. Bioreactors can also provide additional stimuli to support cell proliferation, differentiation and even cell maturation. Regardless of the type of bioreactor, a dynamic environment is achieved with a portfolio of environmental parameters that can be monitored and controlled that includes temperature, pH and gases (O_2, CO_2). This level of control is not possible to be achieved in planar culture vessels. Bioreactors also allow for sterility which is critical for growing mammalian cells. Moreover, the resulting product would be sterile which could result in longer shelf lives for cultivated meat products due to the lack of bacteriological loads obtained through animal slaughter.

Stirred-tank bioreactors (STBs) are a type of bioreactor that uses mechanical means to achieve homogeneity of its contents. STRs are equipped with impellers (agitators) that can be of various shapes and sizes and generate either an axial, radial or mixed flow. For cultivated meat production, mammalian cells will most likely be used for which typical impellers used include pitched-blade or marine that generate a gentler flow to minimise damage (Nienow 2010). This type of bioreactor is a common choice for the production of high value biologics due to its scalability to over 20,000 L, as well as established understanding of operating such systems (Zhong 2011). To achieve the high cell numbers required for meat ($> 2 \times 10^{11}$ cells) (Allan et al. 2019), scalability is critical and as such, STBs are perhaps the most promising type of bioreactor. Moreover, despite initially being optimised for suspension cell types like bacteria or CHO cells used for production of biologics, this type of bioreactor has been successfully adapted to the culture of adherent-dependent cells, and as such, it offers a level of flexibility in regards to the cell culture method to be chosen as it could be used with either free-floating aggregates (Serra et al. 2012, Kropp et al. 2016, Abecasis et al. 2017), microcarriers (deSoure et al. 2016, Gupta et al. 2016, Takahashi et al. 2017, Lawson et al. 2017, Rafiq et al. 2018, Hanga et al. 2020, Bodiou et al. 2020, Hanga et al. 2021) or hydrogel capsules containing cells (Miranda et al. 2010, Tostoes et al. 2010, Jing et al. 2010). Stirred-tank systems are optimal for starter cultures and experimental work due to their flexible operating conditions with easily controlled parameters (impeller RPM, number of impellers, vessel dimensions, etc.), their commercial availability at various scales up to > 1000 L, the efficient gas transfer for growing cells and thorough mixing of the growth medium. However, they are also associated with high power consumption and introduction of high hydrodynamic forces, especially at the proximity of the impeller and the bioreactor walls (Garcia-Ochoa et al. 2011).

While STBs have been extensively and successfully used for the expansion stage, they are less likely applicable for the tissue formation phase for structured cultured meat products, especially when functional scaffolds with anchor points need to be used in that process. Instead, a different type of a mechanical bioreactor could be useful for muscle tissue formation. This bioreactor would incorporate dynamic compression that would provide the mechanical stimuli needed by muscle tissue to achieve maturation and alignment (Meinert et al. 2017).

The differentiation step is an integral part of the production of cultivated meat products. The adipogenic differentiation is easier to achieve as it can be induced using small molecules formulated in the media (Scott et al. 2011, Chen et al. 2017). However, myogenic differentiation has very specific requirements for nutrients and physical environment (typically a mechanical stimulus). To satisfy these requirements, the myogenic differentiation step is typically performed in monolayer using a hydrogel between anchor points to simulate tendons, thus creating a passive tension and the minimal mechanical stimulus required for differentiation and maturation of tissue (Vandenburgh and Karlisch 1989, Morgan et al. 2003). However, this is not a scalable approach. To achieve the mechanical stimulus required in stirred tank bioreactors, several approaches have been tested. Torgan et al. expanded and differentiated satellite cells on microcarriers in a microgravity bioreactor. However, the cells expressed less myogenin, myosin and tropomyosin than satellite cells cultured in a "normal gravity" stirred tank bioreactors (Torgan et al. 2000). Others have reported that the shear stress generated by the stirring alone in stirred tanks used with microcarriers can promote differentiation; however, the shear stress needs to be controlled (Naskar et al. 2017, Wu et al. 2018). For example, high shear stress within the ranges of 5–10 Pa was found detrimental, while a shear stress of 16 mPa resulted in a higher expression of myogenic markers and longer myotubes, while 42 mPa determined a better alignment (Naskar et al. 2017).

Other bioreactor types could be suitable for the production of cultivated meat products. Packed bed bioreactors (PBBs) operate by packing the cells in aggregates or on microcarriers in a tight bed, through which growth medium is perfused. The mixing in this type of bioreactor is achieved through liquid circulation. Cells are retained in the bioreactor to continuously increase cell density while removing spent media without dilution of the cell culture. Very high volumetric cell densities can be achieved this way. However, the system has gas and nutrient diffusion limitations, especially at very high end-point cell densities (Ellis et al. 2005). Fluid flow in packed bed systems can't exceed 3.0×10^{-4} m s^{-1} without relative hydrodynamic forces impacting cell viability (Weber et al. 2010), while stresses exceeding 0.015 Pa have been shown to up-regulate differentiation down osteogenic lineages in human MSCs (Zhao et al. 2007). Scale up of PBBs proves challenging as increasing the depth of the packed bed isn't possible without generating axial gradients of nutrient distribution and heterogeneity within the packed bed; there will be a critical depth below the fluid-gas interface where gas solubility is zero (Mewly et al. 2007). However, once again, the adaptation of suspension cell culture processes to the engineering of a complete, congruous tissue offers, means plateauing at subpar cell densities is inevitable and energy, medium, nutrients, and space requirements will all be far higher relatively to the physiological system a PBB is trying to mimic.

Fluidised bed bioreactors (FBBs) operate on a similar principle to PBBs with the difference that the cell/carrier/aggregate bed is less packed, thus fluidised. This way, pH /DO/nutrient gradients can be more easily avoided, however inevitably compromising the volume. For the bed to be fluidised, a larger bioreactor volume is needed, and, inevitably, the achieved cell densities are lower when compared to the PBBs. However, in such a system, similarly to an STB, continuous addition of fresh microcarriers or encapsulation of cells in small scaffolds/beads can be applied to expand the culture. Usually, fluidisation at a constant or increasing rate is applied from the bottom of the column upwards and a filter membrane is used to prevent any cell loss. The medium is continually pumped around the chamber, and can be aerated either with the help of an external bubble column or by the incorporation of a sparger in the culture column.

Hollow Fibre Bioreactors (HFBs) are another bioreactor type with potential for cultivated meat. HFBs are a type of hydraulic bioreactors meaning that mixing in such a system is achieved through the fluid flow similar to PBB and FBB (Reiss et al. 2021). HFBs have the greatest surface area to volume ratio from all bioreactor types. HFBs are advantageous as they generate a low shear, while provide an increased availability to nutrients due to the perfusion operation mode resulting in high cell densities (Reiss et al. 2021). Although the medium perfusion provides a more efficient distribution of nutrients and gases to the cells when compared to PBBs, at scale, the development of gradients would be inevitable. The maximum perfusion distance allowed for the maintenance of a physiological tissue (i.e., how far can a cell be from a capillary and still survive) is said to be 150–200 μm, which correlates well with the maximum oxygen diffusion distance (Rouwkema et al. 2009). HFBs have been previously used for the successful expansion of skeletal muscle cells (Bettahalli et al. 2011, Yamamoto et al. 2012). HFBs are typically used in the biopharmaceutical industry for the production of biosimilars, but for cultivated meat purpose where the biomass is the product, harvesting from such systems can be challenging particularly as increased cell harvesting yields are required. However, it is conceivable that degradable and/or edible fibre channels can be used for a one-step expansion and differentiation of muscle and fat tissues in the same setup.

The suitability of these types of bioreactors is dependent on the type of cultivated meat product that is targeted. As the industry grows and a diverse range of cultivated meat products will be developed, the continual development and optimisation of the bioreactor platforms to be used for large scale production is a must. It is also important to acknowledge that novel bioreactor designs might also be beneficial for these products.

4.5 Co-culture

Co-culture is defined as two or more cell populations grown with some degree of cell contact between them, primarily to investigate how one cell type affects the proliferative or differentiative capabilities of another cell type. Co-culture can be direct, with cells forming physical bonds, or indirect, throughout a transwell semi-permeable membrane, to gauge the effect of secreted factors

(Goers et al. 2014). Some examples of co-culture that could prove beneficial in the production of cultivated meat include the addition of macrophages and fibroblasts, both used in muscle repair, where macrophages contribute an inflammatory response and fibroblasts have a pro-migratory effect on the myoblasts (Venter and Niesler 2018). Increasing the myoblast motility potential presents an opportunity to enhance structural self-organisation and differentiation through the application of mechanical stimuli via higher contractility. Dendritic cells have also been shown to enhance the proliferative capacity of myoblasts, but don't seem to increase the myofibre number proportionally (Ladislau et al. 2018). This may indicate a delay in the onset of differentiation or the induction of further proliferative cycles, which could be useful in mass cell banking, if a two-stage "proliferation/differentiation" process was to be developed. The further addition of myoglobin, a protein naturally occurring in mammalian skeletal and cardiac muscle cells, has been shown to increase proliferation and metabolic activity of bovine satellite cells, while the addition of haemoglobin can improve colouration of cultivated meat products (Simsa et al. 2019).

To obtain a tissue that resembles the structure and the flavour of livestock meat, co-culture of different cell types (fat, muscle and connective cells) will be necessary particularly for structured cultured meat products. The co-culture of adipocytes with myotubes has been previously reported (Kovalik et al. 2011) with the help of culture inserts. Another co-culture model of adipocytes with cardiomyocytes has been developed by Anan et al. where adipocytes were first mixed with a collagen gel solution in an inner dish which was then solidified at 37°C. The cardiomyocytes were then seeded on the surface of the gel and the inner dish was placed in an outer dish containing complete growth medium (Anan et al. 2011). However, both of these examples of co-culture were two dimensional, did not facilitate scale up and were performed with human cells. Moreover, these models did not support the survival of the two cell types for more than 48 h which is not sufficient for achieving tissue maturation. The co-culture needs to be three-dimensional, scalable and to sustain the growth of all cell types involved. Critically, co-culture of cells in food-grade hydrogel microcapsules would enable a scalable method for production of meat microtissues in bioreactors. The incorporation of hydrogels would also streamline the manufacturing process by eliminating the need to use proteolytic enzymes to detach cells from commercial microcarriers, which are often polystyrene based. Enzymes and polystyrene microcarriers cannot form part of the final consumer product and the presence of such traces could raise safety concerns during the regulatory approval. Grellier et al. have shown that co-immobilisation of human umbilical vein endothelial cells and human osteo-progenitors in alginate hydrogels and their culture under dynamic conditions significantly promoted mineralization over a 3-week culture period (Grellier et al. 2009).

5. Conclusions

Cultivated meat products are highly complex and thus their production is challenging. Several critical challenges need to be overcome in order to make these products affordable and available to the population. Bioprocess development for their production requires a very good understanding of cell requirements including media formulation and environmental conditions. The development of a bioprocess for the scalable production of cultivated meat is even more complex as it requires knowledge and understanding of multiple cell types—a source for muscle and a source for fat cells as a minimum. In addition, their production has multiple stages—expansion, induction, differentiation and even co-culture of multiple cell types if targeting a structured product. It needs to be noted that at each of these stages, the culture requirements are different. Because of that, typical strategies involve separate processes being carried out in parallel and even in multiple bioreactor vessels. Ideally, a simplified bioprocess where subsequent stages such as expansion and differentiation would take place in the same bioreactor would be desirable. However, this one-step bioprocess is highly challenging as it requires flexibility and specific tools such as strategies to address the requirements of differentiation and maturation. A particular challenge is formation of mature muscle fibres due to the mechanical stimulus that needs to be incorporated in the bioprocess.

The general approach to cultivated meat products seems to be adapted from tissue engineering and regenerative medicine applications which typically use highly pure, expensive components. However, for the economic feasibility and affordability of cultivated meat products, a hybrid approach should be taken that incorporates food-grade, cheap components or even uses waste products from other industries as feeds for the cultures in order to achieve sustainability and circularity.

References

Abecasis, B., Aguiar, T., Arnault, E., Costa, R., Gomes-Alves, P., Aspegren, A., et al. 2017. Expansion of 3D human induced pluripotent stem cells aggregates in bioreactors: Bioprocess intensification and scaling-up approaches. J. Biotechnol. 246: 81–93. doi.org/10.1016/j.jbiotec.2017.01.004.

Aguanno, S., Petrelli, C., Di Siena, S., De Angelis, I., Pellegrini, M. and Naro, F. 2019. A three-dimensional culture method of reversibly quiescent myogenic cells. Stem Cells Intl. 2019: 1–12.

Ahmad, K., Shaikh, S., Ahmad, S.S., Lee, E.J. and Choi, I. 2020. Cross-Talk Between Extracellular Matrix and Skeletal Muscle: Implications for Myopathies. Frontiers in Pharmacology; 11.

Allan, S.J., De Bank, P.A. and Ellis, M.J. 2019. Bioprocess design considerations for cultured meat production with a focus on the expansion bioreactor. Front Sustan. Food Syst. 3: 44.

Anan, M., Uchihashi, K., Aoki, S., Matsunobu, A., Ootani, A., Node, K., et al. 2011. A promising culture model for analyzing the interaction between adipose tissue and cardiomyocytes. Endocrinology 152: 1599–1605.

Arora, M. 2013 Cell culture media: a review. Mater Methods 3: 175. Doi.org/10.13070/mm.en.3.175.

Arshad, M.S., Javed, M., Sohaib, M., Saeed, F., Imran, A. and Amjad, Z. 2017. Tissue engineering approaches to develop cultured meat from cells: A mini review. Cogent Food & Agriculture 3(1): 1320814.

Badenes, S.M., Fernandes, T.G., Rodrigues, C.A.V., Diogo, M.M. and Cabral J.M.S. 2015. Scalable expansion of human induced pluripotent stem cells in xeno-free microcarriers. Methods Mol. Biol; 1283: 23–29. Doi:10.1007/7651_2014_106.

Baraniak, P.R. and McDevitt, T.C. 2012. Scaffold-free culture of mesenchymal stem cells spheroids in suspension preserves multilineage potential; Cell Tis Res; 347(3): 701–711.

BBC. 2020. https://www.bbc.co.uk/news/business-55155741. Accessed on 27th of January 2022.

Bell, S.L., Bebbington, C., Scott, M.F., Wardell, J.N., Spier, R.E., Bushell, M.E., et al. 1995. Genetic engineering of hybridoma glutamine metabolism. Enzyme Microbial. Technol. 17(2): 98–106.

Bettahalli, N., Steg, H., Wessling, M. and Stamatialis, D. 2011. Development of poly(lactic-acid) hollow fibre membranes for artifical vasculature in tissue engineering scaffolds. J. Membr. Sci. 371: 117–126.

Bhat, Z.F. and Fayaz, H. 2010 Prospectus of cultured meat—advancing meat alternatives. J. Food Sci. Technol. 48: 125–140.

Bodiou, V., Moutsatsou, P. and Post, M.J. 2020. Microcarriers for upscaling cultured meat production. Front. In Nutrition 7: 10.

Bosnakovski, D., Mizuno, M., Kim, G., Takagi, S., Okumura, M. and Fujinaga, T. 2005. Isolation and multilineage differentiation of bovine bone marrow mesenchymal stem cells. Cell Tissue Res. 319: 243–253.

Bressan, F.F., Bassanezze, V., de Figueiredo Pessôa, L.V., Sacramento, C.B., Malta, T.M., Kashima, S. et al. 2020. Generation of induced pluripotent stem cells from large domestic animals. Stem Cell. Res. Ther. 11: 247.

Bryant, C.J. 2019a. We can't keep meating like this: attitudes towards vegetarian and vegan diets in the United Kingdom. Sustainability 11: 6844.

Bryant, C.J. and Barnett, J.C. 2019b. What's in a name? Consumer perceptions of *in vitro* meat under different names. Appetite 137: 104–113.

Campuzano, S. and Pelling, A.F. 2019. Scaffolds for 3D Cell Culture and Cellular Agriculture Applications Derived From Non-animal Sources. Frontiers in Sustainable Food Systems 17(3): 38.

Chen, A.K-L., Chen, X., Choo, A.B.H., Reuveny, S. and Oh S.K-W. 2011. Critical microcarrier properties affecting the expansion of undifferentiated human embryonic stem cells. Stem cell Res. 7(2): 97–111. doi: 10.1016/j.scr.2011.04.007.

Chen, A.E., Egli, D., Niakan, K., Deng, J., Akutsu, H., Yamaki, M., et al. 2009. Optimal timing of inner cell mass isolation increases the efficiency of human embryonic stem cell derivation and allows generation of sibling cell lines. Cell Stem Cell 4(2): 103–106.

Chen, Q., Shou, P., Zheng, C., Jiang, M., Cao, G., Yang, Q. et al. 2016. Fate decision of mesenchymal stem cells: adipocytes or osteoblasts? Cell Death Differ 23: 1128–1139. https://doi.org/10.1038/cdd.2015.168.

Chen, X.-Y., Chen, J.-Y., Tong, X.-M., Mei, J.G., Chen, Y.-F. and Mou, X.-Z. 2020. Recent advances in the use of microcarriers for cell cultures and their *ex vivo* and *in vivo* applications. Biotech. Lett. 42: 1–10.

Cheng, D., Guo, Y., Li, Z., Liu, Y., Gao, X., Gao, Y. et al. 2012. Porcine induced pluripotent stem cells require LIF and maintain their developmental potential in early stage of embryos. PLoS ONE 7(12): e51778.

Chriki, S. and Hocquette, J.-F. 2020. The myth of cultivated meat: a review. Front. Nutr. 7: 7.

Churchill, W. 1931. Fifty Years Hence. Originally published in Strand Magazine.

Clean Technica. 2019. Mosa Meat: From €250,000 To €9 Burger Patties. Found at: https://cleantechnica.com/2019/09/12/mosa-meat-from-e250000-to-e9-burger-patties/(accessed on 27th of January 2022).

Cultured Meat Market Report. 2019. Found at: https://www.marketsandmarkets.com/Market-Reports/cultured-meat-market-204524444.html. Accessed on 27th of January 2022.

Datar, I., Kim, E. and d'Origny, G. 2016. New Harvest: Building the cellular economy. *In*: Donaldson, B. and Carter, C. (eds.). The future of meat without animals. Rowman and Littlefields International, London. 121–132.

deSoure, A.M., Fernandes Platzgummer, A., daSilva, C.L. and Cabral, J.M. 2016. Scalable microcarrier-based manufacturing of mesenchymal stem/stromal cells. Journal of Biotechnology 236: 88–109. DOI: 10.1016/j.biotec.2016.08.007.

Dillard, C. and Szejda, K. 2019. Consumer Response to Cellular Agriculture Messaging and Nomenclature: A focus group pilot study; The Good Food Institute: Washington, DC, USA.

Ding, S., Swennen, G.M., Messmer, T., Gagliardi, M., Molin, D.G., Li, C. et al. 2018. Maintaining bovine satellite cells stemness through p38 pathway. Sci. Rep. 8 (1): 1–12, 10808.

Dohmen, R.G., Hubalek, S., Melke, J., Messmer, T., Cantoni, F., Mei, A. et al. 2022. Muscle-derived fibro-adipogenic progenitor cells for production of cultured bovine adipose tissue. npj Sci. Food 6: 6. https://doi.org/10.1038/s41538-021-00122-2.

Dufau, J., Shen, J.X., Couchet, M., Barbosa, T.D.C., Mejhert, N., Massier, L. et al. 2021. *In vitro* and *ex vivo* models of adipocytes. Cell Phys. 320(5): C822–C841.

Ellis, M., Jarman-Smith, M. and Chaudhuri, J.B. 2005. Bioreactor systems for tissue engineering: a four-dimensional challenge. *In*: Chaudhuri, J. and Al-Rubeai, M. (eds.). Bioreactors for Tissue Engineering. Springer, Dordrecht. https://doi.org/10.1007/1-4020-3741-4_1.

Eun, Ji Gang, Ju Ah Jeong, Seung Hyun Hong, Soo Han Hwang SWK, Il Ho Yang, Chiyoung Ahn, et al. 2001. Skeletal myogenic differentiation of mesenchymal stem cells isolated from human umbilical cord blood. Stem Cells 19(3): 180–92.

Europa, A.F., Gambhir, A., Fu, P.-C. and Hu, W.-S. 2000. Multiple steady states with distinct cellular metabolism in continuous culture of mammalian cells. Biotech. Bioeng. 67(1): 25–34.

Fang, C.Y.W.C., Fang, C.L., Chen, W.Y. and Chen, C.L. 2017. Long-term growth comparison studies of FBS and FBS alternatives in six head and neck cell lines. PLoS ONE 12: 2017.

FAO. 2006. Livestock's Long Shadow - Environmental Issues and Options. Rome.

FAO. 2012. Alexandratos, N. and Bruinsma, J. World agriculture towards 2030/2050: the 2012 revision. ESA Working Paper. Rome: FAO; 2012. no. 12–03 http://www.fao.org/docrep/016/ap106e/ap106e.pdf.

FAO. 2015. Food wastage footprint & Climate Change. Accessed on 27th of January 2022 at https://www.fao.org/3/bb144e/bb144e.pdf

Fink, T. and Zachar, V. 2011. Adipogenic differentiation of human mesenchymal stem cells. Meth. Mol. Biol. (Clifton, NJ) 698: 243–251.

Fraeye, I., Kratka, M., Vanderburgh, H. and Thorrez, L. 2020. Sensorial and nutritional aspects of cultured meat in comparison to traditional meat: much to be inferred. Frontiers in Nutrition 7: 35.

Francis, G.L. 2010. Albumin and mammalian cell culture: implications for biotechnological applications. Cytotechnol. 62(1): 1–16.

Freund, N.W. and Croughan, M.S. 2018. A simple method to reduce both lactic acid and ammonium production in industrial animal cell culture. Intl. J. Mol. Sci. 19(2).

Garcia-Ochoa, F., Santos, V.E. and Gomez, E. 2011. Stirred Tank Bioreactors. Comprehensive Biotechnology, Second Edition 2: 179–98.

Garg, S.K. and B.N. Johri. 1994. Rennet: current trends and future research. Food Reviews International 10(3): 313–355.

Gerbens-Leenes, P.W., Mekonnen, M.M. and Hoekstra, A.Y. 2013. The water footprint of poultry, pork and beef: a comparative study in different countries and production systems. Water Resources and Industry 1-2: 25–36. Doi.org/10.1016/j.wri.2013.03.001.

Gerber, P.J., Mottet, A., Opio, C.I., Falcucci, A. and Teillard, F. 2015. Environmental impacts of beef production: review of challenges and perspectives for durability. Meat Sci. 109: 2–12. doi: 10.1016/j.meatsci.2015.05.013.

Gershlak, J.R., Hernandez, S., Fontana, G., Perreault, L.R., Hansen, K.J., Larson, S.A. et al. 2017. Crossing kingdoms: Using decellularized plants as perfusable tissue engineering scaffolds. Biomaterials 125: 13–22.

Gholobova, D., Gerard, M., Decroix, L., Desender, L., Callewaert, N., Annaert, P. et al. 2018. Human tissue-engineered skeletal muscle: a novel 3D *in vitro* model for drug disposition and toxicity after intramuscular injection. Scientific Reports 8(1): 1–14.

Grellier, M., Granja, P.L., Fricain, J.C., Bidarra, S.J., Renard, M., Bareille, R. et al. 2009. The effect of the co-immobilization of human osteoprogrenitors and endothelial cells within alginate microscpheres on mineralization in a bone defect. Biomat. 30(19): 3271–3278.

Grossi, G., Goglio, P., Vitali, A. and Williams, A.G. 2019. Livestock and climate change: impact of livestock on climate and mitigation strategies. Animal Frontiers 9(1): 69–76. Doi.org/10.1093/af/vfy034.

Gstraunthaler, G.L.T. and van der Valk, J. 2013. A plea to reduce or replace fetal bovine serum in cell culture media. Cytotechnol. 65: 791–793.

Guhr, A., Kurts, A., Friedgen, K. and Loser, P. 2006. Current state of human embryonic stem cell research: an overview of cell lines and their usage in experimental work; Stem Cells 24: 2187–2191.

Gupta, P., Ismati, M.Z., Verma, P.J., Fouras, A., Jadhav, S., Bellara, J., et al. 2016. Optimization of agitation speed in spinner flasks for microcarrier structural integrity and expansion of induced pluripotent stem cells. Cytotechnol. 68(1): 45–59.

Gyles, D.A., Castro, L.D., Silva, J.O.C. and Ribeiro-Costa, R.M. 2017. A review of the designs and prominent biomedical advances of natural and synthetic hydrogel formulations. European Polymer Journal 88: 373–92.

Hanga, M.P., Ali, J., Moutsatsou, P., de la Raga, F., Hewitt, C.J., Nienow, A.W., et al. 2020. Bioprocess development for scalable production of cultivated meat. Biotech Bioeng; 117: 3029–3039.

Hanga, M.P., de la Raga, F.A., Moutsatsou, P., Hewitt, C.J., Nienow, A. and Wall, I. 2021. Scale-up of an intensified bioprocess for the expansion of bovine adipose-derived stem cells (bASCs) in stirred tank bioreactors. Biotech. Bioeng. 118(8): 3175–3186.

Haraguchi, Y. and Shimizu, T. 2021. Three-dimensional tissue fabrication system by co-culture of microalgae and animal cells for production of thicker and healthy cultured food. Biotechnol. Lett. 43(6): 1117–1129.

Heathman, T.R.J., Glyn, V.A.M., Picken, A., Rafiq, Q.A., Coopman, K., Nienow, A.W., et al. 2015. Expansion, harvest and cryopreservation of human mesenchymal stem cells in a serum-free microcarrier process. Biotechnol. Bioeng. 112: 1696–1707.

Hervy, M., Weber, J.L., Pecheul, M., Dolley-Sonneville, P., Henry, D., Zhou, Y., et al. 2014. Long term expansion of bone marrow-derived hMSCs on novel synthetic microcarriers in xeno-free, defined conditions. PLoS One 9: Article e92120.

Hewitt, C.J., Lee, K., Nienow, A.W., Thomas, R.J., Smith, M. and Thomas, C.R. 2011. Expansion of human mesenchymal stem cells on microcarriers. Biotechnol. Lett. 33: 2325–2335.

Hill, A.B.T., Bressan, F.F., Murphy, B.D. and Gracia, M.J. 2019. Applications of mesenchymal stem cell technology in bovine species. Stem Cell Res. & Therapy. 10, article number 44. Doi:10.1186/s13287-019-1145-9.

Ho, Y.Y., Lu, H.K., Lin, Z.F.S., Lin, H.W., Ho, Y.S. and Ng, S.K. 2021. Applications and analysis of hydrolysates in animal cell culture. Bioresour. Bioprocess 8(1): 93.

Hocquette, J.F. 2016. Is *in vitro* meat the solution for the future? Meat Sci. 120: 167–176.

Hosoyama, T.G., Meyer, M., Krakova, D. and Suzuki, M. 2013. Isolation and *in vitro* propagation of human skeletal muscle progenitor cells form fetal muscle. Cell Biol. Intl. 37: 191–196.

Hsiao, A.Y., Okitsu, T., Teramae, H. and Takeuchi, S. 2016. 3D Tissue formation of unilocular adipocytes in hydrogel microfibers. Adv. Healthc. Mater 5(5): 548–556.

Huang, Y., Song, C., Li, L. and Zhou, Y. 2014. The mechanism and performance of zeolites for ammonia removal in the zeolite packed electrolysis reactor. Electrochem. 82(7): 557–60.

Isern, J., Martín-Antonio, B., Ghazanfari, R., Martín, A.M., Lopez, J.A., Del Toro, R. et al. 2013. Self-renewing human bone marrow mesenspheres promote hematopoietic stem cell expansion. Cell Rep. 3: 1714–1724.

Jeswani, H.K., Figueroa-Torres, G. and Azapagic, A. 2021. The extent of food waste generation in the UK and its environmental impacts. Sustainable Production and Consumption 26: 532–47.

JIACRA III. 2012. Third joint inter-agency report on integrated analysis of consumption of antimicrobial agents and occurrence of antimicrobial resistance in bacteria from humans and food-producing animals in the EU/EEA, accessed on 27th of January 2022 at https://www.ecdc.europa.eu/sites/default/files/documents/JIACRA-III-Antimicrobial-Consumption-and-Resistance-in-Bacteria-from-Humans-and-Animals.pdf.

Jing, D., Parikh, A. and Tzanakakis, E.S. 2010. Cardiac cell generation from encapsulated embryonic stem cells in static and scalable culture systems. Cell Transplantation 19: 1397–1412.

Kalra, K., Banerjee, B., Weiss, K. and Morgan, C. 2019. Developing efficient bioreactor microcarrier cell culture system for large scale production of mesenchymal stem cells (MSCs). Cytotherapy 21(5): S73.

Kim, H., Bae, C., Kook, Y.-M., Koh, W.-G., Lee, K. and Park, M.H. 2019. Mesenchymal stem cell 3D encapsulation technologies for biomimetic microenvironment in tissue regeneration. Stem Cell Research & Therapy 10(1): 51. https://doi.org/10.1186/s13287-018-1130-8.

Kolkmann, A.M., Post, M.J., Rutjens, M.A.M., van Essen, A.L.M. and Moutsatsou, P. 2020. Serum-free media for the growth of primary bovine myoblasts. Cytotechnol. 72: 111–120.

Konakovsky, V., Clemens, C., Müller, M.M., Bechmann, J., Berger, M., Schlatter, S. et al. 2016. Metabolic control in mammalian fed-batch cell cultures for reduced lactic acid accumulation and improved process robustness. Bioengineering 3(1): 5.

Kovalik, J.P., Slentz, D., Stevens, R.D., Kraus, W.E., Houmard, J.A., Nicoll, J.B. et al. 2011. Metabolic remodeling of human skeletal myocytes by cocultured adipocytes depends on the lipolytic state of the system. Diabetes 60(7): 1882–1893.

Krause, R.M., Hamann, M., Bader, C.R., Liu, J.H., Baroffio, A. and Bernheim, L. 1995. Activation of nicotinic acetylcholine receptors increases the rate of fusion of cultured human myoblasts. The Journal of Physiology 489(Pt3): 779.

Kropp, C., Kempf, H., Halloin, C., Robles-Diaz, D., Franke, A., Scheper, T., et al. 2016. Impact of feeding strategies on the scalable expasion of human pluripotent stem cells in single use stirred tank bioreactor. Stem cell Transl Med. 5(10): 1289–1301.

Landers, T.F., Cohen, B., Wittum, T.E. and Larson, E.L. 2012. A review of antibiotic use in food animals: perspective, policy, and potential. Public Health. Rep. 127: 4–22.

Laperle, A., Masters, K.S. and Palecek, S.P. 2015. Influence of substrate composition on human embryonic stem cell differentiation and extracellular matrix production in embryoid bodies. Biotechnology Progress 31(1): 212–219. https://doi.org/https://doi.org/10.1002/btpr.2001.

Lawson, T., Kehoe, D.E., Schnitzler, A.C., Rapiejko, P.J., Der, K.A., Philbrick, K., et al. 2017. Process development for expansion of human mesenchymal stromal cells in a 50 L single-use stirred tank bioreactor. Biochem. Eng. J. 120: 49–62.

Li, B., Wang, X., Wang, Y., Gou, W., Yuan, X., Peng, J., et al. 2015. Past, present and future of microcarrier-based tissue engineering. J. Orthopaedic Transl. 3(2): 51–57.

Li, L., Qin, J., Feng, Q., Tang, H., Liu, R., Xu, L., et al. 2011. Heparin promotes suspension adaptation process of CHO-TS28 cells by eliminating cell aggregation. Mol. Biotechnol. 47(1): 9–17.

Lipsitz, Y.Y., Tonge, P.D. and Zandstra, P.W. 2018. Chemically controlled aggregation of pluripotent stem cells. Biotechnol. Bioeng. 115(8): 2061–2066.

Listrat, A., Lebret, B., Louveau, I., Astruc, T., Bonnet, M., Lefaucheur, L. et al. 2016. How muscle structure and composition influence meat and flesh quality. Sci. World J. 2016: 3182746. doi: 10.1155/2016/3182746.

Lu, Y., West, F.D., Jordan, B.J., Mumaw, J.L., Jordan, E.T., Gallegos-Cardenas, A., et al. 2012. Avian-induced pluripotent stem cells derived using human reprogramming factors. Stem Cell Dev. 21(3): 394–403.

Lu, Y., West, F.D., Jordan, B.J., Beckstead, R.B., Jordan, E.T. and Stice, S.L. 2015. Generation of avian induced pluripotent stem cells. Methods Mol. Biol. 1330: 89–99.

Luo, J., Vijayasankaran, N., Autsen, J., Santuray, R., Hudson, T., Amanullah, A. et al. 2012. Comparative metabolite analysis to understand lactate metabolism shift in Chinese hamster ovary cell culture process. Biotechnol. Bioeng. 109(1): 146–56.

Machalaba, C.C., Loh, E.H., Daszak, P. and Karesh, W.B. 2015. Emerging Diseases from Animals. In State of the World; Island Press.

Manibog, K., Sankar, K., Kim, S.A., Zhang, Y., Jernigan, R.L. and Sivasankar, S. 2016. Molecular determinants of cadherin ideal bond formation: conformation-dependent unbinding on a multidimensional landscape. Proceedings of the National Academy of Sciences of the United States of America 113(39): E5711–20.

Mattick, C.S., Landis, A.E., Allenby, B.R. and Genovese, N.J. 2015. Anticipatory life cycle analysis of *in vitro* biomass cultivation for cultured meat production in the United States. Environ. Sci. Technol. 49: 11941–11949.

Mattick, C.S. 2018. Cellular agriculture: The coming revolution in food production. Bull. At. Sci. 74(1): 32–35.

Meinert, C., Schrobback, K., Hutmacher, D.W. and Klein, T.J. 2017. A novel bioreactor system for biaxial mechanical loading enhances the properties of tissue-engineered human cartilage. Sci. Rep. 7: 1–14.

Mendly-Zambo, Z., Jordan Powell, L. and Newman, L.L. 2021. Dairy 3.0: cellular agriculture and the future of milk. Food, Culture and Society 24(5): 675–693.

Messmer, T., Klevernic, I., Furquim, C., Ovchinnikova, E., Dogan, A., Cruz, H., et al. 2022. A serum-free media formulation for cultured meat production supports bovine satellite cell differentiation in the absence of serum starvation. Nature Food; https://doi.org/10.1038/s43016-021-00419-1.

Meuwly, F., Ruffieux, P.-A., Kadouri, A. and von Stockar, U. 2007. Packed-bed bioreactors for mammalian cell culture: Bioprocess and biomedical applications. Biotechnol. Adv. 25(1): 45–56.

Milburn, J. 2018. Death-free dairy? The ethics of clean milk. J. Agric. Environ. Ethics 31(2): 261–279.

Miranda, J.P., Rodrigues, A., Tostoes, R.M., Leite, S., Zimmerman, H., Carrondo, M.J.T., et al. 2010. Extending hepatocyte functionality for drug-testing applications using high-viscosity alginate–encapsulated three-dimensional cultures in bioreactors. Tis. Eng – Part C; 16(6): 1223–1232.

Molnar, G., Schroed, N.A., Gonda, S.R. and Hartzell, C.R. 1997. Skeletal muscle satellite cells cultured in simulated microgravity. *In vitro* Cell & Dev. Biol. – Animal 33: 386–391. DOI:10.1007/s11626-997-0010-9.

Morgan, J.R., Yarmush, M.L., Vandenburgh, H., Shansky, I., Del Tatto, M. and Chromiak, J. 2003. Organogenesis of skeletal muscle in tissue culture. Tissue Eng. 217–226.

Mueller-Klieser, W. 1987. Multicellular spheroids a review on cellular aggregates in cancer research. J. Cancer Res. Clin. Oncol. 113: 101–22.

Nascar, S., Kumaran, V. and Basu, B. 2017. On the origin of shear stress induced myogenesis using PMMA based lab-on-chip. ACS Biomater Sci. Eng. 3: 1154–1371.

Nienow, A. 2010. Impeller selection for animal cell culture. Encyclopedia of Industrial Biotechnology: Bioprocess, Bioseparation and Cell Technology; edited by M C Flickinger. https://doi.org/10.1002/9780470054581.eib636.

Ng, J.Y., Chua, M.L., Zhang, C., Hong, S., Kumar, Y., Gokhale, R., and Ee, P.L.R. 2020. Chlorella vulgaris extract as a serum replacement that enhances mammalian cell growth and protein expression. Front. Bioeng. Biotechnol. 8: 1068.

Nienow, A.W., Hewitt, C.J., Heathman, T.R.J., Glyn, V.A.M., Fonte, G.N., Hanga, M.P., et al. 2016. Agitation conditions for the culture and detachment of hMSCs from microcarriers in multiple bioreactor platforms. Biochem. Eng. J. 108: 24–29.

Nikolova, M.P. and Chavali, M.S. 2019. Recent advances in biomaterials for 3D scaffolds: a review. Bioactive Materials 4: 271.

Nogueira, D.E.S., Rodrigues, C.A.V., Carvalho, M.S., Miranda, C.C., Hashimura, Y., Jung, S., et al. 2019. Strategies for the expansion of human induced pluripotent stem cells as aggregates in single-use Vertical-Wheel bioreactors. J. Biol. Eng. 13: 74.

OECD/FAO. 2016. MEAT. In: OECD-FAO Agricultural Outlook 2016–2025. Paris: OECD publishing.

Okamoto, Y., Haraguchi, Y., Sawamura, N., Asashi, T. and Shimizu, T. 2020. Mammalian cell cultivation using nutrients extracted from microalgae. Biotechnol. Progress 36(2): e2941.

Okamura, L.H., Cordero, P., Palomino, J., Parraguez, V.H., Torres, C.G. and Peralta, O.A. 2018. Myogenic differentiation potential of mesenchymal stem cells derived from fetal bovine bone marrow. Anim. Biotechnol. 29(1): 1–11.

Oliver, S.P., Murinda, S.E. and Jayarao, B.M. 2011. Impact of antibiotic use in adult dairy cows on antimicrobial resistance of veterinary and human pathogens: a comprehensive review. Foodborne Pathog. Dis. 8: 337–55. doi: 10.1089/fpd.2010.0730.

Park, Y., Chen, Y., Ordovas, L. and Verfaille, C.M. 2014. Hepatic differentiation of human embryonic stem cells on microcarriers. J. Biotechnol. 174: 39–48.

Pillai, V.V., Kei, T.G., Reddy, S.E., Das, M., Abratte, C., Cheong, S.H., et al. 2019. Induced pluripotent stem cells generation from bovine somatic cells indicates unmet needs for pluripotency sustenance. Anim. Sci. J. 90(9): 1149–1160.

Polanco, A., Kuang, B. and Yoon, S. 2020. Bioprocess technologies that preserve the quality of induced pluripotent stem cells. Trends in Biotechnol. 38(10): 1128–1140.

Post, M.J. 2014. Cultured beef: medical technology to produce food. J. Sci. Food Agric. 94(6): 1039–41.

Post, M.J. and Hocquette, J.F. 2017. New sources of animal proteins *in vitro* meat. pp. 425–441. *In*: Purslow, P.P. (ed). New Aspects of Meat Quality. Cambridge: Elsevier Ltd.

Post, M.J., Levenberg, S., Kaplan, D.L., Genovese, N., Fu, J., Bryant, C.J., et al. 2020. Scientific, sustainability and regulatory challenges of cultured meat. Nat Food; 1:403–415. https://doi.org/10.1038/s43016-020-0112-z

Rafiq, Q.A., Brosnan, K.M., Coopman, K., Nienow, A.W. and Hewitt, C.J. 2013. Culture of human mesenchymal stem cells on microcarriers in a 5L stirred-tank bioreactor. Biotechnol. Lett. 35: 1233–1245.

Rafiq, Q.A., Coopman, K., Nienow, A.W. and Hewitt, C.J. 2016. Systematic microcarrier screening and agitated culture conditions improves human mesenchymal stem cell yield in bioreactors. Biotechnol. J. 11: 473–486.

Rafiq, Q.A., Hanga, M.P., Heathman, T.J., Coopman, K., Nienow, A.W., Williams, D.J., et al. 2017. Process development of human multipotent stromal cell microcarrier culture using an automated high-throughput microbioreactor. Biotechnol. Bioeng. 114(10): 2253–2266.

Rafiq, Q.A., Ruck, S., Hanga, M.P., Heathman, T.J., Coopman, K., Nienow, A.W., et al. 2018. Qualitative and quantitative demonstration of bead-to-bead transfer with bone marrow-derived human mesenchymal stem cells on microcarriers: utilising the phenomenon to improve culture performance. Biochem. Eng. J. 135: 11–21.

Reiss, J., Robertson, S. and Suzuki, M. 2021. Cell sources for cultivated meat: applications and considerations throughout the production workflow. Int. J. Mol. Sci. 22: 7513.

Reubinoff, B.E., Pera, M.F., Fong, C.Y., Trounson, A. and Bongso, A. 2000. Embryonic stem cell lines from human blastocysts: somatic differentiation *in vitro*. Nat. Biotechnol. 18: 399–404.

Rischer, H., Szilvay, G.R. and Oksman-Caldentey, K.-M. 2020. Cellular agriculture—industrial biotechnology for food and materials. Current Opinion in Biotech. 61: 128–134.

Rodrigues, A.L., Rodrigues, C.A.V., Gomes, A.R., Vieira, S.F., Badenes, S.M., Diogo, M.M., et al. 2018. Dissolvable microcarriers allow scalable expansion and harvesting of human induced pluripotent stem cells under xeno-free conditions. Biotechnol. J. 14(4).

Rouwkema, J., Koopman, B.F., Blitterswijk, C.A.V., Dhert, W.J. and Malda, J. 2009. Supply of nutrients to cells in engineered tissues. Biotechnol. Genetic Eng. Rev. 26(1): 163–178.

Ryu, A.H., Eckalbar, W.L., Kreimer, A., Yosef, N. and Ahituv, N. 2017. Use antibiotics in cell culture with caution: genome-wide identification of antibiotic-induced changes in gene expression and regulation. Sci. Rep. 7: 7533.

Sart, S., Bejoy, J. and Li, Y. 2017. Characterisation of 3D pluripotent stem cell aggregates and the impact of their properties on bioprocessing. Proc. Biochem. 59(B): 276–288. Doi.org/10.1016/j.procbio.2016.05.024.

Schop, D., Janssen, F.W., van Rijn, D.S., Fernandes, H., Bloem, R.M., de Bruijn, J.D., et al. 2009. Growth, metabolism and growth inhibitors of mesenchymal stem cells. Tis. Eng. Part A. 15(8): 1877–1886.

Schop, D., van Dijkhuizen-Radersma, R., Borgart, E., Janssen, F.W., Rozemuller, H., Prins, H.-J., et al. 2010. Expansion of human mesenchymal stromal cells on microcarriers: growth and metabolism. J. Tis. Eng. Regen. Med. 4(2): 131–140.

Scott, M.A., Nguyen, V.T., Levi, B. and James, A.W. 2011. Current methods of adipogenic differentiation of mesenchymal stem cells. Stem Cell Dev. 20(10): 1793–1804.

Serra, M., Brito, C., Correia, C. and Alves, P.M. 2012. Process engineering of human pluripotent stem cells for clinical applications. Trends Biotechnol. 30(6): 350–359.

Silva, A.K.A., Richard, C., Bessodes, M., Scherman, D. and Merten, O.-W. 2009. Growth Factor Delivery Approaches in Hydrogels. Available from: http://pubs.acs.org.

Skubis, A., Gola, J., Sikora, B., Hybiak, J., Paul-Samojedny, M., Mazurek, U., et al. 2017. Impact of antibiotics on the proliferation and differentiation of human adipose-derived mesenchymal stem cells. Int. J. Mol. Sci. 18: 18.

Smetana, S., Mathys, A., Knoch, A. and Heinz, V. 2015. Meat alternatives: life cycle assessment of most known meat substitutes. Intl. J. Life Cycle Assess 20: 1254–1267.

Specht, E.A., Welch, D.R., Clayton, E.M.R., and Lagally, C.D. 2018. Opportunities for applying biomedical production and manufacturing methods to the development of the clean meat industry. Biochem. Eng. J. 132: 161–168.

Stephens, N., Di Silvio, L., Dunsford, I., Ellis, M., Glencross, A. and Sexton, A. 2018. Bringing cultured meat to market: technical, socio-political, and regulatory challenges in cellular agriculture. Trends in Food Sci & Techn. 78: 155–166.

Stephens, N. and Ellis, M. 2020. Cellular agriculture in the UK: a review. Wellcome Open Research; 5: 12.

Stern-Straeter, J., Bonaterra, G.A., Juritz, S., Birk, R., Goessler, U.R., Bieback, K. et al. 2014. Evaluation of the effects of different culture media on the myogenic differentiation potential of adipose tissue- or bone marrow-derived human mesenchymal stem cells. Int. J. Mol. Med. 33(1): 160–70.

Su, Y., Wang, L., Fan, Z., Liu, Y., Zhu, J., Kaback, D., Oudiz, J., et al. 2021. Establishment of bovine-induced pluripotent stem cells. Int. J. Mol. Sci. 22: 10489.

Szejda, K., Allen, M., Cull, A., Banisch, A., Stuckey, B., Dillard, C., et al. 2019. Meat Cultivation: Embracing the Science of Nature; The Good Food Institute: Washington, DC, USA.

Szejda, K., Bryant, C.J. and Urbanovich, T. 2021. US and UK consumer adoption of cultivated meat: a segmentation study. Foods 10: 1050.

Takahashi, I., Sato, K., Mera, H., Wakitani, S. and Takagi, M. 2017. Effects of agitation rate on aggregation during beads-to-beads subcultivation of microcarrier culture of human mesenchymal stem cells. Cytotechnol. 69(3): 503–509. doi: 10.1007/s10616-016-9999-5.

Takahashi, K. and Yamanaka, S. 2006. Induction of pluripotent stem cells from mouse embryonic and adult fibroblast cultures by defined factors. Cell 126: 663–676.

Takahashi, K. and Yamanaka, S. 2016. A decade of transcription factor-mediated reprogramming to pluripotency. Nat. Rev. Mol. Cell Biol. 17: 183–193.

Tavassoli, H., Alhosseini, S.N., Tay, A., Chan, P.P.Y., Weng Oh, S.K. and Warkiani, M.E. 2018. Large-scale production of stem cells utilizing microcarriers: a biomaterials engineering perspective from academic research to commercialized products. Biomaterials 181: 333–346.

Thakur, A., Sengupta, R., Matsui, H., Lillicrap, D., Jones, K. and Hortelano, G. 2010. Characterization of viability and proliferation of alginate-poly-L-lysine– alginate encapsulated myoblasts using flow cytometry. J. Biomed. Mater Res. B Appl. Biomater. 94(2): 296–304.

Thomson, J.A., Itskovitz-Eldor, J., Shapiro, S.S., Waknitz, M.A., Swiergiel, J.J., Marshall, V.S., et al. 1998. Embryonic stem cell lines derived from human blastocysts. Science 282: 1145–1147.

Tietze, S., Kräter, M., Jacobi, A., Taubenberger, A., Herbig, M., Wehner, R., et al. 2019. Spheroid culture of mesenchymal stromal cells results in morphorheological properties appropriate for improved microcirculation. Adv. Sci. (Weinh); 6(8): 1802104.

Toh, W.S., Lim, T.C., Kurisawa, M. and Spector, M. 2012. Modulation of mesenchymal stem cell chondrogenesis in a tunable hyaluronic acid hydrogel microenvironment. Biomaterials 33(15): 3835–45.

Torgan, C.E., Burge, S.S., Collinsworth, A.M., Truskey, G.A. and Kraus, W.E. 2000. Differentiation of mammalian skeletal muscle cells cultured on microcarrier beads in a rotating cell culture system. Med. Biol. Eng. Computing; 38: 583–590.

Tostoes, R.M., Leite, S.B., Miranda, J.P., Sousa, M., Wang, D.I.C., Carrondo, M.J.T., et al. 2010. Perfusion of 3D encapsulated hepatocytes – a synergistic effect enhancing long-term functionality in bioreactors. Biotech. Bioeng. 108(1): 41–49.

Tsukamoto, S., Shibasaki, A., Naka, A., Saito, H. and Iida, K. 2018. Lactate promotes myoblast differentiation and myotube hypertrophy via a pathway involving MyoD *in vitro* and enhances muscle regeneration *in vivo*. Int. J. Mol. Sci. 19(11): 3649.

Tuomisto, H.L. and de Mattos, M.J.T. 2011. Environmental impacts of cultured meat production. Environ. Sci. Technol. 45(14): 6117–6123. Doi.org/10.1021/es200130u.

Tuomisto, H.L., Scheelbeek, P.F., Chalabi, Z., Green, R., Smith, R.D., Haines, A., et al. 2017. Effects of environmental change on agriculture, nutrition and health: a framework with a focus on fruits and vegetables. Wellcome Open Research 2: 1–31.

United Nations. World Population Prospects 2019: Data Booklet. New York, NY: Department of Economic Social Affairs. p. 1–25.

Upside Foods. 2021. Animal Component Free: UPSIDE's Cell Feed Breakthrough Levels Up the Future of Cultivated Meat. Found at: https://upsidefoods.com/animal-component-free-upsides-cell-feed-breakthrough-levels-up-the-future-of-cultivated-meat/. Accessed on 27th of January 2022.

Van Wezel. 1967. Growth of cell strains and primary cells on microcarriers in homogeneous culture. Nature 216: 64–65.

Van Winkle, A.P., Gates, I.D. and Kallos, M.S. 2012. Mass transfer limitations in embryoid bodies during human embryonic stem cell differentiation. Cells, Tissues, Organs 196(1): 34–47. doi: 10.1159/000330691.

Vandenburgh, H.H. and Karlisch, P. 1989. Longitudinal growth of skeletal myotubes *in vitro* in a new horizontal mechanical cell stimulator. Vitr. Cell Dev. Biol. 25: 607–616.

Verbruggen, S., Luining, D., van Essen, A. and Post, M.J. 2018. Bovine myoblast cell production in a microcarriers-based system. Cytotechnology 70(2): 503–12.

Wang, Y. and Ouyang, F. 1999. Bead-to-bead transfer of Vero cells in microcarrier culture. Cytotechnol. 31(3): 221–224.

Warriss, P.D. and Rhodes, D.N. 1977. Haemoglobin concentrations in beef. J. Sci. Food Agric. 28: 931–934.

Washington, DC, USA, 2015; pp. 105–116.

Weber, C., Pohl, S., Poertner, R., Pino-Grace, P., Freimark, D., Wallrapp, C. et al. 2010. Production process for stem cell based therapeutic implants: expansion of the production cell line and cultivation of encapsulated cells. Adv. Biochem. Eng. Biotechnol. 123: 143–62.

Wei, Y., Li, Y., Chen, C., Stoelzel, K., Kaufmann, A.M. and Albers, A.E. 2011. Human skeletal muscle-derived stem cells retain stem cell properties after expansion in myosphere culture. Exp. Cell Res. 317: 1016–1027.

Wen, Y. and Yang, S.T. 2011. Microfibrous carriers for cell culture: a comparative study. Biotechnology Progress 27(4): 1126–36.

White, L.J., Taylor, A.J., Faulk, D.M., Keane, T.J., Saldin, L.T., Reing, J.E., et al. 2017. The impact of detergents on the tissue decellularization process: a ToF-SIMS study. Acta Biomater. 50: 207–219.

Wu, C.Y., Stoecklein, D., Kommajousula, A., Lin, J., Owsley, K., Ganapathysubramanian, B. et al. 2018. Shaped 3D microcarriers for adherent cell culture and analysis. Microsystems Nanoeng. 4: 21.

Wu, S.-C. 1999. Influence of hydrodynamic shear stress on microcarrier-attached cell growth: cell line dependency and surfactant protection. Bioprocess Eng. 21: 201–206.

Xie, A.W., Binder, B.Y.K., Khalil, A.S., Schmitt, S.K., Johnson, H.J., Zacharias, N.A., et al. 2017. Controlled self-assembly of stem cell aggregates instructs pluripotency and lineage bias. Sci. Rep. 7(1): 140.0. https://doi.org/10.1038/s41598-017-14325-9.

Xu, J., Yu, L., Guo, J., Xiang, J., Zheng, Z., Gao, D., et al. 2019. Generation of pig induced pluripotent stem cells using an extended pluripotent stem cell culture system. Stem Cell Res. Ther. 10: 193.

Yamamoto, Y., Ito, A., JItsunobu, H., Yamaguchi, K., Kawabe, Y., Mizumoto, H., et al. 2012. Hollow fibre bioreactor perfusion culture system for magnetic force-based skeletal muscle tissue engineering. J. Chem. Eng. Jpn. 45: 348–354.

Yang, J., Guertin, P., Jia, G., Lv, Z., Yang, H. and Ju, D. 2019. Large scale microcarrier culture of HEK293T cells and Vero cells in single-use bioreactors. AMB Express 9: 70.

Yoshihara, M., Hayashizaki, Y. and Murakawa, Y. 2017. Genomic Iinstability of iPSCs: challenges towards their clinical applications. Stem Cell Rev. and Rep. 13: 7–16.

Zhao, F., Chella, R. and Ma, T. 2007. Effects of shear stress on 3-D human mesenchymal stem cell construct development in a perfusion bioreactor system: experiments and hydrodynamic modeling. Biotech Bioeng. 96(3): 584–95.

Zhao, Y., Zeng, H., Nam, J. and Agarwal, S. 2009. Fabrication of skeletal muscle constructs by topographic activation of cell alignment. Biotechnol. Bioeng. 102(2): doi:10.1002/bit.22080.

Zhong, J.J. 2011. Bioreactor engineering. Comprehensive Biotechnol. (2nd Ed); 2: 165–177. https://doi.org/10.1016/B978-0-08-088504-9.00097-0.

Zwietering, T.N. 1958. Suspending of solid particles in liquid by agitators. Chem. Eng. Sci. 8: 244–253.

8

Insects as Food
A Sustainable Foodways

Neila Silvia Pereira dos Santos Richards

1. Introduction

Since the beginning of mankind, in the search for food for survival, insects have been part of animal (entomophagy) and human (anthropoentomophagy) food due to their abundance, availability, and ease of collection. Insects have always been used as food throughout human history and are one of the food sources that humans have relied on. Despite the absence of insects in the conventional western diet, they remain popular. Every primate is, to some degree, insectivorous (MacGrew 2014, Soagri et al. 2019, Hunter 2021).

Human evolution has caused man to go through adversity in the search for survival; so, as hunters, they began to collect edible plants, available or cultivated, and realized that if insects fed on the same plants, they could then be also a great source of food (Minas et al. 2016). Eating insects can be considered taboo, an ancient and exclusive tradition of indigenous peoples, and unnatural according to western countries' social precepts and beliefs (Choe and Hong 2018). Eggs, larvae, pupae, and adult insects have always been used by humans as food ingredients, and this trend has continued into modern times. Man was omnivorous in early development and ate insects extensively, and before humans had tools for hunting or agriculture, the insects constituted a critical component of the human diet. What is more, people mostly lived in warm regions where different types of insects were available throughout the year. Insects used to be a welcome source of protein in the absence of vertebrate meat (i.e., prey meat). Arthropods were an important food source for early hominids, especially in the subsistence of females and their offspring (Kourimská and Adàmkovà 2016, Meyer-Rochow and Jung 2020).

The appetite for insects seems to have been the main driver of our ancestors' discovery of tool use. Our closest primate relatives, chimpanzees, are famous for their ability to use tools, and many of them have been used to efficiently collect edible insects (Koops et al. 2015). There is evidence from early human sites that some bone tools were developed for termite extraction. In caves in the USA and Mexico, evidence of insect consumption in human history has been found by analyzing fossils. For example, coprolites found in the caves of Mexico include ants, beetle larvae, lice, ticks,

Federal University of Santa Maria, Av. Roraima, 1000 – Building 42, CCR/DTCA, Santa Maria-RS-Brazil, Postal Code: 97105-900.
Email: neilarichardsprof@gmail.com

and mites, suggesting that the human diet was primarily based on herbivorous animals, including insects, which accounted for 50% of our ancestors' diet. Further evidence has been found in paintings in the Altamira caves in northern Spain, dating back to 9,000–3,000 BC (Hunter 2021). Julie Lesnik found that termites were included in the diet of Plio-Pleistocene hominins (Sun-Waterhouse et al. 2016, Lesnik 2019). In Biblical writings, historical records have cited the 'manna' consumed by the Hebrews during the exodus, with insects responsible for feeding a nation; manna consisted of the secretion of the mealybug *Trabutina mannipara* (Santos and Florêncio 2013). Minas et al. (2016) reported that according to the culinary journey, which dates back to Biblical times, John the Baptist lived in the desert for months whilst living on locusts and honeycomb.

Records of the Aztec culture have unveiled 91 insect species in their diet, which were consumed roasted, fried, in sauces, boiled, or used as a condiment for some dish; some species were stored dry for later use. With the arrival of the Spanish conquistadors, cultural overlap occurred, and many of the insects used as food by the Aztecs were negatively qualified and thus forgotten and/or depreciated (Ramos-Elorduy and Pino 1996, Santos and Florêncio 2013, Soagri et al. 2019). The Romans and Greeks used to dine on beetle and locust larvae; Aristotle, a Greek scientist and philosopher, wrote about collecting cicadas and their use in food, demonstrating that entomophagy or anthropoentomophagy is a historical practice (Minas et al. 2016, Govorushko 2019).

Although little present in the western world, mainly because of prejudice, insect consumption is quite common in Asian and African cultures, in addition to being consumed as a food supplement or as the main constituent of the diet of different peoples in numerous regions of the world, including indigenous people (Kelemu et al. 2015, Gere et al. 2017). In some countries, local people have developed an intricate knowledge of the main species consumed, which is reflected in how insects are prepared, from collection to preparation as food. The intake of some species depends not only on taste and nutritional value but also on customs, ethnic preferences, religion, or prohibitions (e.g., vegan, kosher, and halal) and are an indispensable part of the diet of some ethnic groups (Kourimská and Adámková 2016, Patel et al. 2019).

In Mexico, it is still common to use some types of bed bugs as a condiment, albeit they produce unpleasant or repulsive odorous compounds. Bed bugs are sold alive and consumed roasted and ground, and sometimes, they are used together with pepper to season dishes. On the other hand, the Thais transform the praying mantis into a paste, whose flavor resembles that of shrimp pate with mushrooms; in China, it is believed that the consumption of ants has rejuvenating effects (Santos and Florêncio 2013, Grassi 2014). In central Australia, the aboriginal culture of the Anunta tribe represents six groups of insects through totems; the species they represent are prohibited as food, and feeding on one of these insects is the same as eating one's ancestors. The only member who can eat totemic insects is the head of each group (Grassi 2014).

In Latin America, insect consumption is also widespread among indigenous peoples, such as the Tukano Indians who inhabit the Colombian Amazon, consuming ants and termite soldiers as their only food of animal origin (Lesnik 2019). The culture of eating insects in Brazil has an indigenous origin, predating the Suruí Indians of Pará State, who consumed beetle larvae of the species *Pachymerus cardo*, *Caryobruchus* sp., *Rhynchophorum palmarum*, and *Rhina barbirostris* (Moraes and Fernandes 2018). The population of Pará popularly uses tucumã oil to treat joint pain; the oil is extracted from the beetle larva that grows inside the coconut of the tucumã palm. In general, in the Brazilian states that have the habit of eating insects, four main types of insects are part of the Brazilian diet: the tanajura (female saúvas ant, genus *Atta*), the taquara bug (the larvae of the *Morpheis smerintha* butterfly), palm worms (larva of the beetles *Rhynchophorus palmarum* and *Rhina barbirostris*), and the coconut worm (larva of the beetle *Pachymerus nucleorum*). These species have great nutritional value, and analysis of fried larvae of *R. palmarum* has shown 54.3% protein, 21.1% fat, and 12.7% moisture (Minas et al. 2016). In Pernambuco State (northeastern Brazil), many rural inhabitants enjoy the ovate abdomens of tanajuras with rice and beans, replacing the meat of domestic and wild animals with insects. In the tanajuras trade, the main customers are bar owners, who, during the swarming season, offer a menu of these insects in which they are presented as a delicacy to accompany the *cachaça* (a typical Brazilian alcoholic beverage). In northern and northeastern Brazil and Minas Gerais State (southeastern Brazil), *farofa* (seasoned cassava flour) with tanajura ant (içá, bitú, winged ant or countryside side) is common. Saúva (*Atta cephalotes*) is present in various traditional dishes throughout Brazil, and popular beliefs state that eating tanajura on Santa Luzia's day, on 13 December, is good for the eyes (Costa-Neto 2013, 2015, Minas et al. 2016).

Brazil is recognized globally for its status as a biodiversity hotspot, having a rich cultural diversity, or biosocial diversity, with 222 indigenous ethnic groups as well as other groups, including artisanal fishermen, Amazon *caboclos* (or river-dwellers), and Afro-Brazilians, which are also known as *quilombolas*, among others. A total of 135 edible insect species belonging to nine orders, and 23 families in 14 of Brazil's 26 states, and 95 of them have been identified to the species level and 18 to the genus level, while others are known only by their native names. In Brazil, edible insects are abundantly available, and the most consumed species are in Hymenoptera (63%), Coleoptera with 22 species (16%), and Orthoptera with 9 species (7%) (Costa-Neto 2013, FAO 2013a, Minas et al. 2016). Nevertheless, the exact number of species of edible insects is undervalued. People consume edible species depending on their presence, abundance, and availability (Costa-Neto 2013). In Thailand, Korea, Japan, Botswana, and Mexico, edible insects are sought-after delicacies in local markets and restaurants (Mitsuhashi 2017).

Some insects play positive roles in waste biodegradation, plant pollination, and agricultural pest control and provide valuable products such as honey, wax, silk, food dyes, shellac, etc. Examples include ants, bees, beetle larvae, termites, silkworms, mealybugs, praying mantis, etc., while others are considered pests and can cause considerable damage to plants, animals, and humans, such as maggot larvae, locusts, aphids, whitefly, among others (Sun-Waterhouse et al. 2016, Grassi 2014, Gravel and Doyen 2020). Western cultures have begun to realize that edible insects represent a rich source of biodiversity worldwide, and new organoleptic experiments are being discovered (Fig. 1) (Minas et al. 2016).

In eastern countries, insects are valued as important and delicate foods due to their recognized nutritional and health properties and the low availability of conventional meat sources (including beef, fish, and chicken) (Kim et al. 2019, Lange and Nakamura 2021). Currently, insects are estimated to play a crucial role in the diet of at least two billion people in 140 countries, most of whom are in the southern hemisphere (Acosta-Estrada et al. 2021, Liceaga 2021, FAO 2021a, b).

Edible insects are potential sources of alternative proteins, as they contain essential nutrients such as proteins of high biological value, vitamins, minerals, and fatty acids. They can be used as

Figure 1. Traditional tropical salad in Mexico. Source: Author (personal file).

a food ingredient to increase the nutritional value of different meals worldwide (Acosta-Estrada et al. 2021), contributing to fulfilling Sustainable Development Goals (SDG) 2, 3, and especially 12, since their creation has a low impact on the environment (Schardong et al. 2019). Considering the growing demand for animal protein production due to the world population growth, edible insects are beginning to be considered sustainable protein sources (Sun-Waterhouse et al. 2016).

2. Food Demand and Population

According to the Food and Agriculture Organization of the United Nations (FAO), one in nine people in the world (about 825 million people) do not have enough food to lead a healthy and active life, especially in China and India (FAO 2017, 2018, 2020, 2021a, Raheem et al. 2019). Food insecurity comes from the impossibility of the poorest classes not having access to the food needed for a healthy and balanced diet (Vilela and Pulrolnik 2015, Van Huis et al. 2021). This picture has worsened with the current variability and extreme weather conditions and spread of the COVID-19 pandemic, and it will worsen with conflicts and violence in some parts of the world, including the war between Russia and Ukraine.

It is estimated that the world population is expected to surpass 9.7 billion inhabitants by 2050, reaching 11 billion in 2100, and the demand for animal protein is expected to increase at astounding rates. Even if current trends in productivity growth continue over the next few decades, it will not be enough to meet the demands of a population that grows in wealth at the same time as it grows in numbers (Meyer-Rochow and Jung 2020, FAO 2021b, Ojha et al. 2021); therefore, issues such as production capacity, demand, and resource depletion cannot be ignored (Mishyna et al. 2021).

Nonetheless, population growth, the increase in per capita consumption, the expansion of cities in the coming decades, and the inability to meet the new human needs in terms of the content of proteins, especially those of animal origin, and restrictions on the productivity, primarily cultivable area, exacerbate the issues regarding the inability to supply food in general (Saath and Fachinello 2018, Hunter 2021). In some countries, although there are areas available for agricultural expansion (e.g., Latin America and Sub-Saharan Africa), the demand must be met with increased productivity and/or crop replacement or even adjusting and reassessing eating habits (Orsi et al. 2019).

Countries with the least suitable agricultural land are likely to be the hardest hit by rising food demand. Populations that have already been struggling to obtain food will have an even worse situation in the coming decades, where 75% of their income will be used for food (Ojha et al. 2021). On the other hand, people in developed countries consume about 95 g of protein per day, of which almost 60% is of animal origin; this is a highly concerning piece of evidence, given that the protein intake in developing countries is only 45 g, 15% of which is an animal protein (Abbasi et al. 2016, Lange and Nakamura 2021).

The more significant use of alternative sources of proteins, such as those of plant origin, has been suggested since these products are linked to bioactive compounds that promote human health. However, vegetables pose several limitations, including anti-nutritional factors, amino acid imbalances, and high levels of fiber and non-starch polysaccharides (Mutungi et al. 2019, Hlongwane et al. 2020, Meyer-Rochow and Jung 2020). Food must, in addition to nourish, provide health benefits, and these benefits need to be associated with the presence of bioactive compounds which, when present or incorporated into food, positively affect various biological functions of the body. As alternative protein sources, insects are a potential substrate for obtaining these compounds, especially bioactive peptides (Matos and Castro 2021, Lucas et al. 2021).

In the last 40 years, technological advances and deforestation have allowed food production to grow in tandem with population growth (Saath and Fachinello 2018). However, the overexploitation of land, energy, and water will not only destabilize our ability to produce food but also the impact that the food system has on the environment. Additionally, increasing urbanization rates over time have led to reduced insect numbers due to habitat reduction and damage (Van Huis et al. 2013, Lucas et al. 2020).

One of the biggest threats we face today is climate change, which is exacerbated by excessive deforestation and greenhouse gas (GHG) emissions, and it concerns how adaptation measures could affect the food system (Winsemius et al. 2015, Baiano 2020). In the coming decades, there will be a gradual increase in the demand for high-quality animal protein, and it will be estimated at 72% in meat production in the next 30 years; one of the requirements is that this protein source is of better quality than those currently available. Vegetable products often contain less than 14% protein, whereas some types of edible insects have up to 75% high-quality animal protein, thus being better sources of essential amino acids, interesting amounts of fatty acids such as omega 3 and 6, and other specific nutrients that help in the complex functioning of the human body. When properly produced, insects can be from natural sources, pesticide-free, and locally produced using land more efficiently for food production and, above all, respecting the environment (Dossey et al. 2016, Haloran et al. 2018, Hunter 2021, Van Huis et al. 2021, FAO 2021b, Ojha et al. 2021).

Henchion et al. (2017) estimated that 33% of the proteins consumed in the coming decades will come from alternative sources by 2054, reaching 311 million metric tons of alternative products, and insects will have an 11% share of this market (34 metric tons).

Insects can transform organic matter into high-quality protein and other nutrients more efficiently than conventionally used animals; therefore, they are associated with a small ecological footprint. Insects produce lower levels of GHGs such as methane, carbon dioxide, and nitrous oxide than animals raised for meat production, suggesting they are a greener alternative to animal protein production in terms of these emissions (Abbasi et al. 2016, Gravel and Doyen 2020, Acosta-Estrada et al. 2021). Moruzzo et al. (2021) schematized (Fig. 2) insect farming as a production sector that could contribute to achieving several SDGs.

In western countries, there is a perception that insects are only consumed by people in extreme poverty, which has created an embarrassing factor and caused people who consume insects in tropical countries to not report their preferences. Since 2003, FAO has been working on issues related to food insects in many countries of the world in the following thematic areas: generating and sharing knowledge through publications, expert meetings, and a web portal on edible insects; awareness of the role of insects through the media collaboration (e.g., newspapers, magazines, and TV); providing support to member countries through the field projects; multidisciplinary networks

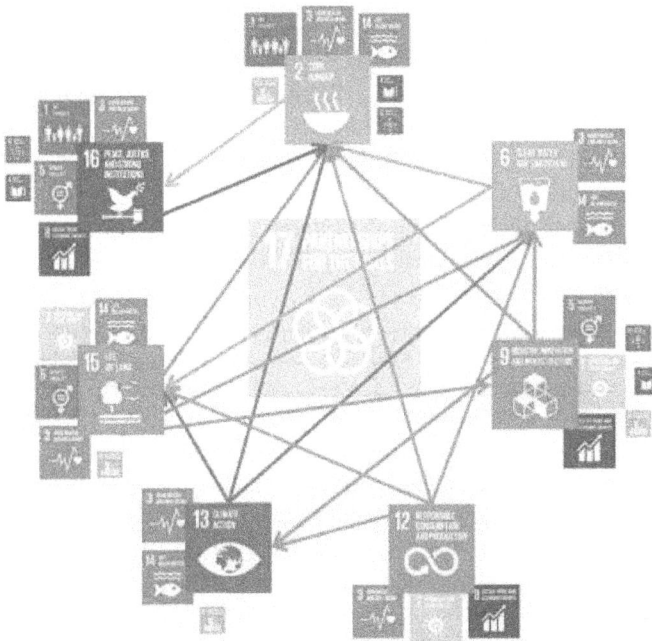

Figure 2. Sustainable Development Goals (SDGs) and links to edible insect farming. Source: Moruzzo et al. (2021).

and interactions (e.g., issues related to nutrition, food, and legislation) with various sectors inside and outside FAO (Van Huis and Dunkel 2017).

3. Food Conversion and Human Nutrition

The feed conversion efficiency for many insect species is much higher than meat from beef animals (cattle, swine, and poultry). Because they are cold-blooded, they have a high feed conversion rate and are efficient in the biotransformation of matter into biomass (Orsi et al. 2019). The nutritional profile of different insect species is variable, with an insect body containing up to 80% protein, which is attributed to controlling the body temperature due to the environment (Orsi et al. 2019, Nowakowski et al. 2021).

Furthermore, 1.7 kg of feed are needed to produce 1 kg of domestic crickets with nutritional proportions superior to beef and with a production cost below that required for conventional meat, which demand 10, 5, and 2.5 kg of food to produce 1 kg of beef, pork, and chicken, respectively. Compared to planting grains to form rations, insect production is 50% higher per hectare (Minas et al. 2016, Raheem et al. 2019, Reis and Dias 2020). Regarding the water footprint, insect production is about 4,341 m^3 per ton, especially for the larvae of *Tenebrio Molitor*, while for cattle, pigs, and chickens, the requirements are 154,115, 5,988, and 4,325 m^3 per ton, respectively (Miglietta et al. 2015).

The area needed to produce the same amount of protein is about 1 ha for mealworms, 3.5 ha for pigs or chickens, and 10 ha for cattle; to produce 1 kg of beef, emissions are equivalent to 14.8 kg of CO_2, and for pigs and chickens, 3.8 and 1.1 kg of CO_2 is necessary. Crickets, larvae, and grasshoppers emit 100 times less GHG emissions and 10 times less ammonia than most beef animals (cattle and pigs) (Dobermann et al. 2017, Reheem et al. 2019, Imathiu 2020).

Roughly one million insect species are known, although it is estimated that there are around 80 million, presenting different stages of development and, consequently, different amino acid and fatty acid profiles (FAO 2019, 2020, 2021a, Reis and Dias 2020). The largest group of edible insects are Coleoptera, followed by Hymenoptera, Orthoptera, and Lepidoptera. The most popular insects for human consumption are beetles (Coleoptera, 31%), caterpillars (Lepidoptera, 18%), bees, wasps, and ants (Hymenoptera, 14%), grasshoppers and crickets (Orthoptera, 13%), cicadas, leafhoppers, mealybugs, and bed bugs (Hemiptera, 10%), termites (Isoptera, 3%), dragonflies (Odonata, 3%), flies (Diptera, 2%), and other orders (5%) (FAO 2021b). Of these species mentioned, 10% are from distinct parts of the world, and the rest are restricted to some geographic zones, of which 12% are aquatic species and 78% terrestrial (Minas et al. 2016, Lange and Nakamura 2021, FAO 2021b).

Today, the most significant consumption of insects is in Africa, Asia, Australia, and Latin America, and they are a typical part of the human diet in some regions with culturally and gastronomically important aspects (Yen 2015, Costa-Neto 2016, Kim et al. 2019). As being considered culturally inappropriate or even taboo, in most European countries, the human consumption of insects is very low, and entomophagy is often considered distasteful (Garofalo et al. 2019, Ordoñez-Araque and Egas-Montenegro 2021).

There is great potential for insect meals in animal feed, mainly because they are an excellent source of protein and present an adequate profile of amino acids depending on the species, diet, and stage of development. The protein concentration of the main insects used in animal feed varies from 46 to 65%, higher than the values found in beans (23.5%), lentils (26.7%), and soybeans (41.1%) (Reis and Dias 2020, Gravel and Doyen 2020). In addition to being a good source of protein of high biological value, insect meals are also good sources of lipids. However, the fat and energy content vary considerably between species and stages of development, ranging from 7 to 77% in dry matter, while energy can vary from 2930 to 7620 kcal per kg of dry matter. Crickets have about 13% lipids, while beetle larvae have 50% dry matter and a low cholesterol concentration compared to products of animal origin, having a favorable proportion of omega 6 and 3 fatty acids and polyunsaturated and saturated fatty acids. Fibers and minerals, such as copper, iron, magnesium, phosphorus, selenium,

Table 1. Nutritional analysis of live and dehydrated edible insects. Source: Schickler (2013).

Insects		Protein (g per 100 g)	Fibers (g per 100 g)	Ashes (g per 100 g)	Calcium (mg per 100 g)	Phoshorus (mg per 100 g)
Alive	Commom Tenébrio	18.72	7.61	1.03	0.05	0.28
	Giant Tenébrio	17.24	7.42	1.04	0.07	0.21
	Black cricket	16.47	10.59	1.42	0.15	0.23
	Cockroach cinerea	20.17	15.58	1.40	0.14	0.25
	Housefly	13.92	8.61	1.08	0.09	0.24
Dehydrated	Commom Tenébrio	47.41	6.45	3.20	0.07	0.52
	Giant Tenébrio	44.03	8.00	3.21	0.12	0.53
	Black cricket	48.76	7.99	4.43	0.19	0.66
	Cockroach cinerea	60.39	21.48	6.92	5.16	0.13
	Housefly	50.25	6.93	4.65	0.39	1.05

and zinc, are present in high concentrations in insect meals, although it is necessary to pay attention to the variation of these minerals in different species (Reis and Dias 2020, Lange and Nakamura 2021).

Nutrient compositions vary with the species, diet, developmental stage, sex, and growth environments of insects, such as temperature, day length, humidity, light intensity, and spectral composition (incident light, reflectance, and spectral transmittance) (Zielinska et al. 2015, Barroso et al. 2017). There are nutrient variations depending on the type of processing, if the insects are alive, dehydrated, cooked, etc., as shown in Table 1.

Humans already consume insects as part of their diet, often without realizing it. For instance, according to the US Food and Drug Administration, an average of 75 or more insect fragments per 50 g are allowed in wheat flour as a defect (AOAC 972.32) (Castro and Chambers 2018).

Studies have shown that the "meat" of insects contains satisfactory amounts of proteins and lipids and is rich in mineral salts, including iron, phosphorus, magnesium, manganese, selenium, zinc, vitamins (A, B1, B2, B5, B6, B7, B9, D, E, and K), carotenoids, and flavonoids (apigenin, quercetin, luteolin, and rutin) and are easily digestible (Belluco et al. 2017, Castro et al. 2018, Hunter 2021, Ojha et al. 2021). For example, the tanajura ant has more protein (42.59%) than chicken (23%) or beef (20%) (Costa-Neto 2013, Fogang et al. 2017). In addition, the composition contains large amounts of fiber in the form of insoluble chitin and important nutrients for the correct functioning of the human body. The chitin is the second most important natural polymer in the world and is generally extracted from marine crustaceans and shrimp, although it is also present in the exoskeleton of many insects (Choi et al. 2016). When partially deacetylated, chitin gives rise to chitosan, a natural polysaccharide used as a pharmaceutical (anti-inflammatory) and food ingredient, which is attributed to a series of benefits, such as being hypocholesterolemic, antimicrobial, antithrombotic, anticoagulant, and with healing properties (Zhang et al. 2014).

Insects have nutritional benefits, including an abundance of proteins, and may contain double or triple protein (> 40%) compared to conventional sources of animal origin. This makes it a good substitute for other animal proteins (Rumpold and Schlüter 2013, Lucas et al. 2020). The protein content of bees, crickets, silkworms, mopane caterpillars that live in Zimbabwe, and palm weevils is higher than beef, chicken, and pork, as demonstrated in Table 2. The mopane caterpillar is traditionally boiled in salted water and dried in the sun; it can last for several months without refrigeration and is important in food shortages (Halloran et al. 2018).

A serving of locusts provides similar amounts of vitamin D as herrings, cooked chicken liver, and egg yolks. Silkworm larvae (*Bombyx mori* L.) feed on mulberry leaves and lettuce and, when processed (ground and freeze-dried), can contain almost 10 mg of vitamin E in 100 g (Tong et al. 2011). Fat is the second-largest component of insects, and they have an unsaturated fatty acid profile similar to poultry and white fish while containing more polyunsaturated fatty acids than poultry or red meat. The fat profile of insects depends on their diets and can range from 7 to 77 g per 100 g of dry weight, which is higher in the larval stages than in adults (Kourimska and Adámkvá 2016,

Table 2. Nutritional value of conventional beef animals and edible insects. Source: Halloran et al. (2018), Hunter (2021), Ojha et al. (2021) (adapted).

Source	kcal per 100 g	Protein (g per 100 g)	Fat (g per 100 g)	Iron (mg per 100 g)	Calcium (mg per 100 g)
Beef	176	20	10.0	1.95	5.0
Chicken	120	22.5	2.62	0.88	8.0
Pork	142	19.8	6.34	0.80	7.0
Cricket	153	25.1	5.06	5.46	104.0
Bee	499	21.0	3.64	18.5	30.0
Silkworm	128	55.6	8.26	1.80	42.0
Mopane caterpillar	409	80.0	15.2	31.0	700.0
Palm weevil	479	42.0	25.3	2.58	39.6
Flour larva	247	19.4	12.3	1.87	42.9

Dobermann et al. 2017). For example, one study showed that eicosapentaenoic acid and docosahexaenoic acid levels could be increased in black soldier flies by feeding them fish offal (Reis and Dias 2020).

4. Consumption Safety

Not all insects are safe to eat, as certain vegetables are. When selecting edible insects, monitoring microbial contamination is the first concern, as microbiological, chemical, physical, toxicological, and allergenic risks must be considered (Sun-Waterhouse et al. 2016, Mishyna and Glumac 2021). The bioavailability and bioaccessibility of insect nutrients are poorly researched topics, and most studies have addressed nutrient content and characterization, although just as any other food, the type of preparation and processing significantly affects the nutrient content that will or will not be bioaccessible (Parada and Aguilera 2007, Megido et al. 2018).

The presence of anti-nutritional components in some insects, such as tannins and phenolic acids, may hinder the bioavailability of some fractions of the compounds (Cardoso et al. 2015). In general, tannins in animal feeds are associated with reduced consumption, growth rate, feed efficiency, metabolizable energy, and protein digestibility, as they form insoluble complexes with protein, thus interfering with bioavailability (Hamerman 2016, Mutungi et al. 2019). In animals, the bioaccessibility and bioavailability of compounds present in insects are variable; however, as in the case of conventional foods, thermal processing can modify the food matrix and increase or decrease the bioaccessibility and bioavailability of nutrients (Barba et al. 2017, Manditsera et al. 2019).

There are natural variations in protein levels and amino acid profiles among different insect species, and the levels of these components are sensitive to preparation methods, including cooking, boiling, frying, and drying (static, lyophilization, air circulation, vacuum, and microwaves). Minerals are less sensitive to processing conditions but can be affected by cooking style as they can be removed during boiling (Zielinska et al. 2015, Manditsera et al. 2019).

The digestibility of edible insect proteins is affected by the preparation and cooking style. For instance, some studies have reported that the roasting and drying of locusts reduces the digestibility of the protein, which does not occur when this process is applied to dehydrated termites. When boiled, Tenebrio increases protein digestibility (Gravel and Diyen 2020, Liceaga 2021).

Some insects naturally have repellents or toxic chemicals, while others contain cross-reacting allergens that can induce allergic reactions similar to those associated with shellfish or molluscs (Barre et al. 2014, Lange and Nakamura 2021). The allergenicity of edible insects may resemble that of shellfish, and the allergenic potential of insects may be associated with so-called arthropod and invertebrate allergy (caused by components such as the hemolymph, cuticle, and parts including exuvia, hair, bristles, and scales) (Broekman et al. 2015, Raheem et al. 2019). Edible insects are somewhat similar in composition to certain seafood and nuts, meaning they are rich in protein, unsaturated fats, fiber, vitamins, minerals, and sterols. The occurrence of allergenic proteins in significant amounts and certain interacting components (e.g., lipid fraction and ions such as Cu2+) can trigger an allergic response in humans (Sun-Waterhouse et al. 2016).

People allergic to shrimp are likely to have allergic reactions to edible insects such as crickets (*Gryllus bimaculatus*), as these insects may contain specific allergens such as hexamerin1B (HEX1B) or even the allergens commonly found in crustaceans (e.g., arginine kinase). The genotoxic and oxidative potentials of edible insects must be evaluated along with analyses of allergenic and toxicogenic substances, elements, and other chemical species (Srinroch et al. 2015, Pyo et al. 2020).

The risk of allergies caused by edible insects varies considerably due to large differences in their biological and ecological characteristics and individual cultivation and processing practices (Broekman et al. 2015). Reverberi (2021) stated that, despite the widespread use of crickets as human food, cases of food allergy are relatively rare. There is a lower risk of transmitting zoonotic infections such as H1N1 and BSE (mad cow disease) to humans in captive breeding compared to beef animals and birds (Raheem et al. 2019). Most species can be eaten whole, while others, such as locusts and grasshoppers, require the legs and wings to be removed; many are processed into powder or paste form. There are still some exceptions involving insects bred mainly to obtain products of commercial value (e.g., silk and honey). Even though some of the common insect species consumed worldwide are known, there is no way to estimate the consumed insects' content (FAO 2021a, Liceaga 2021).

According to Van Huis (2016), food safety is of utmost importance when dealing with new food sources. In the context of edible insects, there are four ways through which food safety risks can arise: (1) the insect itself could be toxic; (2) the insect could have acquired toxic substances or human pathogens from its environment during its life cycle; (3) the insect could become spoiled after harvest; (4) consumers could experience an allergic reaction to the insect.

5. Production and Processing

Insect production can play a significant role in the future, particularly when the circular economy concept is applied, as insects can be raised in secondary organic streams by recycling and cost-effectively transforming agricultural and industrial by-products and/or waste into proteins, promoting food security and minimizing climate change and biodiversity loss, thus contributing to the SDGs (Orsi et al. 2019, Moruzzo et al. 2021).

Insects can be bred to provide an indirect source of food for humans; for example, in the form of food for animals such as pork, poultry, and fish, or even to support the growth of other foods such as *Ophiocordyceps Sinensis* (medicinal mushroom also known as the 'fungus caterpillar' that parasitizes ghost moth larvae through germination in the live larva) (Nowakowski et al. 2021).

Various researchers have referred to the creation of insects as 'mini livestock' because the insects have a more sustainable creation to the animal protein produced and are suitable for industrial production because they have shorter cycles; high rates of reproduction and survival in different locations, both in natural and artificial environments; adulthood reached in days; high density growth (animal welfare); greater efficiency in nutrient conversion (12 to 15 times more efficient than conventional animals); greater use of meat; less need for food and water; use of agricultural waste such as slaughterhouse blood and food waste; less land use; lower GHG emissions (up to 100 times); smaller carbon footprint; low environmental contamination; reduced water pollution (mainly in use for slaughter and processing); lower methane production (only cockroaches, termites, and beetles produce CH_4—a result of bacterial fermentation by Methanobacteriaceae in the upper intestine); faster slaughter and simpler procedures; high profitability and low investment in the implementation of creation (Abbasi et al. 2016, Sun-Waterhaouse et al. 2016, House 2018, Orsi et al. 2019, Patel et al. 2019, Raheem et al. 2019, FAO 2021b, Hunter 2021, Lannag 2021, Pedroso et al. 2021). Sustainable insect breeding can also contribute to using marine resources to reduce overfishing by replacing oil and fish meal in animal feed (Moruzzo et al. 2021).

A key point to be observed in breeding is the collection of insects, and they can be acquired in three ways: wild insect collection, partial domestication, and industrial farming. Currently, collection in nature corresponds to 92% and industrial creation to only 2% (Skotnicka et al. 2021). Insects collected in nature are more susceptible to contamination and may even have intrinsic toxicity, allergenicity, anti-nutritional factors, enzymes, phenolic compounds, and steroids (as part of their defense mechanisms against enemies/predators); physical hazards such as glass, stones, plastic, and metal fragments; chemical hazards that may be naturally present (scombrotoxin, ciguatoxin, tetrodotoxin, mycotoxin, pyrrolizidine, alkaloids, neurotoxins, etc.), synthesized, accumulated, added, transmitted, or leached during their lifetime in the habitat (e.g., allergens, sulfites, nitrites/ nitrates, niacin, pesticides, fertilizers, fungicides, insecticides, antibiotics, growth hormones, lubricants, cleaning products, disinfectants, paint, mercury, copper, lead, and other heavy metals) (Broekman et al. 2015, Sun-Waterhouse et al. 2016).

Insects may contain substances that interfere with nutrient absorption, including thiaminase enzyme, and there may still be different pollutants or contaminants depending on where they were collected, such as aflatoxin, which is carcinogenic, can contaminate edible bed bugs if they are from wooden baskets contaminated with manure or jute bags previously used to store cereals (Musundire et al. 2016).

Preservation and processing techniques are necessary to increase shelf life, preserve quality, and increase the acceptability of food products containing insects; processing procedures are also required to turn insects into protein flour to be used as ingredients in the food industry. Processing can change the nutritional value of insects and improve microbiological safety (Tzompa-Sosa-Fogliano 2017, Melgar-Lalanne et al. 2019).

The processing (thermal or otherwise) can improve the quality (e.g., reduce the allergenic potential of the proteins of insects) and the safety, flavor, and shelf life of the edible insects. However, the species must be observed since uncontrolled processing practices can harm nutritional composition, sensory quality, and the bioaccessibility of nutrients in the human body (Homann et al. 2017, Mutungi et al. 2019, Megido et al. 2018).

Chemical and enzymatic reactions and microbial growth depend on the moisture and water activity of the food; therefore, reducing these parameters will decrease microbial growth and reaction rates, in addition to increasing the shelf life of the product. The main methods used in insect processing are heat treatments with high temperatures (bleaching, dehydration, etc.), low temperatures (chilling and freezing), and fermentation (Liceaga 2021).

Nanotechnology can be used to manipulate the structure of insect proteins to achieve health outcomes and manipulate the processing functionalities of food ingredients and materials to improve nutrient and compound efficiency (Sun-Waterhouse et al. 2006).

6. Food Habits and Consumers

COVID-19 has changed lifestyles, some eating habits, social life, and negatively influenced the world economy and worsened cases of depression, overload, fear, and anguish. Healthy habits are developed in childhood and may persist into adulthood (Al-Awwad et al. 2021, Shaun et al. 2021). Eating habits are normally influenced by personal, cultural, and economic circumstances, as well as internal and external political factors. Ensuring nutritious food for the world's growing population is one of the biggest challenges of this millennium (FAO 2017, 2018, 2019, 2020, 2021a).

Insects can be efficiently cultivated in urban environments. More than 2,140 species are already part of the human diet of up to 3,000 ethnic groups in over 140 countries; 92% of known insects are wild harvested, 6% are semi-domesticated, and 2% are farmed. Among the known wild-harvested edible insect species, 88% are terrestrial in origin, and the remaining are collected from aquatic ecosystems (Jongema 2017). With species diversity, all countries can find edible insects on their land (FAO 2021b, Hunter 2021).

History has shown us that civilizations and peoples who have gone through some food deprivation at some point are those with the greatest spectrum of potential foods that they use in their diets. That is, they are civilizations/peoples that had a greater need to taste new sources of food (Imathiu 2020, Lange and Nakamura 2021).

One example is Mexico, which has many reports in the literature on edible insects, with over 500 species having been recorded. In the case of Brazil, despite little disclosure to the general public, around 100 species of insects are already used as food (Lesnik 2019). The production of insects for human consumption on a large scale is a relatively new concept in society, and one of the main challenges is popularizing this alternative protein source (Patel et al. 2019). Various insects are already part of the cuisine in other countries and even in Brazil, such as the tenébrio (flour beetle), whose larva is used both for feeding reptiles and human consumption, in addition to crickets and some species of cockroaches that are produced for these purposes. In Brazil, despite the National Health Surveillance Agency (ANVISA) not having yet regulated the use of edible insects for human consumption, there are insect breeders spread throughout the country who, artisanally, use insects to produce animal feed and even make artisanal chocolates and cookies for human consumption (Schardong et al. 2019).

The Resolution of the Collegiate Board (RDC) No. 14 of 2014 (ANVISA) establishes tolerable limits of 'foreign matter' in foods indicative of risks to human health and/or indicative of failures in applying good practices in the food and beverage production chain. The RDC lists insects considered part of the food production process (food and beverages), establishing the limits. Insect fragments can only be considered failures in the production process, involving everything from the harvest of the food to the final packaged product (Brasil 2014).

There are no regulatory barriers to their production, marketing, and consumption for countries where eating insects is traditional. Nevertheless, in western countries, regulations represent a significant barrier to the use of insects in feed and food. The European Food Safety Authority (EFSA) has declared that all insect products for human consumption will be considered as 'novel food' and must undergo Novel Food approval, with a 2-year transition period (Dobermann et al. 2017).

In 2021, the EFSA issued the first positive opinion (EU Regulation 2015/2283) for mealworms (EFSA-Q-2018-00262) to be marketed for human consumption of food based on insects. Mealworm larvae can be used as a whole, dried as snack products and ground, powdered in various other food products such as baked goods, energy bars, and pasta (ON-6343), with the only exception that the ingestion of the larvae can cause allergies in some people (EFSA 2021).

Currently, the EFSA is proceeding with eleven applications concerning insect species or certain products made from them, and the following are in the risk assessment stage: dried crickets (*Gryllodes sigillatus*) (EFSA-Q-2018-00263), whole and ground lesser mealworm (*Alphitobius diaperinus*) larvae products (EFSAQ-2018-00282), *Locusta migratoria* (EFSA-Q-2018-00513), *Acheta domesticus* (EFSA-Q-2018-00543), mealworms (*Tenebrio molitor*) (EFSA-Q-2018-00746), whole and ground mealworms (*Tenebrio molitor*) larvae (EFSA-Q-2019-00101), whole and ground grasshoppers (*Locusta migratoria*) (EFSA-Q-2019-00115), whole and ground crickets (*Acheta domesticus*) (EFSA-Q-2019-00121), defatted whole cricket (*Acheta domesticus*) powder (EFSA-Q-2019-00589), *Tenebrio molitor* (mealworm) flour (EFSA-Q-2019-00748), and dried *Acheta domesticus* (EFSA-Q-2020-00748) (EFSA 2021, Skotnicka et al. 2021).

One of the great challenges in accepting edible insects is the aversion to insect-based foods (i.e., neophobia and disgust). Food neophobia is the fear of eating unfamiliar foods, and it is dependent on culture and social rules. Rejection of edible insects can be reduced by increasing familiarity, that is, by making insect-based foods more widely available. Disgust from eating insects has been reported as a reaction resulting from a negative consumption experience, which can be a feeling of discomfort due to the nature, especially filth, of the insects' habitats, fear of contamination and diseases, that is, it is an emotion related food and socialized with group members, although it can be overcome or even eliminated in a generation (Choe and Hong 2018, Gama et al. 2018, Phan and Chambers 2016, Castro and Chambers 2018).

According to Mishyna and Glumac (2021), reasons such as disgust, health risk, lack of familiarity, and sociocultural origins are considered responsible, albeit age-dependent differences have also been observed. Melgar-Lalanne et al. (2019) stated that consumer interest in insect consumption can be increased when insects are used as an ingredient in an unrecognizable form such as powder or flour and introduced into cereal bars, drinks, pasta, cookies, etc. In a survey under review for publication, our research group analyzed two formulations of cookies containing whole and milled larvae of *Tenebrio Molitor* that were developed, characterized, and evaluated for acceptance. The cookies were prepared on the same basis for all formulations and only differed by the presence of yellow mealworms and their form (whole or milled); they were analyzed through an affective test. The evaluators were recruited among students and employees of the Federal University of Santa Maria and consisted of 49 women and 28 men aged between 18 and 60 years. The cookie dough and product after baking are presented in Fig. 3.

The test determined the rate of acceptance of the cookies containing *T. Molitor* (ground and whole), and the attributes were color, smell, texture, taste, and global acceptance using a hedonic scale composed of five points (1 = did not like it much to 5 = liked it a lot). The study had two main challenges, the first being the question of how to conduct sensory analysis during a pandemic, respecting the hygiene and distance protocols without putting the health of the participants at risk, assuming that conducting sensory analysis would traditionally exceed the capacity for people in the laboratory according to the UFSM safety protocols, as well as increasing the possibility of spreading and contamination of the SARS-CoV-2 virus. In parallel, the second challenge was people's resistance to consuming food products containing insects, as reported by several authors (Imathiu 2020, Kim et al. 2019, Orsi et al. 2019, Schardong et al. 2019).

Regarding color, the volunteers did not note a significant difference between the formulations containing insects. The cookies containing ground larvae of *Tenebrio molitor* were more pleasant to the consumers. The results demonstrated the repulsion felt by the consumers related to the ingestion of cookies containing whole insects.

These observations and the results reinforce the statements made by House (2018), Imathiu (2020), Raheem et al. (2019), and Kim et al. (2019), who reported that the 'disgusting barrier' at which consumers are stuck may be broken by modifying the form in which insects are presented to the

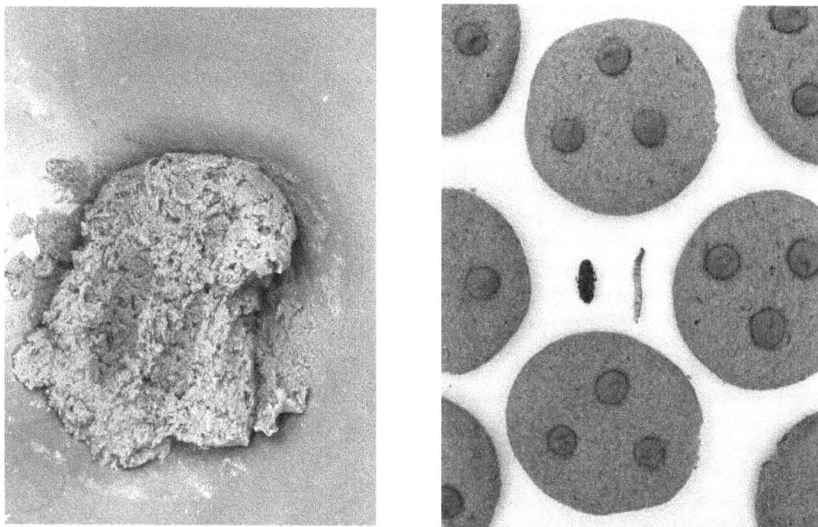

Figure 3. Cookies with *Tenebrio molitor* developed at UFSM. (a) Cookie dough; (b) cookie after provisioning. Source: Author (personal file).

consumer. Furthermore, Schardong et al. (2019) evaluated Brazilian consumers' perception of rated insects and observed that 46% of the 1,619 volunteers preferred consuming insects in the form of flour. The authors pointed out that the geographical factor must be considered besides the form of presentation. The fact that the region of Santa Maria (located in the center of Rio Grande do Sul State, southern Brazil), and where a significant number of the evaluators in this research live, has a high offer of meat at affordable prices (2021), leading the consumers to choose conventional meals instead of trying alternative sources of protein. Interestingly, most Brazilian consumers do not have an opinion on insect safety, although respondents with higher levels of education and familiarity consider consumption safe. Additionally, further evidence has also shown that women are more averse to insects than men (Orsi et al. 2019, Schardong et al. 2019).

In another study, Pedroso et al. (2021) compared the protein content of *Tenebrio Molitor* and *Chrysodeixis includens*, which is identified as a difficult-to-control pest mainly because of its polyphagous characteristic; however, it is believed to have interesting nutritional attributes. *C. includens* caterpillars were bred and slaughtered at the UFSM Entomology Laboratory. The results showed that the caterpillar *C. includens* had the highest protein content (61.95%), while the larva of *T. Molitor* was 50.63%. *T. Molitor* and *C. includens* presented higher values than conventional sources (cattle, swine, sheep, chicken, etc.), thereby confirming that they can be used as alternative sources of protein. Additionally, both demonstrated to have high potential as an alternative source of protein for food, requiring further studies to prove production feasibility through sustainable and low-cost methods.

Castro and Chambers (2018) consulted 630 consumers from 13 countries (except Mexico and Thailand) who were reluctant to accept insects as food. Elhassan et al. (2019) showed that sensory attributes are important in accepting edible insects. The appearance, viscosity, color, odor, texture, and taste of foods containing edible insects can be influenced by particle size. The acceptability of insects as food ingredients largely depends on how they are processed and presented in the final product, including packaging (Naest et al. 2018).

Pyo et al. (2020) showed that ethanolic extracts of *Tenebrio Molitor* and bees exhibited strong hemolytic activity, suggesting that these insects have potential as a food source and can also be used therapeutically. In Brazil, some companies already offer dehydrated insects for food to serve the pet market of exotic animals, such as birds and reptiles. There are no differences in insect production, either for animal or human consumption. Production must follow sanitary requirements, and inspection by the Federal Inspection Service is necessary and the authorization seal from the Ministry of Agriculture, Livestock, and Supply (MAPA), ensuring the quality of edible and non-edible animal products intended for domestic and foreign markets (Moraes and Fernandes 2018).

Edible insects and other foods, to ensure consumer safety and quality, must have regulations and labeling that, in addition to indicating the list of ingredients, also contain potentially allergenic ingredients. The allergen list must be updated constantly, and guidelines harmonized worldwide to avoid potentially life-threatening allergic reactions (Raheem et al. 2019). The use of insects as ingredients in the preparation of food products has helped reduce resistance to their consumption.

In Spain, the supermarket chain Carrefour launched a line of products, including granola and pasta, containing insects in their composition. Other examples are Bugfoundation, a German company, which has invested in the production of hamburgers from larvae, and the company Chirps, which has used crickets as ingredients in savory biscuits sold in the United States (Pedroso et al. 2020).

7. Final Considerations

Ensuring nutritious food for the world's growing population is one of the most significant challenges of this millennium. As pressure on land and water increases, these resources are being depleted at an alarming rate. Thus, efficient ways of producing food in terms of energy and resources are sought. In this regard, insects have proven to be highly promising as their life cycle is short while the turnover rate and biomass conversion are high. From a sustainability point of view, there is no other food source as efficient and effective as insects. If insect-based products can be marketed and commercialized properly, then insects could become a new source of protein. A challenge is the emotional disgust barrier toward insects and their associations (food neophobia, lack of information, etc.). The primary challenges for including insects in our diet include legislation, lowering prices in the acquisition of insects, automation in breeding and cheap substrates, developing insect products that appeal to consumers, and clarifying health benefits. The absence of norms and adequate legislation in countries where entomophagy is a new trend contrasts with countries with a tradition of consumption where norms are well defined. New protein origins cannot be underestimated by outdated legislation any longer, and the possibility of developing sustainable food substitutes is well known, and companies in the food sector must not ignore public health and safety. As new processing technologies emerge, there will be better opportunities for studying the nutritional quality parameters of edible insects for new food product formulations.

Efficient rearing will need to be optimized in the future, so there is a need to evaluate insect rearing facilities before they become available. This will provide the basis for a more informed discussion of sustainability and food security. The regulation represents security for the consumer and greater business opportunities for the producer. Insect breeding allows producers in small communities to increase their income and improve the social conditions of the families involved. What is more, insects are an alternative to enrich the diets of these communities. The enrichment of diets meets the issues related to hunger and malnutrition that affect poor communities around the world and encompasses the proposals of the 2030 Agenda and the Sustainable Development Goals that emphasize the need for transformative measures in search of real sustainable development for humanity. In addition, insect breeding must be promoted and encouraged as a socially inclusive activity. Insect farming requires minimal technical knowledge and capital investment, and since it does not require access to or ownership of land, it is within reach of even the poorest and most vulnerable members of society.

In the future, with rising prices of conventional animal proteins, insects may become a cheaper source of protein than conventionally produced meat and sea-caught fish. This will require significant technological innovation, changes in consumer preferences, food and feed legislation covering insects, and more sustainable food production. Consumers need to be sure that eating insects is not just good for their health, it is good for the planet.

Furthermore, the insect breeding sector is expected to increase the number of jobs from a few hundred to a few thousand by 2025, thus contributing to the economy. Employment opportunities are also linked to job creation to support the sector, such as specialized retail, administration, logistics, and research. Much attention is being paid to the potential of edible insects to diversify diets and improve food security in many parts of the world. Insect farming can help solve two of the world's biggest problems simultaneously: food insecurity and the climate crisis. Land use efficiency is pivotal to achieving food security for a growing population, and the nutritional composition of insects combined with more efficient use of natural resources and a smaller environmental footprint contribute to SDG 2 (zero hunger), SDG 3 (health and well-being), and SDG 12 (responsible consumption and production). The differential will be achieved with government, industry, and academia collaboration. Edible insects have an exceptionally high potential to contribute to more sustainable and socially just global food security.

References

Abbasi, T., Abbasi, T. and Abbasi, S.A. 2016. Reducing the global environmental impact of livestock production: the minilivestock option. J. Clean Prod. 112: 1754–1766.

ABPA - Associação Brasileira de Proteína Animal (ABPA). 2021. Relatório Anual. Available in: https://abpa-br.org/relatorios/. Access: Dez. 13, 2021.

Acosta-Estrada, B., Reis, A., Rosell, C.M., Rodrigi, D. and Ibarra-Herrera, C.C. 2021. Benefits and challenges in the incorporation of insects in food products. Front. Nutr. 30: 687712.

Al-Awwad, N.J., Al-Sayyed, H.F., Zeinah, Z.A. and Tayyem, R.F. 2021. Dietary and lifestyle habits among university students at different academic years. Clin. Nutr. ESPEN 44: 236–242.

Baiano, A. 2020. Edible insects: An overview on nutritional characteristics, safety, farming, production technologies, regulatory framework, and socio-economic and ethical implications. Trends Food Sci. Tech. 100: 35–50.

Barba, F.J., Mariutti, L.R., Bragagnolo, N., Mercadante, A.Z., Barbosa Canovas, G.V. and Orlien, V. 2017. Bioaccessibility of bioactive compounds from fruits and vegetables after thermal and nonthermal processing. Trends Food Sci. Tech. 67: 195–206.

Barre, A., Caze-Subra, S., Gironde, C., Bienvenu, F., Bienvenu, J. and Rouge, P. 2014. Entomophagy and the risk of allergy. Rev. Fr. Allergol. 54: 315–321.

Barroso, F.G., Sanchez-Muros, M.J., Segura, M., Morote, E., Torres, A., Ramos, R., et al. 2017. Insects as food: enrichment of larvae of *Hermetia illucens* with omega 3 fatty acids by means of dietary modifications. J. Food Compos. Anal. 62: 8–13.

Belluco, S., Halloran, A. and Ricci, A. 2017. New protein sources and food legislation: the case of edible insects and EU law. Food Sec. 9: 803–814.

BRASIL. Ministério da Saúde. Agência Nacional de Vigilância Sanitária (ANVISA). (2014). Resolução da Diretoria Colegiada – RDC n° 14, de 28 de março de 2014, dispõe sobre matérias estranhas macroscópicas e microscópicas em alimentos e bebidas, seus limites de tolerância e dá outras providências. Available in: https://bvsms.saude.gov.br/bvs/saudelegis/anvisa/2014/rdc0014_28_03_2014.pdf. Access: Ago. 13, 2021.

Broekman, H., Knulst, A., Jager, S.D.H., Gaspari, M., Jong, G.D., Houben, G. et al. 2015. Shrimp allergic patients are at risk when eating mealworm proteins. Clin. Transl. Allergy 5: 77.

Cardoso, C., Afonso, C., Lourenço, H., Costa, S. and Nunes, M.L. 2015. Bioaccessibility assessment methodologies and their consequences for the risk-benefit evaluation of food. Trends Food Sci. Tech. 41: 5–23.

Castro, M. and Chambers IV, E. 2018. Willingness to eat an insect based product and impact on brand equity: a global perspective. J. Sens. Stud. 34: e12486.

Castro, R.J.S., Ohara, A., Aguilar, J.G.S. and Domingues, M.A.F. 2018. Nutritional, functional and biological properties of insect proteins: processes for obtaining, consumption and future challenges. Trends Food Sci. Tech. 76: 82–89.

Choe, S.-Y. and Hong, J.H. 2018. Can information positively influence familiarity and acceptance of a novel ethnic food? A case study of Korean traditional foods for Malaysian consumers. J. Sens. Stud. 33: e12327.

Choi, J.H., Kim, S.J. and Kim, S. 2016. A novel anticoagulant protein with antithrombotic properties from the mosquito *Culex pipiens pallens*. Int. Biol. Macromol. 93: 156–166.

Costa-Neto, E.M. 2013. Insect as human food: an overview. Rev Antropol. 5: 562–582.

Costa-Neto, E.M. 2015. Anthropo-entomophagy in Latin America: an overview of the importance of edible insects to local communities. J. Insect. Food Feed. 1: 17–23.

Dobermann, D., Swift, J.A. and Field, L.M. 2017. Opportunities and hurdles of edible insects for food and feed. Nutrition Bulletin. 42: 293–308.

Dossey, A.T., Morales-Ramos, J.S. and Rojas, M.G. 2016. Insects as sustainable food ingredients. Academic Press, London, United Kingdom.

EFSA. European Food Safety Authority. Panel on Nutrition, Novel Foods and Food Allergens (NDA), Turck, D., Castenmiller, J., De Henauw, S., Hirsch-Ernst, K.I., Kearney, J., Maciuk, A., et al. 2021. Safety of dried yellow mealworm (*Tenebrio molitor larva*) as a novel food pursuant to Regulation (EU) 2015/2283. EFS2 19. Available online: https://doi.org/10.2903/j.efsa.2021.6343. Access: Jan. 13, 2022.

EFSA. European Food Safety Authority. Register of Questions out of Service. Available online: https://www.efsa.europa.eu/en/ register-of-questions. Access: Feb. 03, 2022.

Elhassan, M., Wendin, K., Olsson, V. and Langton, M. 2019. Quality aspects of insects as food – nutritional, sensory, and related concepts. Foods 8: 1–14.

FAO. Food and Agricultural Organization of the United Nations. 2009. How to feed the world in 2050. High level expert forum. Available online: https://www.fao.org/wsfs/forum2050/wsfs-background-documents/hlef-issues-briefs/en/. Access: Dez. 13, 2021.

FAO. Food and Agricultural Organization of the United Nations. 2013a. Edible insects. Future prospects for food and feed security. Available online: https://www.fao.org/3/i3253e/i3253e.pdf . Access: Dez. 13, 2021.

FAO. Food and Agricultural Organization of the United Nations. 2013b. Fao statistical yearbook 2013 world food and agriculture. Available online: https://www.fao.org/3/i3107e/i3107e00.htm. Access: Abr. 11, 2015.

FAO. Food and Agricultural Organization of the United Nations. 2014. The state of food insecurity in the world 2014. Available online: http://www.fao.org/ publications/sofi/en/. Access: Ago. 12, 2015.

FAO. Food and Agricultural Organization of the United Nations. 2017. El estado de la seguridad alimentaria y la nutrición en el mundo. Available online: https://www.fao.org/3/i7695s/i7695s.pdf. Access: Dez. 13, 2021.

FAO. Food and Agricultural Organization of the United Nations. 2018. El estado de la seguridad alimentaria y la nutrición en el mundo. Available online: https://www.fao.org/3/I9553ES/i9553es.pdf. Access: Dez. 13, 2021.

FAO. Food and Agricultural Organization of the United Nations. 2019. El estado de la seguridad alimentaria y la nutrición en el mundo. Available online: https://reliefweb.int/sites/reliefweb.int/files/resources/ca5162es.pdf. Access: Dez. 13, 2021.

FAO. Food and Agricultural Organization of the United Nations. 2020. El estado de la seguridad alimentaria y la nutrición en el mundo. Available online: https://www.fao.org/3/ca9699es/ca9699es.pdf. Access: Dez.13, 2021.

FAO. Food and Agricultural Organization of the United Nations. 2021a. El estado de la seguridad alimentaria y la nutrición en el mundo. Available online: https://reliefweb.int/sites/reliefweb.int/files/resources/SOFI2021_InBrief_SP_web.pdf. Acesso: Jan. 10, 2022.

FAO. Food and Agricultural Organization of the United Nations. 2021b. Looking at edible insects from a food safety perspective. Challenges and opportunities for the sector. FAO, Rome. Available online: http://www.fao.org/3/cb4094en/cb4094en.pdf. Access: Feb. 03, 2022.

Fogang, A.R., Kansci, G., Viau, M., Hafnaoui, N., Meynier, A., Demmano, G., et al. 2017. Lipid and amino acid profiles support the potential of *Rhynchophorus phoenicis* larvae for human nutrition. J. Food Compos. Anal. 60: 64–73.

Gama, A.P., Adhikari, K. and Hoisington, D.A. 2018. Factors influencing food choices of Malawian consumers: a food choice questionnaire approach. J. Sens. Stud. 33: e12442.

Gere, A., Székely, G., Kovács, S., Kókai, Z. and Sipos, L. 2017. Readiness to adopt insects in Hungary: a case study. Food Qual Prefer. 59: 81–86.

Govorushko, S. 2019. Global status of insects as food and feed source: A review. Trends Food Sci. Tech. 91: 436–445.

Grassi, M.K. 2014. Let's Eat Bugs! A Thought-Provoking Introduction to Edible Insects for Adventurous Teens and Adults (2nd Edition) (Let's Eat Bugs). Kindle Edition.

Gravel, A. and Doyen, A. 2020. The use of edible proteins in food: challenges and issues related to their functional properties. Innov. Food Sci. Emerg. 59: 102272.

Halloran, A., Flore, R., Vantomme, P. and Roos, N. 2018. Edible Insects in Food sustainable Food Systems. Springer, New York, USA.

Hamerman, E.J. 2016. Cooking and disgust sensitivity influence preference for attending insect-based food events. Appetite 96: 319–326.

Hlongwane, Z.T., Slotow, R. and Munyai, T.C. 2020. Nutritional composition of edible insects consumed in Africa: a systematic review. Nutrients 12: 1–28.

Homann, A.M., Ayieko, M.A., Konyole, S.O. and Roos, N. 2017. Acceptability of biscuits containing 10% cricket (*Acheta domesticus*) compared to milk biscuits among 5-10-year-old Kenyan schoolchildren. J. Insects Food Feed. 3: 95–103.

House, J. 2018. Insects as food in the Netherlands: production networks and the geographies of edibility. Geoforum. 94: 82–93.

Hunter, G.L. 2021. Edible Insetcts. A Global History. Reaktion Books, London, United Kingdom.

Imathiu, S. 2020. Benefits and food safety concerns associated with consumption of edible insects. NFS Journal 18: 1–11.

Jongema, Y. 2017. List of edible insects of the world. Available online: https://www.wur.nl/en/Research-Results/Chair-groups/Plant-Sciences/Laboratory-of-Entomology/Edible-insects/Worldwide-species-list.htm. Access: Ago. 12, 2021.

Kelemu, S., Niassy, S., Torto, B., Fiaboe, K., Affognon, H., Tonnang, H. et al. 2015. African edible insects for food and feed: inventory, diversity, commonalities and contribution to food security. J. Insect Food Feed. 1: 103–119.

Kim, T., Yong, H.I., Kim, Y., Kim, H. and Choi, Y.S. 2019. Edible insects as a protein source: a review of public perception, processing technology, and research trends. Food Sci. Anim. Resour. 39: 521–540.

Kourimská, L. and Adámková, A. 2016. Nutritional and sensory quality of edible insects. NFS Journal 4: 22–26.

Lange, K.W. and Nakamura, Y. 2021. Edible insects as future food: chances and challenges. J. Future Foods. 1: 38–46.

Lesnik, J.J. 2019. Edible Insects and Human Evolution. University Press of Florida, Pensacola, USA. Kindle Edition.

Liceaga, A.M. 2021. Processing insects for use in the food and feed industry. Curr Opin Insect Sci. 48: 1–5.

Lucas, A.J.S., Oliveira, L.M., Rocha, M. and Prentice, C. 2020. Edible insects: an alternative of nutritional, functional and bioactive compounds. Food Chem. 311: 126022.

Manditsera, F.A., Luning, P.A., Fogliano, V. and Lakemond, C.M. 2019. Effect of domestic cooking methods on protein digestibility and mineral bioaccessibility of wild harvested adult edible insects. Food Res. Int. 121: 404–411.

Matos, F.M. and Castro, R.J.S. 2021. Edible insects as potential sources of proteins for obtaining bioactive peptides. Braz. J. Food Technol. 24: e2020044.

McGrew, W.C. 2014. The "other faunivory" revisited: insectivory in human and non-human primates and the evolution of human diet. J. Human Evolution 71: 4–11.

Megido, R.C., Poelaert, C., Ernens, M., Liotta, M., Blecker, C., Danthine, S. et al. 2018. Effect of household cooking techniques on the microbiological load and the nutritional quality of mealworms (*Tenebrio molitor* L. 1758). Food Res. Int. 106: 503–508.

Melgar-Lalanne, G., Hernández-Álvarez, A.J. and Salinas-Castro, A. 2019. Edible insects processing: tradicional and innovative technologies. Compr. Rev. Food Sci. F. 18: 1166–1191.

Melse-Boonstra, A. 2020. Bioavailability of micronutrients from nutrient-dense whole foods: zooming in on dairy, vegetables, and fruits. Front Nutr. 7: 10.

Meyer-Rochow, V. and Jung, C. 2020. Insects used as food and feed: isn't that what we all need? Foods 9: 1003.

Miglietta, P., De Leo, F., Ruberti, M. and Massari, S. 2015. Mealworms for food: a water footprint perspective. Water 7: 6190–6203.

Minas, R.S., Kwiatkowski, A., Klein, S., Oliveira, R.F. and Diemer, O. 2016. Antropoentomofagia e Entomofagia: insetos, a salvação nutricional da humanidade. Editora Kiron, Brasília, Brasil.

Mishyna, M. and Glumac, M. 2021. So different, yet so alike Pancrustacea: health benefits of insects and shrimps. J. Funct. Foods. 76: 104316.

Mishyna, M., Keppler, J.K. and Chen, J. 2021. Techno-functional properties of edible proteins and effects of processing. Curr. Opin. Colloid In. 56: 101508.

Mitsuhashi, J. 2017. Edible insects of the world. CRC Press, Boca Raton, USA.

Moraes, B. and Fernandes, L. 2018. O promissor mercado de insetos comestíveis. Avaliable online: https://medium.com/@leodemf/o-promissor-mercado-de-insetos-comest%C3%ADveis-131a4572a98f. Access: Ago. 13, 2021.

Moruzzo, R., Mancini, S. and Guidi, A. 2021. Edible insects and sustainable development goals. Insects. 12: 557.

Mutungi, C., Irungu, F.G., Nduko, J., Mutua, F., Affognon, H., Nakimbugwe, D. et al. 2019. Postharvest processes of edible insects in Africa: a review of processing methods, and the implications for nutrition, safety and new products development. Crit. Rev. Food Sci. Nutr. 59: 276–298.

Næs, T., Varela, P. and Berget, I. 2018. Individual differences in descriptive sensory data (DA). pp. 25–55. *In*: Næs, T., Varela, P. and Berget, I. (eds.). Individual differences in Sensory and Consumer Science. Woodhead Publishing, Duxford, United Kingdom.

Nowakowski, A.C., Miller, A.C., Miller, M.E., Xiao, H. and Wu, X. 2021. Potential health benefits of edible insects. Crit. Rev. Food Sci. Nutr. 1–10.

Ojha, S., Bußler, S. and Schlüter, O.K. 2020. Food waste valorisation and circular economy concepts in insect production and processing. Waste Manage 118: 600–609.

Ojha, S., Bekhit, A.E.D., Grune, T. and Schlüter, O. 2021. Bioavailability of nutrients from edible insects. Current Opinion in Food Science 41: 240–248.

Ordoñez-Araque, R. and Egas-Montenegro, E. 2021. Edible insects: A food alternative for the sustainable development of the planet. International Journal of Gastronomy and Food Science 23: 100304.

Parada, J. and Aguilera, J.M. 2007. Food microstructure affects the bioavailability of several nutrients. J. Food Sci. 72: R21–R32.

Patel, S., Suleira, H.A.R. and Rauf, A. 2019. Edible insects as innovative foods: nutritional and functional assessments. Trends Food Sci. Tech. 86: 352–359.

Pedroso, M.A.P., Schardong, I.S., Jiménez, M.S.E., Boff, J., Bernardi, O. and Richards, N.S.P.S. 2021. Edible insects as potential source of alternative protein. *In*: Simpósio Latino-americano de Ciências de Alimentos – SLACA, 14, Campinas. Anais eletrônicos... Campinas: UNICAMP. Available online: https://proceedings.science/slaca-2021/papers/edible-insects-as-potential-source-of-alternative-protein, Access: Dez. 13, 2021.

Phan, U.X.T. and Chambers IV, E. 2016. Application of an eating motivation survey to study eating occasions. J. Sens. Stud. 31: 114–123.

Pyo, S.J., Kang, D.G., Jung, C. and Sohn, H.Y. 2020. Anti-thrombotic, anti-oxidant and hemolysis activities of six edible insect species. Foods 9: 401.

Raheem, D., Raposo, A., Oluwole, O.B., Nieuwland, M., Saraiva, A. and Carrascosa, C. 2019. Entomophagy: nutritional, ecological, safety and legislation aspects. Food Res. Inter. 126: 108672.

Ramos-Elorduy, J. and Pino, M.M. 1996. El consumo de insectos entre los Aztecas. pp. 89–101. *In*: Long, J. (ed.). Conquista y comida. Consecuencias del encuentro de dos mundos. UNAM, México.

Reis, T.L. and Dias, A.C.C. 2020. Farinha De Insetos Na Alimentação De Não-Ruminantes, Uma alternativa alimentar. Vet Zootec. 27: 001-017. Avaliable online: https://www.researchgate.net/publication/341193012_FARINHA_DE_INSETOS_NA_ALIMENTACAO_DE_NAO_RUMINANTES_UMA_ALTERNATIVA_ALIMENTAR. Access: Dez. 13, 2021.

Reverberi, M. 2021. The new packaged food products containing insects as an ingredient. J. Insect. Food Feed. 7: 901–908.

Rumpold, B.A. and Schlüter, O.K. 2013. Nutritional composition and safety aspects of edible insects. Mol. Nut. Food Res. 57: 802–823.

Saath, K.C.O. and Fachinello, A.L. 2018. Crescimento da demanda mundial de alimentos e restrições do fator terra no Brasil. RESR. 56: 195–212.

Santos, C.A.B. and Florêncio, R.R. 2013. Breve histórico das relações homen-ambiente presentes na entomofagia e entomoterapia. Polêm!ca. 12: 786–798.

Schardong, I.S., Freiberg, J.A., Santana, N.A. and Richards, N.S.P.S. 2019. Brazilian consumers' perception of edible insects. Ciência Rural. 49: e20180960.

Schickler, G. 2013. Insetos na Alimentação. Available online: http://www.nutrinsecta.com.br/artigos/artigo-completo-para-a-revista-passarinheiros-e-cia-edicao-72/. Access: Dez. 13, 2021.

Shaun, M.M.A., Nizum, M.W.R., Munny, S., Fayeza, F., Mali, S.K., Abid, M.T. et al. 2021. Eating habits and lifestyle changes among higher studies students post-lockdown in Bangladesh: a web-based cross-sectional study. Heliyon 7: e07843.

Skotnicka, M., Karwowska, K., Kłobukowski, F., Borkowska, A. and Pieszko, M. 2021. Possibilities of the development of edible insect-based foods in Europe. Foods. 10: 766.

Soagri, G., Mora, C. and Menozzi, D. 2019. Edible insects in the food sector. Springer, New York, USA.

Srinroch, C., Srisomsap, C., Chokchaichamnankit, D., Punyarit, P. and Phiriyangkul, P. 2015. Identification of novel allergen in edible insect, *Gryllus bimaculatus* and its cross-reactivity with *Macrobrachium* spp. allergens. Food Chem. 184: 160–166.

Tong, L., Yu, X. and Lui, H. 2011. Insect food for astronauts: gas exchange in silkworms fed on mulberry and lettuce and the nutritional value of these insects for human consumption during deep space flights. Bulletin of Entomological Research 101: 613–622.

Tzompa-Sosa, D.A. and Fogliano, V. 2017. Potential of insect-derived ingredients for food applications. pp. 215–231. *In*: Shields, V.D. (ed.). Insect physiology and ecology, IntechOpen Limited, London, United Kingdom.

Van Huis, A., Van Itterbeeck, J., Klunder, H., Mertens, E., Halloran, A., Muir, G. et al. 2013. Edible insects: future prospects for food and feed security. FAO, Rome, Italy.

Van Huis, A. 2016. Edible insects are the future? Proceedings of the Nutrition Society, 75: 294–305.

Van Huis, A. and Dunkel, F.V. 2017. Edible insects: A neglected and promising food source. *In*: Nadathur, S.R., Wanasundara, J.P.D. and Scanlin, L. (eds.). Sustainable Protein Sources (Chap. 21, 341–355) London: Academic Press.

Van Huis, A. 2018. Insects as human Food. pp. 195–213. *In*: Alves, R.R.N. and Albuquerque, I.P. (eds.). Ethnozoology: Animals in our lives. Academic Press, Cambridge, United Kingdom.

Van Huis, A., Rumpold, B., Maya, C. and Roos, N. 2021. Nutritional qualities and enhancement of edible insects. Annu. Rev. Nutr. 11: 551–576.

Vilela, L. and Pulrolnik, K. 2015. Segurança alimentar e a sustentabilidade. Opiniões 12: 12–14.

Yen, A.L. 2015. Insects as food and feed in the Asia Pacific region: current perspectives and future directions. J. Insect. Food Feed. 1: 33–55.

Zhang, Z., Gao, L., Shen, C., Rong, M., Yan, X. and Lai, R. 2014. A potent anti-thrombosis peptide (vasotab TY) from horsefly salivary glands. Int. J. Biochem. Cell Biol. 54: 83–88.

Zielinska, E., Baraniak, B., Karas, M., Rybczynska, K. and Jakubczyk, A. 2015. Selected species of edible insects as a source of nutrient composition. Food Res. Int. 77: 460–466.

9

Mealworm-based Foods

Lee Seong Wei,[1,2,*] *Zulhisyam Abdul Kari,*[1,2] *Wendy Wee,*[3]
Noor Khalidah Abdul Hamid[4] *and Mahmoud A.O. Dawood*[5,6,*]

1. Introduction

The global human population is expected to increase rapidly from the current total population of 7.6 billion to 9.8 billion in 2050 (United Nations 2017). Therefore, a more sustainable food supply is needed to produce along with the exponential growth of the human population. As income rises, lifestyle changes and meat consumption is growing, especially in developed countries (Gerber et al. 2013). The high demand for animal protein from consumers catalyzed the livestock industry into intensification, and it has burdened the environment as the industry needs enormous energy tto support operations such as animal feed production, transportation, and processing (Boer et al. 2006). Furthermore, overreliance on animals for protein sources can disrupt world sustainability programs, and all sustainability-related activities may come to an impasse. Hence, alternative food sources need to be explored aggressively for environmental betterment. One of the alternative food sources is insects, and the global proposal concerning the topic was made in 1975 (Meyerrochow 1975).

Primates are known to consume insects as a source of protein (Raubenheimer and Rothman 2013). Entomophagy, or insect-eating, has been practised by many people in different parts of the globe for generations (Evans et al. 2015). Consumption of insects by humans has been extensively documented in the literature. Statistically, more than 2 billion people are recorded consuming insects as a protein source, and 2000 insect species are edible (Jongema 2015). For instance, Van Huis (2003) reported that communities in the African continent consumed grasshoppers and nymphs of the beetle as the source of protein, whereas grasshopper, ant larvae and giant water bugs are popular as local delicacies in Thailand (Tan et al. 2015). Mealworm is one of many insects that humans and

[1] Department of Agricultural Sciences, Faculty of Agro-Based Industry, Universiti Malaysia Kelantan, Jeli Campus, 17600 Jeli, Kelantan, Malaysia.
   Email: zulhisyam.a@umk.edu.my
[2] Advanced Livestock and Aquaculture Research Group, Faculty of Agro-Based Industry, Universiti Ma-laysia Kelantan, Jeli Campus, 17600 Jeli, Kelantan, Malaysia.
[3] Center of Fundamental and Continuing Education, Universiti Malaysia Terengganu, 21030 Kuala Nerus, Terengganu.
   Email: wendy@umt.edu.my
[4] School of Biological Sciences, Universiti Sains Malaysia, 11800 Minden, Pulau Pinang, Malaysia.
   Email: khalidah.hamid@usm.my
[5] Animal Production Department, Faculty of Agriculture, Kafrelsheikh University, Kafr El-Sheikh 33516, Egypt.
[6] The Center for Applied Research on the Environment and Sustainability, The American University in Cairo, 11835, Cairo, Egypt.
* Corresponding authors: leeseong@umk.edu.my; mahmoud.dawood@agr.kfs.edu.eg

animals have consumed. It is commonly sold in pet shops to feed pet companions such as birds, reptiles, amphibians, and ornamental fish. In Thailand, live mealworm is served in various cuisines or snacks (Tan et al. 2015). Consumption of insects is ingrained in the diets and cultures of insect-eating communities such as Thailand and the African continent. However, insect consumption has become a global trend and is expected to grow in popularity in the coming years due to concerns about sustainability.

The scientific report about mealworms was found as early as in 1939 by Evans and Goodliffe (1939), who reported the nutritional value of mealworm larvae. Martin et al. (1976) claimed that mealworms had been used to feed captivity animals such as pottos, *Perodicticus potto*, tree shrews, *Tupaia belangeri* and mouse *Microcebus murinus*. Recent studies suggested that mealworm could be sustainable food for the future because it is rich in nutrients. However, the nutrient profile of mealworms is influenced by the preparation methods (Ramos-Elorduy 2009). There is no doubt that the nutrient profile of mealworms can compete with other human conventional food as a protein source. For example, protein in fresh mealworm larvae is around 18% (Oonincx et al. 2015), whereas the protein in mutton is about 23% (Kenya 2018), 22% in beef and 23% in pork (Smil 2002). Hence, mealworms can become an alternative protein source in the future. Mealworm is starting to get attention from all around the world. This insect was reported easy to raise and the most reared in Europe (Bordiean et al. 2020). The advantages of rearing mealworms are cost-effective and can sustain the agricultural industry, foraging agricultural wastes such as wheat bran for growth (Oonincx et al. 2015). Furthermore, the European Commission has approved the mealworm production for food and feed in the European Union (European Commission 2015, 2017). Besides, it was reported that mealworm can be used as a biodegradable agent in degrading plastic-based products in a recent study (Brandon et al. 2018). The biodegradable activity of mealworms can be enhanced with the presence of probiotics such as *Lactobacillus* and *Mucispirillum* in the gut of mealworms (Lou et al. 2021).

2. Source, forms, and Derivatives of Mealworm

Mealworms consist of three species: yellow mealworm, *Tenebrio molitor* L., giant mealworm, *Zophobas atratus* and lesser mealworm, *Alphitobius diaperinus panzer*. These mealworms can be found in many countries as they are nasty pests to grains and flours (Ramos-Elorduy et al. 2002). Furthermore, these pests can survive in very tough environments and are challenging to eliminate, as reported by Renault et al. (2003). However, nutrient, mineral and trace elements-rich mealworms (Grau et al. 2017) are farmed commercially to cater to high demand of the item as pet food for birds (Bańbura et al. 2013), amphibians (Pasmans et al. 2012), reptiles (Gregorovičová and Černíková 2016), ornamental fish, cats (Bosch et al. 2014) and dogs (Bosch et al. 2016).

A commonly farmed mealworm is *T. molitor*, which has four life stages: adult (bettle), egg, larva, and pupa (Li et al. 2013). Mating will occur during the beetle life stage, and female beetle will produce eggs. Subsequently, the eggs hatch and transform into larva and pupa. Mealworms require food during its life stages as bettle and larva, but not during its egg and pupa life stages (Fig. 1). The diet of mealworm varies and comprises vegetables, fruits, wheat bran, rice bran, and many more (Van Broekhoven et al. 2016). Mealworm prefers warm temperatures ranging between 25 and 28°C at any life stages (Kim et al. 2015, Siemianowska et al. 2013). It also favors high humidity, especially at the beetle stage (Hardouin and Mahoux 2003) and dark environments (Oonincx et al. 2015). In a commercial farm, mealworms are fed with wheat bran, wheat, maize, rapeseed, sunflower, soybean, and sugar beet pulp (Sauvant et al. 2004). Thévenot et al. (2018) reported that mealworms would be harvested and processed into meal and oil at the larvae stage. These mealworm products will be used in animal feed production and human use (Son et al. 2020). Few studies suggested that mealworm farming has less impact on the environment than other animals and ingredients for feed production. For instance, Oonincx and De Boer (2012) reported that dairy products and meat production have a higher environmental impact than mealworms. Another study by Thévenot et al.

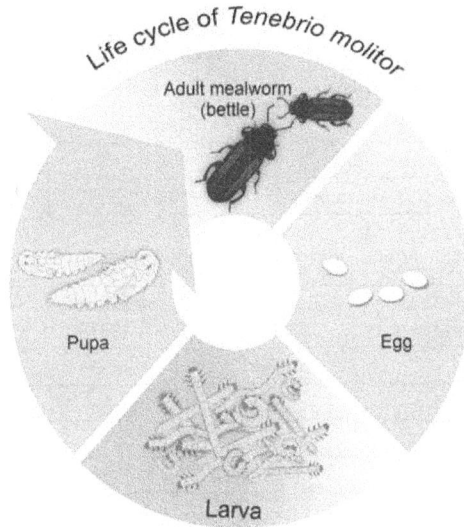

Figure 1. Life cycle of the mealworm, *Tenebrio molitor*.

(2018) revealed that mealworm production has a lesser environmental impact than soybean meal and fishmeal production.

3. Nutritional Composition of Mealworm

Mealworm nutritional content has been extensively researched and recorded in the literature. The nutritional content of mealworms can undoubtedly compete with that of other conventional protein sources on the market. Mealworms has a comparatively high crude protein composition and are on par with conventional animal feed ingredients such as soybean meal and fishmeal. Soybean meal has about 40% crude protein (Bovera et al. 2015) whereas it has about 60% in fishmeal (Wang et al. 2005). Live mealworm is also rich in other nutrients to cater to the nutrition needed for many animal species, including humans (Table 1). However, the nutrient profile of mealworms reported is inconsistent and varies depending on the studies and may also be due to food given to the larvae (Table 2). For example, the highest protein value of mealworm (53%) was reported in the study of Khan et al. (2018) and Mariod (2020), whereas the lowest protein value of mealworm recorded in the study of Hussain et al. (2017) was 45.8%. Another example of the variation of mealworm nutrient profile is fat content. The lowest fat content value was recorded as 3.1% whereas the highest was 34.2%. A similar pattern also reported on the amino acid profile (Table 3). For instance, Finke (2002) claimed that methionine in mealworms is about 0.6% while Bovera et al. (2015) showed methionine in mealworms is 1.6%. At present, commercial essential amino acids for animal uses are derived from the plant through the fermentation process (Kari et al. 2022, 2021, 2020, Zulhisyam et al. 2020). Although the amino acids profile of the consumer is not reflective of the food, some essential amino acids will display alteration respective to the input. DL-Methionine, L-lysine, L-threonine, and L-tryptophan are essential amino acids required for commercially farmed animal and are frequently supplied in synthetic form into the feed formulation (Karau and Grayson 2014). These amino acids are present in reasonable amounts in mealworms. Thus, the inclusion of mealworms in animal feed could serve as an organic source of essential amino acids.

Another important nutrient in mealworms is fatty acids such as palmitic acid, oleic acid and linoleic acid (Ravzanaadii et al. 2012). These fatty acids are also crucial in animal feed production. Furthermore, Jeon et al. (2016) claimed that mealworm oil has fatty acids that stand at high temperatures. Kim et al. (2013) claimed that mealworm has vast potential to be used as a feed additive for animal. In addition, entomopathogenic fungus, *Beauveria bassiana*, was introduced to

Table 1. Nutrient profile of live mealworm in the study of Mariod (2020).

Nutrient	Composition (%)
Protein	20
Fat	13
Fiber	2
Moisture	62

Table 2. Comparison of the nutrient profile of dried mealworm among studies in the literature.

Nutrient/References	Ravzanaadii et al. (2012)	Bovera et al. (2015)	Khan et al. (2018)	Hussain et al. (2017)	Mariod (2020)
Protein %	46.4	51.9	53	45.8	53
Fat %	32.7	21.6	3.6	34.2	28
Fiber %	4.6	7.2	3.1	4	6
Moisture %	5.3	Not provided	Not provided	5.8	5

Table 3. Comparison of essential amino acids (EAAs) in dried mealworm among studies in the literature.

EAAs/References	Finke (2002)	Bovera et al. (2015)	Khan et al. (2018)	Hussain et al. (2017)
Methionine %	0.6	1.6	1.5	1.3
Lysine %	2.7	4.5	5.4	4.5
Threonine %	2.5	2.7	4.0	1.6
Tryptophan %	0.4	1.7	0.6	Not provided

infect *T. molitor* to enhance green fluorescent protein in the worm. Nevertheless, study by Liu et al. (2020) claimed that mealworm given feed consisting of high antioxidant materials such as orange, carrot and red cabbage, did not contribute to antioxidant properties in the mealworm. Many factors can influence the nutrient values of mealworms, especially in dried mealworm preparation, such as temperature, storage, and many more (Zhao et al. 2016). Therefore, it is a must to establish a mealworm processing method in order to maintain the quality of mealworm consistently.

Mealworm is rich in glucosamine compounds and can become functional food for human uses (Son et al. 2020). Both chitin and chitosan are glucosamine compounds found in mealworms and are very useful for maintaining human health (Son et al. 2021). Chitin from mealworms can be further processed through deacetylation to produce chitosan (Shin et al. 2019). At present, chitin and chitosan are widely used in the healthcare industry, such as antimicrobial agents, cosmetics, wound healing, and many more (Elieh-Ali-Komi and Hamblin 2016). The commercial chitin and chitosan were produced from crustacean exoskeletons, mainly shrimp and crab (Nag et al. 2021). The current finding showed that the quality of chitin and chitosan derived from mealworms are equally similar to that of shrimp (Son et al. 2021). Thus, industry players have more options in selecting the source of ingredients for chitin and chitosan production.

Many studies have reported the medicinal properties of mealworms. For example, Ding et al. (2021) reported that mealworm larvae aqueous extract up-regulated the expression of caspases, i.e., caspace-8 and-9, followed by the activation of caspase-3 which induced apoptosis in cancer cells. Dávalos Terán et al. (2020) discovered that strategic processing of yellow mealworm proteins is able to produce dipeptidyl peptidase (DPP)-IV inhibiting peptides. This DPP can be used in diabetes treatment. Other medicinal properties of mealworm were reported in the study of Song et al. (2018). The study claimed that mealworm whole body and exuvium are good chitin and chitosan sources. The extension study of Shin et al. (2019) showed that chitosan extracted from mealworm possesses antimicrobial activities against *Bacillus cereus*, *Listeria monocytogenes*, *Escherischia coli* and *Staphylococcus aureus*. Other usefulness of mealworms is its larvae exuviae that can be served as prebiotic to mice at a dose of 20% of diet (Kwon et al. 2020). In conclusion, huge nutrient composition in mealworms can benefit both animal production and human health.

4. Mealworm and Food Safety

Mealworm is commonly reported to infest grain products and lower quality products (Vigneron et al. 2019). Many studies were documented methods to get rid of this pest from grain products such as essential oil. The essential oil was reported to possess insecticidal activity against mealworm larvae. For example, Szołyga et al. (2014) claimed that *Thuja occidentalis* and *Tanacetum vulgare* essential oil could work against lesser mealworm, *Alphitobius diaperinus*. Other essential oils such as *Origanum vulgare* and *Artemisia dracunculus* also had similar inhibitory activity against mealworm (Szczepanik et al. 2018). In addition, enthomopathogenic fungi such as *Metarhizium anisopliae* could be used as a biocontrol agent against the lesser mealworm (Gindin et al. 2009). Although mealworm has a defence mechanism against pathogen and fungal infection (Silva et al. 2016), some pathogen may associate with the worm. Crippen et al. (2018) reported that beetle mealworms become a carrier of *Salmonella* and are able to spread the bacteria among broiler chicken. This was supported by a study by Crippen et al. (2012), which claimed that *Salmonella* can stay with a mealworm in its all-life cycle. Hence, mealworms can be a reservoir of *Salmonella*. Besides, mealworm was also reported to become a reservoir of parasite, protozoa and other harmful pathogens (Vigneron et al. 2019). Despite the facts, mealworm at the larval stage possesses antibacterial property that can reduce *Salmonella* population in mealworm feed and larvae itself within seven days as claimed in the study of Wynants et al. (2019). Thus, the pathogen on mealworm larvae and its surroundings will reduce over time. Another study by Stoops et al. (2016) suggested that mealworms need to be disinfectant properly to eliminate harmful pathogens colonized on mealworms before utilizing it.

Other safety issues in using mealworms as an alternative protein source for animals and human uses are antibiotic residues (Osimani et al. 2018), heavy metals and pesticide residues (Poma et al. 2017), allergens (Nebbia et al. 2019) and toxin compounds (Poma et al. 2017) in mealworm. Antibiotic-resistance microorganisms is a primary issue in food production for human uses. The incidence of antibiotic resistance cases among pathogenic microorganisms posed a threat to public health and the environment. The study of Osimani et al. (2018) revealed that high antibiotic resistance cases were reported among microorganisms isolated from laboratory reared mealworms. However, no antibiotic residues were detected in the feed given to the mealworm. The study concluded that antibiotic resistance cases would happen naturally in mealworms. The presence of heavy metals such as cadmium, lead, arsenic and mercury in mealworms has become an issue in food safety. Many European countries have banned feed ingredients contaminated with heavy metals exceeding certain levels that can pose a threat to public health. Different insect meal is proned to accumulate specific heavy metals. For example, cadmium accumulation in black soldier fly, whereas arsenic can remain in mealworms (Schrögel and Wätjen 2019). Therefore, the frequency of monitoring insect-based products should be increased to avoid the flow of heavily contaminated products into the market.

According to the study by Broekman et al. (2015), consumption of mealworm proteins has a likelihood of causing allergic reactions. The study revealed that shrimp allergic patient has similar allergic reactions to mealworm. This suggested that people with shrimp allergies should avoid taking mealworm-based products. On the other hand, study by Han et al. (2014) concluded that feeding freeze-dried powder of mealworm larvae to the Sprague-Dawley rats under experimental condition did not develop genotoxicity. In another study by Han et al. (2016), no adverse effect was observed on Sprague-Dawley rats which received freeze-dried powdered mealworm at a dose of 3000 mg/kg/day. Therefore, mealworm is safe and does not harm a host's genomic property and health. Nevertheless, more studies are required to address the gaps and constraints on food safety issues before this protein source is introduced commercially.

5. Mealworm Meal in Animal Nutrition

Many small-scale mealworm farming has been carried out to cater to demand from the market to feed pet companions such as birds, mice, reptiles, and amphibians. Mealworm is available in live, frozen, dried and powder form. Live, frozen, and dried mealworms can be consumed directly, whereas powdered mealworms is used as feed additives or protein replacement in animal feed formulation. Hence, mealworm farming becomes an industry where many people earn a living. Therefore, the impact of mealworms on animal nutrition, economic and social, is huge. This section will discuss and summarise the roles and impacts of mealworms for animal production, namely livestock, poultry, and aquaculture.

5.1 Livestock

Based on the literature review survey, most of the studies only focused on the viability of mealworms as a protein source in monogastric livestock, i.e., pigs and rabbits (Table 4). A recent study by Volek et al. (2021) showed that mealworm could be used as a feed additive in rabbit feed. Although diet containing mealworm did not improve growth performance significantly when compared to the control group, the inclusion also had no adverse effects on the rabbits. Besides, it was found that mealworm can be used as a feed additive and protein replacement of fishmeal in pig feed. For instance, mealworms was utilized as a feed additive to promote growth and enhance the health of pigs, as claimed in the study of Jin et al. (2016). Moreover, studies by Meyer et al. (2020) documented that the application of mealworm as a feed additive in pig feed has no adverse effect on pigs' growth performance and health. Mealworm performed promising finding as protein replacement of fish meal in pig feed. Yoo et al. (2019) revealed that mealworms could replace 10% of fishmeal in the diet of growing pig without any adverse effect on the pig's growth performance and health. Mealworm can also replace fishmeal in pig feed as high as 50% in the study of Ko et al. (2020). Another study by Ao et al. (2020) claimed that mealworms could only reduce fish meal proportion in pig feed but cannot 100% replace fishmeal as the protein source in pig feed. Overall, replacing mealworms in pig feed is promising and can reduce the reliance on conventional protein sources such as fishmeal in pig feed formulation.

Table 4. Impact of mealworm on livestock farming.

Livestock	Dosage + duration	Impact	References
Weaner piglet	1.5–6 % of diet + 35 days	+ Growth performance + Health	Jin et al. (2016)
Weaner piglet	2% of diet + 28 days	± Growth performance ± Health	Jonas-Levi and Martinez (2017)
Grower pig	10% of fish meal replacement + 28 days	+ Can digest mealworm	Yoo et al. (2019)
Weaner piglet	50% of fish meal replacement + 28 days	+ Average daily gain	Ko et al. (2020)
Weaner piglet	5–10% of diet + 28 days	± Growth performance ± Health	Meyer et al. (2020)
Rabbit	30 g/kg diet + 47 days	± Growth performance – Nitrogen in urine	Volek et al. (2021)

± = no significant difference in comparison to the control group;
+ = positive response with a significant difference in comparison to the control group;
– = negative response with a significant difference in comparison to the control group.

Poultry

Mealworm can only be used as feed additive in poultry farming, as described in many studies in the literature. Mealworm was reported to be viable as a feed additive without any adverse effect on growth performance and health of the broiler (Biasato et al. 2019, Bovera et al. 2016, Ramos-Elorduy et al. 2002) and layer (Stastnik et al. 2021). In the meantime, many studies revealed the huge potential of mealworms as a feed additive in poultry feed formulation to promote growth performance, increase flesh quality, enhance the immune system and stimulate disease resistance of poultry. Islam and Yang (2017) claimed that mealworm combined with two probiotics, namely lactic acid bacteria, *Lactobacillus plantarum* and yeast, *Saccharomyces cerevisiae* can stimulate disease-resistant of Ross male broiler chick against *Salmonella enteritidis* and *Escherichia coli*. Furthermore, Hussain et al. (2017) revealed that mealworms could help broiler chick stimulate disease resistance to Newcastle disease. Sedgh-Gooya et al. (2021) found that mealworms can significantly contribute to broiler weight gain and reduce *E. coli* in the broiler. The study of mealworms' role in poultry farming is well documented and established. Appropriate dosage of mealworm as a feed additive in poultry feed can increase poultry production.

Table 5. Impact of mealworm on poultry farming.

Poultry species	Dosage + duration	Impact	References
Broiler	5–10% of diet + 15 days	± Growth performance	Ramos-Elorduy et al. (2002)
Broiler	0.5–10% of diet + 3 weeks	+ Growth performance	Ballitoc and Sun (2013)
Free range chicken	75 g/kg diet + 97 days	+ Safe to use in poultry diet	Biasato et al. (2016)
Broiler	20 g/kg diet + 32 days	± Growth performance ± Flesh quality	Bovera et al. (2016)
Ross male broiler chick	0.4% of diet + 7 days	+ Disease resistance to *Salmonella* and *Escherichia coli*	Islam and Yang (2017)
Broiler chick	1–3 g/kg diet + 6 weeks	± Feed intake + Growth performance + Disease resistance to Newcastle disease	Hussain et al. (2017)
Broiler	5 g/kg diet + 5 weeks	+ FCR + Flesh quality	Khan et al. (2018)
Male broiler	7.5% of diet + 52 days	± Growth performance ± Flesh quality	Biasato et al. (2019)
Broiler	2.5–5 % of diet + 42 days	+ FCR + Body weight gain + Eliminate *E. coli*	Sedgh-Gooya et al. (2021)
Ross 308 male broiler chick	4% of diet + 42 days	+ Growth performance	Elahi et al. (2020)
Broiler	5% of diet + 38 days	+ FCR + Health	Bellezza Oddon et al. (2021)
Lohmann Brown classic hen, layer	2.5–5% of diet + 25 weeks	± Growth performance ± Flesh quality	Stastnik et al. (2021)

± = no significant difference in comparison to the control group;
+ = positive response with a significant difference in comparison to the control group;
– = negative response with a significant difference in comparison to the control group.

5.2 Aquaculture

The application of fishmeal in aquaculture species feed has been debated for a long time. As aquaculture is a fast-growing industry, more fishmeal is needed in the near future (Olsen and Hasan 2012). Thus, the aquaculture industry faces its sustainability issue as the demand for fishmeal for aquaculture species feed is expected to increase. Therefore, alternative protein sources such as mealworms should be introduced into aquaculture feed formulation. A recent study by Quang Tran et al. (2022) suggested that insect meal, especially mealworm, can be used as a feed additive

Table 6. Impact of mealworm on aquaculture species farming.

Aquaculture species	Dosage + duration	Impact	References
African catfish, *Clarias gariepinus*	20–80% replacement + 7 weeks	+ Growth performance	Ng et al. (2001)
Rainbow trout, *Oncorhynchus mykiss*	25–50% replacement + 3 weeks	– Growth performance – Flesh quality	Belforti et al. (2015)
Pacific white shrimp, *Litopenaeus vannamei*	25–100% + 6 weeks	± Growth performance ± Survival rate	Panini et al. (2017)
Gilthead seabream (*Sparus aurata*)	25% + 163 days	± Growth performance ± Market acceptance	Piccolo et al. (2017)
Blackspot sea bream (*Pagellus bogaraveo*)	25–50% fishmeal replacement + 131 days	± Growth performance – Flesh quality	Laconisi et al. (2017)
Juvenile rockfish (*Sebastes schlegeli*)	< 16% > 16 % replacement + 8 weeks	+ Growth performance – Growth performance	Khosravi et al. (2018)
European sea bass, *Dicentrarchus labrax*	25–50% replacement of fish meal + 6 weeks	+ Immune system + Disease resistant to bacteria and parasites	Henry et al. (2018)
Juvenile Mandarin fish, *Siniperca scherzeri*	10–20% fish meal replacement > 20%–30% fish meal replacement	+ Growth performance – Growth performance	Sankian et al. (2018)
Black Sea trout (*Salmo trutta labrax*)	2% of diet + 50 days	– Growth performance + Egg quality	Gelinçek and Yamaner (2020)
Rainbow trout, *Oncorhynchus mykiss*	7 – 14% protein replacement + 8 weeks > 14% protein replacement	+ Growth performance + FCR – Growth performance	Jeong et al. (2020)
Sea trout (*Salmo trutta m. trutta* L.)	10–40 % replacement + 8 weeks	+ Growth performance + Feed utilization + Health	Hoffmann et al. (2021)
Narrow-clawed crayfish (*Pontastacus leptodactylus*)	50% replacement + 80 days	+ Growth performance	Mazlum et al. (2021)
Juvenile olive flounder, *Paralichthys olivaceus*	40% replacement + 8 weeks	+ Growth performance + Feed utilization + Health	Jeong et al. (2021)
Juvenile of Black porgy, *Acanthopagrus schlegelii*	60% protein replacement + 12 weeks	+ Growth performance + Weight gain + Health	Jeong et al. (2022)
Juvenile largemouth bass, *Micropterus salmonids*	4% of diet + 8 weeks	± Growth performance + Health	Gu et al. (2022)
Rainbow trout, *Oncorhynchus mykiss*	50% of replacement + 21 days	+ Growth performance + Immune system	Melenchón et al. (2022)

± = no significant difference in comparison to the control group;
+ = positive response with a significant difference in comparison to the control group;
– = negative response with a significant difference in comparison to the control group.

and protein replacement in aquaculture species feed formulation. Most of the studies documented that mealworm can improve growth performance, enhance health, and stimulate disease resistance of aquaculture species. For instance, Henry et al. (2018) claimed that dietary mealworm meal could stimulate anti-parasitic defense of European sea bass, *Dicentrarchus labrax* because both exoskeletons of the mealworm and fish parasite share common composition that possibly serving as an immunostimulant which enhance the fish immune system. Despite the fact that the chitin content of mealworm meal was anticipated to enhance resistance to bacterial diseases in fish, the addition of exogenous proteases considerably decreased the antibacterial efficacy against *E. coli*, in the same study. However, Panini et al. (2017) found that mealworm lacks amino acid methionine which is, essential for aquaculture species. Therefore, the additional commercial amino acid in aquaculture species formulation is needed. The study on the impact of using mealworms as a feed additive and protein replacement in aquaculture is extensive. Many studies revealed the promising result of using mealworms in aquaculture species feed formulation. Hence, mealworm has a tremendous prospect of replacing or reducing fishmeal's overreliance in aquaculture species feed formulation. An appropriate dosage of mealworm to use in aquaculture species feed formulation will enhance the viability of this insect meal as an alternative protein source for now and future uses.

6. Conclusion and Future Remarks

The intensification of animal farming requires the use of sustainable raw materials to support the expansion of food production required to feed the world's rapidly growing population. At present, mealworm is the best candidate for the new sustainable raw materials for animal production as it has a lesser ecological footprint. This insect is easy to cultivate by using agricultural wastes as feed, and it could sustain the current agricultural industry. Being rich in nutrient content such as protein, amino acid, fatty acid, and medicinal properties allowed mealworms to compete with other conventional protein sources. However, several constraints hinder the application of mealworms as alternative protein sources. A reservoir of harmful pathogens, antibiotic, heavy metals and pesticides residues, allergens and toxin compounds are several issues that needed to be tackled before utilizing mealworms as a feed additive, and the source of protein replacement for soybean meal and fishmeal in the animals' production. Although some studies highlighted the setback of using mealworms, many showed promising findings in utilizing mealworms in livestock, poultry, and aquaculture. Overall, mealworm as an alternative protein source is in the pipeline. Extension works with the support of blockchain technology in providing transparency, traceability, and efficiency throughout the supply chain of mealworm are highly recommended.

References

Ao, X., Yoo, J., Wu, Z. and Kim, I. 2020. Can dried mealworm (*Tenebrio molitor*) larvae replace fish meal in weaned pigs? Livestock Science 239: 104103.

Ballitoc, D.A. and Sun, S. 2013. Ground yellow mealworms (*Tenebrio molitor* L.) feed supplementation improves growth performance and carcass yield characteristics in broilers. Open Sci. Repos. Agric, e23050425.

Bańbura, J., Babura, M., Gldalski, M., Kaliński, A., Marciniak, B., Markowski, M., et al. 2013. Consequences of experimental changes in the rearing conditions of Blue Tit Cyanistes caeruleus and Great Tit Parus major nestlings. Acta Ornithologica 48: 129–139.

Belforti, M., Gai, F., Lussiana, C., Renna, M., Malfatto, V., Rotolo, L., et al. 2015. *Tenebrio molitor* meal in rainbow trout (*Oncorhynchus mykiss*) diets: effects on animal performance, nutrient digestibility and chemical composition of fillets. Italian Journal of Animal Science 14: 4170.

Bellezza Oddon, S., Biasato, I., Imarisio, A., Pipan, M., Dekleva, D., Colombino, E., et al. 2021. Black soldier fly and yellow mealworm live larvae for broiler chickens: effects on bird performance and health status. Journal of Animal Physiology and Animal Nutrition 105: 10–18.

Biasato, I., De Marco, M., Rotolo, L., Renna, M., Lussiana, C., Dabbou, S., et al. 2016. Effects of dietary *Tenebrio molitor* meal inclusion in free-range chickens. Journal of Animal Physiology and Animal Nutrition 100: 1104–1112.

Biasato, I., Ferrocino, I., Grego, E., Dabbou, S., Gai, F., Gasco, L., et al. 2019. Gut microbiota and mucin composition in female broiler chickens fed diets including yellow mealworm (*Tenebrio molitor*, L.). Animals 9, 213.

Boer, J., Helms, M. and Aiking, H. 2006. Protein consumption and sustainability: diet diversity in EU-15. Ecological Economics 59: 267–274.

Bordiean, A., Krzyżaniak, M., Stolarski, M.J., Czachorowski, S. and Peni, D. 2020. Will yellow mealworm become a source of safe proteins for Europe? Agriculture 10: 233.

Bosch, G., Zhang, S., Oonincx, D.G. and Hendriks, W.H. 2014. Protein quality of insects as potential ingredients for dog and cat foods. Journal of Nutritional Science 3.

Bosch, G., Vervoort, J. and Hendriks, W. 2016. *In vitro* digestibility and fermentability of selected insects for dog foods. Animal Feed Science and Technology 221: 174–184.

Bovera, F., Piccolo, G., Gasco, L., Marono, S., Loponte, R., Vassalotti, G., et al. 2015. Yellow mealworm larvae (*Tenebrio molitor*, L.) as a possible alternative to soybean meal in broiler diets. British Poultry Science 56: 569–575.

Bovera, F., Loponte, R., Marono, S., Piccolo, G., Parisi, G., Iaconisi, V., et al. 2016. Use of *Tenebrio molitor* larvae meal as protein source in broiler diet: effect on growth performance, nutrient digestibility, and carcass and meat traits. Journal of Animal Science 94: 639–647.

Brandon, A.M., Gao, S.-H., Tian, R., Ning, D., Yang, S.-S., Zhou, J., et al. 2018. Biodegradation of polyethylene and plastic mixtures in mealworms (larvae of *Tenebrio molitor*) and effects on the gut microbiome. Environmental Science & Technology 52: 6526–6533.

Broekman, H., Knulst, A., Den Hartog Jager, S., Gaspari, M., De Jong, G., Houben, G., et al. 2015. Shrimp allergic patients are at risk when eating mealworm proteins. Clinical and Translational Allergy 5: 1–1.

Crippen, T., Zheng, L., Sheffield, C., Tomberlin, J., Beier, R. and Yu, Z. 2012. Transient gut retention and persistence of Salmonella through metamorphosis in the lesser mealworm, *Alphitobius diaperinus* (Coleoptera: Tenebrionidae). Journal of Applied Microbiology 112: 920–926.

Crippen, T., Sheffield, C., Beier, R. and Nisbet, D. 2018. The horizontal transfer of Salmonella between the lesser mealworm (*Alphitobius diaperinus*) and poultry manure. Zoonoses and Public Health 65: e23–e33.

Dávalos Terán, I., Imai, K., Lacroix, I.M., Fogliano, V. and Udenigwe, C.C. 2020. Bioinformatics of edible yellow mealworm (*Tenebrio molitor*) proteome reveal the cuticular proteins as promising precursors of dipeptidyl peptidase-IV inhibitors. Journal of Food Biochemistry 44: e13121.

Ding, Q., Wu, R.A., Shi, T., Yu, Y., Yan, Y., Sun, N., et al. 2021. Antiproliferative effects of mealworm larvae (*Tenebrio molitor*) aqueous extract on human colorectal adenocarcinoma (Caco-2) and hepatocellular carcinoma (HepG2) cancer cell lines. Journal of Food Biochemistry 45: e13778.

Elahi, U., Wang, J., Ma, Y.-b., Wu, S.-g., Wu, J., Qi, G.-h., et al. 2020. Evaluation of yellow mealworm meal as a protein feedstuff in the diet of broiler chicks. Animals 10: 224.

Elieh-Ali-Komi, D. and Hamblin, M.R. 2016. Chitin and chitosan: production and application of versatile biomedical nanomaterials. International journal of Advanced Research 4: 411.

Evans, A. and Goodliffe, E. 1939. The utilisation of food by the larva of the mealworm *Tenebrio molitor* L.(Coleopb.), Proceedings of the Royal Entomological Society of London. Series A, General Entomology. Wiley Online Library, pp. 57–62.

Evans, J., Alemu, M.H., Flore, R., Frøst, M.B., Halloran, A., Jensen, A.B., et al. 2015. 'Entomophagy': an evolving terminology in need of review. Journal of Insects as Food and Feed 1: 293–305.

Finke, M.D. 2002. Complete nutrient composition of commercially raised invertebrates used as food for insectivores. Zoo Biology: Published in Affiliation with the American Zoo and Aquarium Association 21: 269–285.

Gelinçek, İ. and Yamaner, G. 2020. An investigation on the gamete quality of Black Sea trout (*Salmo trutta labrax*) broodstock fed with mealworm (*Tenebrio molitor*). Aquaculture Research 51: 2379–2388.

Gerber, P.J., Steinfeld, H., Henderson, B., Mottet, A., Opio, C., Dijkman, J., et al. 2013. Tackling climate change through livestock: a global assessment of emissions and mitigation opportunities. Food and Agriculture Organization of the United Nations (FAO).

Gindin, G., Glazer, I., Mishoutchenko, A. and Samish, M. 2009. Entomopathogenic fungi as a potential control agent against the lesser mealworm, *Alphitobius diaperinus* in broiler houses. BioControl 54: 549–558.

Grau, T., Vilcinskas, A. and Joop, G. 2017. Sustainable farming of the mealworm *Tenebrio molitor* for the production of food and feed. Zeitschrift für Naturforschung C 72: 337–349.

Gregorovičová, M. and Černíková, A. 2016. Reactions of leopard geckos (*Eublepharis macularius*) to defensive secretion of *Graphosoma lineatum* (*Heteroptera Pentatomidae*): an experimental approach. Ethology Ecology & Evolution 28: 367–384.

Gu, J., Liang, H., Ge, X., Xia, D., Pan, L., Mi, H., et al. 2022. A study of the potential effect of yellow mealworm (*Tenebrio molitor*) substitution for fish meal on growth, immune and antioxidant capacity in juvenile largemouth bass (*Micropterus salmoides*). Fish & Shellfish Immunology 120: 214–221.

Han, S.-R., Yun, E.-Y., Kim, J.-Y., Hwang, J.S., Jeong, E.J. and Moon, K.-S. 2014. Evaluation of genotoxicity and 28-day oral dose toxicity on freeze-dried powder of *Tenebrio molitor* larvae (Yellow Mealworm). Toxicological Research 30: 121–130.

Han, S.-R., Lee, B.-S., Jung, K.-J., Yu, H.-J., Yun, E.-Y., Hwang, J.S., et al. 2016. Safety assessment of freeze-dried powdered *Tenebrio molitor* larvae (yellow mealworm) as novel food source: Evaluation of 90-day toxicity in Sprague-Dawley rats. Regulatory Toxicology and Pharmacology 77: 206–212.

Hardouin, J. and Mahoux, G. 2003. Zootechnie d'insects-Breeding and use for the benefit of man and certain animals. Bureau for the exchange and distribution of information on mini-livestock (BEDIM) 164pp.

Henry, M., Gasco, L., Chatzifotis, S. and Piccolo, G. 2018. Does dietary insect meal affect the fish immune system? The case of mealworm, *Tenebrio molitor* on European sea bass, *Dicentrarchus labrax*. Developmental & Comparative Immunology 81: 204–209.

Hoffmann, L., Rawski, M., Nogales-Mérida, S., Kołodziejski, P., Pruszyńska-Oszmałek, E. and Mazurkiewicz, J. 2021. Mealworm meal use in sea trout (*Salmo trutta m. trutta*, L.) fingerling diets: effects on growth performance, histomorphology of the gastrointestinal tract and blood parameters. Aquaculture Nutrition 27: 1512–1528.

Hussain, I., Khan, S., Sultan, A., Chand, N., Khan, R., Alam, W., et al. 2017. Meal worm (*Tenebrio molitor*) as potential alternative source of protein supplementation in broiler. Int. J. Biosci. 10: 225–262.

Islam, M.M. and Yang, C.-J. 2017. Efficacy of mealworm and super mealworm larvae probiotics as an alternative to antibiotics challenged orally with *Salmonella* and *E. coli* infection in broiler chicks. Poultry Science 96: 27–34.

Jeon, Y.-H., Son, Y.-J., Kim, S.-H., Yun, E.-Y., Kang, H.-J. and Hwang, I.-K. 2016. Physicochemical properties and oxidative stabilities of mealworm (*Tenebrio molitor*) oils under different roasting conditions. Food Science and Biotechnology 25: 105–110.

Jeong, S.-M., Khosravi, S., Mauliasari, I.R. and Lee, S.-M. 2020. Dietary inclusion of mealworm (*Tenebrio molitor*) meal as an alternative protein source in practical diets for rainbow trout (*Oncorhynchus mykiss*) fry. Fisheries and Aquatic Sciences 23: 1–8.

Jeong, S.-M., Khosravi, S., Yoon, K.-Y., Kim, K.-W., Lee, B.-J., Hur, S.-W., et al. 2021. Mealworm, *Tenebrio molitor*, as a feed ingredient for juvenile olive flounder, *Paralichthys olivaceus*. Aquaculture Reports 20: 100747.

Jeong, S.-M., Khosravi, S., Kim, K.-W., Lee, B.-J., Hur, S.-W., You, S.-G., et al. 2022. Potential of mealworm, *Tenebrio molitor*, meal as a sustainable dietary protein source for juvenile black porgy, *Acanthopagrus schlegelii*. Aquaculture Reports 22: 100956.

Jin, X., Heo, P., Hong, J., Kim, N. and Kim, Y. 2016. Supplementation of dried mealworm (*Tenebrio molitor larva*) on growth performance, nutrient digestibility and blood profiles in weaning pigs. Asian-Australasian Journal of Animal Sciences 29: 979.

Jonas-Levi, A. and Martinez, J.-J.I. 2017. The high level of protein content reported in insects for food and feed is overestimated. Journal of Food Composition and Analysis 62: 184–188.

Jongema, Y. 2015. World list of edible insects. Wageningen University 75.

Karau, A. and Grayson, I. 2014. Amino acids in human and animal nutrition. Biotechnology of Food and Feed Additives, 189–228.

Kari, Z.A., Kabir, M.A., Razab, M.K.A.A., Munir, M.B., Lim, P.T. and Wei, L.S. 2020. A replacement of plant protein sources as an alternative of fish meal ingredient for African catfish, *Clarias gariepinus*: a review. Journal of Tropical Resources and Sustainable Science 8: 47–59.

Kari, Z.A., Kabir, M.A., Mat, K., Rusli, N.D., Razab, M.K.A.A., Ariff, N.S.N.A., et al. 2021. The possibility of replacing fish meal with fermented soy pulp on the growth performance, blood biochemistry, liver, and intestinal morphology of African catfish (*Clarias gariepinus*). Aquaculture Reports 21: 100815.

Kari, Z.A., Kabir, M.A., Dawood, M.A., Razab, M.K.A.A., Ariff, N.S.N.A., Sarkar, T., et al. 2022. Effect of fish meal substitution with fermented soy pulp on growth performance, digestive enzyme, amino acid profile, and immune-related gene expression of African catfish (*Clarias gariepinus*). Aquaculture 546: 737418.

Khan, S., Khan, R., Alam, W. and Sultan, A. 2018. Evaluating the nutritive profile of three insect meals and their effects to replace soya bean in broiler diet. Journal of Animal Physiology and Animal Nutrition 102: e662–e668.

Khosravi, S., Kim, E., Lee, Y.S. and Lee, S.M. 2018. Dietary inclusion of mealworm (*Tenebrio molitor*) meal as an alternative protein source in practical diets for juvenile rockfish (*Sebastes schlegeli*). Entomological Research 48: 214–221.

Kim, J.S., Choi, J.Y., Lee, S.J., Lee, J.H., Fu, Z., Skinner, M., et al. 2013. Transformation of *Beauveria bassiana* to produce EGFP in *Tenebrio molitor* for use as animal feed additives. FEMS Microbiology Letters 344: 173–178.

Kim, S.Y., Park, J.B., Lee, Y.B., Yoon, H.J., Lee, K.Y. and Kim, N.J. 2015. Growth characteristics of mealworm *Tenebrio molitor*. Journal of Sericultural and Entomological Science 53: 1–5.

Ko, H., Kim, Y. and Kim, J. 2020. The produced mealworm meal through organic wastes as a sustainable protein source for weanling pigs. Journal of Animal Science and Technology 62: 365.

Kwon, G.T., Yuk, H.-G., Lee, S.J., Chung, Y.H., Jang, H.S., Yoo, J.-S., et al. 2020. Mealworm larvae (*Tenebrio molitor* L.) exuviae as a novel prebiotic material for BALB/c mouse gut microbiota. Food Science and Biotechnology 29: 531–537.

Laconisi, V., Marono, S., Parisi, G., Gasco, L., Genovese, L., Maricchiolo, G., et al. 2017. Dietary inclusion of *Tenebrio molitor* larvae meal: Effects on growth performance and final quality treats of blackspot sea bream (*Pagellus bogaraveo*). Aquaculture 476: 49–58.

Li, L., Zhao, Z. and Liu, H. 2013. Feasibility of feeding yellow mealworm (*Tenebrio molitor* L.) in bioregenerative life support systems as a source of animal protein for humans. Acta Astronautica 92: 103–109.

Liu, C., Masri, J., Perez, V., Maya, C. and Zhao, J. 2020. Growth performance and nutrient composition of mealworms (*Tenebrio molitor*) fed on fresh plant materials-supplemented diets. Foods 9: 151.

Lou, Y., Li, Y., Lu, B., Liu, Q., Yang, S.-S., Liu, B., et al. 2021. Response of the yellow mealworm (*Tenebrio molitor*) gut microbiome to diet shifts during polystyrene and polyethylene biodegradation. Journal of Hazardous Materials 416: 126222.

Mariod, A.A. 2020. African edible insects as alternative source of food, oil, protein and bioactive components. Springer Nature.

Martin, R., Rivers, J. and Cowgill, U. 1976. Culturing mealworms as food for animals in captivity. International Zoo Yearbook 16: 63–70.

Mazlum, Y., Turan, F. and Bircan Yıldırım, Y. 2021. Evaluation of mealworms (*Tenebrio molitor*) meal as an alternative protein source for narrow-clawed crayfish (*Pontastacus leptodactylus*) juveniles. Aquaculture Research 52: 4145–4153.

Melenchón, F., de Mercado, E., Pula, H.J., Cardenete, G., Barroso, F.G., Fabrikov, D., et al. 2022. Fishmeal dietary replacement Up to 50%: a comparative study of two insect meals for rainbow trout (*Oncorhynchus mykiss*). Animals 12: 179.

Meyer, S., Gessner, D.K., Braune, M.S., Friedhoff, T., Most, E., Höring, M., et al. 2020. Comprehensive evaluation of the metabolic effects of insect meal from *Tenebrio molitor* L. in growing pigs by transcriptomics, metabolomics and lipidomics. Journal of Animal Science and Biotechnology 11: 1–19.

Meyerrochow, V.B. 1975. Can Insects Help to Ease Problem of World Food Shortage 6: 261–262.

Nag, M., Lahiri, D., Mukherjee, D., Banerjee, R., Garai, S., Sarkar, T., et al. 2021. Functionalized chitosan nanomaterials: a jammer for quorum sensing. Polymers 13: 2533.

Nebbia, S., Lamberti, C., Giorgis, V., Giuffrida, M.G., Manfredi, M., Marengo, E., et al. 2019. The cockroach allergen-like protein is involved in primary respiratory and food allergy to yellow mealworm (*Tenebrio molitor*). Clinical and Experimental Allergy: Journal of the British Society for Allergy and Clinical Immunology 49: 1379–1382.

Ng, W.K., Liew, F.L., Ang, L.P. and Wong, K.W. 2001. Potential of mealworm (*Tenebrio molitor*) as an alternative protein source in practical diets for African catfish, *Clarias gariepinus*. Aquaculture Research 32: 273–280.

Olsen, R.L. and Hasan, M.R. 2012. A limited supply of fishmeal: impact on future increases in global aquaculture production. Trends in Food Science & Technology 27: 120–128.

Oonincx, D.G. and De Boer, I.J. 2012. Environmental impact of the production of mealworms as a protein source for humans–a life cycle assessment. PloS one 7: e51145.

Oonincx, D.G., Van Broekhoven, S., Van Huis, A. and van Loon, J.J. 2015. Feed conversion, survival and development, and composition of four insect species on diets composed of food by-products. PloS one 10: e0144601.

Osimani, A., Milanović, V., Cardinali, F., Garofalo, C., Clementi, F., Ruschioni, S., et al. 2018. Distribution of transferable antibiotic resistance genes in laboratory-reared edible mealworms (*Tenebrio molitor* L.). Frontiers in Microbiology, 2702.

Panini, R.L., Freitas, L.E.L., Guimarães, A.M., Rios, C., da Silva, M.F.O., Vieira, F.N., et al. 2017. Potential use of mealworms as an alternative protein source for Pacific white shrimp: Digestibility and performance. Aquaculture 473: 115–120.

Pasmans, F., Janssens, G.P., Sparreboom, M., Jiang, J. and Nishikawa, K. 2012. Reproduction, development, and growth response to captive diets in the Shangcheng stout salamander, *Pachyhynobius shangchengensis* (Amphibia, Urodela, Hynobiidae). Asian Herpetol. Res. 3: 192–197.

Piccolo, G., Iaconisi, V., Marono, S., Gasco, L., Loponte, R., Nizza, S., et al. 2017. Effect of *Tenebrio molitor* larvae meal on growth performance, *in vivo* nutrients digestibility, somatic and marketable indexes of gilthead sea bream (*Sparus aurata*). Animal Feed Science and Technology 226: 12–20.

Poma, G., Cuykx, M., Amato, E., Calaprice, C., Focant, J.F. and Covaci, A. 2017. Evaluation of hazardous chemicals in edible insects and insect-based food intended for human consumption. Food and Chemical Toxicology 100: 70–79.

Quang Tran, H., Van Doan, H. and Stejskal, V. 2022. Environmental consequences of using insect meal as an ingredient in aquafeeds: a systematic view. Reviews in Aquaculture 14: 237–251.

Ramos-Elorduy, J., González, E.A., Hernández, A.R. and Pino, J.M. 2002. Use of *Tenebrio molitor* (Coleoptera: Tenebrionidae) to recycle organic wastes and as feed for broiler chickens. Journal of Economic Entomology 95: 214–220.

Ramos-Elorduy, J. 2009. Anthropo-entomophagy: cultures, evolution and sustainability. Entomological Research 39: 271–288.

Raubenheimer, D. and Rothman, J.M. 2013. Nutritional ecology of entomophagy in humans and other primates. Annual Review of Entomology 58: 141–160.

Ravzanaadii, N., Kim, S.-H., Choi, W.-H., Hong, S.-J. and Kim, N.-J. 2012. Nutritional value of mealworm, *Tenebrio molitor* as food source. International Journal of Industrial Entomology 25: 93–98.

Renault, D., Hervant, F. and Vernon, P. 2003. Effect of food shortage and temperature on oxygen consumption in the lesser mealworm, *Alphitobius diaperinus* (Panzer) (Coleoptera: Tenebrionidae). Physiological Entomology 28: 261–267.

Sankian, Z., Khosravi, S., Kim, Y.-O. and Lee, S.-M. 2018. Effects of dietary inclusion of yellow mealworm (*Tenebrio molitor*) meal on growth performance, feed utilization, body composition, plasma biochemical indices, selected immune parameters and antioxidant enzyme activities of mandarin fish (*Siniperca scherzeri*) juveniles. Aquaculture 496: 79–87.

Sauvant, D., Perez, J.-M. and Tran, G. 2004. Tables of composition and nutritional value of feed materials: pigs, poultry, cattle, sheep, goats, rabbits, horses and fish. Wageningen Academic Publishers.

Schrögel, P. and Wätjen, W. 2019. Insects for food and feed-safety aspects related to mycotoxins and metals. Foods 8: 288.

Sedgh-Gooya, S., Torki, M., Darbemamieh, M., Khamisabadi, H., Karimi Torshizi, M.A. and Abdolmohamadi, A. 2021. Yellow mealworm, *Tenebrio molitor* (Col: Tenebrionidae), larvae powder as dietary protein sources for broiler chickens: effects on growth performance, carcass traits, selected intestinal microbiota and blood parameters. Journal of Animal Physiology and Animal Nutrition 105: 119–128.

Shin, C.-S., Kim, D.-Y. and Shin, W.-S. 2019. Characterization of chitosan extracted from Mealworm Beetle (*Tenebrio molitor, Zophobas morio*) and Rhinoceros Beetle (*Allomyrina dichotoma*) and their antibacterial activities. International journal of Biological Macromolecules 125: 72–77.

Siemianowska, E., Kosewska, A., Aljewicz, M., Skibniewska, K.A., Polak-Juszczak, L., Jarocki, A., et al. 2013. Larvae of mealworm (*Tenebrio molitor* L.) as European Novel Food.

Silva, F.W., Araujo, L.S., Azevedo, D.O., Serrão, J.E. and Elliot, S.L. 2016. Physical and chemical properties of primary defences in *Tenebrio molitor*. Physiological Entomology 41: 121–126.

Smil, V. 2002. Eating meat: evolution, patterns, and consequences. Population and Development Review 28: 599–639.

Son, Y.-J., Choi, S.Y., Hwang, I.-K., Nho, C.W. and Kim, S.H. 2020. Could defatted mealworm (*Tenebrio molitor*) and mealworm oil be used as food ingredients? Foods 9: 40.

Son, Y.-J., Hwang, I.-K., Nho, C.W., Kim, S.M. and Kim, S.H. 2021. Determination of carbohydrate composition in mealworm (*Tenebrio molitor* L.) larvae and characterization of mealworm chitin and chitosan. Foods 10: 640.

Song, Y.S., Kim, M.W., Moon, C., Seo, D.J., Han, Y.S., Jo, Y.H., et al. 2018. Extraction of chitin and chitosan from larval exuvium and whole body of edible mealworm, *Tenebrio molitor*. Entomological Research 48: 227–233.

Stastnik, O., Novotny, J., Roztocilova, A., Kouril, P., Kumbar, V., Cernik, J., et al. 2021. Safety of mealworm meal in layer diets and their influence on gut morphology. Animals 11: 1439.

Stoops, J., Crauwels, S., Waud, M., Claes, J., Lievens, B. and Van Campenhout, L. 2016. Microbial community assessment of mealworm larvae (*Tenebrio molitor*) and grasshoppers (*Locusta migratoria migratorioides*) sold for human consumption. Food Microbiology 53: 122–127.

Szczepanik, M., Walczak, M., Zawitowska, B., Michalska-Sionkowska, M., Szumny, A., Wawrzeńczyk, C., et al. 2018. Chemical composition, antimicromicrobial activity and insecticidal activity against the lesser mealworm *Alphitobius diaperinus* (Panzer) (Coleoptera: Tenebrionidae) of *Origanum vulgare* L. ssp. hirtum (Link) and *Artemisia dracunculus* L. essential oils. Journal of the Science of Food and Agriculture 98 767–774.

Szołyga, B., Gniłka, R., Szczepanik, M. and Szumny, A. 2014. Chemical composition and insecticidal activity of *Thuja occidentalis* and *Tanacetum vulgare* essential oils against larvae of the lesser mealworm, *Alphitobius diaperinus*. Entomologia Experimentalis et Applicata 151: 1–10.

Tan, H.S.G., Fischer, A.R., Tinchan, P., Stieger, M., Steenbekkers, L. and van Trijp, H.C. 2015. Insects as food: exploring cultural exposure and individual experience as determinants of acceptance. Food Quality and Preference 42: 78–89.

Thévenot, A., Rivera, J.L., Wilfart, A., Maillard, F., Hassouna, M., Senga-Kiesse, T., et al. 2018. Mealworm meal for animal feed: environmental assessment and sensitivity analysis to guide future prospects. Journal of Cleaner Production 170: 1260–1267.

Van Broekhoven, S., Bastiaan-Net, S., de Jong, N.W. and Wichers, H.J. 2016. Influence of processing and *in vitro* digestion on the allergic cross-reactivity of three mealworm species. Food Chemistry 196: 1075–1083.

Van Huis, A. 2003. Insects as food in sub-Saharan Africa. International Journal of Tropical Insect Science 23: 163–185.

Vigneron, A., Jehan, C., Rigaud, T. and Moret, Y. 2019. Immune defenses of a beneficial pest: the mealworm beetle, *Tenebrio molitor*. Frontiers in Physiology 10: 138.

Volek, Z., Adámková, A., Zita, L., Adámek, M., Plachý, V., Mlček, J., et al. 2021. The effects of the dietary replacement of soybean meal with yellow mealworm larvae (*Tenebrio molitor*) on the growth, nutrient digestibility and nitrogen output of fattening rabbits. Animal Feed Science and Technology 280: 115048.

Wang, D., Zhai, S.W., Zhang, C.X., Bai, Y.Y., An, S.H. and Xu, Y.N. 2005. Evaluation on nutritional value of field crickets as a poultry feedstuff. Asian-australasian Journal of Animal Sciences 18: 667–670.

Wynants, E., Frooninckx, L., Van Miert, S., Geeraerd, A., Claes, J. and Van Campenhout, L. 2019. Risks related to the presence of *Salmonella* sp. during rearing of mealworms (*Tenebrio molitor*) for food or feed: survival in the substrate and transmission to the larvae. Food Control 100: 227–234.

Yoo, J., Cho, K., Hong, J., Jang, H., Chung, Y., Kwon, G., et al. 2019. Nutrient ileal digestibility evaluation of dried mealworm (*Tenebrio molitor*) larvae compared to three animal protein by-products in growing pigs. Asian-Australasian Journal of Animal Sciences 32: 387.

Zhao, X., Vázquez-Gutiérrez, J.L., Johansson, D.P., Landberg, R. and Langton, M. 2016. Yellow mealworm protein for food purposes-extraction and functional properties. PloS one 11: e0147791.

Zulhisyam, A.K., Kabir, M.A., Munir, M.B. and Wei, L.S. 2020. Using of fermented soy pulp as an edible coating material on fish feed pellet in African catfish (*Clarias gariepinus*) production. Aquaculture, Aquarium, Conservation & Legislation 13: 296–308.

10

Microalgae as Food Source

Marcele Leal Nörnberg, Pricila Pinheiro Nass, Patricia Acosta Caetano,
Luísa Chitolina Schetinger and *Leila Queiroz Zepka**

1. Introduction

Meeting the growing demand of the world population for nutritious diets is a global challenge this century. To address the risks of famine and worldwide food insecurity, contemporary food policy discourses center on the role of food systems in improving human nutrition (Chaudlhary et al. 2018, Goden et al. 2021).

Based on this understanding, microalgal products represent an opportunity for the global food system to nourish the world without exceeding planetary boundaries (Kusmayadi et al. 2021). It is estimated that the industrial relevance of microalgae as food ingredients or products is full commercial expansion. As a result, this market presents around 150 new companies each year (Deprá et al. 2020).

However, despite projections of promising growth, conventional large-scale production is still questionable. These barriers are attributed to (i) economic viability, (ii) low production capacities, (iii) legislative and regulatory issues, and (iv) sensory characteristics (Lafarga 2019).

Despite the numerous challenges to be overcome, the production of microalgae-based foods is a reality. Microalgae have been highlighted for their nutritional potential associated with the presence of bioactive chemical specialties (Souza et al. 2019, Jacob-Lopes et al. 2019, Rahman 2020, Nörnberg et al. 2022b,d).

Examples of microalgae-based products include plant-based foods, functional beverages, dry powders, supplements, capsules, and tablets, which are marketed worldwide (Lafarga et al. 2019).

Based on the above, the purpose of this chapter was to provide an overview of microalgae-based products. Here, we cover topics related to cultivation systems, nutritional composition and bioactivity, food products enriched with microalgae biomass, and, finally, regulatory framework and legislation.

2. Cultivation Strategies for Microalgae Production

Historically, circular ponds and open raceway ponds are classic types of photobioreactors that have been practiced at a large scale for the commercial production of microalgae (Fu et al. 2021,

Department of Food Science and Technology, Federal University of Santa Maria (UFSM), Roraima Avenue, 1000, 97105-900, Santa Maria, RS, Brazil.
* Corresponding author: zepkaleila@yahoo.com.br

Chew et al. 2018, Lam et al. 2018, Jacob-Lopes and Franco 2010). These are characterized by their cost-effectiveness and simplicity of operation. However, the major bottlenecks are the external agent's contaminations and the vulnerability of seasonal intervention, which directly inhibits their productivity (Mehariya et al. 2021). Table 1 summarizes the main culture systems currently used for this purpose.

Despite the numerous challenges to be overcome, open raceway ponds are undeniably well established in the industrial production of microalgae species such as *Nannochloropsis* sp., *Chlorella* sp., *Tetraselmis* sp., *Arthrospira platensis*, *Dunaliella salina*, *Scenedesmus* sp., *Haematococcus pluvialis* and *Phaeodactylum tricornutum* (Kumar et al. 2021).

For example, open raceway ponds represent a promising model for industrial production of astaxanthin, found substantially in species such as *Haematococcus pluvialis*. In practice, average astaxanthin yields are 2.10 g 100 g^{-1} dry wt in a 100 m^2 open system (Zhang et al. 2009, Yu et al. 2022).

On the other hand, closed-loop photobioreactors provide a system with controlled conditions (Fu et al. 2021). In particular, tubular closed-loop photobioreactors have been highlighted for their industrial relevance in microalgae production systems (Lam et al. 2018, Yu et al. 2022).

As a result, various tubular photobioreactors have been designed, including horizontal, vertical, conical, near horizontal, inclined types made of plastic or glass tube in which microalgae cultures were re-circulated either with a pump or airlift technique (Kumar et al. 2021).

The improved design allows a controlled environment, high biomass production (up to 10 times more than the open pond), low evaporation rate, reduced risks of contaminations, higher photosynthetic efficiency, and efficient transfer of nutrients (Kumar et al. 2021).

By way of comparison, studies by Benavides et al. (2013) evaluated the biomass productivity by *Phaeodactylum tricornutum* in consolidated systems. The assessed scenario presents higher closed-loop photobioreactors' (1.0 g L^{-1}) values than open raceway ponds (0.6 g L^{-1}). However, the critical point is the high capital and operating costs that limit the viability of this system. Also, closed photobioreactors face scaling issues. This is because enlarging the photobioreactors will affect light availability and, consequently, trigger effects on microalgal productivity (Yu et al. 2022).

Despite the photosynthetic nature of microalgae, some species have the potential to grow in the dark. Heterotrophic systems allow the use of conventional fermenters (i.e., stirred tank bioreactors), whose operation and maintenance are considered more straightforward, reducing the production costs (Perez-Garcia et al. 2011, Hu et al. 2018). When using this cultivation condition, the productive capacities are superior to the photoautotrophic systems in microalgae biomass production (Jacob-Lopes et al. 2020). The main interests associated with this condition are related to the substantial content of lipids. Especially the species *Crypthecodinium cohnii* has been cultured in a heterotrophic environment for commercial production of long-chain unsaturated fatty acids (Chew et al. 2018, Lam et al. 2018). However, although it is an apparent attractive advantage, only a few species can successfully grow heterotrophically.

Given these aspects, it is important to mention that microalgae cultivation still faces barriers attributed to technological limitations. Finally, the scale of production needs to be expanded at reasonable costs (Torres-Tiji et al. 2018, Jacob-Lopes et al. 2020, Fu et al. 2021).

3. Nutritional Composition and Bioactivity of Microalgae Compounds

For microalgae to be considered as potential new food sources, a crucial fact is their nutritional composition, which varies between species, and even within the same species, it can also vary significantly based on the form of cultivation. Important nutritional components to consider are proteins, lipids, carbohydrates, and pigments' content, all of which are known to positively impact human health (Torres-Tiji et al. 2020).

Table 1. Industrial microalgae producers and technological routes used.

Cultivation systems	Species	Products	Company
Photosynthetic growth in open ponds	*Dunaliella salina*	Whole dried biomass	Plankton Australia Pty Ltd
			Algalimento SL, Spain
			Seagrass Tech Private Limited, India
			Fuqing King Dnarmsa *Spirulina* Co., Ltd., China
		Dried extract β-carotene	BASF, Australia
	Chlorella sp.	Whole dried biomass	Seagrass Tech Private Limited, India
			Fuqing King Dnarmsa *Spirulina* Co., Ltd., China
			Gong Bih Enterprise Co., Ltd., Taiwan
			Sun *Chlorella* Corporation, Japan
			Parry Nutraceuticals, India
		Tablets	Fuqing King Dnarmsa Spirulina Co., Ltd., China
			Gong Bih Enterprise Co., Ltd., Taiwan
			Sun *Chlorella* Corporation, Japan
		Chlorella extract	Sun *Chlorella* Corporation, Japan
	Spirulina sp.	Organic pasta	Algosud, France
		Gluten-free organic pasta	Algosud, France
		Whole dried biomass	Seagrass Tech Private Limited, India
			Parry Nutraceuticals, India
			Algosud, France
			Cyanotech, Hawaii
		Tablets	Fuqing King Dnarmsa *Spirulina* Co., Ltd., China
			Cyanotech, Hawaii
			Algosud, France
		Dried extract phycocyanin	Fuqing King Dnarmsa *Spirulina* Co., Ltd., China
			Parry Nutraceuticals, India
	Nannochloropsis sp.	Whole dried biomass	Monzón Biotech S.L, Spain
		Tablets	Monzón Biotech S.L, Spain
		Paste	Monzón Biotech S.L, Spain
	Crypthecodinium cohnii	DHA	Yaeyama Shokusan Co., Ltd., Japan
	Haematococcus pluvialis	Whole dried biomass	Cyanotech, Hawaii
			Atacama Bio Natural Products, Chile
		Astaxanthin oil	Atacama Bio Natural Products, Chile
		Dried extract astaxanthin	Atacama Bio Natural Products, Chile
		Capsules	Atacama Bio Natural Products, Chile
Photosynthetic growth in closed photobioreactors	*Haematococcus pluvialis*	Astaxanthin oil	Algatechnologies, Israel
			BGG, China
			Algalif, Iceland
			Algamo, Czech Republic
		Dried extract astaxanthin	Algamo, Czech Republic
			BGG, China
		Capsules	Algamo, Czech Republic
			BGG, China
		Whole dried biomass	Algamo, Czech Republic
	Phaeodactylum tricornutum	Fucoxanthin oil	Algatechnologies, Israel

Table 1 contd. ...

...*Table 1 contd.*

Cultivation systems	Species	Products	Company
Heterotrophic growth in bioreactors	*Haematococcus pluvialis*	Astaxanthin oil	Shaanxi Rebecca Bio-Tech Co., LTD, China
			AstaReal, Sweden, the USA and Japan
		Dried extract astaxanthin	AstaReal, Sweden, the USA and Japan
		Capsules	AstaReal, Sweden, the USA and Japan
		Whole dried biomass	AstaReal, Sweden, the USA and Japan

Arthrospira plantesis, *Chlorella vulgaris*, *Dunaliella salina*, *Haematococcus pluvialis*, *Crypthecodinium cohnii*, and *Nitzschia laevis* are the main strains commercially used for production of microalgae-based ingredients (Souza et al. 2019).

In this context, microalgae have great potential to meet the dietary needs of the population, generating wide interest in these microorganisms due to their nutritional composition as well as the presence of bioactive compounds, physiologically active substances with functional properties at the metabolic level. These biocompounds, such as polysaccharides, sterols, pigments (carotenoids, chlorophylls, and phycobiliproteins), volatile organic compounds, polyphenols, vitamins, minerals, fatty acids, and amino acids (proteins/enzymes) are changing their status in the field of microalgae research, and gaining space in relevant research for the application of pharmacological, nutraceutical and functional products, holding countless opportunities for the development of innovative and sustainable products in the food industries. These microalgae biomolecules have shown essential activities at the level of health regulation and disease prevention, such as antioxidant, antitumor, anti-inflammatory, antivirus, and antithrombotic activity, among other important potentialities (Jacob-Lopes et al. 2019, Wu et al. 2021, Nörnberg et al. 2022c).

3.1 Proteins

The great advantage of microalgae proteins lies in the constitution of amino acids in their structures, as they have a complete profile of essential amino acids that are often not found in plants (Koyande et al. 2019), which cannot be synthesized by the human body, requiring consumption through food. These essential compounds are histidine, isoleucine, leucine, lysine, methionine, phenylalanine, threonine, tryptophan, and valine. Other amino acids such as arginine, cysteine, glutamine, glycine, proline, tyrosine, and aspartic acid also appear in the protein fraction of some species (Torres-Tiji et al. 2020).

Some microalgae have a high percentage of protein in their dry biomass (40–70%). Species such as *Spirulina* are reported to have up to 70% in their biomass, one of the richest sources of protein in nature. Other examples of different microalgae with the potential to obtain proteins include *Aphanizomenon flos-aquae* (60%), *Chlorella pyrenoidosa* (55%), *Dunaliella salina* (55%), and *Scenedesmus obliquus* (50%), which is also superior when compared to other plant sources like soy (38%) or even animal sources like milk (4%), and eggs (13%) (Souza et al. 2019, He et al. 2020, Torres-Tiji et al. 2020, Boukid et al. 2021).

3.2 Lipids

Lipids are indispensable components of cells and are precursors of many essential molecules. Its adequate intake is crucial for human nutrition. Some species can accumulate up to 70% of lipids in dry biomass (Torres-Tiji et al. 2020). However, in average terms most species accumulate levels of 5 to 15% (Barkia et al. 2019).

The lipid profile of microalgae consists mainly of triacylglycerol molecules (Tang et al. 2020). Microalgae triacylglycerols are formed by a mixture of monounsaturated fatty acids, such as palmitoleic (16:1) and oleic (18:1), polyunsaturated, such as linoleic (18:2) and linolenic (18:3), and saturated fatty acids, such as palmitic (16:0) and stearic (18:0) (Souza et al. 2019).

The PUFA profile in microalgae includes omega-6 fatty acids such as linoleic acid (LNA, 18:2n-6), γ-linolenic acid (GLA, 18:3n-6), and arachidonic acid (ARA, 20:4n-6), as well as omega-3 fatty acids that include α-linolenic acid (ALA, 18:3n-3), docosapentaenoic acid (DPA, 22:5n-3), docosahexaenoic acid (DHA, 22:6n-3), and eicosapentaenoic acid (EPA, 20:5n-3) (Morais et al. 2015, Katiyar and Arora 2020).

Apart from essential amino acids, there are fatty acids that are also essential, including α-linolenic acid and linoleic acid. In addition, there are certain lipids that have been proven to have a positive impact on human health. One of the most prominent examples is omega-3 fatty acids, DHA and EPA, which are generally introduced into the human diet through the consumption of fish or through supplements. The traditional source of such nutrients in the human diet has been cold-water fish and seafood in general. However, fish are enriched with these omega-3 fatty acids because they consume plankton and algae as part of their diet, and it is these algae that actually produce these polyunsaturated fatty acids (PUFA). In this context, microalgae such as *Phaeodactylum tricornutum* can accumulate up to 30% to 40% of the total fatty acids produced as EPA, and other species, such as *Schizochytrium* sp. can accumulate about 50% of the cell's total lipids as DHA. Therefore, microalgae can be an effective substitute for fish oil supplements, providing health-beneficial fatty acids (Torres-Tiji et al. 2020, Moradi and Saidi 2022).

3.3 Carbohydrates

Microalgae can have approximately 25% of their dry weight as carbohydrates, depending on the species. The polysaccharides present can modulate the immune system, and inflammatory reactions, making them favorable to act as sources of active molecules in food products and as natural therapeutic agents (Souza et al. 2019).

One such example is the β-glucans, a group of polysaccharides composed of D-glucose joined through β bonds found in the cell walls of plants, fungi, and bacteria. β-glucans act physiologically as soluble fiber, with effects on reducing LDL cholesterol and the risk of cardiovascular diseases. They also have recognized anti-inflammatory, anticancer, immunomodulatory effects, implications for the reduction of blood glucose, as well as prebiotic properties. This type of molecule is produced by many green microalgae, thus increasing the nutritional value of products containing them (Torres-Tiji et al. 2020, Nörnberg et al. 2022a), with *Chlorella* sp. recognized as the microalgae that have the highest content of β-glucans in its composition (Souza et al. 2019).

According to Niccolai et al. (2019), high total dietary fiber intake has beneficial health effects, such as lowering cholesterol and blood glucose, increased volume of fecal bolus and decreased intestinal transit time, entrapment of substances that can be dangerous to the human body (mutagenic and carcinogenic agents), and stimulation of the proliferation of beneficial intestinal flora. In a study carried out by the same authors, they showed that, in general, the total dietary fiber content of the microalgae examined was significantly higher (between 4.4% and 17%) compared to some cooked cereals, such as white rice (0.3%) and oats (1.7%), raw vegetables such as tomatoes (1.3%) and lettuce (1.0%) and raw fruits such as bananas (1.8%) and pineapples (1.5%).

3.4 Pigments

The main pigments (~ 1–9%) present in microalgae are carotenoids, chlorophylls, and, in some cases, phycobiliproteins. Microalgal pigments have found a market in the food industry due to the current search for natural ingredients that stand out in relation to synthetic ones for being non-toxic, having been used as natural dyes, and food supplements, in addition to being added as ingredients in foods. Chlorophyll, carotenoids, and phycobiliproteins have colors ranging from green, yellow, and brown to red. These colors vary with each microalgae. For example, the blue pigment of *Spirulina* (provided by phycocyanins), yellow pigment of *Dunaliella* (due to the presence of β-carotene), and yellow to red pigment of *Haematococcus* (due to astaxanthin), which, in addition to pigmentation, have anti-inflammatory, antihypertensive, anticancer, antioxidant, antidepressant, and antiaging properties (Souza et al. 2019, Vendruscolo et al. 2021).

3.4.1 Carotenoids

Microalgae are considered the major sources of obtaining natural carotenoids. Carotenoids stand out as a diverse group of compounds with potential health benefits. These microorganisms can synthesize complex mixtures of carotenoids, ranging from structures found in conventional plants, such as lutein, β-carotene, and zeaxanthin, to specific carotenoids with potentiated bioactive abilities, such as echinenone, astaxanthin, and canthaxanthin. The growing interest in this class of compounds results from its effectiveness in promoting human health, which is why its use has potentially been directed to several biomedical applications. Carotenoids play a remarkable role as antioxidants, as precursors to vitamin A, and in eye health. In addition, scientific evidence suggests its role in reducing cardiovascular disease, obesity, diabetes, cancer, and protecting neurons. The biomedical contribution of these isoprenes is made possible in part by their structural properties that orchestrate essential biological activities, mainly antioxidants and provitamin A, related to production, chemical structure, relationships between structural and biological activity, and challenges for use (Nörnberg et al. 2021, Nascimento et al. 2022).

3.4.2 Chlorophylls

Chlorophylls are the major pigments involved in the metabolism of photosynthetic organisms. Originally, four types of chlorophyll were known, named a (from higher plants), b (from higher plants and green algae), c (from diatoms), and d (only related to red algae). Recently, a new type of chlorophyll has been discovered, the type f, found in cyanobacteria (Chen et al. 2012).

Chlorophylls can be used as naturally sourced food colorants (Timberlake and Henry 1986). Additionally, studies suggest that chlorophylls can have health benefits, acting like an antitumor agent (Vesenick et al. 2012), and presenting anti-inflammatory effects (Subramoniam et al. 2012).

4. Food Products Enriched with Microalgal Biomass

The interest in the nutritional composition and bioactivity of the high-value compounds of microalgae increases the common sense that health care and the environment are essential factors for maintaining a good quality of life. In this sense, the use of microalgae as ingredients/functional food products and/or nutraceuticals has gained a booming market in recent decades (Fernandes et al. 2021a,b, Dias et al. 2022). This incorporation is feasible to increase the nutritional and functional value in most products that are consumed regularly (Tang et al. 2020). One of the main advantages of using microalgae in nutrition is the possibility of offering several compounds of interest simultaneously. Furthermore, microalgae can be used to increase the shelf life of the product and act as a source of natural dyes (Souza et al. 2019).

Spirulina and *Chlorella* were the first microalgae to be marketed as functional foods, as they are recognized as safe (GRAS); therefore, they are among the most commercially exploited species in terms of profile and characterization of new bioactive compounds (Fernandes et al. 2020, Markou et al. 2021).

Spirulina is one of the main trends in the food industry (Lafarga 2019), among other microalgae such as *Chlorella*. Most of the *Spirulina* that is being produced today is consumed as a nutritional supplement promoted as a "superfood" and marketed as a dry powder, flakes, or capsules. Also, aware of the richness of its bioactive compounds, *Spirulina* supplementation (1 g daily) was highlighted for its high hypolipidemic effect, especially triglyceride concentration, in dyslipidemic individuals (Mazokopakis et al. 2014).

Numerous other microalgae combinations can be found on the global market as nutritional supplements, such as in the form of pills, lozenges, and liquids. They can also be incorporated into food products (e.g., bakery items, cheeses, yogurts, vegetable creams, ice creams, meat analogs, beverages, sweets, etc.), providing health benefits (Benavalte-Valdes 2021). Some of these products are mentioned in Table 2, most of them being produced with concentrations of 1–2% of the biomass,

Table 2. Commercialized products containing microalgae.

Products	Additive microalgae	Company	Country
Mint chocolate-covered coconut cookies	*Spirulina*	Emmy's Organics	USA
Chocolate cookies	*Spirulina*	Kookie Cat	Bulgaria
Spirulina and cranberry biscuits	*Spirulina*	Casino	France
Algencracker rosmarin and meersalz	*Chorella*	Helga	Austria
Algencracker sesam and leinsamen			
Organic algae drink			
Spirulina Filled Crackers	*Spirulina*	Lee Biscuits	Malaysia
Bio-matcha and *Spirulina* biscuits	*Spirulina*	Próvida Produtos Naturais	Portugal
Oat flakes with coconut and *Spirulina*			
Spirulina and lemon protein bar	*Spirulina*	Sottolestelle	Italy
Gomasio with *Spirulina*			
Penne pasta with *Spirulina*			
Protein *Spirulina* and lemon bar	*Spirulina*	Roobar	Bulgaria
Protein chia and *Spirulina* bar			
Plant-based bar free from gluten, grains, and soy	*Spirulina* and *Chlorella*	OHi Foods	USA
Smoked Scottish seaweed and sea salt flavored organic puffs	*Chlorella*	SC Honest Fields Europe	Romania
Thyme and lime-flavored organic puffs			
Matcha and lime-flavored organic puffs			
Granola with *Spirulina* and orange	*Spirulina*	Healthy Tradition	Ukraine
Instant spinach and *Spirulina* soup	*Spirulina*	Nature et Aliments	France
Vegan eggs	*Spirulina*	Follow Your Heart	Earth Island
Blue cheese			
Pack natural food colors	*Spirulina*	Color Garden	USA
Green sugar crystals			
Blue sugar crystals			
Cashew kind of blue cheese	*Spirulina*	Pulse Kitchen	Canada
Cheese super blue	*Spirulina*	Nuts for cheese	Canada
White corn and *Spirulina* pasta	*Spirulina*	EATTIAMO	Italy
Wheat noodles with *Spirulina*	*Spirulina*	Bionsan	Spain
Algae noodles	*Chlorella*	Van der Moolen Foodgroup B.V.	Netherlands
Spirulina waffle	*Spirulina*	Damhert Nutrition	Belgium
Spirulina burger			
Chocolate hazelnut flavored spread	*Spirulina*	Mars Wrigley Confectionery	Ireland
Bubble Gum Pops	*Spirulina*	Tree Hugger	USA
Honey, acerola and *Spirulina*	*Spirulina*	Meltonic	France
Organic fudge with *Chlorella*	*Chlorella*	Majami	Poland
Mint flavored pastille	*Spirulina*	Ferrero Ibérica	Spain
Balsamic *Spirulina* algae pastille	*Spirulina*	Cesare Carraro	Italy
Chocolate hazelnut flavored spread	*Spirulina*	Nestlé	Canada
Lemon chocolate truffles with *Spirulina*	*Spirulina*	Lubs	Germany

Table 2 contd. ...

...Table 2 contd.

Products	Additive microalgae	Company	Country
Green tea and *Spirulina* candies	*Spirulina*	Incap	Italy
Tender creamy chocolate with almond and *Spirulina*	*Spirulina*	Lovechock B.V.	Netherlands
Spirulina ice cream	*Spirulina*	Nuts	USA
Chlorella pistachio ice cream	*Chlorella*	Organic burst	England
Blue *Spirulina* cacao bar	*Spirulina*	Benjamissimo	Bulgaria
Green fruit smoothie with *Spirulina* and *Chlorella*	*Spirulina* and *Chlorella*	Happy Planet Foods	Canada
Chlorella and *Spirulina* tea	*Spirulina* and *Chlorella*	Smart Organic	Europe
Hot shot *Spirulina*	*Spirulina*	Urban Remedy	Brazil
Blue colored cashew milk containing *Spirulina*	*Spirulina*	Urban Remedy	USA
Kombucha with Spirulina and peppermint	*Spirulina*	Lökki	France
Bio Shot Ginseng Chokeberry with *Chlorella* Algae	*Chlorella*	Viva Maris GmbH	Germany
Fresh *Chlorella*	*Chlorella*	Duplaco B.V.	Netherlands
Spirulina powder	*Spirulina*	Dragon Superfoods	Bulgaria
		NOW	Canada
		Unilife Vitamins	Brazil
		Próvida Produtos Naturais	Portugal
		Label Spiruline	France
		Sottolestelle	Italy
Softgel capsules astaxanthin	*Haematococcus pluvialis*	NOW	Canada
Softgel capsules β-carotene	*Dunaliella salina*	NOW	Canada
Chlorella powder	*Chlorella*	NOW	Canada
		Dragon Superfoods	Bulgaria
		Duplaco B.V.	Netherlands
		Helga	Austria
Chlorella tablets	*Chlorella*	Organic burst	England
		NOW	Canada
		Duplaco B.V.	Netherlands
		Dragon Superfoods	Bulgaria
Spirulina tablets	*Spirulina*	Organic burst	England
		Dragon Superfoods	Bulgaria
		NOW	Canada
		Nutrex Hawai	USA
Apogen Children supplement concentrated	*Spirulina*	Far east BIO-TEC CO., LTD.	Taiwan
Chlorella and *Spirulina* capsules	*Spirulina* and *Chlorella*	Unilife Vitamins	Brazil
Astaxanthin capsules	*Haematococcus pluvialis*	Unilife Vitamins	Brazil
DHA capsules	*Schizochytrium* sp.	Unilife Vitamins	Brazil
Nannochloropsis powder	*Nannochloropsis* sp.	Monzón Biotech S.L	Spain
Nannochloropsis tablets			
DHA	*Crypthecodinium cohnii*	Yaeyama Shokusan Co. Ltd.	Japan
Fucoxanthin oil	*Phaeodactylum tricornutum*	Algatechnologies	Israel

but also produced with lower (below 0.1%) and higher (up to 3%) concentrations (Bernaerts et al. 2021).

In addition to nutrient enrichment, microalgal biomass presents interesting technological features in food products, which include sensory properties (taste, texture, and color). Some microalgae such as *Spirulina*, *Chlorella*, *Dunaliella*, and *Scenedesmus*, when processed correctly, have an attractive flavor and can be incorporated into various types of food (Souza et al. 2019).

Gouveia et al. (2007) successfully used *Chlorella vulgaris* for the green coloring of traditional shortbread cookies. Furthermore, Graça et al. (2018), using the same strain, demonstrated that the addition of wheat flour improved the viscoelastic and rheological parameters of the bread dough.

Marti-Quijal et al. (2018, 2019) elaborated hybrid meat analogs, one with turkey breast and the other with pork, respectively, added with *Spirulina* and *Chlorella* microalgae, in addition to other plant sources. In both studies, there was a significant difference between treatments in the amino acid profile, which was improved with the addition of *Spirulina*, with a higher content of essential amino acids.

Microalgae proteins, in addition to having essential amino acids, have physicochemical properties such as solubility, emulsification, gelation, and foaming that are adjustable for different applications. Generally, the protein digestibility coefficient of some microalgae species such as *Chlorella* sp. and *Spirulina* sp. is better than wheat gluten and soy protein (Fu et al. 2021). Thus microalgae-based products may become widely appreciated and consumed as functional foods in the near future. However, despite their potential, the success of these products still depends on regulatory approval.

5. Regulatory Framework and Legislation for Microalgae-based Food Products

Microalgae-based systems emerge as a successful model in the commercial production of food raw materials. Associated with this perspective, food safety is the most crucial parameter for ingredients' selection (Fu et al. 2021, Markou et al. 2021).

In principle, in European Union, the European Food and Safety Authority (EFSA) considers it safe to consume *Arthrospira platensis*, *Arthrospira maxima*, *Chlorella pyrenoidosa*, *Chlorella luteoviridis*, and *Chlorella vulgaris*, due to their consumption significantly, without the need to comply with Regulation (EU) 2015/2283 on novel foods.

On the other hand, *Aurantiochytrium limacinum*, *Euglena gracilis*, *Odontella aurita*, and *Tetraselmis chui* are subjected to the regulations on novel foods. Besides, compounds coming from microalgae as β-carotene produced from *Dunaliella salina*, astaxanthin-rich oleoresin extracted from *Haematococcus pluvialis*, DHA and EPA acquired from *Schizochytrium* sp., oil-rich in PUFA obtained from *Ulkenia* sp., and oil-rich in EPA derived from *Phaeodactylum tricornutum* are approved by regulation (EU) 2470/2017 and labeled as "novel food" (E.F.S.A. 2020, Torres-Tiji et al. 2020, Matos 2019).

In that line, in Canada, the license for food-grade microalgae is issued under 'novel foods' status by Health Canada. Among the species considered for use include *Arthrospira platensis*, *Chlorella vulgaris*, *Chlorella sorokiniana*, *Chlorella regularis*, *Dunaliella salina*, and *Euglena gracilis* (CFIA 2020).

The Food and Drug Administration (FDA) declared the GRAS status (Generally Recognized As Safe) in the USA. Yet, *Arthrospira platensis*, *Chlorella vulgaris*, *Dunaliella Salina*, *Haematococcus pluvialis*, and *Phaeodactylum tricornutum* are authorized as foods and extracts. Moreover, *Chlamydomonas reinhardtii*, *Auxenochlorella prototothecoides*, *Dunaliella bardawil*, and *Euglena gracilis* also have GRAS status (Torres-Tiji et al. 2020, Jacob-Lopes et al. 2019, Matos 2019).

In China, *Arthrospira platensis*, *Arthrospira maxima*, *Chlorella pyrenoidesa*, *Dunaliella Salina*, and *Haematococcus pluvialis* are considered safe to consume as dried biomass. On the other hand,

Crypthecodinium cohnii, Ulkenia amoeboida, and *Schizochytrium* sp. are only approved to be used as DHA algal oil (Fu et al. 2021, National Health and Family Planning Commission of PRC 2016).

Finally, *Arthrospira* sp. (*Spirulina*), *Chlorella* sp., *Dunaliella* sp., and *Haematococcus* pluvialis are the common species widely consumed as food ingredients across continents (Fu et al. 2021, Torres-Tiji et al. 2020, Jacob-Lopes et al. 2019).

6. Conclusion

Boosted by global changes in food consumption patterns, microalgae-based products have been gaining prominence due to the potential of their biocompounds, providing healthier and safer diets. Besides, these bioproducts are rich in proteins (essential amino acids), lipids (polyunsaturated fats, mainly from the omega-3 family), carbohydrates (with the presence of polysaccharides such as β-glucans), and pigments (carotenoids and chlorophylls). Many of these compounds have antioxidant behavior and important bioactivity. Among the various species and technologies for food production, *Spirulina* has been the most explored so far. Although some obstacles still need to be overcome in the face of bottlenecks such as increasing biomass yield, problems of scale, being economically viable, in addition to the sensory properties that are vital for successful product commercialization, this is a growing market and indicates that microalgae are indeed a future food source.

References

Barkia, I., Saari, N. and Manning, S.R. 2019. Microalgae for high-value products towards human health and nutrition. Marine Drugs 17(5): 304.

Benavente-Valdés, J.R., Méndez-Zavala, A., Hernández-López, I., Carreón-González, B.A., Velázquez-Arellano, M.E., Morales-Oyervides, L., et al. 2021. Unconventional microalgae species and potential for their use in the food industry. In Cultured Microalgae for the Food Industry (pp. 49–71). Academic Press.

Bernaerts, T.M. and Van Loey, A.M. 2021. Microalgae as structuring ingredients in food. In Cultured Microalgae for the Food Industry (pp. 265–286). Academic Press.

Boukid, F., Rosell, C.M., Rosene, S., Bover-Cid, S. and Castellari, M. 2021. Non-animal proteins as cutting-edge ingredients to reformulate animal-free foodstuffs: present status and future perspectives. Critical Reviews in Food Science and Nutrition, 1–31.

CFIA. 2020. https://www.canada.ca/en/health-canada/services/food-nutrition/genetically-modified-foods-other-novel-foods.html.

Chaudhary, A., Gustafson, D. and Mathys, A. 2018. Multi-indicator sustainability assessment of global food systems. Nature Communications 9(1). https://doi.org/10.1038/s41467-018-03308-7.

Chen, M., Li, Y., Birch, D. and Willows. R.D. 2012. A cyanobacterium that contains chlorophyll f—a red-absorbing photopigment. FEBS Lett. 586: 3249–3254.

Chew, K.W., Chia, S.R., Show, P.L., Yap, Y.J., Ling, T.C. and Chang, J.S. 2018. Effects of water culture medium, cultivation systems and growth modes for microalgae cultivation: a review. Journal of the Taiwan Institute of Chemical Engineers 91: 332–344. https://doi.org/10.1016/j.jtice.2018.05.039.

Deprá, M.C., Severo, I.A., dos Santos, A.M., Zepka, L.Q. and Jacob-Lopes, E. 2020. Environmental impacts on commercial microalgae-based products: sustainability metrics and indicators. Algal Research 51: 102056. https://doi.org/10.1016/j.algal.2020.102056.

Dias, R.R., Deprá, M.C., Severo, I.A., Zepka, L.Q. and Jacob-Lopes, E. 2022. Smart override of the energy matrix in commercial microalgae facilities: a transition path to a low-carbon bioeconomy. Sustainable Energy Technologies and Assessments 52: 102073.

EFSA. 2020. https://ec.europa.eu/food/safety/novel_food/catalogue/search /public/index.cfm ?ascii=N.

EN. 2009. https://cdn.standards.iteh.ai/samples/29302/799c23fe6a6a 4e8cb2dd72ae1050e81f/SIST-EN-15763-2010. pdf.

FAO. 1997. http://www.fao.org/3/w7241e/w7241e0h.htm#chapter% 206%20%20oil%20production.

Fernandes, A.S., Petry, F.C., Mercadante, A.Z., Jacob-Lopes, E. and Zepka, L.Q. 2020. HPLC-PDA-MS/MS as a strategy to characterize and quantify natural pigments from microalgae. Current Research in Food Science 3: 100–112.

Fernandes, A.S., Nascimento, T.C., Pinheiro, P.N., Vendruscolo, R.G., Wagner, R., de Rosso, V.V., et al. 2021a. Bioaccessibility of microalgae-based carotenoids and their association with the lipid matrix. Food Research International 148: 110596.

Fernandes, A.S., Nascimento, T.C., Pinheiro, P.N., de Rosso, V.V., de Menezes, C.R., Jacob-Lopes, E., et al. 2021b. Insights on the intestinal absorption of chlorophyll series from microalgae. Food Research International, 140: 110031.

Fu, Y., Chen, T., Chen, S.H.Y., Liu, B., Sun, P., Sun, H., et al. 2021. The potentials and challenges of using microalgae as an ingredient to produce meat analogues. Trends in Food Science and Technology 112: 188–200. https://doi.org/10.1016/j.tifs.2021.03.050.

Golden, C.D., Koehn, J.Z., Shepon, A., Passarelli, S., Free, C.M., Viana, D.F., et al. 2021. Aquatic foods to nourish nations. Nature 598(7880): 315–320. https://doi.org/10.1038/s41586-021-03917-1.

Gouveia, L., Batista, A.P., Miranda, A., Empis, J. and Raymundo, A. 2007. *Chlorella vulgaris* biomass used as colouring source in traditional butter cookies. Innovative Food Sci. Emerg. Technol. 8: 433–436.

Graça, C., Fradinho, P., Sousa, I. and Raymundo, A. 2018. Impact of *Chlorella vulgaris* on the rheology of wheat flour dough and bread texture, LWT – Food Science and Technology 89: 466–474.

He, J., Evans, N.M., Liu, H. and Shao, S. 2020. A review of research on plant-based meat alternatives: driving forces, history, manufacturing, and consumer attitudes. Comprehensive Reviews in Food Science and Food Safety, 1–18.

Hu, J., Nagarajan, D., Zhang, Q., Chang, J.S. and Lee, D.J. 2018. Heterotrophic cultivation of microalgae for pigment production: a review. Biotechnology Advances 36(1): 54–67. https://doi.org/10.1016/j.biotechadv.2017.09.009.

Jacob-Lopes, E., Maroneze, M.M., Depra, M.C., Sartori, R.B., Dias, R.R. and Zepka, L.Q. 2019. Bioactive food compounds from microalgae: an innovative framework on industrial biorefineries. Current Opinion in Food Science 25: 1–7. https://doi.org/10.1016/j.cofs.2018.12.003.

Jacob-Lopes, E., Santos, A.B., Severo, I.A., Deprá, M.C., Maroneze, M.M. and Zepka, L.Q. 2020. Dual production of bioenergy in heterotrophic cultures of cyanobacteria: process performance, carbon balance, biofuel quality and sustainability metrics. Biomass and Bioenergy, 142. https://doi.org/10.1016/j.biombioe.2020.105756.

Katiyar, R. and Arora, A. 2020. Health promoting functional lipids from microalgae pool: a review. Algal Research 46: 101800.

Koyande, A.K., Chew, K.W., Rambabu, K., Tao, Y., Chu, D.T. and Show, P.L. 2019. Microalgae: a potential alternative to health supplementation for humans. Food Science and Human Wellness 8(1): 16–24.

Kumar, B.R., Mathimani, T., Sudhakar, M.P., Rajendran, K., Nizami, A.S., Brindhadevi, K., et al. 2021. A state-of-the-art review on the cultivation of algae for energy and other valuable products: application, challenges, and opportunities. Renewable and Sustainable Energy Reviews 138: 110649. https://doi.org/10.1016/j.rser.2020.110649.

Kusmayadi, A., Leong, Y.K., Yen, H.W., Huang, C.Y. and Chang, J.S. 2021. Microalgae as sustainable food and feed sources for animals and humans – Biotechnological and environmental aspects. Chemosphere 271: 129800. https://doi.org/10.1016/j.chemosphere.2021.129800.

Lafarga, T. 2019. Effect of microalgal biomass incorporation into foods: Nutritional and sensorial attributes of the end products. Algal Research 41: 101566.

Lafarga, T., Rodríguez-Bermúdez, R., Morillas-España, A., Villaró, S., García-Vaquero, M., Morán, L., et al. 2021. Consumer knowledge and attitudes towards microalgae as food: the case of Spain. Algal Research 54: 102174.

Lam, T.P., Lee, T.M., Chen, C.Y. and Chang, J.S. 2018. Strategies to control biological contaminants during microalgal cultivation in open ponds. Bioresource Technology 252: 180–187. https://doi.org/10.1016/j.biortech.2017.12.088.

Markou, G., Chentir, I. and Tzovenis, I. 2021. Microalgae and cyanobacteria as food: legislative and safety aspects. In Cultured Microalgae for the Food Industry (pp. 249–264). Academic Press.

Marti-Quijal, F.J., Zamuz, S., Galvez, F., Roohinejad, S., Tiwari, B.K., Gómez, B., et al. 2018. Replacement of soy protein with other legumes or algae in turkey breast formulation: changes in physicochemical and technological properties. Journal of Food Processing and Preservation e13845: 1–9.

Marti-Quijal, F.J., Zamuz, S., Tomašević, I., Gómez, B., Rocchetti, G., Lucini, L., et al. 2019. Influence of different sources of vegetable, whey and microalgae proteins on the physicochemical properties and amino acid profile of fresh pork sausages. LWT – Food Science and Technology 110: 316–323.

Matos, A.P. 2019. Microalgae as a potential source of proteins. *In*: Galanakis, C. (ed.). Proteins: sustainable source, processing and applications (pp., ed C. Galanakis, 63–93). Academic Press, Cambridge.

McCarty, M.F., Barroso-Aranda, J. and Contreras, F. 2008. Genistein and phycocyanobilin may prevent hepatic fibrosis by suppressing proliferation and activation of hepatic stellate cells. Medical Hypotheses 72(3): 330–332.

Mehariya, S., Goswami, R.K., Karthikeysan, O.P. and Verma, P. 2021. Microalgae for high-value products: a way towards green nutraceutical and pharmaceutical compounds. Chemosphere 280: 130553. https://doi. org/10.1016/j.chemosphere.2021.130553.

Moradi, P. and Saidi, M. 2022. Biodiesel production from *Chlorella vulgaris* microalgal-derived oil via electrochemical and thermal processes. Fuel Processing Technology 228: 107158.

Nascimento, T.C., Nass, P.P., Fernandes, A.S., Nörnberg, M.L., Zepka, L.Q. and Jacob-Lopes, E. 2022. Microalgae carotenoids: an overview of biomedical applications. Algal Biotechnology, 409–425.

National Health and Family Planning Commission of PRC. 2016. The lists of novel food ingredients and common food ingredients. http://www.nhc.gov.cn/sps/pztq/2.

Niccolai, A., Chini Zittelli, G., Rodolfi, L., Biondi, N. and Tredici, M.R. 2019. Microalgae of interest as food source: biochemical composition and digestibility. Algal Research 42: 101617.

Nörnberg, M.L., Nass, P.P., Nascimento, T.C., Fernandes, A.S., Jacob-Lopes, E. and Zepka, L.Q. 2021. Carotenoids profile of *Desertifilum* spp. in mixotrophic conditions. Brazilian Journal of Development 7(3): 33017–33029.

Nörnberg, M.L., Bortolotti, C.M., Minella, E. and Nörnberg, J.L. 2022a. Use of barley flour as a source of biocompounds in bakery products. Brazilian Journal of Development 8(2): 10334–10353.

Nörnberg, M.L., Caetano, P.A., Nass, P.P., Vieira, K.R., Wagner, R., Jacob-Lopes, E., et al. 2022b. Limonene production in microalgal photoautotrophic cultivation. Brazilian Journal of Development 8(2): 10241–10254.

Nörnberg, M.L., Nass, P.P., Nascimento, T.C., Fernandes, A.S., Jacob-Lopes, E. and Zepka, L.Q. 2022c. Production of microalgae biocompounds in different cultivation conditions. Brazilian Journal of Development 8(2): 10226–10240.

Nörnberg, M.L., Nass, P.P., Nascimento, T.C., Fernandes, A.S., Nörnberg, M.F.B.L., Jacob-Lopes, E., et al. 2022d. Bioactive compounds in butters: carotenoids and fatty acids. Brazilian Journal of Development 8(2): 10270–10288.

Perez-Garcia, O., Escalante, F.M.E., de-Bashan, L.E. and Bashan, Y. 2011. Heterotrophic cultures of microalgae: Metabolism and potential products. Water Research 45(1): 11–36. https://doi.org/10.1016/j.watres.2010.08.037.

Rahman, K.M. 2020. Food and high value products from microalgae: market opportunities and challenges. *In*: Alam, M.A., Xu, J.-L. and Wang, Z. (eds.). Microalgae Biotechnology for Food, Health and High Value Products, 483.

Shu, M.H., Appleton, D., Zandi, K. and AbuBakar, S. 2013. Anti-inflammatory, gastroprotective and anti-ulcerogenic effects of red algae *Gracilaria changii* (*Gracilariales, Rhodophyta*) extract. BMC Complementary and Alternative Medicine 13. https://doi.org/10.1186/1472-6882-13-61.

Souza, M.P., Hoeltz, M., Gressler, P.D., Benitez, L.B. and Schneider, R.C.S. 2019. Potential of microalgal bioproducts: general perspectives and main challenges. Waste and Biomass Valorization 10: 2139–2156.

Subramoniam, A., Asha, V.V., Nair, S.A., Sasidharan, S.P., Sureshkumar, P.K., Rajendran, K.N., et al. 2012. Chlorophyll revisited: anti-inflammatory activities of chlorophyll a and inhibition of expression of TNF-α gene by the same. Inflammation 35: 959–966.

Tang, D.Y.Y., Khoo, K.S., Chew, K.W., Tao, Y., Ho, S.-H. and Show, P.L. 2020. Potential utilization of bioproducts from microalgae for the quality enhancement of natural products. Bioresource Technology 304: 122997.

Timberlak, C.F. and Henry, B.S. 1986. Plant pigments as natural food colours. Endeavour. 10: 31–36.

Tokuşoglu, Ö. and Üunal, M.K. 2003. Biomass nutrient profiles of three microalgae: *Spirulina platensis*, *Chlorella vulgaris*, and *Isochrisis galbana*. Journal of Food Science 68(4): 1144–1148.

Torres-Tiji, Y., Fields, F.J. and Mayfield, S.P. 2020. Microalgae as a future food source. Biotechnology Advances 41(107536): 1–13. https://doi.org/10.1016/j.biotechadv.2020.107536.

Veerabadhran, M., Natesan, S., MubarakAli, D., Xu, S. and Yang, F. 2021. Using different cultivation strategies and methods for the production of microalgal biomass as a raw material for the generation of bioproducts. Chemosphere 285: 131436. https://doi.org/10.1016/j.chemosphere.2021.131436.

Vendruscolo, R.G., Fernandes, A.S., Fagundes, M.B., Zepka, L.Q., Menezes, C.R., Jacob–Lopes, E., et al. 2021. Development of a new method for simultaneous extraction of chlorophylls and carotenoids from microalgal biomass. Journal of Applied Phycology 33(4): 1987–1997.

Vesenick, D.C., Paula, N.A., Niwa, A.M. and Mantovani, M.S. 2012. Evaluation of the effects of chlorophyllin on apoptosis induction, inhibition of cellular proliferation and mRNA expression of CASP8, CASP9, APC and b-catenin. Current Research Journal of Biological Sciences 4: 315–322.

Wu, J., Gu, X., Yang, D., Xu, S., Wang, S., Chen, X., et al. 2021. Bioactive substances and potentiality of marine microalgae. Food Science and Nutrition 9(9): 5279–5292.

11

3D Food Printing

Fundamental Concepts, Crucial Factors, and A Critical Discussion about Acceptance and Sustainable Contribution

Jaqueline Souza Guedes,[1] *Bruna Sousa Bitencourt,*[1]
Pedro Augusto Invernizzi Sponchiado,[2] *Pedro Esteves Duarte Augusto,*[3]
Ana Paula Ramos[2] and *Bianca Chieregato Maniglia*[4,*]

1. Introduction

Three-dimensional (3D) printing is a widespread type of additive manufacturing technology with potential application in several fields as civil engineering, textile, packaging, medical, veterinary, and odontology. In the food industry, this technique enables the creation of food by adding the ingredients layer-by-layer to build edible 3D structures from digital designs (Zhao et al. 2021).

The primary advantage of this technique compared to the traditional manufacturing process is the possibility of creating products with complex geometries and controlled internal structures. Furthermore, 3D printing can be successfully applied in the personalized-food industry to manufacture foods with bespoke texture, taste, and nutritional content, to customize diets with individualized nutrition for health benefits. This technique can also enhance visual appeal increasing food acceptability and reducing food waste due to the efficient use of the ingredients. Other positive aspects of this technology are the possibility of using functional ingredients on demand, improved food experience, automation, labor savings, lower energy and transport costs, easier supply chain, and food manufacturing closer to consumers (Jiang et al. 2019, Le-Bail et al. 2020, Nachal et al. 2019).

---

[1] Departamento de Agroindústria, Alimentos e Nutrição da ESALQ-USP (LAN), Colégio de Agricultura Luiz de Queiroz (ESALQ), Universidade de São Paulo (USP), Piracicaba, SP, Brasil.

[2] Departamento de Química, Faculdade de Filosofia, Ciências e Letras (FFCLRP), Universidade de São Paulo (USP), Ribeirão Preto, SP, Brasil.

[3] Université Paris-Saclay, Centrale Supélec, Laboratoire de Génie des Procédés et Matériaux, Centre Européen de Biotechnologie et de Bioéconomie (CEBB), Pomacle, France.

[4] Instituto de Química de São Carlos (IQSC), Universidade de São Paulo (USP), São Carlos, SP, Brasil.

* Corresponding author: biancamaniglia@iqsc.usp.br

The implementation of 3D printing technology in the food industry depends on the following aspects:

- Ingredients: Choice of ingredients by the in-depth understanding of their physicochemical and rheological properties. In the food field, these materials have been named food ink. Different food ingredients such as sugar, chocolate, gelatin, and flours have been used to create a designed shape based on the layer-by-layer method. In addition, 3D printing has made possible the use of alternative ingredients such as insect flours, algae, microorganisms, and agri-food residues.

- 3D printing technology: The choice of technology is based on the material's physicochemical and rheological properties, applicability, and post-processing requirements. Diversified techniques can be used to print food, such as extrusion, selective laser sintering, and inkjet.

- 3D model and formats: A huge diversity of software for designing the printed construction is available, from beginners to advanced (e.g., SketchUp, 3D Builder, OnShape, and Tinkercad). The design is then converted to a *.stl* file and "sliced" by the slicing software (e.g., PrusaSlicer, Repetier, Simplify3D, and Cura). Then, a G code that consists of the commands necessary to guide the printhead under predetermined conditions of flow (i.e., temperature and speed) is generated. This step is crucial as the success of the printing procedure is closely tied to selecting the optimal plan for guiding the object's design.

- Evaluation of the printing parameters and quality of the printed object: The printing parameters and the quality of the printed material can be evaluated with regards to its form fidelity (compared to the original design) and mechanical properties.

Figure 1 shows a schematic representation of the steps involved in 3D printing, from CAD design to the final 3D printed construction (Mantihal et al. 2020).

In this context, this chapter brings the background knowledge on (1) 3D techniques for food printing, (2) major food ingredients for 3D printing, (3) crucial factors that influence the 3D food printing, and a discussion about (4) acceptance of 3D printed foods, (5) sustainable contributions, and (6) perspectives on 3D technology for food printing.

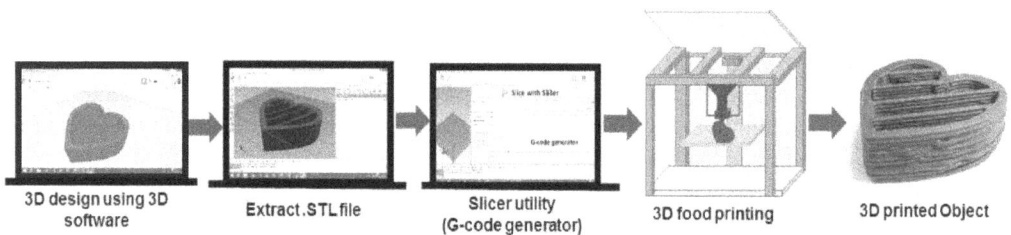

| 3D design using 3D software | Extract .STL file | Slicer utility (G-code generator) | 3D food printing | 3D printed Object |

Figure 1. Schematic representation of the steps involved in 3D printing, from CAD design to the final 3D printed material. The figure was extracted from (Mantihal et al. 2020). Copyright 2020, with permission from Elsevier.

2. 3D Technology for Food Printing

2.1 Techniques and the Machines

Several 3D printing techniques can be used for the production of food. They are classified according to either the materials used (paste, powder, solutions, etc.) or based on the mechanisms used to confer structure and stability to the final products (Godoi et al. 2016). This part of the chapter will address three different 3D printing techniques for food production and their peculiarities. Figure 2 illustrates the main techniques: inkjet, selective sintering, and extrusion printing.

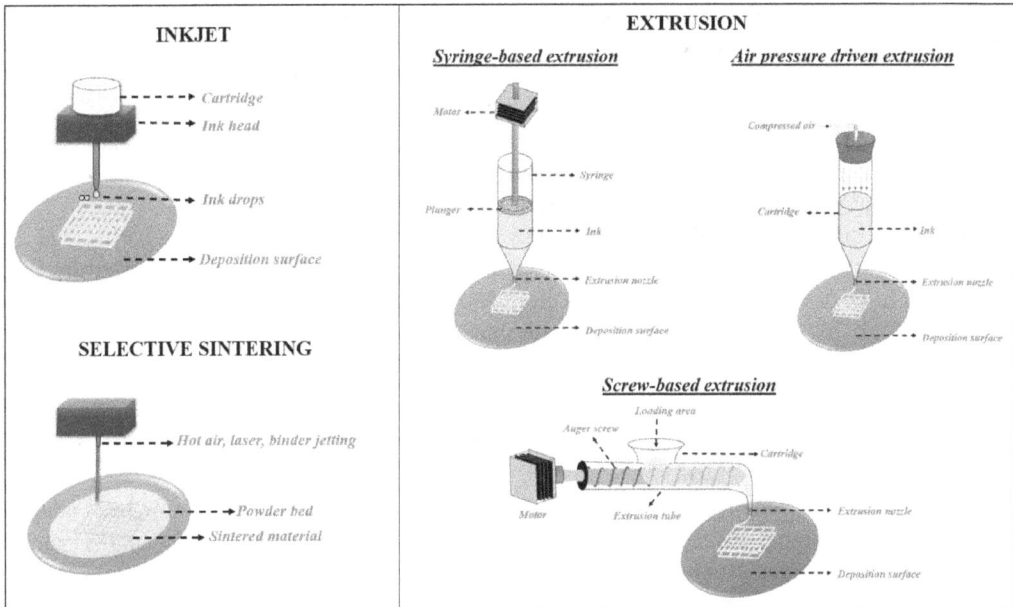

Figure 2. Illustration of the main 3D printing techniques for food production: inkjet, selective sintering, and extrusion printing.

2.1.1 Inkjet Printing

Inkjet printing uses a pneumatic membrane nozzle-jet which lays little drops of liquid-food-ink to form 2D and 3D patterns onto a moving object by a computer-aided design (Fig. 2). Inkjet printing can deposit a wide range of liquid foods in different ways and it can be carried out continuously or by drop-on-demand. In both methods, fluid is pushed through a nozzle, which splits into a stream of droplets of the same volume (Godoi et al. 2019). Low viscosity ranging from 2.8 to 6 cP is preferable since the deposition of small drops is required by this technique. It includes aqueous solutions (Ebrahimi et al. 2020, Bilbao et al. 2021) containing small molecules and nanoparticles' dispersion (Chou et al. 2021).

The driving force for droplets' extrusion may be achieved by either heating the liquid inside the nozzle to expand and squeeze it out through an electrical heating element (Godoi et al. 2019) or installing a piezoelectric ceramic element in the printhead to generate energy through its deformation (Huang et al. 2020).

The inks' fluidity, viscosity, and surface tension are the most important properties to get a high printing fidelity (Soleimani-Gorgani et al. 2016, Voon et al. 2019). Although inkjet printing shows the best results in pharmaceutical applications focused on the printing of drugs (Eleftheriadis and Fatouros 2021), it has also shown some interesting results in the food field. This technique has mainly been applied to decorate and cover food substrates, as well as to fill food cavities (Jiang et al. 2021). The FoodJet is a typical example of the commercial application of inkjet printing (FoodJet 2021).

2.1.2 Selective Sintering Printing

This technique can be classified depending on the type of sintering, such as selective laser sintering (SLS) and selective hot air and fusion sintering (SHASAM). Both the methods use a sintering source to melt the powder particles and to form solid layers (Fig. 2) (Sun et al. 2015). A roller distributes a layer of powder onto the printer bed (or the previous layer) while the sintering process is controlled by the CAD (computer-aided design) model, allowing the structural stability of the 3D material to be achieved.

In addition to selective laser sintering (SLS) and selective melt sintering (SHASAM), the binder jetting technique can also be cited. In this method, coating layers of the powdered materials are bonded together using a liquid binder. The powder is spread using a counter-rotating roller. Then, an inkjet print head deposits the binding agent onto the powders to create the desired 3D pattern (Ziaee and Crane 2019). Although it does not use a sintering source to melt the powder particles, this method uses the principle of deposition of the powder layer on the printer and the binder jet flow is controlled by a CAD process like in the sintering-based techniques (Mostafaei et al. 2021).

The main advantages of these technologies are the non-requirement of post-curing and the support structure, although post-processing is required to remove excess powder from the final products (Liu et al. 2017). However, they have the disadvantage of higher residue production (Mantihal et al. 2020). Also, due to its nature, this technology is limited to powdered materials which hinders its application in the food industry using fruit, vegetables, meat, or fish as inks (Le-Bail et al. 2020).

2.1.3 Extrusion-based Printing

This technology was initially developed for modeling plastics; however, it has now been adapted and has become the most used 3D printing technique in the food sector (Lee 2021). In this printing process, the ink is loaded into an extruder (cylinder) before being extruded through a nozzle with constant pressure to build up the food layer by layer according to a template (Fig. 2). This process is repeated until the shape of the desired object is achieved (Mazzanti et al. 2019). This type of printing technology is similar to conventional fused deposition modeling (FDM); nevertheless, the inks used in the extrusion-based printing can be either solid or pasty (soft) with low viscosity, while thermoplastic materials are used for the FDM.

The material can be extruded through the nozzle tip from different mechanisms: air pressure-driven extrusion, screw-based extrusion, and syringe-based extrusion (Dick et al. 2019, Sun et al. 2018) (Fig. 2). Each type of extrusion requires inks with viscosities suitable for processing. Air extrusion is best suited for low viscosity inks. Screw extrusion has been used when there is interest in mixing ingredients and then continuously depositing the material, avoiding air entrapment. Syringe-based extrusion is more suitable for high viscosity inks (Dick et al. 2019, Sun et al. 2015). In this sense, the viscosity is a critical parameter in extrusion since it has to be low enough to allow the ink to exit through the nozzle but high enough so that the structure formed does not collapse after extrusion (Jiang et al. 2019). The challenge lies in the choice of the ingredients able to meet the rheological parameters for the preparation of the ink to ensure fluidity and self-sustainability. Usually, a food additive or hydrocolloid is added to facilitate the inks' printability, flowability, and solidification. Printability refers to the facility by which the material can be controlled and deposited by a 3D printer, including shape retention after deposition. Materials with good printability can be used to fabricate complex structures (Wang et al. 2021). Different foods based on doughs, pastes, gels, chocolates, cheeses among others have been printed using the extrusion technique (Tan et al. 2018, Mantihal et al. 2020). In this chapter, special attention will be dedicated to extrusion-based printing, since it has been most used in the food sector.

2.2 Ingredients for Extrusion-based 3D Food Printing

Knowing the constituents of raw materials and how their properties will influence 3D printing technology is mandatory to ensure the quality of the final product (Godoi et al. 2016). In this sense, this section will cover the application of carbohydrate, protein, and lipid-based foods for food printing and the role of these components during the printing process.

2.2.1 Carbohydrates

Carbohydrate-based materials such as starch (Yang et al. 2018a, Liu et al. 2018, Maniglia et al. 2019, 2020), pectin (Derossi et al. 2018), and hydrocolloids (Kim et al. 2017) in which carbohydrate is used as a structuring ingredient have been studied for application in 3D printing.

Starch has been applied due to its rheological characteristics and its wide variety of natural and modified sources, making it a promising ingredient for 3D application. Some studies have been carried out using native starches as a gelling agent for 3D printing. Yang et al. (2018b) investigated the 3D printing of a lemon juice gel using native potato starch. The study revealed that the lemon gel containing 15 g/100 g of starch presented better printability. Chen et al. (2019a) studied the rheological properties of different starch sources in their native form. They reported that high concentrations of starch prevented the material from extruding smoothly and those low concentrations (10–15 wt.%), especially for maize starch, deformed immediately, resulting in poor printing resolution and sagging. Optimized conditions were dependent on the starch source.

Modified starches have also been masterfully applied in 3D printing technology. Studies have already been carried out using ozone in cassava starch (Maniglia et al. 2019), dry heat treatment (DHT) in cassava (Maniglia et al. 2020b), and wheat (Maniglia et al. 2020a) starches, and pulsed electric field (PEF) in cassava starches and wheat (Maniglia et al. 2021). The use of modified starches, in special DHT and ozone, resulted in 3D products with well-defined geometries.

One fruit-based food formula was used for the development of 3D printed snacks nutritionally designed for children (Derossi et al. 2018). Pectin solution was added to the fruit-based formula to advance the rheological behavior of the ink and to prevent phase separation between water and fruit formula during deposition. Vancauwenberg et al. (2019) studied pectin-based food ink formulation for 3D printing of customizable porous food simulators. The authors observed that food objects can be printed using different pectin-based food inks and that the concentration is the determinant for the firmness and strength of the printed objects. A higher concentration of sugar and pectin affected the build quality and increased the viscosity. The authors emphasized that future research should address the incorporation of additional ingredients to add color and flavor to printed products.

The use of hydrocolloids such as hydroxypropyl methylcellulose (MC) and others has also been described for the production of 3D printed foods. Kim et al. (2017a) established an underlying technical classification system to evaluate 3D printing quality using hydrocolloids as reference materials. MC was selected as suitable reference material, as it presented better texture and viscosity results. Different MC concentrations can simulate the deformation behavior in a series of foods, being the most suitable for 3D printing of foods. Figure 3 brings photographs of several 3D-printed foods showing the effect of the MC concentration from 5 to 14 wt.%, on the dimensional stability.

2.2.2 Proteins

Proteins are composed of amino acids with different functional groups bearing positive or negative charges, dependent on the pH. The distribution and the amount of each amino acid in the primary structure will dictate the isoelectric point of the protein. The printing of protein-based materials is highly influenced by protein aggregation, which in turn depends on the isoelectric point (IP) and the working pH (Portanguen et al. 2019).

There are few studies on the use of animal protein for 3D printing of food, as it is still a great challenge to use meat products or an alternative mixture of protein sources for the elaboration of meat products without the use of additives, and that presents a rheological property that guarantees a good printability and resolution of the final printed food. Liu et al. (2017) pointed out that accurate 3D structures can be better obtained by the addition of texturizers such as hydrocolloids or gellable proteins to the inks. Dick et al. (2019) produced 3D recombined meats and appetizing tender meat products by using a multi-component ink.

Lipton et al. (2010) studied the effect of transglutaminase as a food additive for the building of complex meat products. The authors postulated that transglutaminase allowed the meat to be printed

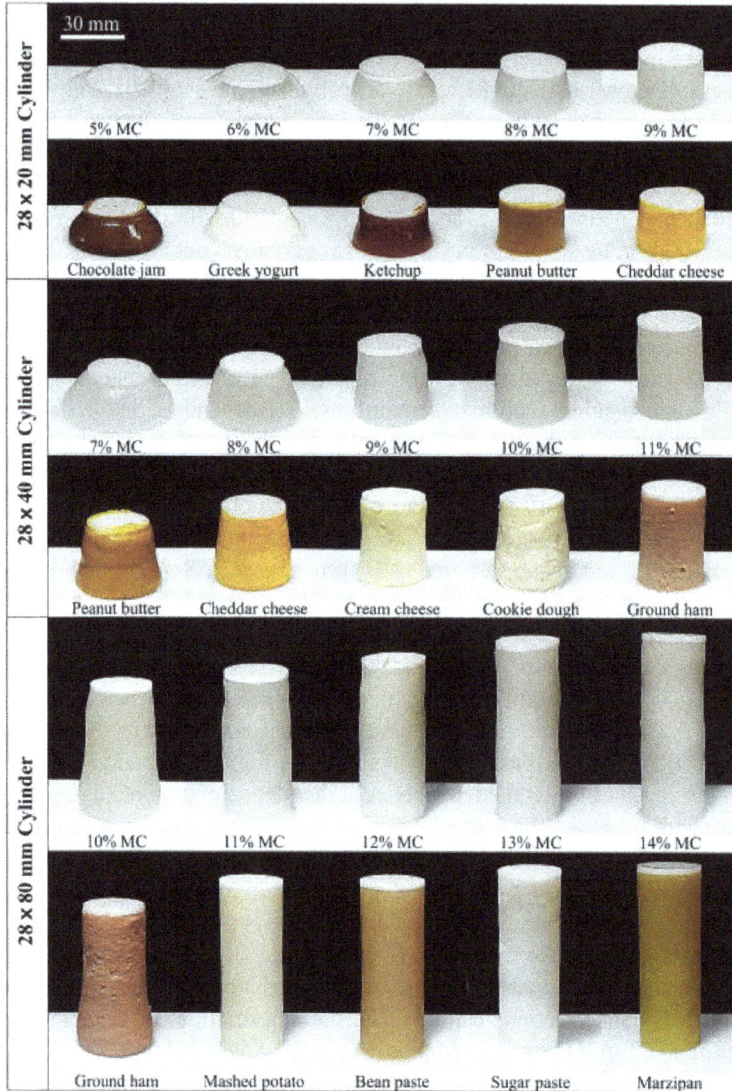

Figure 3. Photographs of 3D printed hydroxypropyl methylcellulose (MC)-reference materials, revealing different deformation behavior as a function of the concentration compared to experimental foods. The scale bar corresponds to 30 mm. Reprinted with permission from (Kim et al. 2017). Copyright 2017, with permission from Elsevier.

directly in 3D. However, the use of additives can be a problem, particularly for consumers who tend to prefer clean label products containing as few additives as possible.

Wang et al. (2018) investigated the influence of sodium chloride on the rheological properties, water holding capacity, gel strength, and other parameters on the printability of surimi mixtures. The improved rheological properties of the mixtures due to the addition of sodium chloride resulted in better flow of the ink through the mouthpiece and maintenance of the shape of the printed material. This was related to the salt-in effect that led to solubilization of myofibrillar proteins and possible protein structural changes. Finally, it was suggested that surimi gels are promising inks for the production of 3D printed food with complex shapes.

Severini et al. (2018a) designed 3D printed snacks with a cylindrical geometry using wheat flour dough enriched with ground Yellow mealworms (*Tenebrio molitor*) larvae as a protein source (Fig. 4). The authors observed that the addition of insect ground up to 20 g/100 g (d.b.) resulted in a softer dough, besides improving other final quality parameters in the snacks. In this sense, the

Insect enrichment (%)

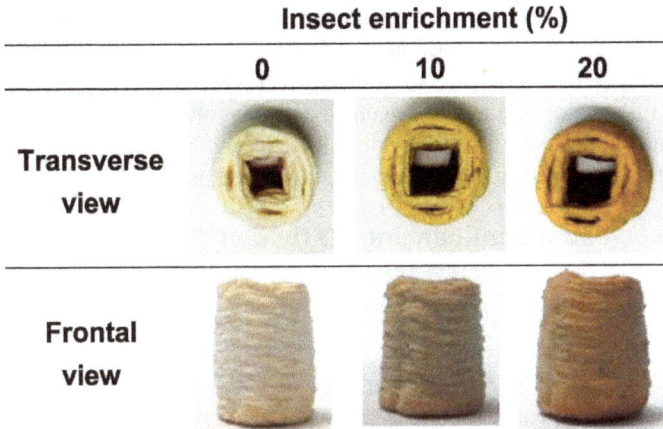

Figure 4. 3D printed snacks using wheat flour dough enriched with ground yellow mealworms' (*Tenebrio molitor*) larvae as a protein source, baked at 200°C for 22 min. Reprinted with permission from Severini et al. (2018a). Copyright 2018, with permission from Elsevier.

insects could be a nutritional option, which validates the use of 3D printing as a technology for the manufacture of innovative snacks without impacting technological quality.

Powdered milk has also been studied as a potential source of protein for the production of 3D printed food (Lille et al. 2018). A high viscous ink was produced using 50 wt.% of skimmed milk powder, which hindered printing due to stickiness. On the other hand, 60 wt.% of semi-skimmed milk performed better as ink which may have contributed to its fat and carbohydrate content.

2.2.3 Lipids

The structure and composition of triglycerides can modify the properties of inks and the functional properties of the 3D printed products, as melting point, crystalline structure, and solid fat index (Godoi et al. 2016).

Chocolate is the ingredient with the highest lipid composition used in 3D food printing (Lanaro et al. 2019, Mantihal et al. 2020). The reasons for choosing chocolate as an ingredient are related to its ability to behave as a viscous fluid that can be easily solidified facilitating the demolding and packaging of the product after a cooling process (Hao et al. 2010). However, the use of high-quality chocolate for 3D printing is a challenge due to its softness and highly complex structure, which includes six crystalline phases that must be controlled throughout the printing process to guarantee the quality of the final product. Furthermore, its non-Newtonian behavior can make difficult the control of the material's flow properties (Hao et al. 2010).

Le Tohic et al. (2018) sought to understand the influence of the 3D printing process on the properties of processed cheeses. The results indicated that the melting and extrusion processing promoted by the 3D printing affected the properties of the cheese-melted and printed cheeses had lower hardness and higher meltability compared to the untreated ones. It was assigned to changes in the microstructure due to protein phase separation and the presence of fat globules.

Lipton et al. (2010) used bacon fat as a flavor enhancer to print turkey puree using transglutaminase as an additive. The same authors also used butterfat in a traditional pasta recipe to prevent the melting of the printed structure when roasted. Another interesting aspect related to the use of lipids for 3D printing is that they can act as a plasticizer or lubricant, which can help the extrusion, facilitating the flow of the ink, as observed by Lille et al. (2018).

2.3 Crucial Factors that Influence Extrusion-based 3D Food Printing

Many factors are essential for obtaining high-quality 3D printed foods, such as the properties of food inks like rheology, texture, ingredient concentration, and composition; and external factors such as the 3D printing equipment, printing parameters, processing techniques, and printing design, as summarized in Fig. 5. Therefore, for a more efficient 3D printing job, all these factors must be considered and adapted to the specific conditions of each study.

Table 1 shows the major findings of selected studies involving 3D food printing by extrusion. The results presented in these studies will be discussed during the development of this chapter, with a focus on the factors that influence extrusion-based 3D food printing.

Factors influencing 3D food printing

Food ink

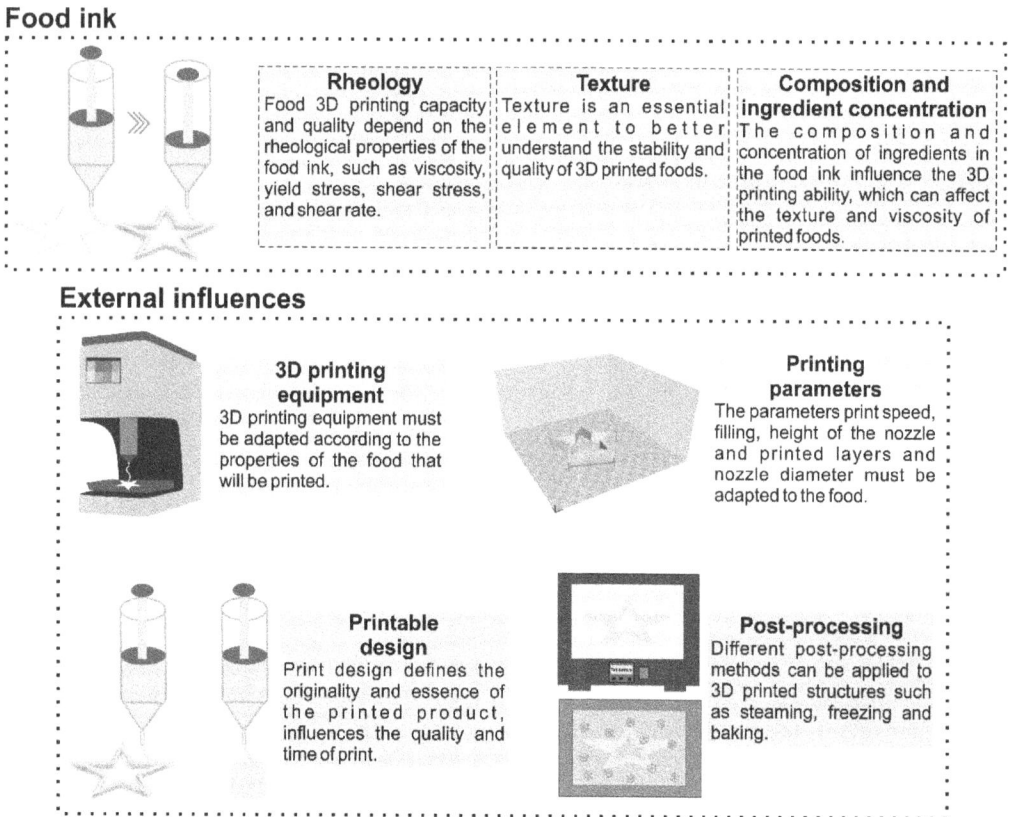

Rheology
Food 3D printing capacity and quality depend on the rheological properties of the food ink, such as viscosity, yield stress, shear stress, and shear rate.

Texture
Texture is an essential element to better understand the stability and quality of 3D printed foods.

Composition and ingredient concentration
The composition and concentration of ingredients in the food ink influence the 3D printing ability, which can affect the texture and viscosity of printed foods.

External influences

3D printing equipment
3D printing equipment must be adapted according to the properties of the food that will be printed.

Printing parameters
The parameters print speed, filling, height of the nozzle and printed layers and nozzle diameter must be adapted to the food.

Printable design
Print design defines the originality and essence of the printed product, influences the quality and time of print.

Post-processing
Different post-processing methods can be applied to 3D printed structures such as steaming, freezing and baking.

Figure 5. Summary of the factors that influence 3D food printing.

2.3.1 Food Ink

2.3.1.1 Rheology

Food inks used for 3D printing must present some specific characteristics to ensure sufficient stability of the printed object, such as ease of extrusion and printing and the ability to maintain its structure. All these characteristics are related to the rheology of the material (Godoi et al. 2016). Therefore, food 3D printing capability and quality depend on the rheological behavior of the food ink, on the properties of viscosity, yield stress, shear stress, and shear rate (Zhang et al. 2021, Sweeney et al. 2017). For example, hydrogels used as food inks must be viscous enough to flow through the nozzle, maintaining shape during and after printing (Kyle et al. 2017).

The food inks can be classified into viscoelastic and pseudoplastic and can behave as both Newtonian and non-Newtonian fluids. Newtonian fluids exhibit shear-independent viscosity while non-Newtonian fluids have shear rate-dependent viscosity. The shear rate of the inks is important information for 3D printing applications since it can be associated with the material's ability to

Table 1. Major findings of selected works involving extrusion-based 3D food printing.

Ink composition	Printing parameters	Major findings	References
Mashed potatoes added to 0%, 1%, 2%, 4% (w/w) potato starch	Nozzle height: 3.0 mm	• The inks showed low yield stress (σ_0: 195.90 Pa) and the printed objects were easily deformed. • The addition of 4% potato starch resulted in better shape retention, as the ink showed higher σ_0 (370.33 Pa) and adequate G'.	Liu et al. (2018)
Lemon juice added to different corn starch content (10, 12.5, 15, 17.5, 20 g/100 g)	Nozzle diameter: 0.5, 1.0, 1.5, and 2.0 mm Nozzle speed: 15, 20, 25, 30, and 35 mm/s Extrusion rate: 20, 24, and 28 mm³/s Printing temperature: 35°C	• 15 g/100 g of starch was the optimal concentration (apparent viscosity: 8079.3 Pa.s, G': 4924.2 Pa; G'': 760.8 Pa and tanδ: 0.155). • The parameters that presented the best-printed lines were nozzle diameter (1 mm), the extruded rate (24 mm³/s), and the nozzle speed (30 mm/s)	Yang et al. (2018b)
Wheat flour and the insect's mealworm powder (100:0, 90:10, and 80:20 w/w, d.b.)	Nozzle diameter: 0.84 mm Nozzle speed: 30 mm/s Travel speed: 50 mm/s Nozzle height: 0.5 mm Infill density: 15%	• The softer dough was obtained with ground insects to 20 g/100 g (d.b.). • A significant increase in total essential amino acids was observed in snacks printed with 20% ground insects (from 32.5 to 41.3 g/100 g of protein).	Severini et al. (2018a)
Soy protein (20%); Soy protein (20%) + Alginate (2%); Soy protein (20%) + Alginate (2%) + Gelatin (2, 6, 10%)	Nozzle diameter: 1.55 mm Nozzle speed: 20 mm/s Flow rate: 80% Fill density: 100% Printing temperature: 35°C	• The addition of alginate increased the G' and G'' values of the inks and the hardness and chewiness of the 3D printed geometry. • Good flowability with lower viscosity and higher tan σ and G' was obtained for soy protein mixtures with 6% and 10% gelatin.	Chen et al. (2019b)
Rice, cassava, potato starch suspensions at 5, 10, 15, 20, 25, and 30% (w/w, dry basis)	Nozzle diameter: 0.8 mm Nozzle height: 2.0 mm Nozzle speed: 20 mm/s Extrusion rate: 30 mm/s Printing temperature: 70, 75, 80°C	• Rice starch: 15–25% (w/w) at 80°C, cassava starch: 20–25% (w/w) at 75°C, potato starch: 15–20% (w/w) at 70°C possessed appropriate, τ f (140–722 Pa), τ γ (32–455 Pa), and G' (1150–6909 Pa) values, which were preferable, showed excellent printability, shape retention, and resolutions.	Chen et al. (2019a)
Cassava starch native and modified by ozone processing (15 and 30 min) Inks: 10 g starch/ 100 g (d.b.)	Nozzle diameter: 0.8 mm Nozzle speed: 20 mm/s Nozzle height: 18 mm Extrusion rate: 30 mm/s Printing temperature: 20°C	• At the gelatinization temperature of 65°C, ink gels based on native starches and modified starches (30 min) showed good printability, but up to this temperature, only modified starch (30 min) produced gels with good printability.	Maniglia et al. (2019)

Table 1 contd. ...

...Table 1 contd.

Ink composition	Printing parameters	Major findings	References
Mushroom powder at 5, 10, 15, 20, and 25% was mixed with wheat flour	Nozzle speed: 200, 400, 600, 800, and 1000 mm/min Nozzle diameter: 1.28 and 0.82 mm	• The formulation with 20% mushroom powder printed at a speed of 800 mm/min, nozzle diameter of 1.28 mm, extrusion motor speed of 300 rpm at 4 bar of pressure, presented the best stability and printing accuracy (78.13%).	Keerthana et al. (2020)
Cookie dough formulations with different types of fat (butter and shortening), flour (wheat, rice, and tapioca), non-fat milk (32.5 or 65 g/100 g flour), and sugar (37.5 or 55 g/100 g flour)	Nozzle speed: 300 mm/min Nozzle diameter: 2.4 mm	• The formulation based on 37.5 g sugar, 62.5 g butter, 100 g tapioca flour, and 32.5 g milk showed the best visual printing and the smallest structural deformation after baking	Pulatsu et al. (2020)
Fresh vegetables (garden pea, carrot, and bok choy) added to hydrocolloids 0–2% (w/w) (xanthan gum, kappa carrageenan, and locust bean gum)	Nozzle diameter: 0.84 mm Nozzle speed: 25 mm/s Nozzle height: 5 mm Flow rate: 80% Infill density: 85%	• Smaller amounts of hydrocolloids are needed in formulations with higher starch content and a lower percentage of water in vegetables.	Pant et al. (2021)
Peanut butter, cream cheese, fractionated animal fat, or 20% rice-starch gel	Nozzle diameter: 1.50, 083, and 0.41 mm Extrusion speeds: 10 and 30 mm/s	• The best printing conditions were observed using fractional animal fat and 20% rice starch gel. • Non-uniformity of the filament width and the relevance of the extrusion multiplier to adjust the different rheology between food materials were highlighted.	Nijdam et al. (2022)

deform itself by its weight and construction (extrusion speed, shape, and geometry) (Zhang et al. 2021). Viscoelastic fluids can be characterized by the storage module (G'), which consists of the sample's ability to store energy elastically; and the loss modulus (G"), which is the ratio between viscosity and stress, and yield stress (σ_0: the crossover point between G' and G"), will provide information related to the shear conditions in which irreversible deformation occurs, causing the material not to return to its original shape when the shear load is removed (Tabilo-Munizaga and Barbosa-Cánovas 2005). Tanδ (loss factor) defines the G"/G', and this factor indicates the shape retention of printed products after extrusion (Oliveira et al. 2020). According to Jiang et al. (2019), when G' is greater than G", the food has an elastic gel-like structure with higher structural stability and print resistance. Studies have reported success in printing using inks with G' and G" values in the range of 100 Pa to 10,000 Pa, for both the properties (Wang et al. 2018, Yang et al. 2018b, Lille et al. 2018).

The pseudoplastic materials usually behave in a non-Newtonian way. Pseudoplastic foods present a reduction of shear viscosity (Feng et al. 2019). Pseudoplastic foods are ideal for 3D printing as the decrease in viscosity allows for easier extrusion into the mouthpiece. Upon completion of printing, the printed food can recover to a viscous state that will maintain structural integrity (Feng et al. 2019, Jiang et al. 2019). Pseudoplastic foods that have been successfully printed include baking dough, mashed potatoes, agar, vegemite, and marmite (Yang et al. 2018a, Liu et al. 2018, Feng et al. 2019, Hamilton et al. 2018). Researchers have reported successful 3D printing with viscosities ranging from 1000 to 10 Pa, corresponding to shear rates of 0.1 in 500 s^{-1} (Hamilton et al. 2018, Liu et al. 2018, Yang et al. 2018b). The vast range of values demonstrated that shear roughing principles applied directly to 3D food result in the ability to deliver a variety of successfully printed materials.

Chen et al. (2019a) evaluated the effect of the rheological properties of different starches on 3D printing by hot extrusion. The authors observed that the storage modulus of food inks is essential to support the various layers deposited during 3D printing and to maintain shapes after printing. The results indicated that rice starch of 15–25 wt% at 80°C, cassava starch of 20–25 wt.% at 75°C, and potato starch of 15–20 wt% at 70°C bears appropriate G' (1150–6909 Pa) values, which were preferable for 3D food printing with great printability, resolutions, and shape quality. In addition, the flow tension influenced the smoothness of the food inks extrusion, ensuring uniformity and high resolution of the printed objects.

Liu et al. (2018) investigated the impact of rheological properties on 3D printing of inks based on mashed potatoes mixed with potato starch at different concentrations (i.e., 1, 2, 3, and 4%). The authors concluded that highly desirable materials for 3D food printing must not only have an adequate modulus of elasticity (G') to hold the printed shapes but also have proper viscosities to be easily extruded from the nozzle in an extrusion-based type printer. The mashed potato added with 2% potato starch showed excellent extrudability and printability, i.e., thinning behavior (σ_0: 312.16 Pa and adequate G'). Although the mashed potato mixed with 2% potato starch represented good shape retention (σ_0: 370.33 Pa and adequate G'), the low extrudability made printing difficult due to the high viscosity.

Maniglia et al. (2020a) reported that hydrogels based on modified wheat starches by dry heating treatment showed higher G' and yield stress (G': 430 Pa and σ_0 54.56 Pa) than the hydrogels based on native wheat starch (G': 175 MPa and σ_0: 12.74 Pa). These rheological parameters predicted the behavior observed for the 3D printed materials based on these hydrogels. The modified starch hydrogels showed better printability than native starch hydrogels.

Therefore, the different properties that determine the rheology of the food inks must be carefully considered and evaluated to ensure the success of the 3D food prints. The rheological parameters determine the strength of the food ink to maintain the printed form, the smoothness of extrusion and the ability to adhere to previously deposited layers.

2.3.1.2 Texture

3D printing enables the production of foods with adjustable textures, which guarantees the access of a broad range of the population, including people with masticatory diseases. For instance, the texture is linked to the structural profile of foods, so textural analysis is an essential element to better understand the stability of 3D printed foods, as well as to indicate the most appropriate sensorial characteristics. Therefore, the texture is one of the main parameters used to evaluate the quality of 3D printed food, since this attribute has a great influence on the sensorial acceptance of the product (Mantihal et al. 2020, Zhang et al. 2018). Texture can also influence both the stability after printing and the material extrusion out of the syringe.

Reported texture profile analysis (TPA) parameters include hardness, chewiness, stickiness, cohesion, resilience, and springiness (Nishinari et al. 2019). Toughness, resilience, cohesion, and springiness are good indicators of structural integrity, as they represent the resistance and resilience to external forces. Chewability and stickiness can provide information about the quality and mouthfeel of 3D printed food (Zhang et al. 2021).

Keerthana et al. (2020) used 3D printing technology to produce fiber-enriched snacks from different foods. It was observed that the formulation based on wheat dough added to 20% mushroom showed the lowest hardness (53.17 N) and adhesiveness (–197.80 N.s) and it was easier to extrude. Pant et al. (2021) produced 3D printed food of fresh vegetables (pea, carrot, and Chinese cabbage) using food hydrocolloids for dysphagic patients. The authors related a wide texture range observed for the printed food with hardness ranging from 60 to 320 g, adhesiveness 1.7–6.7 mJ, and gumminess of 27–230 g, depending on the type of vegetable, concentration, and type of hydrocolloids added. In general, the printed samples showed textural parameters suitable for dysphagic patients. Lin et al. (2020) presented the FoodFab system that allows users to modify the texture and internal structure of foods via two 3D printing parameters: infill pattern and infill density. The authors evaluated how to modify the chewing time by changing the infill pattern (honeycomb, rectilinear, and Hilbert) and density (25%), and thus control perceived satiety. In addition, they observed that the higher bite force was necessary for the snacks with Hilbert infill pattern, and the consequent longer chewing time affect people's sense of satiety.

In general, it has been reported in the literature that successful 3D printed food has values from 4.5 to 60.0 N for hardness and springiness ratio below 1 (Zhang et al. 2021). Foods with a springiness ratio above 1 showed are mechanically fragile and present poor structural stability (Kim et al. 2017)

2.3.1.3 Composition and Ingredient Concentration

The composition and concentration of food ink ingredients, such as the composition of nutrients and the use of additives, significantly influence the printability which can directly affect the viscosity and texture of the final product (Zhang et al. 2021).

Food inks can be prepared from a single to a multicomponent composition, achieving different levels of effectiveness when printed. Chen et al. (2019b) prepared inks using a mixture of soy protein isolate (SPI) and water with poor printability and low structural integrity, related to an inadequate rheological behavior (hardness 1.18 N). The authors evaluated the addition of alginate (0.5% wt.) and gelatin (0, 2, 6, and 10% wt.) to the SPI-based inks. The best synergistic effect of materials was reported when SPI was mixed with alginate and gelatin (2–10%). The hardness increased up to 15.66 N using the mixtures.

Yang et al. (2018a) highlighted in their study the quality of 3D printing of dough for baking. They observed that the rheology of the ink and the physical properties of the product were affected by the concentration of water, sucrose, flour, butter, and egg. The gel strength, elasticity, and viscosity increased while the ductility was reduced with an increased concentration of sucrose, butter, and flour.

However, when the concentration of these same ingredients exceeded a certain value, the extrusion capacity of the material was compromised. The formulation containing water (29.0 g), butter (6.0 g), flour (48.0 g), sucrose (6.6 g), and egg (10.4 g) showed the best printing capacity, elasticity, and lower ductility.

Dick et al. (2019) studied the impact of the fat content (0, 1, 2, 3 fat layers within a structure) and fill density (50, 75, and 100%) on print and product properties. For the 3D printing of the rectangular prism design, a food ink was prepared using a base of meat [meat (85%) and water (15%)], NaCl (1.5%), and guar gum (0.5%). The chewability, hardness, and moisture retention parameters were proportionally influenced by the filling density; however, this same factor inversely affected the cohesion and shrinkage of the printed material. Fat content inversely affected chewiness, fat retention, hardness, and moisture retention, proportionally affecting shrinkage, cooking loss, and cohesiveness.

2.3.2 External Factors that Influence 3D Printing

3D food printing is also affected by external aspects, such as the type of equipment, processing parameters, techniques, print design, among others. External influences can affect print accuracy and final product quality. In this section, we will discuss how these external factors affect 3D food printing.

2.3.2.1 The 3D printers

The use of 3D printing technology for food applications is one more recent development when compared to the 3D printing technology involving the mechanism for plastic filaments. In this way, in previous works, many of the 3D printers used have been modified or customized to print food effectively, including the addition of a motorized system for food extrusion and a syringe for food intake. For example, Reprap Pro Ormerod 1, Felix 3.0, and Delta 2040 printers were originally designed for plastic materials but were modified and reconfigured with a custom-designed syringe that was appropriate for food extrusion (Le Tohic et al. 2018, Chen et al. 2019a, Derossi et al. 2018).

In this scenario, what we observe is that 3D printers have varying capabilities, specifications, and design differences that make standardization and reproducibility difficult. Thus, there is a need for further clarification of the configuration, modifications, and customizations of 3D printers for developing universal printers for food 3D printing.

2.3.2.2 Printing Parameters

The quality of 3D printed food is directly related to the optimization of the printing parameters, namely nozzle height, diameter and movement speed, extrusion rate, and filling (Fig. 6).

The distance between the nozzle tip of the extruder and the deposited top layer is called the nozzle height. Under ideal conditions, without shrinkage or expansion of the lines, the extruded product has the same diameter as the nozzle; ideally, smaller nozzle heights provide better print quality, as they ensure that the material can be fixed to the previous layer avoiding inaccuracies caused by delay (Yang et al. 2018b) in deposition. However, under non-ideal conditions, a critical nozzle height can be calculated (Eq. 1).

$$hc = \frac{V_d}{v_n D_n} \tag{1}$$

where:
- h_c: critical nozzle height (mm),
- V_d: volume of the extruded rate (mm³/s),
- v_n: nozzle movement speed (mm/s),
- D_n: nozzle diameter (mm).

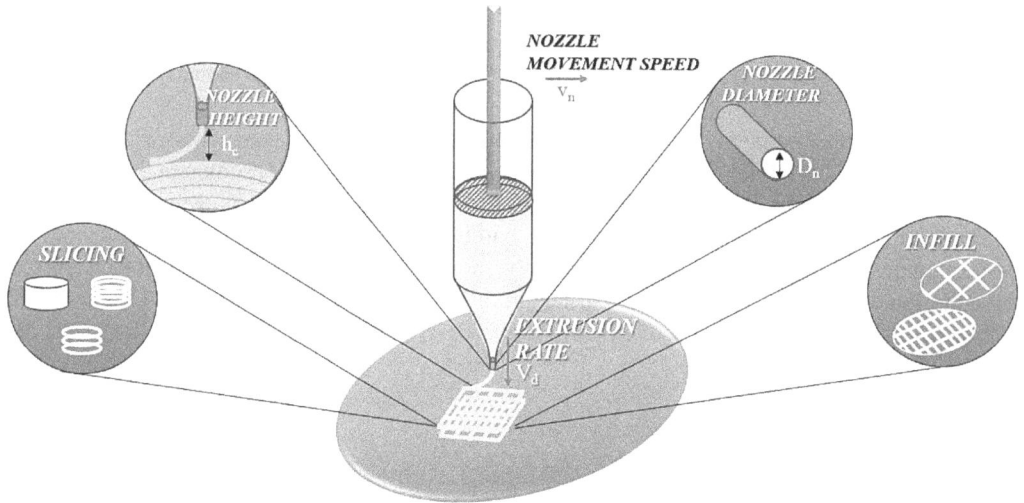

Figure 6. Schematic diagram of the 3D printing process parameters: nozzle height, nozzle diameter, extrusion rate, nozzle movement speed, and filling.

The literature describes that the nozzle height greatly influences the geometric shape of the printed samples and that when using the critical height value, it provides the desired shapes with greater fidelity and a good connection between the layers (Wang et al. 2018).

The diameter (D_n) of the ink outlet nozzle is largely responsible for the thickness of the line to be deposited. Depending on the properties of the inks, the effect of high pressure in the flow channel in a nozzle with reduced diameters causes the material to store elastic energy and, when released after extrusion, promotes swelling of the extrudate. Therefore, a suitable diameter is required to provide smooth extrusion without a considerable increase in the system pressure. Yang et al. (2018b) produced corn starch gels added with lemon juice and evaluated the influence of nozzle diameter (0.5, 1.0, 1.5, and 2.0 mm) on the quality of printed materials. The authors noted that the larger diameter nozzle lines resulted in relatively thick and poor lines, while the smaller diameter nozzle resulted in relatively delicate and fine lines, ensuring a better quality of the printed food. Huang et al. (2019) evaluated the effect of nozzle diameter (0.84 mm, 1.20 mm, 1.56 mm) on the dimensional properties of 3D printed samples based on brown rice. These authors stated that the smaller diameter (0.84 mm) of the nozzle resulted in printed samples with dimensions closer to the design project. In addition, the nozzle diameter was able to change the void rate and the number of layers deposited, directly affecting the texture characteristics (hardness and gumminess). The authors also noted that the larger nozzle size (1.56 mm) significantly reduced print time but resulted in dimensional deviations. Nijdam et al. (2022) evaluated the validity of different filament extrusion models typically used in open-source software and commercial slicers such as Slicer. The authors evaluated different inks based on peanut butter, cream cheese, fractionated animal fat or 20% rice-starch gel, different nozzle diameters (0.41, 083, and 1.5 mm), and extrusion speeds (10 and 30 mm/s). The model that most corresponded to the behavior observed by the lines made with the inks was of oblong cross-section and provided accurate predictions of the filament width under the nozzle diameter and print speed conditions evaluated in this work. Fractionated animal fat and rice starch gel presented ideal conditions for 3D printing, forming smooth, continuous, and stable filaments, considering the wide range of printer operating conditions evaluated in this work.

The nozzle movement speed and extrusion rate parameters simultaneously affect 3D printing. In this sense, according to Yang et al. (2019), the nozzle movement speed and extrusion rate can be determined considering the nozzle height and diameter (Eq. 2).

$$V_d = \frac{\pi}{4} v_n D_n^2 = \frac{\pi}{4} v_n h_c^2 \qquad (2)$$

where:
- V_d: volume of the extruded rate (mm³/s),
- v_n: nozzle movement speed (mm/s),
- D_n: nozzle diameter (mm),
- h_c: nozzle height.

Although there is a positive relationship between nozzle movement speed and extrusion rate, there is a limit to printing efficiency when changing these parameters. High speeds can cause a drastic increase in the extrusion rate of the material and cause the strands of the extruded product to be dragged and thick wavy lines to form. On the other hand, low extrusion rates can lead to broken lines, which result in collapse and a considerable difference between the defined objective and the actual product. Therefore, it is necessary to determine a suitable extrusion rate to print a smooth line with a uniform diameter (Yang et al. 2018b).

Theagarajan et al. (2020) evaluated the parameters' nozzle speed (800 to 2200 mm/min) and extrusion rate based on motor speed (120 to 240 rpm) for 3D printing of rice starch gel inks. The authors observed that for high nozzle speed (up to 2400 mm/min), the printed samples showed serious structural defects; for intermediate nozzle speed (1500 to 800 mm/min), the samples showed better appearance, and for low nozzle speed (below 800 mm/min), the samples showed zones with excessive material deposition. Furthermore, these authors observed that the best printability was obtained for values from 180 to 240 rpm for the engine speed.

Another crucial parameter in the final characteristics of the printed food, mainly in its three-dimensional stability, is the infill which can be regulated by the infill level, patterns, and perimeter of the printed sample. The fill level is the percentage of the inside of the printed object that is filled with printing ink. The fill pattern is the pattern that the nozzle draws to fill the object (Liu et al. 2018). The perimeter consists of the amount of layer on the outside (Huang et al. 2019). Liu et al. (2018b) investigated the effect of different filling levels (10, 40, and 70%) and filling patterns (honeycomb, rectilinear, honeycomb, and Hibert curve) on the texture of printed foods based on mashed potatoes. It was noted that increasing the filling percentage had a positive effect on the hardness, firmness, and rigidity of the printed products, while the infill pattern had no significant effect.

2.3.2.3 Print Design

The choice of the design desired for the final product is essential to ensure the originality and essence of the 3D object. The complexity of the design will influence the quality and printing time. In this sense, a compromise between the complexity and printing time must be reached.

3D printing is a valuable technology in food prototyping as it allows the development of personalized foods based on the needs of individuals. In this sense, choosing the right and personalized print design is essential to attract the individual's attention, presenting food with a pleasing appearance. Print design can also influence the quality of 3D printed foods. For instance, Zhang et al. (2018) performed 3D printing of cereal-based food structures containing probiotics and applied the post-processing process that consisted of baking the structures in an oven at different temperatures, followed by a cooling step. The authors reported that a print design with a higher volume could influence the survival of probiotics, thus improving the final quality of the developed product.

In general, printing even simple structures, like stars, can be challenging due to the fine adjustment of angles and shapes required for the process.

2.3.2.4 Post-processing

Raw materials and ingredients used as food inks for 3D printing may require post-processing techniques, which can cause significant changes in the physicochemical properties of the product, also altering the nutritional value and sensory acceptance (Keerthana et al. 2020).

Different methods have been applied in the post-processing of 3D printed food, such as steaming (Thangalakshmi et al. 2021), freezing, baking (Severini et al. 2018a, Pulatsu et al. 2020), hot air drying, refractive window drying, and microwave drying (Yoha et al. 2021). The post-processing method applied depends on the specific needs of each food. For example, Jo et al. (2021) applied freeze-drying technology after 3D printing of functional snacks as an alternative to traditional cooking methods applied in the post-processing of 3D printed foods. The authors reported that post-processing methods based on freeze-drying can result in 3D printed snacks with a texture similar to commercial products. Furthermore, Yoha et al. (2021) evaluated the effect of different post-processing methods on the stability of encapsulated and 3D printed probiotics. It was observed that 3D printing had no negative impact on the viability of the encapsulated probiotics and that the freeze-drying method was the most efficient method for the survival of the probiotics.

3. Acceptance of 3D Printed Foods

Much is said about the potential of 3D printing technology to improve the acceptance of food products. We do agree with it, and we think it is particularly important to consider food products for special needs. However, surprisingly, there are limited studies in the literature assessing the consumer acceptance of 3D printed foods products. Among different reasons for that, the difficulties associated with this kind of study seem to be an important factor.

Severini et al. (2018b) produced food formulas constituted by a fruit and vegetable paste made of carrots, kiwi, pears, avocado, and broccoli raab leaves, which was gelled with fish gelatin (1%) and printed in a pyramidal shape. The product was evaluated by 30 untrained panelists about the overall appearance, color, odor, off-odor, taste, off-flavor, comparing the 3D printed formula with the initial paste. As expected, the 3D printed food analog presented higher scores about "overall appearance" in comparison with the paste, while both products presented similar scores for the other sensorial parameters. The authors highlight this result supports the hypothesis that 3D printing technology can improve the visual aspect of food, positively affecting their acceptance. Those results are relevant to start this discussion, but it is important to inquire about their representativeness. To go further in this evaluation, the acceptance of actual consumers of a given food proposal must be considered, as well as comparing the printed food with other alternatives of meals. It is worth mentioning how difficult it is to conduct this kind of experiment considering specific audiences, such as children and elderly consumers.

In this context, Caulier et al. (2020) studied the acceptance of snack bars produced by 3D printing through Dutch soldiers, considering the possibility of personalization of a product they already consume after training sections. By focusing on a specific public, the authors gave an important step toward understanding the potential of 3D printing technology for personalization. The panelists were 12 men average of 32 years old, that consumed and evaluated a variety of sweet and savory flavors for both snacks' fillings and doughs. This study provided an increased customization option over four weeks, starting with no choice (the benchmark bar) and finalizing with 4 types of dough and 13 types of filling, considering different textures, tastes, and ingredients. It is interesting to notice that even considering so many personalization options and information, the final result shows the panelists scored the printed snacks equally or slightly lower than the benchmark product (control treatment), with scores from 5.3–6.8 on a scale up to 9. In addition, they pointed to the "interest of health" as the most important interest for this technology in its context (scoring this importance as 5.8/7). This work showed the potential of 3D food printing technology for personalized nutrition for soldiers on the battlefield.

Theagarajan et al. (2020) studied different post-processes to the 3D printing of rice starch gels: blanching, steaming, microwaving cooking, roasting, shallow-frying, and deep-frying. The sensorial acceptability was evaluated by a semi-trained panel with 20 members aged between 24 and 45 years, comparing each product obtained after post-processing with the recently obtained gel, considering nine attributes. This approach is interesting to select a post-processing method for a given food analog, but it does not assess the actual acceptability of that meal, specially considering other possible alternatives.

Chow et al. (2021) evaluated the 3D printability, physical, and sensory properties of lemon mousse produced with whipping cream, sugar, eggs, lemon zest, porcine gelatin (1–2%), citric acid (0.9–1.5%), and whey protein isolate (8–18%). They printed cylinders up to 6 layers, whose sensorial properties were evaluated through two assays. First, the sensory profiling was evaluated through descriptive analysis by a trained panel from the University of Copenhagen (Denmark) containing 10 panelists (9 women, 1 man). Then, a consumer test was conducted with 32 panelists (21 women, 32 men, 35 years old as average). They evaluated six desserts, including two produced by 3D printing, through 22 pre-selected sensory attributes. The 3D-printed desserts were scored ranging from 9.1 to 10.6 in a range of 15 and considering attributes of appearance, taste, texture, desire to eat, salivating, and overall liking, considering geometries obtained with one or three layers. Although those results are interesting and important to foster our knowledge about the consumer perception concerning the 3D printing technology, they were obtained with a limited number of panelists, having a focus on understanding the importance of each ingredient on the final product. Unfortunately, the work did not present the comparison of the printed dessert with other desserts prepared by conventional methods, for example.

Consequently, there is still a need to understand how different groups of consumers (and specially those listed as a possible focus of this technology) accept, think, and feel about the 3D printing technology for different objectives of food processing. For instance, people's perception about a dessert printed to be a different or fun product would be different from consumers with special needs evaluating a whole meal.

Two important points to be considered are the so-called "food technology neophobia" (the consumers' fears of novel food technologies (Cox and Evans 2008)) and the rising number of consumers against processed foods (Augusto 2020). Both aspects are highly influenced by the consumer generation, age, social context, and previous knowledge.

To summarize, the 3D printing technology is in an interesting stage of development, where some products (such as chocolate, dessert, and pasta) are already offered in selected markets, while basic knowledge is still being developed. However, although there are current studies on equipment and process conditions, as well as ingredients for this process, consumer perception of this technology still needs to be understood. Consequently, future studies in this field are highly necessary.

4. Sustainable Contribution of 3D Printing to the Food Sector

It is well known that one of the biggest challenges in the food processing sector is the losses and generation of waste. The introduction of Industry 4.0 technologies in the food sector can contribute to the reduction of losses, such as the use of 3D printers. 3D printing has stood out in recent years in different industrial sectors, including food production. According to Attaran (2017), minimizing food waste is another advantage of 3D printing, as this technology only uses the materials needed for consumption. Moreover, this innovative technology outperforms traditional food processing, as explained in previous sections, as well as being potentially important for sustainable development (Gebler et al. 2014).

For example, the startup Upprinting Food has been using 3D printing to recover food waste and create a more sustainable and nutritious food economy (Upprinting Food 2022). Other researches have also addressed the use of 3D printing focusing on sustainability, for example, the work developed by Jagadiswaran et al. (2021) which used 3D printing to develop functional cookies using grape pomace. This approach has advantages not only from the point of view of food production with customized formats but also helps to overcome the consumer's conception of purchasing foods developed from by-products. Furthermore, it is known that the use of agro-industrial by-products is an alternative to the sustainable development of food, contributing to lower environmental impacts.

3D printing has sparked interest in several applications, ranging from food for people with special needs to its use in the military and space field (Le-Bail et al. 2020). NASA (USA) has also demonstrated interest in this technology, mainly due to the ease of producing foods with adequate nutritional requirements, acceptability, and food safety (Lipton 2017).

Another interesting approach is the application of this technology in the development of foods for specific groups of individuals, as this technology can facilitate and motivate people to have control over food intake through personalized diets (Le-Bail et al. 2020), for example, as already mentioned, people with specific needs related to food composition (such as nutritional intolerances and allergies) or its physical properties (such as texture and chewing and swallowing problems, such as dysphagia).

Many authors have studied the development of 3D printed foods with different textures, nutritious and suitable for people with dysphagia. Dick et al. (2019) incorporated hydrocolloids into pork to prepare a 3D printed pork paste with modified textures suitable for people with chewing and swallowing difficulties. Xing et al. (2022) developed 3D printed foods based on black fungi and added gums as a diet for dysphagia, offering insights into the development of visually appealing foods suitable for people with dysphagia using 3D printing.

Moreover, the potential uses of 3D food printing encompass home or small-scale 3D printing, resulting in easier food acquisition and promoting better food experiences. In fact, in the course of developing 3D printing, one of the first goals was the idea of home 3D printing. In addition, small-scale 3D food production allows for the customization of food products, for example, the confectionery and bakery industry (Lipton et al. 2010).

The printing of food changes our view about the manufacture and preparation of food, since, in the future, an individual could get the food ready in moments. All of this results in less work involved and, consequently, lower food cost, and food becomes more versatile as an individual can make their food at home (Tran 2019). Of course, 3D food printing will still face several obstacles to be completely suitable for all the purposes for which the application is possible. As discussed throughout this chapter, many factors directly influence 3D food printing, making it necessary to optimize processes for each specific demand. Despite this, the technology has great advantages to be applied in the food sector, as presented so far, the use of this technology will revolutionize the industrial scenario as this technology can provide customized solutions, with nutritional appeal and built to measure for people who even now have been excluded from certain markets due to the most diverse conditions, such as health. There is still a long path to further progress, development, and application on a large scale, and although these are difficult challenges, they are not impossible to overcome.

5. Perspectives and Conclusion

This chapter highlighted 3D food printing as a revolutionary and promising technology to transform conventional food manufacturing, but this futuristic technology presents many challenges, from the materials to be processed, to the processing, and post-processing parameters. Among the different techniques, extrusion-based 3D printing has been the most explored in the food sector. In addition, different food ingredients, based on carbohydrates, proteins, and lipids, have been used to create foods based on the layer-by-layer method. The 3D printing technology has made significant contributions to the food sector, offering new possibilities for the application of alternative ingredients such as insect meals, algae, microorganisms, and agro-food residues, efficient use of food ingredients with minimal or no waste. Moreover, this technology opens avenues for the increased acceptance of foods that can contribute to better use of underutilized food. The inclusion of on-demand functional ingredients, resulting in a better food experience, automation, labor savings, lower energy, and transportation costs, and easier supply chain are other examples of the advantages of 3D printed foods. Therefore, it is undeniable how 3D printing can contribute to a sustainable future.

In conclusion, 3D printing is at a stage of development involving the search for basic knowledge of this type of technology, but already with products such as printed pasta, chocolates, and desserts already being offered in selected markets. Furthermore, although there are current studies on equipment and process conditions, as well as ingredients for this process, consumers' perception of this technology still needs to be understood. Thus, future studies in this area both in terms of technology and market are highly necessary.

Acknowledgments

The authors are grateful to the São Paulo Research Foundation (FAPESP, Brazil) for funding the projects n° 2019/25054-2, 2019/05043-6 and, 2020/08727-0, for finance the BC Maniglia Young Investigator grant (2021/05947-2) and the JS Guedes Ph.D. scholarship (2021/06398-2). The authors are also grateful to the National Council for Scientific and Technological Development (CNPq, Brazil) for funding the productivity grant of PED Augusto (306557/2017-7) and for financing the PAI Sponchiado Master scholarship (162334/2021-4). In addition, they are grateful to the "Coordenação de Aperfeiçoamento de Pessoal de Nível Superior – Brasil (CAPES)" – for financing the BS Bitencourt PhD scholarship (88887.636998/2021-00). Finally, the authors would also like to thank Dr. Fernanda Condi de Godoi for her confidence in indicating us to write this chapter.

References

Attaran, Mohsen. 2017. The rise of 3-D printing: the advantages of additive manufacturing over traditional manufacturing. Business Horizons 60(5): 677–688. doi:10.1016/j.bushor.2017.05.011.

Augusto, Pedro E.D. 2020. Challenges, trends, and opportunities in food processing. Current Opinion in Food Science 35(October): 72–78. doi:10.1016/j.cofs.2020.03.005.

Bilbao, Emanuel, Sunil Kapadia, Verónica Riechert, Javier Amalvy, Fabricio N. Molinari, Mariano M. Escobar, et al. 2021. Functional aqueous-based polyaniline inkjet inks for fully printed high-performance ph-sensitive electrodes. Sensors and Actuators B: Chemical 346 (November): 130558. doi:10.1016/j.snb.2021.130558.

Caulier, Sophie, Esmée Doets and Martijn Noort. 2020. An exploratory consumer study of 3D printed food perception in a real-life military setting. Food Quality and Preference 86 (December): 104001. doi:10.1016/j.foodqual.2020.104001.

Chen, Huan, Fengwei Xie, Ling Chen and Bo Zheng. 2019a. Effect of rheological properties of potato, rice and corn starches on their hot-extrusion 3D printing behaviors. Journal of Food Engineering 244(March): 150–158. doi:10.1016/j.jfoodeng.2018.09.011.

Chen, Jingwang, Taihua Mu, Dorothée Goffin, Christophe Blecker, Gaëtan Richard, Aurore Richel, et al. 2019b. Application of soy protein isolate and hydrocolloids based mixtures as promising food material in 3D food printing. Journal of Food Engineering 261. Elsevier: 76–86.

Chou, Wai-Houng, Alexander Gamboa and Javier O. Morales. 2021. Inkjet printing of small molecules, biologics, and nanoparticles. International Journal of Pharmaceutics 600 (May): 120462. doi:10.1016/j.ijpharm.2021.120462.

Chow, Ching Yue, Camilla Doris Thybo, Valeska Farah Sager, Reisya Rizki Riantiningtyas, Wender L.P. Bredie, et al. 2021. Printability, stability and sensory properties of protein-enriched 3D-printed lemon mousse for personalised in-between meals. Food Hydrocolloids 120(November): 106943. doi:10.1016/j.foodhyd.2021.106943.

Cox, D.N. and Evans, G. 2008. Construction and validation of a psychometric scale to measure consumers' fears of novel food technologies: the food technology neophobia scale. Food Quality and Preference 19(8): 704–710. doi:10.1016/j.foodqual.2008.04.005.

Derossi, A., Caporizzi, R., Azzollini, D. and Severini, C. 2018. Application of 3D printing for customized food. A case on the development of a fruit-based snack for children. Journal of Food Engineering 220(March): 65–75. doi:10.1016/j.jfoodeng.2017.05.015.

Dick, Arianna, Bhesh Bhandari and Sangeeta Prakash. 2019. Post-processing feasibility of composite-layer 3D printed beef. Meat Science 153 (July). Elsevier: 9–18. doi:10.1016/j.meatsci.2019.02.024.

Ebrahimi, Amir, Francisco J. Tovar-Lopez, James Scott and Kamran Ghorbani. 2020. Differential microwave sensor for characterization of glycerol–water solutions. Sensors and Actuators B: Chemical 321 (October): 128561. doi:10.1016/j.snb.2020.128561.

Eleftheriadis, Georgios, K. and Dimitrios G. Fatouros. 2021. Haptic evaluation of 3D-printed braille-encoded intraoral films. European Journal of Pharmaceutical Sciences 157 (February): 105605. doi:10.1016/j.ejps.2020.105605.

Feng, Chunyan, Min Zhang and Bhesh Bhandari. 2019. Materials properties of printable edible inks and printing parameters optimization during 3D Printing: a review. Critical Reviews in Food Science and Nutrition 59(19): 3074–3081. doi:10.1080/10408398.2018.1481823.

FoodJet. 2021. FoodJet Printing Systems - Web Page. https://www.foodjet.com/.

Gebler, Malte, Anton J.M. Schoot Uiterkamp and Cindy Visser. 2014. A global sustainability perspective on 3D printing technologies. Energy Policy 74 (November): 158–167. doi:10.1016/j.enpol.2014.08.033.

Godoi, Fernanda C., Sangeeta Prakash and Bhesh R. Bhandari. 2016. 3D printing technologies applied for food design: status and prospects. Journal of Food Engineering 179 (June). Elsevier: 44–54. doi:10.1016/j.jfoodeng.2016.01.025.

Godoi, Fernanda C., Bhesh R. Bhandari, Sangeeta Prakash and Min Zhang. 2019. An introduction to the principles of 3D food printing. In Fundamentals of 3D Food Printing and Applications, 1–18. Elsevier. doi:10.1016/B978-0-12-814564-7.00001-8.

Hamilton, Charles Alan, Gursel Alici and Marc in het Panhuis. 2018. 3D printing vegemite and marmite: redefining 'Breadboards'. Journal of Food Engineering 220 (March): 83–88. doi:10.1016/j.jfoodeng.2017.01.008.

Hao, L., Mellor, S., Seaman, O., Henderson, J., Sewell, N. and Sloan, M. 2010. Material characterisation and process development for chocolate additive layer manufacturing. Virtual and Physical Prototyping 5(2): 57–64. doi:10.1080/17452751003753212.

Huang, Jida, Luis Javier Segura, Tianjiao Wang, Guanglei Zhao, Hongyue Sun and Chi Zhou. 2020. Unsupervised learning for the droplet evolution prediction and process dynamics understanding in inkjet printing. Additive Manufacturing 35 (October): 101197. doi:10.1016/j.addma.2020.101197.

Huang, Meng-sha, Min Zhang and Bhesh Bhandari. 2019. Assessing the 3D printing precision and texture properties of brown rice induced by infill levels and printing variables. Food and Bioprocess Technology 12(7): 1185–1196. doi:10.1007/s11947-019-02287-x.

Jagadiswaran, Bhagya, Vishvaa Alagarasan, Priyadharshini Palanivelu, Radhika Theagarajan, J.A. Moses and C. Anandharamakrishnan. 2021. Valorization of food industry waste and by-products using 3D printing: a study on the development of value-added functional cookies. Future Foods 4 (December): 100036. doi:10.1016/j.fufo.2021.100036.

Jiang, Hao, Luyao Zheng, Yanhui Zou, Zhaobin Tong, Shiyao Han and Shaojin Wang. 2019. 3D food printing: main components selection by considering rheological properties. Critical Reviews in Food Science and Nutrition 59(14): 2335–2347. doi:10.1080/10408398.2018.1514363.

Jiang, Xiaoxiao, Rui Zhai and Mingjie Jin. 2021. Increased mixing intensity is not necessary for more efficient cellulose hydrolysis at high solid loading. Bioresource Technology 329 (June): 124911. doi:10.1016/j.biortech.2021.124911.

Jo, Gyu Han, Woo Su Lim, Hyun Woo Kim and Hyun Jin Park. 2021. Post-processing and printability evaluation of red ginseng snacks for three-dimensional (3D) printing. Food Bioscience 42 (August): 101094. doi:10.1016/j.fbio.2021.101094.

Keerthana, K., Anukiruthika, T., Moses, J.A. and Anandharamakrishnan, C. 2020. Development of fiber-enriched 3Dprinted snacks from alternative foods: a study on button mushroom. Journal of Food Engineering 287 (May). Elsevier Ltd: 110116. doi:10.1016/j.jfoodeng.2020.110116.

Kim, Hyun Woo, Hojae Bae and Hyun Jin Park. 2017. Classification of the printability of selected food for 3D printing: development of an assessment method using hydrocolloids as reference material. Journal of Food Engineering 215 (December): 23–32. doi:10.1016/j.jfoodeng.2017.07.017.

Kyle, Stuart, Zita M. Jessop, Ayesha Al-Sabah and Iain S. Whitaker. 2017. 'Printability' of candidate biomaterials for extrusion based 3D printing: state-of-the-art. Advanced Healthcare Materials 6(16): 1700264. doi:10.1002/adhm.201700264.

Lanaro, Matthew, Mathilde R. Desselle and Maria A. Woodruff. 2019. 3D printing chocolate: properties of formulations for extrusion, sintering, binding and ink jetting. In Fundamentals of 3D Food Printing and Applications, 151–173. Elsevier.

Le-Bail, Alain, Bianca Chieregato Maniglia and Patricia Le-Bail. 2020. Recent advances and future perspective in additive manufacturing of foods based on 3D printing. Current Opinion in Food Science 35 (October): 54–64. doi:10.1016/j.cofs.2020.01.009.

Le Tohic, Camille, Jonathan J. O'Sullivan, Kamil P. Drapala, Valentin Chartrin, Tony Chan, Alan P. Morrison, et al. 2018. Effect of 3D printing on the structure and textural properties of processed cheese. Journal of Food Engineering 220 (March): 56–64. doi:10.1016/j.jfoodeng.2017.02.003.

Lee, Jinyoung. 2021. A 3D food printing process for the new normal era: a review. Processes 9(9): 1495. doi:10.3390/pr9091495.

Lille, Martina, Asta Nurmela, Emilia Nordlund, Sini Metsä-Kortelainen and Nesli Sozer. 2018. Applicability of protein and fiber-rich food materials in extrusion-based 3D printing. Journal of Food Engineering 220 (March): 20–27. doi:10.1016/j.jfoodeng.2017.04.034.

Lin, Ying-Ju, Parinya Punpongsanon, Xin Wen, Daisuke Iwai, Kosuke Sato, Marianna Obrist, et al. 2020. FoodFab: creating food perception illusions using food 3D printing. In Proceedings of the 2020 CHI Conference on Human Factors in Computing Systems, 1–13. New York, NY, USA: ACM. doi:10.1145/3313831.3376421.

Lipton, Jeffrey, Dave Arnold, Franz Nigl, Nastassia Lopez, D.L. Cohen, Nils Norén, et al. 2010. Multi-material food printing with complex internal structure suitable for conventional post-processing. In Solid Freeform Fabrication Symposium, 809–815.

Lipton, Jeffrey I. 2017. Printable food: the technology and its application in human health. Current Opinion in Biotechnology 44 (April): 198–201. doi:10.1016/j.copbio.2016.11.015.

Liu, Zhenbin, Min Zhang, Bhesh Bhandari and Yuchuan Wang. 2017. 3D printing: printing precision and application in food sector. Trends in Food Science & Technology 69. Elsevier: 83–94.

Liu, Zhenbin, Min Zhang, Bhesh Bhandari and Chaohui Yang. 2018. Impact of rheological properties of mashed potatoes on 3D printing. Journal of Food Engineering 220 (March): 76–82. doi:10.1016/j.jfoodeng.2017.04.017.

Maniglia, Bianca C., Dâmaris C. Lima, Manoel D. Matta Junior, Patricia Le-Bail, Alain Le-Bail and Pedro E.D. Augusto. 2019. Hydrogels based on ozonated cassava starch: effect of ozone processing and gelatinization conditions on enhancing 3D-printing applications. International Journal of Biological Macromolecules 138 (October): 1087–1097. doi:10.1016/j.ijbiomac.2019.07.124.

Maniglia, Bianca C., Dâmaris C. Lima, Manoel da Matta Júnior, Anthony Oge, Patricia Le-Bail, Pedro E.D. Augusto, et al. 2020a. Dry heating treatment: a potential tool to improve the wheat starch properties for 3D food printing application. Food Research International 137 (November): 109731. doi:10.1016/j.foodres.2020.109731.

Maniglia, Bianca C., Dâmaris C. Lima, Manoel D. Matta Junior, Patricia Le-Bail, Alain Le-Bail and Pedro E.D. Augusto. 2020b. Preparation of cassava starch hydrogels for application in 3D printing using Dry Heating Treatment (DHT): a prospective study on the effects of DHT and gelatinization conditions. Food Research International 128 (February): 108803. doi:10.1016/j.foodres.2019.108803.

Maniglia, Bianca Chieregato, Gianpiero Pataro, Giovanna Ferrari, Pedro Esteves Duarte Augusto, Patricia Le-Bail and Alain Le-Bail. 2021. Pulsed Electric Fields (PEF) treatment to enhance starch 3D printing application: effect on structure, properties, and functionality of wheat and cassava starches. Innovative Food Science & Emerging Technologies 68 (March): 102602. doi:10.1016/j.ifset.2021.102602.

Mantihal, Sylvester, Rovina Kobun and Boon-Beng Lee. 2020. 3D food printing of as the new way of preparing food: a review. International Journal of Gastronomy and Food Science 22 (December): 100260. doi:10.1016/j.ijgfs.2020.100260.

Mazzanti, Valentina, Lorenzo Malagutti and Francesco Mollica. 2019. FDM 3D printing of polymers containing natural fillers: a review of their mechanical properties. Polymers 11(7): 1094. doi:10.3390/polym11071094.

Mostafaei, Amir, Amy M. Elliott, John E. Barnes, Fangzhou Li, Wenda Tan, Corson L. Cramer, et al. 2021. Binder jet 3D printing—process parameters, materials, properties, modeling, and challenges. Progress in Materials Science 119 (June): 100707. doi:10.1016/j.pmatsci.2020.100707.

Nachal, N., Moses, J.A., Karthik, P. and Anandharamakrishnan, C. 2019. Applications of 3D printing in food processing. Food Engineering Reviews 11(3): 123–141. doi:10.1007/s12393-019-09199-8.

Nijdam, Justin J., Deepa Agarwal and Ben S. Schon. 2022. An experimental assessment of filament-extrusion models used in slicer software for 3D food-printing applications. Journal of Food Engineering 317 (March): 110711. doi:10.1016/j.jfoodeng.2021.110711.

Nishinari, Katsuyoshi, Yapeng Fang and Andrew Rosenthal. 2019. Human oral processing and texture profile analysis parameters: bridging the gap between the sensory evaluation and the instrumental measurements. Journal of Texture Studies 50(5): 369–380. doi:10.1111/jtxs.12404.

Oliveira, Sara M., Luiz H. Fasolin, António A. Vicente, Pablo Fuciños and Lorenzo M. Pastrana. 2020. Printability, microstructure, and flow dynamics of phase-separated edible 3D inks. Food Hydrocolloids 109 (December): 106120. doi:10.1016/j.foodhyd.2020.106120.

Pant, Aakanksha, Amelia Yilin Lee, Rahul Karyappa, Cheng Pau Lee, Jia An, Michinao Hashimoto, et al. 2021. 3D food printing of fresh vegetables using food hydrocolloids for dysphagic patients. Food Hydrocolloids 114 (May): 106546. doi:10.1016/j.foodhyd.2020.106546.

Portanguen, Stéphane, Pascal Tournayre, Jason Sicard, Thierry Astruc and Pierre-Sylvain Mirade. 2019. Toward the design of functional foods and biobased products by 3D printing: a review. Trends in Food Science & Technology 86 (April): 188–198. doi:10.1016/j.tifs.2019.02.023.

Pulatsu, Ezgi, Jheng-Wun Su, Jian Lin and Mengshi Lin. 2020. Factors affecting 3D printing and post-processing capacity of cookie dough. Innovative Food Science & Emerging Technologies 61 (May): 102316. doi:10.1016/j.ifset.2020.102316.

Severini, C., Azzollini, D., Albenzio, M. and Derossi, A. 2018a. On printability, quality and nutritional properties of 3D printed cereal based snacks enriched with edible insects. Food Research International 106 (April): 666–676. doi:10.1016/j.foodres.2018.01.034.

Severini, C., Derossi, A., Ricci, I., Caporizzi, R. and Fiore, A. 2018b. Printing a blend of fruit and vegetables. New advances on critical variables and shelf life of 3D edible objects. Journal of Food Engineering 220 (March): 89–100. doi:10.1016/j.jfoodeng.2017.08.025.

Soleimani-Gorgani, Atasheh, Izdebska, J. and Thomas, S. 2016. 14. Inkjet Printing. Printing on Polymers, William Andrew Publishing, New York, USA, 231–246.

Sun, Jie, Weibiao Zhou, Dejian Huang, Jerry Y.H. Fuh and Geok Soon Hong. 2015. An overview of 3D printing technologies for food fabrication. Food and Bioprocess Technology 8(8): 1605–1615. doi:10.1007/s11947-015-1528-6.

Sun, Jie, Weibiao Zhou, Liangkun Yan, Dejian Huang and Lien-ya Lin. 2018. Extrusion-based food printing for digitalized food design and nutrition control. Journal of Food Engineering 220 (March). Elsevier: 1–11. doi:10.1016/j.jfoodeng.2017.02.028.

Sweeney, Michael, Loudon L. Campbell, Jeff Hanson, Michelle L. Pantoya and Gordon F. Christopher. 2017. Characterizing the feasibility of processing wet granular materials to improve rheology for 3D printing. Journal of Materials Science 52(22): 13040–13053. doi:10.1007/s10853-017-1404-z.

Tabilo-Munizaga, Gipsy and Gustavo V. Barbosa-Cánovas. 2005. Rheology for the Food Industry. Journal of Food Engineering 67(1-2): 147–156. doi:10.1016/j.jfoodeng.2004.05.062.

Tan, Cavin, Wei Yan Toh, Gladys Wong and Lin Li. 2018. Extrusion-based 3D food printing—materials and machines. International Journal of Bioprinting 4(2): 1–13. doi:10.18063/IJB.v4i2.143.

Thangalakshmi, S., Vinkel Kumar Arora, Barjinder Pal Kaur and Santanu Malakar. 2021. Investigation on rice flour and jaggery paste as food material for extrusion-based 3D printing. Journal of Food Processing and Preservation 45(4). doi:10.1111/jfpp.15375.

Theagarajan, Radhika, Moses, J.A. and Anandharamakrishnan, C. 2020. 3D extrusion printability of rice starch and optimization of process variables. Food and Bioprocess Technology 13(6): 1048–1062. doi:10.1007/s11947-020-02453-6.

Tran, Jasper L. 2019. Safety and labelling of 3D printed food. In Fundamentals of 3D Food Printing and Applications, 355–371. Elsevier. doi:10.1016/B978-0-12-814564-7.00012-2.

Upprinting Food. 2022. UPPRINTING FOOD Transforms Food Waste into Edible 3D Printed Snacks.

Vancauwenberghe, Valérie, Victor Baiye Mfortaw Mbong, Els Vanstreels, Pieter Verboven, Jeroen Lammertyn and Bart Nicolai. 2019. 3D printing of plant tissue for innovative food manufacturing: encapsulation of alive plant cells into pectin based bio-ink. Journal of Food Engineering 263: 454–464. doi:10.1016/j.jfoodeng.2017.12.003.

Voon, Siew Li, Jia An, Gladys Wong, Yi Zhang and Chee Kai Chua. 2019. 3D food printing: a categorised review of inks and their development. Virtual and Physical Prototyping 14(3): 203–218. doi:10.1080/17452759.2019.1603508.

Wang, Lin, Min Zhang, Bhesh Bhandari and Chaohui Yang. 2018. Investigation on fish surimi gel as promising food material for 3D printing. Journal of Food Engineering 220 (March): 101–108. doi:10.1016/j.jfoodeng.2017.02.029.

Wang, Mingshuang, Dongnan Li, Zhihuan Zang, Xiyun Sun, Hui Tan, Xu Si, et al. 2021. 3D food printing: applications of plant-based materials in extrusion-based food printing. Critical Reviews in Food Science and Nutrition, April, 1–15. doi:10.1080/10408398.2021.1911929.

Xing, Xuebing, Bimal Chitrakar, Subrota Hati, Suya Xie, Hongbo Li, Changtian Li, et al. 2022. Development of black fungus-based 3D printed foods as dysphagia diet: effect of gums incorporation. Food Hydrocolloids (February): 107173. doi:10.1016/j.foodhyd.2021.107173.

Yang, Fan, Min Zhang, Sangeeta Prakash and Yaping Liu. 2018a. Physical properties of 3D printed baking dough as affected by different compositions. Innovative Food Science & Emerging Technologies 49 (October): 202–210. doi:10.1016/j.ifset.2018.01.001.

Yang, Fan, Min Zhang, Zhongxiang Fang and Yaping Liu. 2019. Impact of processing parameters and post-treatment on the shape accuracy of 3D-printed baking dough. International Journal of Food Science and Technology 54(1): 68–74. doi:10.1111/ijfs.13904.

Yang, Fanli, Min Zhang, Bhesh Bhandari and Yaping Liu. 2018b. Investigation on lemon juice gel as food material for 3D printing and optimization of printing parameters. LWT 87 (January): 67–76. doi:10.1016/j.lwt.2017.08.054.

Yoha, Kandasamy Suppiramaniam, Thangarasu Anukiruthika, Wilson Anila, Jeyan Arthur Moses and Chinnaswamy Anandharamakrishnan. 2021. 3D printing of encapsulated probiotics: effect of different post-processing methods on the stability of *Lactiplantibacillus Plantarum* (NCIM 2083) under Static *in vitro* digestion conditions and during storage. LWT 146 (July): 111461. doi:10.1016/j.lwt.2021.111461.

Zhang, John Y., Janam K. Pandya, David Julian McClements, Jiakai Lu and Amanda J. Kinchla. 2021. Advancements in 3D food printing: a comprehensive overview of properties and opportunities. Critical Reviews in Food Science and Nutrition, February, 1–18. doi:10.1080/10408398.2021.1878103.

Zhang, Lu, Yimin Lou and Maarten A.I. Schutyser. 2018. 3D printing of cereal-based food structures containing probiotics. Food Structure 18 (August): 14–22. doi:10.1016/j.foostr.2018.10.002.

Zhao, Zilong, Qian Wang, Bowen Yan, Wenhua Gao, Xidong Jiao, Jianlian Huang, et al. 2021. Synergistic effect of microwave 3D print and transglutaminase on the self-gelation of surimi during printing. Innovative Food Science and Emerging Technologies 67. Elsevier Ltd: 102546. doi:10.1016/j.ifset.2020.102546.

Ziaee, Mohsen and Nathan B. Crane. 2019. Binder jetting: a review of process, materials, and methods. Additive Manufacturing 28 (August): 781–801. doi:10.1016/j.addma.2019.05.031.

Part III
Policy, Social, Economic, and Environmental Aspects in Food Industries

12

Smart Food Policy as a Response to a New Sustainable Food System Agenda

Susanna Kugelberg[1],* and *Fabio Bartolini*[2]

1. Introduction

Intensive conventional agriculture and poor food environments have led to severe ecological damages, climate change and contributed to a growing burden of non-communciable diseases (Herrero et al. 2010). In response, there is an increasing call for a reorientation of food system policies from a progressive, neo-liberal and top-down approach to an inclusive governance approach towards a sustainable development agenda (Kugelberg et al. 2021).

The nature and urgency of a food system transition require a simultaneous implementation of short-term policies to reduce the negative effects of the current regimes and long-term ones to promote transformative changes in society by supporting sustainable niches (Geels 2002). In parallel, disruptive technologies, such as AI and blockchain in general, are increasingly reshaping governance structures and pave the way for making organisational and governance processes run more efficiently (Baralla et al. 2019, Awan et al. 2020).

As a response to the growing complexity and need for systemic change, "smart policy" is gaining ground as a governance heuristics for transitioning the food system towards sustainability. Although the concept of smart policy is not well conceptualised, it emerges as a response to new policy paradigms and as a governance tool to solve wicked problems at a systemic level (Christensen 2009).

We argue that capacity-building across the food system—both at the socio-technological and governance level—will contribute to a shift towards a sustainable food system. As a result, this short chapter aims to contribute to an understanding of smart policies as essentially a governance heuristic for improving policy design and implementation. Based on a snapshot of the literature on "smart policy", we aim to highlight that it has essentially been applied as a governance approach for dealing with complex problems and induce policy change.

Lastly, we present a conceptual framework based on the seminal work of Christensen (2009) on smart policy as a way to illustrate capacities for governing a food system that can contribute to food

---

[1] Research, Innovation and Organization (RIO) group, Department of Organization, Copenhagen Business School.
[2] Agricultural Economics, Via L Borsari, 46 - 44121 Ferrara.
  Email: fabio.bartolini@unife.it
* Corresponding author: sku.ioa@cbs.dk

security while ensuring prosperity and equity of natural, social and human capital. By doing so, we highlight a need for further capacity-building both at governance and the broader system level for a transition towards a sustainable food system.

2. A Snapshot of "Smart Policy" from the Literature

The concept of "smart policy" is not new in the literature, but it has in recent years attracted more attention as a policy design and management tool for dealing with complex policy problems. The notion of smart policy has its roots in the seminal work of Christensen (2009), who conceptualised smart policy from an instrumental-technical perspective to implement the "new public administration" approach. Since then, the term has been increasingly applied in a diversity of studies focusing mainly on the linkages between, on the one hand, policy design, organisation, governance, and on the other hand, effective and efficient outcomes (Kugelberg et al. 2021). Since Christensen's seminal work (Christensen 2009), the term smart policy appears in 85 papers on Scopus, with a significant increase in the last decade. One-fourth of these papers belong to social science literature. Still, the term smart policy has been found in other subject areas, including environmental studies, computers, economics, engineering, energy, and health studies.

One of the first articles on smart policy focuses on technological innovation in the health care system and the introduction of Smart Cards in the health system (Alpert 1993). The paper shows how new governance and management structures can address tensions between principles of dignity and privacy to be applied in health care strategies. In the following years, O'Keefe (2000) proposes a smart policy approach to face climate change, while Sherman (2001) applies a smart policy framework to reduce gun violence from a public health perspective. These authors focus on the effective use of science in policy-making to reduce uncertainties in decision-making. This can also be ensured by making scientific results available for policy-making and increasing access to non-academic scientific literature.

Forte et al. (2006) claim that smart policy is conducive to support innovative business solutions for urban development. The authors describe smart policy both as an evidence-based approach (i.e., multicriteria analysis based on sustainable indicators) and as a new governance structure (support the integration of the private and public sector) as a way to face complex and unsettled issues in the urban planning. The proposed smart policy represents a way to integrate the private and public sectors and support the adoption of shared strategies.

Leow et al. (2014) conceptualised smart policy as a coherence framework for intervention among several sectoral policies to combine education, nutrition, and health in public schools. Despite high ambition from governments to increase coherence, the authors observe that smart policies have not resulted in new practice for intervention but rather a legitimisation of the existing ones. Similar approaches to smart policy talk about Smart City Policy. Caragliu and Del Bo (2016) have noticed that smart policies are more likely to be implemented in context with high human and social capital, in areas with well-established ICT infrastructure, or finally, where the public engagement exercises in policy decision making are in place.

Studies have also approached smart policy as a way to respond to issues with high conflicts. These studies tend to focus on policy design as a way of resolving conflicts; for example, Grovermann et al. (2017) describes the use of evidence-based "smart policy", based on a modelling approach to design pesticide regulation, to balance trade-off between farmers, consumers and inhabitants. Other authors have stressed the introduction of a smart policy in situations with powerful stakeholders and weak institutions (see, for example, the case of illegal deforestation in developing countries (Blum et al. 2022)).

Yet, more policy analysis-oriented studies focus on smart policy within the frame of good governance and focus on the use of an inclusive approach in the implementation of evidence-based policy (see Matthews (2021) for a review). (Hawkes et al. 2015), in a paper on smart food environments, suggest that public health nutrition policies should assess food preference using the theory of change to better describe causal mechanisms. Smart food policies are those interventions that are effective in changing food behaviour based on robust evidence without or little trade-offs over scales and groups. Other research within food policy literature has advocated smart food policy as linked with tailored intervention. Turnwald and Crum (2019) advocated the need to develop smart food policies to improve dietary intake using the tailored intervention.

3. Capacity-building of Smart Food Policy

As the above section reveals, the increasing scholarly interest in "smart policy" as a governance heuristic is most likely a consequence of three factors: (a) the emergence of complex and wicked problems (Rittle and Webber 1973) such as climate change, public health, SDGs, (b) the recognition for a pluralism approach to deal with these challenges (Pascual et al. 2021a), and (c) the availability of advanced instrumental-technical tools and innovations for designing and implementing policy (Rijswijk et al. 2021). Smart food policy can enhance the capacity and performance of the food system. We argue that capacity-building at governance and the broader system level is necessary for a transition towards a sustainable food system that delivers efficiency, effectiveness, and Sustainable policy performance (Fig. 1).

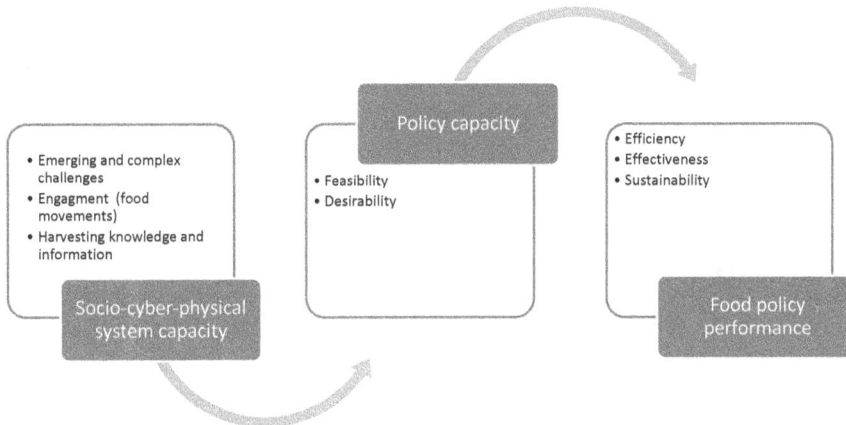

Figure 1. Capacity-building for increasing food policy performance.

3.1 Socio-Cyber-Physical System Capacity

Smart policy applied to the food system is highly linked to the innovation capacity of the socio-cyber-physical system. Private and public sectors' innovation capacity is a key driver of economic and social growth and can contribute as disruptive innovations for complex challenges (Gray et al. 2014). As a way to boost innovation capacity and innovation diffusion, public actors can engage food system actors in policy and innovation laboratories in the real world (McCrory et al. 2020). Following (Gergen et al. 2004), the innovation moves beyond the co-creation of products towards the co-creation of new worlds and desirable futures, which also includes reflexive space to change overall societal systems, which include social innovation (Wittmayer et al. 2019), governance innovation (Griffin 2010) and responsible innovation.

In addition, to increase the capacity for innovation within business and food producers, the role of technologies (i.e., data acquisition tools and new data processing platforms) is also a gamechanger for impact assessments and inclusive governance, as technology can create "global participatory platforms" or enhance surveillance and monitoring (Seele and Lock 2017). For example, De Gennaro et al. (2016) claimed that big data could be an enabling factor for supporting synergies of private/public investments in planning. Other authors propose using a global platform integrating qualitative information for assessing policy acceptability, based on simulation results from general and partial equilibrium models. Other authors (Hübler and Pothen 2021) advocate integrating knowledge among "distant" stakeholders through a global data processing platform as a driver of inclusive green growth. This approach can support the coalition formulation or strategic alliances from a "distant" stakeholder group.

Therefore, the capacity-building of the socio-cyber-physical system (Rijswijk et al. 2021) can be through sharing infrastructure (i.e., platform) or technology that allows a better-informed decision on food choices (i.e., blockchain), as well as ICT tools used along the supply chain (i.e., precision farming), which represent a source of new governance and responsible innovation (Lin and Darnall 2015).

Addressing complex issue also calls for a deeper reflexivity by scientists, policy makers and practitioners to include knowledge and information from a diversity of perspectives in planning (Pascual et al. 2021b). Multidisciplinary studies and sustainable value creation through VCN as a normative concept represent the basis for coordination of public private initiative for developing innovative business opportunities.

Complex environmental and social concerns (i.e., biodiversity or global pollution) have partly undetermined regulatory trajectories as well as unclear political responses, as these concerns can affect distant groups and countries. Therefore, increasing the capacity of the socio-cyber physical system to create real world laboratories, based on inclusive and plural approaches offers an opportunity to improve policy making process through the development of smart food policy.

3.2 Policy Capacity

In the global demands for food system transition, policy-makers are increasingly confronted with demands to become more responsive and efficient, especially given budgetary constraints, and the growing pressure for keeping low food prices and the push for taking action on climate change and public health. In particular, policy-makers are supposed to learn new capabilities and practices in order to adapt to system innovation, digitalisation and changing policy goals often labeled as "smart" (Lovell 2007). Therefore, the success of governments to implement smart food policies depend partly on their existing policy capacity—approached here as a specific type of capacity for smart policy-making. Policy capacity can largely be understood as the resources, skills and abilities that these actors brings to the table (Bali and Ramesh 2018).

Combining management ideas to organisational solutions (Christensen 2009), has identified two main criteria for smart policies related to policy capacity: feasibility and desirability. Feasibility broadly concerns "the quality of organisational thinking and the potential for controlling the reform process and its implementation" (p. 453). Hence, feasibility is both about the extent to which the choice of policy instruments is strategically aligned with long-term policy goals and the efficiency of these instruments in terms of implementation (Howlett et al. 2017). Bettini et al. (2015) point to the need to develop public actors' policy capacity in terms of detecting problematic features of the current system and to strategically analyse the system for targeting leverage points (Bettini et al. 2015). Hence, it is also linked to government's capability for collective envisioning, experiments and cycles of learning and adaptation (Kemp et al. 2007, Voß et al. 2009). In particular, this translates to governments' ability to engage, execute and interpret results from foresights and scenario planning studies (Howlett 2015, Allain et al. 2020).

The second criteria of desirability is related to the wider society and political-administrative system and points to the importance of policies' fit with the wider context (Howlett 2009, Chindarkar 2017). As such, desirability is closely related to the overall fit between smart food policy and capacity of socio-cyber-technical system. Here, policy-makers are confronted with the task to assess stakeholders' barriers and opportunities. In particular, this demands attention to formative evaluation and continuous engagement processes (Allain et al. 2020). The issue at stake is how to design policy mixes—integrating and supporting new social and technological innovations—that increase desired outcomes while reducing the number of negative side effects (Kugelberg et al. 2021). In addition to organisational studies, research on innovation and socio-technological systems are paying an increased attention on how to involve stakeholders in policy learning so that implementation barriers can be reduced and simultaneously accelerate adaptive capacity (Pahl-Wostl 2009).

4. Discussion and Conclusion

We have shown that smart policy can largely be understood as a governance heuristic for dealing with complex issues across a range of topics, such as climate change, public health and digitalisation, and SDGs. We have also shown that smart policy claims a multi-actor approach to deal with these challenges, and there is a need to develop skills and increase access to advanced instrumental-technical tools and innovations for designing, implementing and evaluating policy.

First, we argue that capacity-building at the socio-cyber-physical system level and at the policy-making level is crucial for increasing the overall performance of the food system to become more efficient, effective and sustainable. In particular, we link policy capacity to the specific resources, instruments and abilities that governments have at hand for implementing smart food policy. We put forward the need (i) to increase resources for digitalisation and future-oriented tools, e.g., foresight studies; (ii) that governmental actors acquire appropriate skills for translating results from complex system modelling and; (iii) of deliberative processes and integration of anticipatory knowledge in the governance of food system towards sustainability. Secondly, we argue that abilities and resources linked to engaging societal actors in continuous formative evaluations are crucial for understanding and tackling grand challenges. Moreover, this type of knowledge needs deliberation in wide stakeholder platforms for increasing social learning. This type of reflexive knowledge also needs to be coupled with the policy-process to increase the overall policy capacity for smart food policy-making (Kugelberg et al. 2021).

We conclude that a transition to a sustainable food system is related to its overall capacity at the socio-cyber-technological system and policy level. We argue that capacity-building efforts directed at these two levels can help to increase performance of the food system for dealing with complex issues and grand challenges. In particular, we put forward a set of resources and tools that can enhance the diffusion of innovations and readiness for change, feasibility of collective envisions of long-term food system goals and desirability of courses of actions and disruptive change.

References

Allain, S., Plumecocq, G. and Leenhardt, D. 2020. Linking deliberative evaluation with integrated assessment and modelling: a methodological framework and its application to agricultural water management. Futures 120. doi:10.1016/j.futures.2020.102566.

Alpert, S. 1993. Smart Cards, Smarter Policy Medical Records, Privacy, and Health Care Reform. The Hastings Center Report 23(6): 13. doi:10.2307/3562918.

Awan, S.H., Nawaz, A., Ahmed, S., Khattak, H.A., Zaman, K. and Najam, Z. 2020. Blockchain based Smart Model for Agricultural Food Supply Chain. In 2020 International Conference on UK-China Emerging Technologies, (UCET) (pp. 1–5). IEEE.

Bali, A. and Ramesh, M. 2018. Policy capacity: a design perspective. Routledge, pp. 331–344.

Baralla, G., Ibba, S., Marchesi, M., Tonelli, R. and Missineo, S. 2019. A blockchain based system to ensure transparency and reliability in food supply chain. In Euro-Par 2018: Parallel Processing Workshops: Euro-Par 2018 International Workshops, Turin, Italy, August 27–28, 2018, Revised Selected Papers 24 (pp. 379–391). Springer International Publishing.

Bettini, Y., Brown, R.R., de Haan, F.J. and Farrelly, M. 2015. Understanding institutional capacity for urban water transitions. Technological Forecasting and Social Change, 94: 65–79.

Blum, D., Aguiar, S., Sun, Z., Müller, D., Alvarez, A., Aguirre, I., et al. 2022. Subnational institutions and power of landholders drive illegal deforestation in a major commodity production frontier. Global Environmental Change, 74: 102511.

Caragliu, A. and Del Bo, C.F. 2016. Do smart cities invest in smarter policies? Learning from the past, planning for the future. Social Science Computer Review 34(6): 657–672. doi:10.1177/0894439315610843.

Christensen, T. 2009. Smart Policy? In The Oxford Handbook of Public Policy. Oxford Uni, pp. 448–468.

Forte, F., Girard, L.F. and Nijkamp, P. 2006. Smart policy, creative strategy and urban development. Studies in Regional Science, pp. 947–963. doi:10.2457/srs.35.947.

Gergen, M.M., Gergen, K.J. and Barrett, F. 2004. Appreciative inquiry as dialogue: generative and transformative. pp. 3–27. *In*: Cooperrider, D.L. and Avital, M. (eds.). Constructive Discourse and Human Organization. Emerald Group Publishing Limited (Advances in Appreciative Inquiry). doi:10.1016/S1475-9152(04)01001-4.

Gray, M., Mangyoku, M., Serra, A., Sánchez, L. and Aragall, F. 2014. Integrating design for all in living labs. Technology Innovation Management Review, 4(5): 50–59.

Griffin, L. 2010. Governance innovation for sustainability: exploring the tensions and dilemmas. Environmental Policy and Governance 20(6): 365–369. doi:https://doi.org/10.1002/eet.555.

Grovermann, C., Schreinemachers, P., Riwthong, S. and Berger, T. 2017. 'Smart' policies to reduce pesticide use and avoid income trade-offs: An agent-based model applied to Thai agriculture. Ecological Economics, 132: 91–103.

Hawkes, C., Smith, T.G., Jewell, J., Wardle, J., Hammond, R.A. and Friel, S. 2015. Smart food policies for obesity prevention. Lancet [Internet]. 385(9985): 2410–21.

Herrero, M., Thornton, P.K., Notenbaert, A.M., Wood, S., Msangi, S., Freeman, H.A. et al. 2010. Smart investments in sustainable food production: revisiting mixed crop-livestock systems. Science, 327(5967): 822–825.

Howlett, M. 2009. Governance modes, policy regimes and operational plans: a multi-level nested model of policy instrument choice and policy design. Policy Sciences 42(1): 73–89.

Howlett, M. 2015. Policy analytical capacity: The supply and demand for policy analysis in government. Policy and Society, 34(3-4): 173–182.

Howlett, M., Mukherjee, I. and Rayner, J. 2017. The elements of effective program design: a two-level analysis. Edward Elgar Publishing, pp. 129–144.

Hübler, M. and Pothen, F. 2021. Can smart policies solve the sand mining problem? PLoS ONE, 16(4 April), pp. 1–15. doi:10.1371/journal.pone.0248882.

Kemp, R., Loorbach, D. and Rotmans, J. 2007. Transition management as a model for managing processes of co-evolution towards sustainable development. International Journal of Sustainable Development and World Ecology 14(1): 78–91. doi:10.1080/13504500709469709.

Kugelberg, S., Bartolini, F., Kanter, D.R., Milford, A.B., Pira, K., Sanz-Cobena, A. et al. 2021. Implications of a food system approach for policy agenda-setting design. Global Food Security, 28: 100451.

Leow, A.C.S., Macdonald, D., Hay, P. and McCuaig, L. 2014. Health-education policy interface: The implementation of the Eat Well Be Active policies in schools. Sport, Education and Society, 19(8): 991–1013.

Lin, H. and Darnall, N. 2015. Strategic alliance formation and structural configuration. Journal of Business Ethics, 127(3): 549–564. doi:10.1007/s10551-014-2053-7.

Matthews, A. 2021. The contribution of research to agricultural policy in Europe. Bio-based and Applied Economics, 10(3): 185–205. doi:10.36253/bae-12322.

McCrory, G., Schäpke, N., Holmén, J. and Holmberg, J. 2020. Sustainability-oriented labs in real-world contexts: An exploratory review. Journal of Cleaner Production, 277: 123202.

N Chindarkar, M.H.M.R. 2017. Introduction to the special issue: conceptualizing effective social policy design: Design spaces and capacity challenges. Public Administration and Development 37(1): 3–14.

O'Keefe, W.F. 2000. The bridge from cold facts and hot rhetoric to rational climate policy. ACS Division of Fuel Chemistry, Preprints 45(1): 147–150.

Pahl-Wostl, C. 2009. A conceptual framework for analysing adaptive capacity and multi-level learning processes in resource governance regimes. Global Environmental Change 19(3): 354–365. doi:10.1016/j.gloenvcha.2009.06.001.

Pascual, U., Adams, W.M., Díaz, S., Lele, S., Mace, G.M. and Turnhout, E. 2021. Biodiversity and the challenge of pluralism. Nature Sustainability, 4(7): 567–572.

Pelling, M., High, C., Dearing, J. and Smith, D. 2008. Shadow spaces for social learning: a relational understanding of adaptive capacity to climate change within organisations. Environment and Planning A, 40(4): 867–884.

Rijswijk, K., Klerkx, L., Bacco, M., Bartolini, F., Bulten, E., Debruyne, L., et al 2021. Digital transformation of agriculture and rural areas: A socio-cyber-physical system framework to support responsibilisation. Journal of Rural Studies, 85: 79–90.

Seele, P. and Lock, I. 2017. The game-changing potential of digitalization for sustainability: possibilities, perils, and pathways. Sustainability Science 12(2): 183–185. doi:10.1007/s11625-017-0426-4.

Strasser, T., de Kraker, J. and Kemp, R. 2019. Developing the transformative capacity of social innovation through learning: a conceptual framework and research agenda for the roles of network leadership. Sustainability (Switzerland) 11(5). doi:10.3390/su11051304.

Turnwald, B.P. and Crum, A.J. 2019. Smart food policy for healthy food labeling: leading with taste, not healthiness, to shift consumption and enjoyment of healthy foods. Preventive Medicine 119(November 2018): 7–13. doi:10.1016/j.ypmed.2018.11.021.

Voß, J.P., Smith, A. and Grin, J. 2009. Designing long-term policy: rethinking transition management. Policy Sciences 42(4): 275–302. doi:10.1007/S11077-009-9103-5/TABLES/1.

Wittmayer, J.M., Backhaus, J., Avelino, F., Pel, B., Strasser, T., Kunze, I., et al. 2019. Narratives of change: How social innovation initiatives construct societal transformation. Futures, 112: 102433.

13

Food Safety in the Sustainable Food Industry

Ioana Mihaela Balan,[1],* *Teodor Ioan Trasca,*[1] *Tiberiu Iancu,*[1] *Nastasia Belc,*[2] *Isidora Radulov*[1] and *Camelia Tulcan*[1]

1. Food safety in circular economy

Scientific knowledge and technology development—both related to food safety—have grown considerably in the last two centuries with direct consequences in the significant decrease of foodborne disease (FBD) outbreaks in developed countries. Nowadays, people are expecting that all the food they buy is safe. Unfortunately, there are still several cases of large FBD outbreaks caused by food of vegetable and/or animal origin. Food safety affects everyone worldwide and will remain a global challenge for human health in the foreseeable future requiring the rapid, sensitive, efficient and inexpensive detection of food contaminants. Due to this very large food safety issue, a huge quantity of food is being taken from the market and trade and destroyed.

In the near future, due to the rapid growth of the world's population, the food systems will face a very high demand for food and now our duty is to minimize the food waste which occurs along the food chain, especially during processing when it is about food safety issues.

For a global sustainable development and a healthy population, food safety is one of the most relevant factors, being an integral part of food and nutrition security.

The Sustainable Development Goals (SDGs) set up in 2015 by the United Nations General Assembly call for numerous actions. Ending famine, improving nutrition and encouraging sustainable food systems act to attain food security. Food industry plays an important role in achieving several SDGs such as: SDG2—Zero Hunger, SDG3—Good Health and Well-Being, SDG12—Responsible Production and Consumption and SDG13—Climate Action.

The food industry significantly affects the quality of the environment. Water depletion and land use directly influence climate change.

The European Union's Research and Innovation Policy Food 2030, launched in 2016, aims to change the food systems and to ensure that everyone has enough nutritious food for a healthy life.

[1] University of Life Sciences "King Mihai I" from Timisoara, Romania.
[2] National Research and Development Institute for Food Bioresources, Bucharest, Romania.
* Corresponding author: ioanabalan@usvt.ro

Between the 10 pathways for action, developed by the Food 2030 Research and Innovation Policy for sustainable and healthy food systems, four of them—Alternative proteins and dietary shift; Food waste and resource efficiency; Healthy, sustainable and personalized nutrition and Food safety systems of the future are very related with food safety and circularity (European Commission 2020a).

The Standing Committee on Agricultural Research (SCAR) Foresight Expert Group led the 5th SCAR Foresight Exercise which was focused on the question: Wherewith to reach "a safe and just operating space" for society, by a better management of natural food systems and resources? The question stems from prior research around the SDGs, the 21st session of the Conference of the Parties (COP21) to the United Nations Framework Convention on Climate Change (UNFCCC), the Green Deal initiative and the European Commission (EC) "Farm to Fork Strategy" for fair, healthy and environmentally friendly food systems. Three main goals or three transition-related priorities were identified: sustainable and healthy diets for all as a social imperative, diversifying food systems as a key to resilience and a circular bio-economy as a road to sustainability (European Commission 2020b).

In order to feed a growing number of people and to increase food production, while assuring a healthy and regenerable environment, food systems should take into account reducing CO_2 emissions, energy and water consumption, food loss and waste.

In terms of food safety, new challenges arise along the food systems due to the increase of microbial resistance, changes in raw materials (waste become new raw materials in some cases), and producing new food products. A better resource efficiency of food systems can impact food safety, for example increased sustainability through changes in food systems via more energy efficient and sustainable freezing and cooling systems and/or by reducing utilization of water in cleaning procedures, but also re-circulate waste as new raw materials may compromise food safety and subsequently produce another waste.

It was estimated that almost 80% of the economic costs of a product are defined during the product design stage and around 80% of the environmental impact of a product are set at the design phase. Therefore, there is potential to reduce environmental and economic costs, and accordingly improve sustainability performance, by assessing products proactively before their manufacture. Prevention approach is desired instead of using a reactive approach to minimize impacts of a product already designed (Cooper and Chew 1996, DG Enterprise & Industry and DG Energy European Commission 2014, McAloone and Bey 2009, Garcia-Garcia et al. 2020).

At the level of global food system, some general measures should be taken such as: searching the nutritional profile of material resources and vary them for the food industry; re-engineering the technological processes in the sense of efficiency and sustainability; recovery waste and by-products in foods with high added value; optimizing food consumption in terms of quantity and quality, encouraging a sustainable eating behavior.

Food safety and security are two complementary elements of our resilience and for better and sustainable future.

In order to assess trade-offs correctly and achieve sustainability, in the near future, it will be necessary to approach the whole situation by taking into consideration the One Health concept (Boqvist et al. 2018).

Several foodborne diseases are zoonotic, which means that between animals and people diseases can be transmitted easily. But some pathogens can change their host, or their geographical location and even their impact, and therefore we have to deal with new and emerging diseases (for example, bovine spongiform encephalopathy or new strains of highly pathogenic avian influenza). A large population may be affected by the emerging diseases that have the ability to trigger pandemics. Other problems about production and consumption of food, related and linked with food safety include food allergies, non-communicable diseases like obesity, food fraud, bioterrorism, antimicrobial drug resistance and food waste.

Emerging and re-emerging foodborne pathogens are continuously changing their epidemiology of foodborne diseases.

More than 200 diseases are also caused by contaminated food with human pathogens or chemical substances, for example heavy metals. This increasing public health problem causes significant socioeconomic impact on health-care systems and also harms tourism and trade (Fig. 1).

These diseases contribute significantly to the global burden of disease and mortality.

Contaminants can occur at any stage of the food production, delivery and consumption chain because of environmental contamination including water, soil or air pollution, as well as unsafe methods for food processing and storage.

In order to achieve SDG2 – Zero Hunger in terms of eradication of famine—to feed 10 billion people by 2050, it is necessary to get the trade-offs right between food security, food safety, sustainability, and a better use of food already produced.

Sustainable food security requires: (a) sufficient food production and availability in the market, (b) ability to buy food, (c) nutritionally correct recipes with valuable food products including energy, proteins and micronutrients as well as safety, and (d) the steadiness and foreseeing ability of these conditions (Helland and Sörbö 2014).

Therefore, feeding 10 billion people in a sustainable way will probably require disruptive changes of the food supply chains during the next 20 years. According to some studies, the total food production should increase by 60%, while other studies have shown that the solution is not to increase production but to improve food distribution, change diet pattern and reduce food waste. Moreover, minimizing the food loss and food waste are part of the solution. Another challenge is the swoop of edible crops (e.g., corn, sugar cane) to biofuel production. There are data that show that converting their crops to biofuel could provide food for 400 million people (Kearney 2019, EC DG for Research and Innovation and Bioeconomy Directorate 2016, Helland and Sörbö 2014).

The EAT-Lancet Commission on healthy diets from sustainable food production showed that numerous dietary changes can allow feeding up to 10 billion people in a resilient way. In short, the

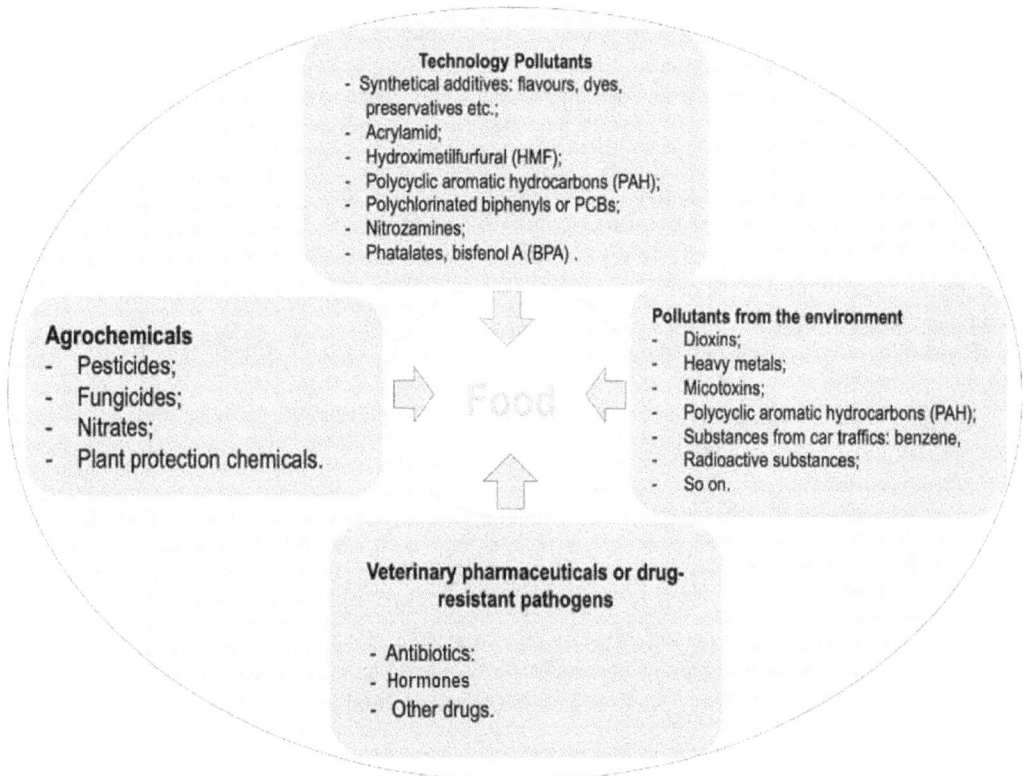

Figure 1. Example of chemical contaminants in food. Source: Original design by authors.

recommendation was to double the consumption of fruits, vegetables, nuts, and legumes and to halve the consumption of red meat and sugar. There are several health and environmental benefits if we can enforce a diet rich in plant-based foods and with fewer animal source foods. The sustainability and resilience of food systems are directly correlated with the land use, biodiversity, fresh water use, greenhouse gas emissions as well as nitrogen and phosphorus cycling. In this view, it appears that doubling of animal food production will need expansion of cereal and vegetable production beyond sustainable levels (Willett et al. 2019, Kearney 2019).

Everyone involved in the food industry, starting with the farmers and continuing with the ones who are preparing and delivering the food, can help to implement durable practices.

There are several examples which show that, instead of the big effort and time spent to implement sustainable practices along the supply chain, some points are not completely clarified and there are issues which should be solved in order to achieve the main goal: a safe and sustainable product in the market.

In 2017, the European Parliament adopted the resolution for reducing food waste by 30 to 50% in 2025 and 2030, respectively, through two approaches to make food donations easier and make "best before" and "use by" labels less confusing.

Food loss occurs because of low quality of vegetal or animal origin raw materials, problems due to ingredients, foods and steps along the food chain starting with improper weather conditions, continuing with improper practices along the food chain, such as cold chain and inadequacies of infrastructure. Food waste is represented by the differences between production and needs buying with the balance tilted towards production and stocks. Here, foods that are close to, at or beyond the "best-before" dates which are thrown up by retailers and consumers and large quantities of food that are often discarded from hospitality industry (HoReCa) and households could be included; additionally, beyond retailer and producer recall of foods.

One of the main challenges in addressing sustainability is avoiding or minimizing food loss and waste as much as possible, in parallel with the aims of resource efficiency and reducing environmental footprints. Food companies and retailers have a huge challenge in avoiding food waste and stock-outs that occur when the sales are not appropriately predicted which leads to an incorrect ordering of products (Arunraj and Ahrens 2015).

Food waste is about loss of resources and it is a sustainability issue that exacerbates climate and environmental impact of food, and impair efforts to reduce food insecurity. Food waste is generally assessed in comparison to the total amount of food produced for a particular food group (Corrado et al. 2020, Vanham et al. 2015, Bene et al. 2019, FAO 2011a, 2014, Springmann et al. 2018, HLPE 2014, Chaudhary et al. 2018, Fanzo et al. 2020).

The increasing interest for food waste results from two big challenges: stabilization, and increase of food security and reduction in greenhouse gas emissions (Willett et al. 2019, WRI 2018).

The food systems consume extensive resources of land, water, energy, fertilizer and other inputs, meanwhile engendering environmental adversities, e.g., biodiversity and habitat loss, soil and water degradation, and greenhouse gas emissions. The resources are sacrificed for no clear inputs and the negative impacts on environment are in vain if the food is wasted. With the growing food demand amid climate change and dwindling resources, enhancing the utilization of food produced and cutting down on waste is a necessity (Foley et al. 2011).

While not yet globally applied, food waste and food safety are progressively acknowledged as key areas of worry, specifically in the most recent frameworks. The global-level impacts of diseases such as campylobacteriosis, salmonellosis and Avian Influenza are a testament to the importance of considering food safety and zoonotic diseases linked to intensified animal production. Assessments firstly lean on data provided by the WHO that specifies the burden of foodborne illnesses (EFSA and ECDC 2019, Leibler et al. 2017, Chaudhary et al. 2018).

A new alternative to the linear take-make-consume-dispose economic model is represented by the circular economy model, which is a no waste and pollution industrial system, restorative or regenerative by intention and design.

Some key characteristics define a circular economy such as: efficiency of using natural resources and inputs, including energy and water which should be minimized and optimized; efficient exploitation of raw materials and ingredients; reducing emissions and loss/waste; keep values of waste in the food system by re-using and recycling;; increasing share of recyclable resources and renewable energy (EC 2016).

CE has the main aim to preserve resource's value such as raw materials, water and energy in a regenerative way. These resources should be used for as long as possible by food systems, while also taking into consideration sustainable production and consumption through resources efficiency, environmental protection and consumer awareness.

A new strategy that supports the transition to an EU CE was launched in December 2015 by EC with an Action Plan, which consists in a CE Package that adopted several legislative proposals including, among others, directives on waste, on packaging waste and on landfill. The restraint has started initiating a call for a major shift towards sustainable food production, supply, and consumption system based on CE approach (Fei et al. 2020, Borsellino et al. 2020, Ibn-Mohammed et al. 2021).

The breakdown of agri-food chains as a result of the ongoing COVID-19 pandemic has generated the need to have a durable monitoring assessment system that will make up for possible pandemic scenarios in the near future (Rowan and Galanakis 2020).

A CE report in agroecosystems is focused on a new and reliable prevalent linear economy approach of "take-make-waste" by reducing the amount of external agricultural inputs, and closing nutrient cycles. This system approach also has the potential to mitigate unfavorable environmental impacts through the elimination of pollution from fertilizers, runoff contamination, excess nutrient load, eutrophication, and food wastage. Indeed, the CE offers optimal use and reuse of agricultural raw materials with a great emphasis on assessing and mitigating environmental impacts that may result in unfavorable climate change (Tseng et al. 2019, Barros et al. 2020).

CE can be defined in general terms as a model of production and consumption that focuses on the sharing, leasing, repairing, refurbishing, reusing, and recycling of available products and materials and reduction of generated wastes (Bahn-Walkowiak et al. 2019).

CE was defined by Ghisellini et al. to be an industrial economy with a focus on achieving sustainability via restorative objects and design (Ghisellini et al. 2014).

As earlier stated, global food availability is being threatened by demographic, economic, and climate change factors and CE has been reported to provide an effective framework for achieving a closed-loop system aiming at combating the issues mentioned before (Kirchherr et al. 2017a, Tseng et al. 2019).

CE offers potential applications in improving food security and sustainability in the food systems. CE concept has found applications in nutrient cycling and inputs in the agri-food industry (Verger et al. 2018, van der Wiel et al. 2020, Billen et al. 2019).

Many challenges occur at the intersection between food security, safety and sustainability. For designing circular food systems where nutrients are recycled, it could happen to have also an accumulation of contaminants (chemical, microbiological, even pathogens) (Fig. 2).

Food safety (hazards and risks) has to be taken into consideration in the development and production of healthy and sustainable food products through CE. Circularity should take into account any resource: water, energy, raw material, and food contact material while food safety is definitively assured (Table 1).

Global sourcing and production, with multiple stakeholders across the food value-chain, requires more-advanced monitoring and hazard assessment. The presence of toxins and environmental pollutants, and the transformation towards new, more sustainable marine and plant-based sources, further highlights the importance of new research in this field.

Advances in processing techniques, preservation, and packaging have enabled the food industry to consistently supply consumers with a wide array of healthy and fresh products all year round. Food packaging can improve food safety by reducing bacterial contamination, prolonging shelf life,

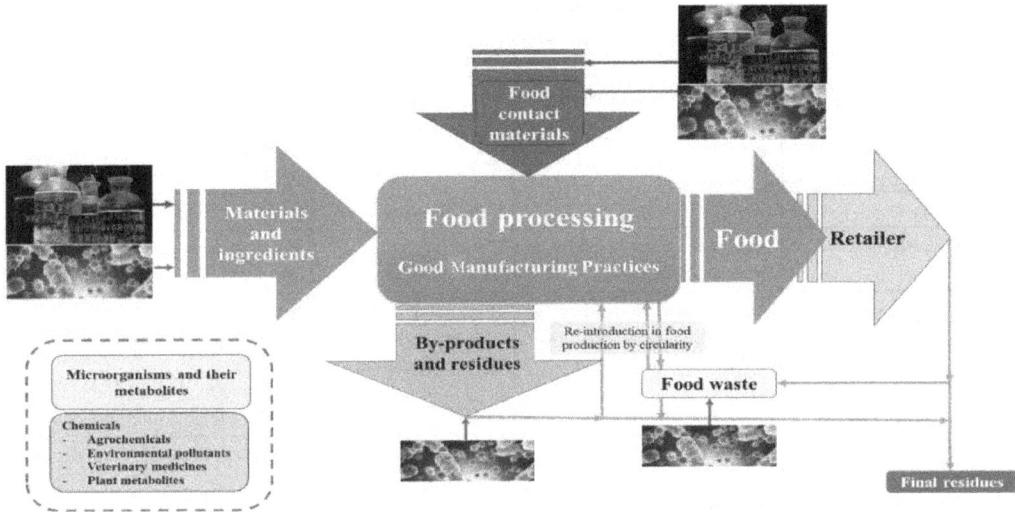

Figure 2. Hazard in circular food processing. Source: Original design by authors.

ensuring convenience in distribution and handling. On the other hand, food contact materials can transfer chemicals to food with partly unknown effects.

So, one of the food safety issue can come from food contact materials which also have a very big impact on the environment. The main intention of packaging material is to get food safely from producers to consumers' cupboards and refrigerators. Also, food packaging improves food safety by reducing bacterial contamination. It has been proposed for example that increased use of packaging for fresh produce could prevent contamination with *Salmonella* spp., a leading cause of foodborne diseases.

Plastic food contact materials protect food against contamination but because of a huge amount of plastic spread out on the environment, several states have restrictions on using plastic bags, through taxes and bans. But, how to protect better food against the environment contaminants or against cross-contamination? This worrying situation has led over the years to the development of various green, sustainable technologies, which can contribute to not only reduce packaging of food but also to reduce the energy consumption that goes into the production and processing such as non-thermal technologies: pulsed electric field, high-pressure processing, cold plasma, ozone, and electrolyzed water. Research on bio-degradable materials are to be taken into consideration (Kearney 2019).

The development of new labeling and traceability systems, new assessment methodologies, will support risk assessors, innovators, policy makers and consumers, and will address related challenges and concerns. A new food safety ecosystem should be designed on the new models of collaborating to foster communities and food systems' sectors. Innovative methodologies and technologies will also aim to make an efficient food waste management.

The hazard potential involved in the use of side streams from agriculture and fisheries also needs further investigation, particularly the possible increased risk during food processing, but also the technological possibilities to purify side streams.

Blockchain technology appears to offer exciting opportunities here. Alongside the risks of environmental pollutants, approaches should also consider contamination during production or via packaging material, and associated issues such as authenticity and traceability.

Research should also be developed on predictive toxicology, such as the safety assessment of new processes and materials, the interactions between food contaminants and the microbiota and the internal exposure to environmental toxins released from fat tissue while dieting.

Table 1. Examples on Circular Economy in food system.

Action on:	Process/product	Where/Who	Effect/Food safety issues
Agricultural waste	Fertilizer by methanisation	France, Danone	The methaniser is producing 1,485,000 m^3 and is reducing GHG emissions per year by approx. 2,000 to equivalent CO_2
Food waste	Fertilizer and energy	Food industry	Reducing food waste
Food by-products	Animal feed and as inputs for other industries		Food safety issues: microbiologic and chemical contamination
FoodWARD project	Analysis of food waste composition from milk, meat and canning sectors	Italy, Food Industry Federation (Federalimentare)	Training of horticulturists; Increasing awareness of waste issues; Reducing food waste and building up a sustainable food system.
Cut-offs from potatoes processing	Potatoes based products such as potato flakes, hash browns' purées, etc.	Potato processors	Increase sources of raw material
Wet starch by-product in cutting process potatoes	Starch industries and bioplastics industry		Reducing the plastic production
Rejected fruits and vegetables for juice production and the remaining pulp	Sauce producing	The Netherlands, Provalor	Food safety issues: microbiologic contamination
Old bread	Sourdough bread produced by fermentation	The Netherlands, ingredient producer Sonneveld	Food safety issues: microbiologic contamination
Kellogg operations in Europe donated 1,110 ton of food	Food redistribution charities and food banks	15 European countries/Kellogg	Reducing food waste
Whey	Protein, animal feed, food and feed supplements	Dairy industry	Reducing food waste
	10% for Biogas and 90% on pig farms as wet feed	Croatian dairy factories Belje's	Company achieved 70,000 Eur of savings per year by re-using the whey
Waste from confectionery factories and mill and mixes—e.g., oat husks from milling operations	90% recycled as renewable energy and clean water through anaerobic digestion	Finland, Fazer company partnership with Lahti Energia	Reducing food waste
Surplus food is redirected	Feed people or animals	HKScan - Nordic meat producer	Prevention food waste
Using parts of the animals that are not sold	Food and food ingredients, pharmaceuticals, pet food and animal feed, renewable energy		
Cow stomach and waste water sludge	Electricity and heating, biogas, renewable energy, organic fertilizers		
Recyclable PET plastic bottle	Reducing by 10% the usage of raw materials used in the processing of food contact materials	Food contact material industry	Reducing plastics
Improve the recyclability of PET package material	Increasing the use of recycled material in packaging sector		

Table 1 contd. ...

...Table 1 contd.

Action on:	Process/product	Where/Who	Effect/Food safety issues
Approx. 70% of waste is recycled or reused: - 57% from animal by-products; - 43% the rest	Biogas Animal protein and fat, pet food	PIK Vrbovec, Croatian meat processor	It is an example of industrial symbiosis by which annual savings of 330,000 Eur are achieved
Food contact materials and innovative packaging	As a replacement for cardboard used in fresh meat industry, re-usable plastic boxes were introduced		CO_2 emissions were reduced
Site canteen waste	Vermicomposting	Unilever's factory	More vegetables are growing
Re-use of packaging materials for transport	On the inbound packaging to use a weaker postage tape	Belgium, Brussels, Unilever tea factory	Modifying the packs in order to be easier to be reused
Up to 4 ton of solid waste residues (every day) from the confectionery processes are used for producing renewable energy and clean water	Anaerobic digestion of a 'chocolate soup' obtained from a mixture of rejected chocolates, sweets, leftover residues of starch and sugar using bacteria that break down the biodegradable material and converts it into biogas	UK, Fawdon, Nestlé's factory	The biogas obtained represents 5–8% of the factory energy needs
Using fatty acid by-product from ice cream for alternative source of energy	Alternative fuel	Swedish ice cream company, Ben & Jerry's	A boiler from a Swedish factory is powered with this fuel
Waste from mayonnaise production	Bio-diesel	Hellmann's mayonnaise	New source of energy
First PET bottle made from plant materials presented at the World Expo, Milan, 2015	A patented technology that converts natural sugars from plant into a compound which is used for obtaining recyclable PET bottles (PlantBottle* packaging™). For 7 years, Coca-Cola has distributed, in nearly 40 countries, more than 35 billion bottles, using the new PET bottle (up to 30 percent plant-based recyclable materials)	Coca-Cola	PlantBottle* packaging™: – functions, looks, and recycles like traditional PET with lower footprint; – can produce a bottle from renewed and recycled materials; – can save 315,000 ton of CO_2 (the equivalent of annual emissions)
Increase by 24% of water use efficiency, implementing vulnerability assessments and source water protection plans in all facilities.	Replenishing or balancing the water for an approx. 94% of the water used in beverages (year 2014)		The Coca-Cola system has replenished an estimated 153.6 billion liters of water back to communities and nature
Reducing the raw material used in the manufacture of primary and secondary packaging by 10%	"2015 Naturally"	Voluntary commitment between Spanish Ministry of Agriculture, Food and ANEABE Association	Preventing the generation of packaging waste; Improving PET recyclability – using 5% of recycled material.

Table 1 contd. ...

...Table 1 contd.

Action on:	Process/product	Where/Who	Effect/Food safety issues
Recovery of water	Reverse osmosis, an innovative water treatment technology, reuses water for different dairy industry operations after recycling wastewater over it by recycling waste	EU Dairy Association	Reducing the impact of water basins and re-using the water from the waste water
Steam waste resulted by burning facility household is transferred by underground pipes of 1,2 km of distance to the chocolate factory	The heat is used for heating buildings but mainly for melting chocolate for M&M's producing	Mars Chocolate France	It has reduced the factory's emissions from energy consumption by 60% and CO_2 emissions by 8,700 ton; annually, 90% of the steam requirement is represented by the green steam system
Capture and use of heat as an energy source. An advanced heat pump system was installed on the waste water treatment system for waste water cooling down tanks to the needed temperature, and to recover the energy to heat water used for cleaning	Project initiated by Wrexham factory to recover wasted heat from the cooker's exhaust systems for using it, and further, to preheat the water which is entering into the boiler	Kellogg factories in Wrexham and Manchester	Reducing gas consumption; 25% of the factory's hot water demand is assured by the heat pump; 24% reduction of GHG emissions at Manchester factory; 17% reduction of GHG emissions at Kellogg in the last 10 years
Water condensate is wasted. This heat and wasting water can be used in the oil and margarine processing	Water condensate obtained is pumped back into the heating facility, and after slightly heating at the steam stage, it will return back to process	The largest producer of edible oils in Croatia, Zvijezda	Re-using the water and heat, the company savings are approx. 80,000 Eur on an annual basis
The International Food Waste Coalition, launched 2015	In several EU countries, pilot projects in schools are to identify, in the different steps of the food chain, where food waste is happening	PepsiCo, Unilever Food Solutions, McCain, WWF and Sodexo	Efficient solutions to build up awareness among students and teachers
Awareness campaign: including tea bags as part of food waste	Education, motivation and changing behavior of UK tea drinkers	UK tea brand - Unilever	Reducing the amount of waste consisting in tea bags that are sent for disposal at landfill
An interactive online game about food waste reduction ("JEUX NE GASPILLE PAS!")	Consumer's tips about how to keep food in the best conditions and where to store food in the fridge	ANIA - French Association for Food and Drink Industry	Raise awareness about food waste
A behavioral change initiative 'Fresher for Longer' was launched in March 2013 under the Waste & Resources Action Programme's (WRAP) Love Food Hate Waste campaign	Demonstrations about how better use of packaging and the information from label can lead to help consumers to reduce the food waste at home, saving money and environment	UK Food and Drink Federation	Reducing food waste

Source: Adapted by authors, FoodDrinkEurope 2016

2. Impact of Food Safety on Environmental Sustainability

Environmental sustainability is part of the 17 SDGs, including social and economic sustainability, as well as food security and safety. A sustainable food system is one that supports food security through the optimal use of human and natural resources, that protects biodiversity and ecosystems for present and future generations, that is accessible and culturally acceptable, economically viable and environmentally friendly, and that provides the consumer with nutritious, safe, healthy and convenient food.

Strictly speaking, sustainability involves the use of resources at rates that do not exceed the Earth's ability to replace them. As regards food, a sustainable system must ensure security of food supply, health, safety, accessibility, food quality, employment and, at the same time, a sustainable, adaptable environment to climate change, preserving biodiversity and ensuring the quality of water and soil.

The level of environmental impact of food production refers to where and how food is produced and the local availability of natural resources such as water and soil. There are often trade-offs between environmental factors and, to date, there is no simple set of principles to determine whether one food is more environmentally sustainable than another.

Food security affects us all, because food production, trade, impact of agriculture on the environment, threat of climate change and factors determining food prices are largely of global nature—there is no single solution that a country can adopt to ensure access to affordable prices, sustainable, safe and nutritious food for all. It is predicted that the global food system will go through pressure in the next 40 years, when the population will grow to 9 billion people. There will be a huge demand for protein-rich foods as a major challenge to maintaining sustainable production (FAO 2011a, Fésüs et al. 2008).

As confirmed by the World Food Summit in 1996, access to safe and nutritious food is a right for everyone. Ensuring food safety has a cost and excessive food safety requirement may impose constraints on production, storage and distribution systems, which may lead to trade barriers or hinder competitiveness (FAO 2009, Vågsholm et al. 2020).

The growth of the global population, the increase in pressure on natural resources and global warming will lead to the need for a new framework at national and international level. Population ageing in Europe is also an additional challenge. All these factors have profound implications for agriculture and rural areas. Global demand for food is increasing, urbanization is accelerated, input prices are rising, the vulnerability of crops and livestock to climate change and the pressure on water resources is growing and, all these will limit food production.

The world's population is expected to grow from 7 billion, as it is currently, to 9 billion by the middle of this century, and 90% of this increase will be in the least developed countries (in 50 of the least developed countries worldwide). Global income growth will mostly be associated with urbanization (70% of the world's population is expected to live in urban areas by 2050, compared to 49% currently) and rapid economic growth in some of the countries with the largest population (e.g., Brazil, China, India and Russia) (Slow Food Foundation 2015).

Climate change will have an increasing impact on the food security in general. Global warming is expected to generate mixed and unevenly distributed effects across the world.

Although agriculture has always focused on food security and has been a positive engine of economic growth, society's growing concern for the environment in recent decades has affected agricultural policies worldwide. However, despite the common framework, the European model of agriculture takes different forms. For example, France has followed a path of competitiveness, driven first of all by the strong integration of the value chain and the strengthening of farms, which has helped it to become the largest wheat producer and exporter in Europe. France has increasingly focused on its efforts to become a leader in high value-added products bearing Europe's flagship quality labels (e.g., traditional products, controlled designation of origin products or protected geographical indication). Austria has developed a vision based on respect for the environment. As a result, its agriculture is focused on multifunctionality and specializes in agri-food products with high added value, mostly organic products.

Food safety should also be addressed from the perspective of food consumption, e.g., changes generated by the change of diet. High protein consumption, such as meat-based diet, leads to the use of larger land areas for livestock compared to the areas used for agricultural crops. Human diets undissolved environment and environmental sustainability and have the potential to feed both. Traditional diets (usually rich in herbal foods) have passed a "Western food model", characterized by high calorie consumption, very processed foods (refined carbohydrates, added sugars, sodium and unhealthy fats) and large amounts of animal products. Together with the negative effects on human health associated with this nutritional transition, this dietary model is also unsustainable. Current food production already determines climate change, biodiversity loss, drastic pollution and changes in land and water use (Willett et al. 2019, Harvard School of Public Health).

More than 4 billion people depend on three types of basic cultures—rice, corn and wheat. These three cultures provide us with two-thirds of our energy consumption. Although there are over 50,000 edible plant species, only a few hundred species contribute to food intake.

In 2050, we will need to feed three times more people than a century ago. The choices we make in terms of production and consumption of food already have direct or indirect climate consequences, resource use such as water and soil, and also the ability of people to feed and have a decent living. Climate change adds pressure on food safety, and some countries feel the pressure more than others. Drought, fires or floods directly make production capacity. Unfortunately, climate change often affects countries that are more vulnerable and may not have the means of adaptation.

According to Our World In Data publication, the food industry is responsible for 26% of the total carbon emissions in a year worldwide. Maybe it is hard to believe that this industry can cause a quarter of total greenhouse gases. If we analyze the whole process at the microscopic level, starting with production and ending with product distribution to the customer, we will notice that there are many parts involved in this process. From this point of view, each entity is responsible for a certain percentage of the total of 26% (Ritchie and Roser 2020).

The food industry contributes to:

- global warming more than all machines, planes and trains taken together;
- eutrophication of aquatic media due to pesticides and fertilizers;
- accelerated loss of biodiversity;
- deforestation and desertification.

Achieving a production level that responds to our needs, from a natural resource base already exhausted, will be impossible without profound changes in our food and agricultural systems. As food production systems turns to adapt to current conditions, it is necessary to consider the impact on food safety to assess the optimal ways of approaching potential risks. As a consequence, agriculture is confronted with unprecedented confluence of pressures that cause profound changes in the food industry (cultures, animals, forestry, fisheries and aquaculture). The industrial agri-food systems involve the extension of arable land and increased yields per hectare by irrigation, intensive use of agricultural fertilizers, the development and widespread use of selected plant hybrids, breeds of commercial animals and genetically modified organisms and production concentration (agricultural societies and far greater farms). The impact—on the environment, society and human health—of this system proves to be more and more devastated. The consequences are measured in terms of air pollution and groundwater, soil degradation, ocean acidification, reduction of energy resources, loss of biological and cultural biodiversity and damage to ecosystems (EEA 2020).

The FAO's vision of sustainable food and agriculture is one in which food is nutritious and accessible to everyone and where natural resources are managed in a way that maintains ecosystem functions to support current as well as future human needs. Intensive animal husbandry, according to the FAO (linked to rising levels of meat consumption), produces about 14% of greenhouse gas emissions, taking into account the entire production chain, from feed cultivation to final consumption. The intensification of livestock production in its current form will have an impact on human health through the contamination of the environment and as a consequence of it, through food contamination. The increase in humidity, precipitation amounts and temperature favors fungal contamination and the accumulation of mycotoxins, contributing to the transmission of zoonotic diseases and chemical contaminants from animals to plants or other animals, and even to humans. The influence of climate change on the prevalence of foodborne diseases must be limited by introducing effective adaptation and intervention strategies. Antimicrobial resistant organisms from livestock production continue to threaten food safety. In order to avoid unintended consequences for food safety, it is necessary to reduce emissions and increase production without antimicrobial growth promoters, but also to evaluate the use of new additives in feed, such as prebiotics, probiotics and bioactive substances to improve immunity (Harvard School of Public Health).

Sustainable agriculture tends to increase production by limiting the over-application of pesticides and chemical fertilizers. The application of organic fertilizers resulting from industry, as well as irrigation with wastewater, can contribute to contamination by heavy metals and various microbial contaminants of agricultural soils and crops. Climate change is contributing to the intensification of the attack of diseases and pests of crops, which implies an excessive use of pesticides, their residues being found in crop plants. Eutrophication, as a result of nutrient accumulations in aquatic environments, is a source of contamination of food of aquatic origin as a result of the development of algae that produce harmful toxins. Hydroponics and vertical agriculture, although they prolong the vegetation period and preserve water resources, will increase the potential for the spread of contaminants and pathogens in food.

The entire global agri-food system has to be readjusted so as to allow simultaneously:

- remedying climate change causes, reducing the impact of agriculture on the climate, reducing carbon dioxide and nitrogen oxide emissions;

- mitigating, in other words, making farmers less economically, socially and environmentally vulnerable by reducing the climate change impact on agriculture;

- adaptation, i.e., improving farmers' ability to respond to climate change by giving priority to local management practices that protect biodiversity and ecosystems.

The sustainable agri-food systems preserve natural balances, are against monocultures, promote diversification and the multitude of plant varieties and local animal breeds. They also promote lowest possible consumption of fossil fuels, pesticides and chemical fertilizers. They use techniques that preserve soil fertility and retain its moisture, increasing carbon storage capacity, slow down the desertification process, protecting the land from erosion; they support animal husbandry systems that use indigenous breeds, adapted to local pedoclimatic conditions, as well as respecting animal welfare.

One of the objectives of sustainable agriculture is to limit the cultivation of the same type of plant in the same place, year after year, because it depletes the soil of nutrients and leaves it more vulnerable to erosion. Beyond this, the loss of fertilizers and herbicides used in the monoculture agricultural system increases greenhouse emissions and damages wildlife, leading to a decrease in the number of important birds, insects and fish. Also, the monoculture is more susceptible to the attack of diseases and pests, which has an impact both on the quality and quantity of the harvest but also economically.

Traditional agriculture uses large amounts of resources. Common agricultural practices consume large amounts of fossil fuels used to transport food and contribute significantly to greenhouse gas emissions. Sustainable agriculture uses alternative energy sources such as wind, solar, used vegetable oil and more, and focuses on reducing pollution by optimizing water, fuel and land consumption.

The productivity of many agricultural and food production systems can be improved by changing current practices. A sufficient supply of food and other agricultural products in the future implies a further increase in productivity, but at the same time the expansion of agricultural land must be limited and environmental protection must be ensured. This is the focus of the changes needed to ensure the sustainability of food and agricultural systems. In the past, productivity efficiency has been expressed mainly by yield (kg per hectare of production), but further productivity growth must take into account several dimensions. As water scarcity increases and agriculture needs to look for ways to reduce greenhouse gas emissions, smart water management and energy production systems will become increasingly important. This will also have an effect on the use of fertilizers and other agricultural inputs.

Animal husbandry and fish farming are responsible for 31% of greenhouse gas emissions from agricultural activities and the food industry. Ruminants, especially cattle, produce methane through their digestive process. The management of manure, pastures, as well as the fuel consumed by fishing vessels are other causes of greenhouse gases' production. 31% of emissions do not include those from feed production.

Plant production is responsible for 27% of greenhouse gas emissions, of which 21% comes from plant production for human consumption and 6% from feed production. They are direct emissions from agricultural production, including nitrogen oxides from mineral and organic fertilizers, methane from rice production and carbon dioxide from agricultural machinery.

24% of emissions come from land use, 16% from land used for animal husbandry, and 8% from land used for agricultural crops (Ritchie and Roser 2020).

Supply chains account for 18% of emissions from the food and agriculture industries. Food processing, transportation, packaging and retail require energy inputs and resources. Although local consumption is considered to reduce emissions due to the transport of agri-food products, these emissions represent a very small percentage of total emissions—only 6% globally. Emissions from food waste are high: a quarter of emissions (3.3 billion to of CO_2) from food production are either waste or losses along the supply chain (Ritchie and Roser 2020).

The EU's Common Agricultural Policy aligns agriculture with the European Green Pact, which aims to create an inclusive, competitive and environmentally friendly future.

Farmers, agri-food businesses, foresters and rural communities play a key role in some of the main policy areas of the Green Pact, including:

- building a sustainable food system through the "From farm to consumer" strategy;
- integration into the new biodiversity strategy by protecting and enhancing the variety of plants and animals in the rural ecosystem;
- contributing to action to combat climate change under the European Green Pact to achieve the EU's zero emissions target by 2050;
- supporting the updated forestry strategy by maintaining healthy forests;
- contribution to a "zero pollution" action plan by protecting natural resources such as water, air and soil.

Food waste also puts pressure on the environment. FAO has defined food losses as losses along the food chain between the producer and the market, while safe and nutritious food is thrown away in the form of waste. About a third of the food produced is lost. A significant amount of food is disposed of as waste, especially in developed countries, and this also means eliminating the resources used to produce food. In the EU alone, 90 million tons of food or 180 kg per person are disposed of as waste every year, much of which is still suitable for human consumption. About a third of globally produced food is lost or discarded. In total, this means that 1.3 billion tons of food is not consumed annually. Taken together, this food waste produces 3.6 gigatons of carbon dioxide, according to FAO estimates. It is incredible that food waste occurs in industrialized countries that account for only 15% of the world's population, but consume most of the world's resources (FAO 2011b, FAO 2021).

Food waste is produced in each phase of the production and supply chain, as well as in the consumption phase. Some food waste is produced by law enforcement, which is often implemented to protect human health. Another part may be related to consumer preferences and habits. All the different phases and reasons must be analyzed and established as necessary to reduce food waste. Losses along the food chain as well as food waste are caused by:

- problems due to processing, handling, packaging, transport and retail;

- reduction of quality, such as damage to packaging, fruit or vegetables;

- quality defects—for example, fresh products that deviate from what is considered optimal in terms of shape, size and color, small or irregularly shaped vegetables and fruits being discarded during sorting operations;

- foods on the expiry date are discarded by retailers and consumers;

- large amounts of unhealthy edible healthy food are thrown away from households and food establishments;

- lack of time—society is increasingly busy, often preferring to go shopping less often and store food;

- the variety of products on the market—we live in a society where the image of the product sells, and out of curiosity we prefer to buy the same product from different brands only under the pretext of trying it;

- promotional offers—we meet more and more often the so-called promotional products, that are quite tempting such as the 2×1, which urges us to buy a larger number of products.

3. Standardization in Food Industry

The EU as well as all other world communities adopted various policies and strategies in regard to international trade, being a complex mix of worldwide guidelines, agreements, national laws and requirements. Of these, food quality standards are perceived for their effectiveness throughout the entire process chain and are a fundamental device for ensuring safe and comparable processes in the agri-food industry. As a rule, due to standardization, agribusiness systems are transformed into an organized food framework, prompting competition not only between individual organizations in the food chain but also between supply chains and organizations. However, the protection of the environment and the economy of sustainable development are the most current trends these days, and additionally, there is an expansion in consumers' demand on food safety and its usefulness. Consequently, food production systems have to operate in a sustainable manner (Pietrzyck et al. 2021).

In the last two decades, standards have proliferated in the worldwide food supply chains, overwhelming world agri-food trade, the requirements on product quality being exceptionally high (Swinnen et al. 2015).

According to Kotsanopoulos and Arvanitoyannis, food safety and quality assurance systems can take many forms (Kotsanopoulos and Arvanitoyannis 2017):

- international quality assurance standards (e.g., ISO 9000);

- national farm-level assurance;

- proprietary quality assurance systems.

Concerning food quality, international and national governments, just as private actors, have imposed new guidelines and prerequisites targeting quality and safety, as well as environmental protection, animal welfare and employment conditions. A number of standards are set by international institutions (e.g., Codex Alimentarius for food safety, International Plant Protection Convention (IPPC) regarding plant health and World Organization for Animal Health (OIE) concerning animal health) (Swinnen et al. 2015).

In the food industry, different audit formats are at presently utilized, including standards that do not intend to support a specific group or country over another, such as the International Organization for Standardization (ISO) with 9001, 22000 and 14001 standards. The ISO 9000 group of guidelines considers a variety of quality management aspects and includes a part of ISO's most generally known requirements.

ISO 22000:2005 characterizes the necessity of an effective food safety management system and can be used to exhibit an association's capacity to control food safety hazards. This standard can and must be implemented by associations of all sizes that are engaged with any part of the supply chain. ISO 22000 empowers organizations to embrace a food chain approach for the development, implementation and improvement of efficient and effective food safety management systems. This standard can likewise be implemented with ISO 9001 and the other standards that ensure effective Quality Management System (QMS) guidelines (Kotsanopoulos and Arvanitoyannis 2017).

Standardization in the food industry can influence both the supply (composition standards) and the demand (food labeling standards) of food. A crucial part of standardization is that it can ensure both efficiency and equity. However, the costs of compliance with standards might be high, especially for developing countries due to the lack of infrastructure, technical and scientific capacity with respect to food quality, safety management and the disparity between the global guidelines for public food quality and wellbeing. The empirical evidence shows that the implementation of certain standards can lead to very high costs, yet can likewise provide significant advantages. Many authors have raised the concern that more standards can pose barriers to emerging countries' incorporation in worldwide food markets (Lawrence et al. 2019, Swinnen 2017, 2015).

When discussing the costs for the implementation of a food safety program, it varies generally based on the segment of the industry (Table 2).

Food safety essential projects like Good Manufacturing Practices (GMPs) and Sanitation Standard Operating Procedures (SSOPs) provide the starting point to more extensive food safety programs like Hazard Analysis Critical Control Points (HACCP). The expanded concern for food handling safety led the United States Food and Drug Administration (FDA) to create the Food Safety Modernization Act (FSMA). The underlying cost for the first year of FSMA implementation is assessed to be somewhere in the range of $520 and $860 million. However, trust in the food and beverage industry has declined by two points since 2019 according to the latest Edelman Trust Barometer (Hessing et al. 2020, Global Report: Edel-man Trust Barometer 2020).

4. Consumer Ethics in the Context of Food Safety

Food safety in a sustainable environment and, implicitly, with a sustainable food industry is the goal of any human community. To achieve this requires development of CE, control of negative effects of food safety measures on the environment, as well as legal and effective regulations to ensure consumer health, doubled by a high-level standard on consumer ethics. Many consumers who are concerned about food safety are, at the same time, concerned about environment sustainability, animal welfare, protection and safety. Moreover, they accept food waste as a necessary behavior for assurance of food safety but are, at the same time, concerned about environmental protection and about the reduction of waste of natural resources. But aren't reciprocally exclusive? Are these concerns sincere? Are these concerns that really comes to protect both the environment and future

Table 2. Initiatives to stimulate sustainable practices.

Practices in:		
Food Supply Chain	**Food Quality Standards**	**Global Trade**
Development of a contingency plan (2021) for ensuring the food supply and security in times of crisis	Development of a contingency plan (2021) for ensuring the food supply and security in times of crisis	Proposal for a revision of the EU legislation on food contact materials (food safety, environmental footprint; 2002)
Develop an EU code and monitoring framework for responsible business and marketing conducted in the FSC (2021)	Stimulate reformulation of standards in processed food (2021/2020)	Proposal to require origin indication for certain products (2022)
Revision to EU marketing standards for agricultural, fishery, and aquaculture products to ensure the uptake and supply of sustainable products (2021–2022)	Proposal for the revision of the EU legislation on food contact materials (food safety, environmental footprint; 2002)	Proposal of global transitions caused due to international cooperation
Proposal for a sustainable food labeling framework to empower consumers to make sustainable food choices (2024)	Work through international standard-setting bodies (e.g., Codex Alimentarius)	Inclusion of ambitious sustainability chapter, including food, in all EU bilateral trade agreements
Proposal to require origin indication for certain products (2022)	Environmental aspects considered when assessing requests for import tolerances (e.g., standards for pesticides)	Environmental aspects considered when assessing requests for import tolerances (e.g., standards for pesticides)
Promotion for appropriate labeling schemes – to ensure that food imported into the EU is gradually produced in a sustainable manner	Promotion for appropriate labeling schemes – to ensure that food imported into the EU is gradually produced in a sustainable manner	Promotion for appropriate labeling schemes – to ensure that food imported into the EU is gradually produced in a sustainable manner

Source: Adapted by authors after Pietrzyck et al. (2021). Exploring sustainable aspects regarding the food supply chain, agri-food quality standards, and global trade: An empirical study among experts from the European Union and the United States, based on Westhoek (DG SANTE)[1]

generations? Maybe if the answers are positive, then the food waste would decrease in all its aspects, such as:

- fewer requirements of quality standards, often unnecessary and unjustified;
- fewer purchase by consumers of excess food, followed by their disposal after the expiration of the validity;
- minimum outdated methods and technologies in some countries, which generate high technological loss along the entire food chain;
- minimalization of food overconsumption.

From all of these, food waste due to overconsumption is the one that can be interpreted as the most selfish, although it is more accessible to all to be reduced, being clearly in contradiction with the care for the population that is in a state of starvation, and at the same time care for the protection of animals. Most people also waste meat, not only vegetables, but their care for animals is declared, official and effective, these consumers being often involved in concrete actions to protect the environment, as fauna and flora.

---

[1] Westhoek, H.; DG SANTE, European Commission. EU Farm to Fork Strategy: Promoting Sustainable Food Consumption and Facilitating the Shift to Healthy, Sustainable Diets. Presentation at Online-Webinar by Valumics: Putting Solutions on the Table. Successful Approaches and Interventions to Support more Sustainable Food Consumption Behaviors in the EU. 16 July 2020. Available online: https://valumics.eu/events/(accessed on 5 April 2021)

According to Härtel and Yu, in 2013 around 842 million people were affected by absolute starvation. Presently, 795 million individuals face starvation and chronic malnutrition. World hunger expanded further in 2020 to 720–811 million individuals, exacerbated by the effect of the COVID-19 situation. The reduction of this number represents a great challenge nowadays for the world community particularly in terms of food safety and security. In expansion to these issues, an increasing number of individuals are suffering from the so-called "hidden hunger", which is an acute nutritional deficiency, in particular in vitamins, minerals and micronutrients (Härtel and Yu 2018, Chichaibelu et al. 2021, FAO 2021).

The primary goal is to prioritize the eradication of poverty and end hunger in all its forms; nonetheless, the worldwide conjecture of hunger indicates that the world isn't on target to accomplish this objective by 2030. Estimated studies from 2020 projected that the number of people affected by hunger will outperform 10% of the worldwide population. The world is additionally not on track to accomplish the 2030 targets for child malnutrition. Over the recent years, the number of malnutrition cases in children up to the age of 5 years has reached a percentage of 21.3% of the global population, representing, in numbers, 144 million children. Until this year, the number logically expanded to 149.2 million. Chichaibelu et al. affirmed that the combined impact of the expansion in wasting and a 25 percent decrease in the coverage of nutrition and health services due to COVID-19 will cause 128,605 additional deaths in children (UNICEF 2020, Chichaibelu et al. 2021, Headey et al. 2020).

In contrast to this constant requirement for food resources, which an ever-increasing number of individuals are deprived of, there is an excessive process of food wasting. According to a study carried out by the University of Stuttgart, each year nearly 11 million ton of foodstuffs are discarded as waste by industry, trade, wholesalers, and private households (Hafner et al. 2012).

Härtel and Yu emphasize the possibility that the reasons behind the food-wasting are multi-layered and they lie particularly in the fact that there are strict legal provisions of the law for food hygiene just as for food shelf-life and the combination in the worth chain. They declared that the central issue of overall food squander has additionally been perceived at the worldwide political level within the setting of food security (Härtel and Yu 2018).

Around the world, there are several projects that aim to guarantee decency with respect to the way of life, particularly regarding the feeding of an ever-growing population while preserving the environment and natural resources.

Of these, One Health (Fig. 3) is the concept in which the health of humans, animals, and environment are inseparably connected. This methodology can be applied to food safety, sustainability, and ecological stewardship (Boqvist et al. 2018).

Since millions of individuals suffer from food instability and most of them live in outrageous poverty, acquiring less than $2 per day, sustainable agricultural production is one of the bases to accomplish food security, reduce malnutrition and alleviate poverty. Additionally, eradication of hunger is the main objective of the SDGs. It is referenced that to feed 10 billion people by 2050, we need to get the trade-offs right directly between sustainability, food security, food safety and make better use of food already produced (FAO 2015, Garcia et al. 2020, Vågsholm et al. 2020).

Food safety and security are two complementing elements, directing the quality and consumption of food. However, there is a requirement for future novel solutions with regard to food security and safety without compromising sustainability. Vågsholm et al. brought up the possibility that as a rule, measures to guarantee safety and quality of food can reduce the amount of food accessibility, and subsequently enhance food scarcities. They exemplified: consumers interpret best before dates as food being noxious from there on dates, consequently expanding food waste and threatening food security (Vågsholm et al. 2020).

Regarding safety and security in the food industry, there are many causes that can influence these two concepts. Some of these are presented in Fig. 4.

Perhaps one of the main aspects of the current century with regards to the production and trade of agricultural products and foodstuffs is addressed by the expanding globalization and market integration.

ONE HEALTH

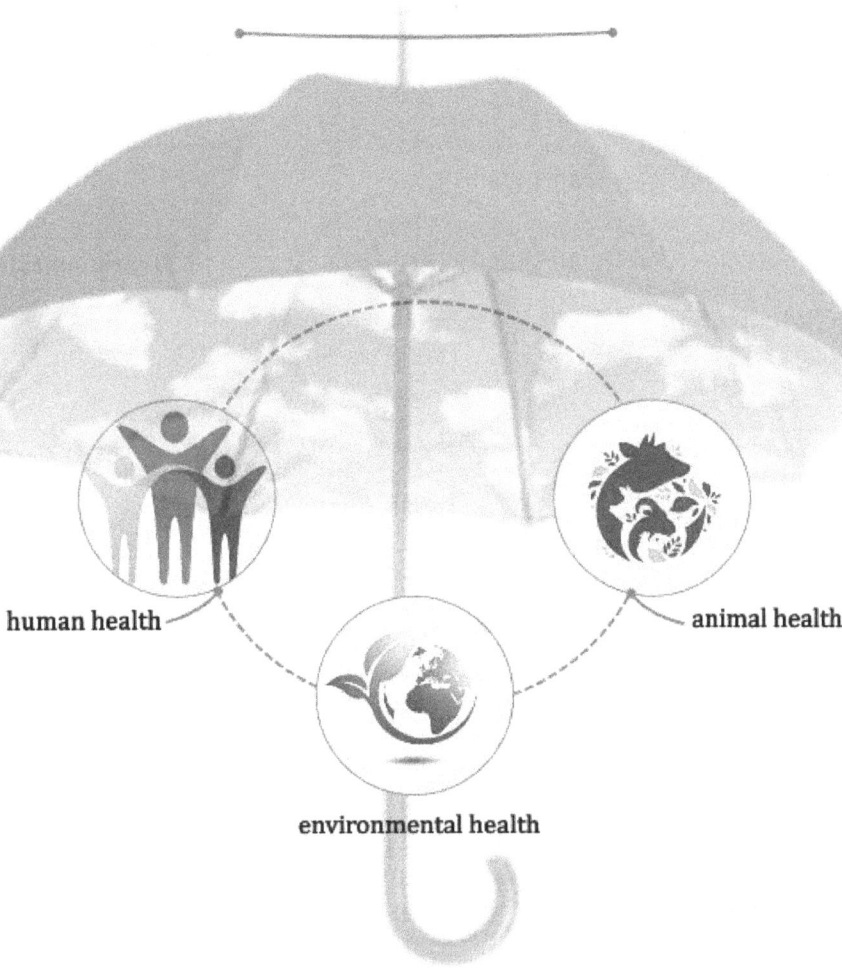

Figure 3. The "*One Health*" concept regarding the human, animal, and environmental health. Source: Original design by authors.

In this context, it is noteworthy that, as Gomez points out, a global cocoa crisis is looming, due to the fact that in 2021 the demand of chocolate and implicitly the demand for cocoa is higher than the supply of cocoa. This is happening, on the one hand, in the current situation in which a large number of children in developed countries suffer from obesity and overweight, and a large part of doctors, parents and authorities are concerned about this issue and try to reduce the amount of sweets, and implicitly chocolate, in children's nutrition, and on the other hand, in the context that it is known that African countries, where most of the cocoa production comes from, use children as labor, even from an early age of 5 years, and the same parents and authorities condemn the exploitation of minors, in any form. At the same time, the intensification of cocoa production will seriously affect the environment in these countries, by introducing fertilization and modern exploitation systems (Gomez 2020).

Then, the question comes back, from another point of view: when is the human population sincere? When overconsuming and wasting food or when it is concerned about environmental sustainability?

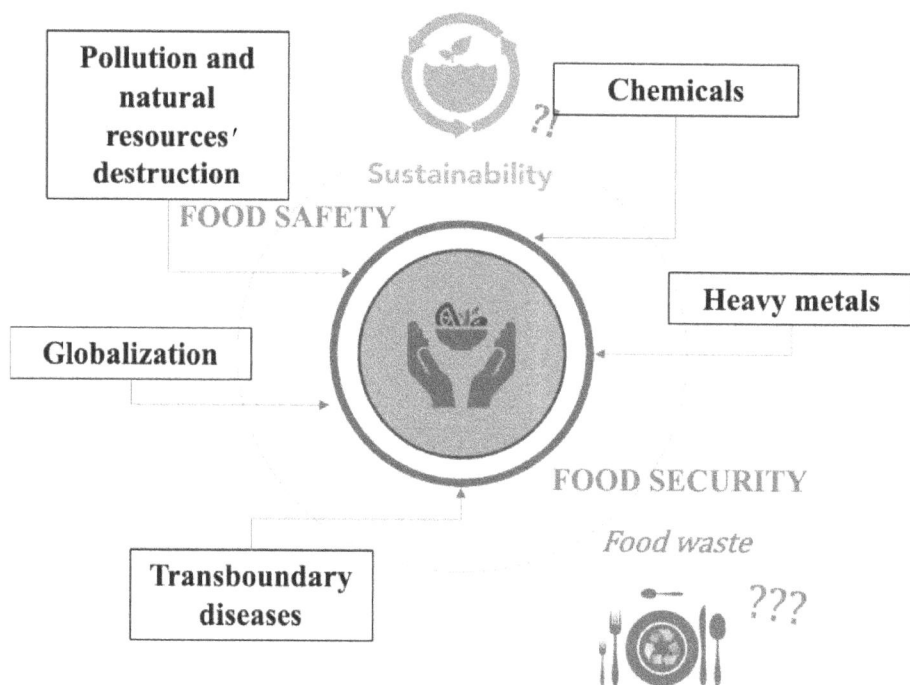

Figure 4. Affecting factors of food safety and security, with implications in sustainability and excessive production of food waste. Source: Original design by authors.

References

Arunraj, N.S. and Ahrens, D. 2015. A hybrid seasonal autoregressive integrated moving average and quantile regression for daily food sales forecasting. Int. J. Prod. Econ. 170: 321–335. doi: 10.1016/j.ijpe.2015.09.039

Bahn-Walkowiak, B., Wilts, H., Reimer, W. and Lee, M. 2019. Overview Report on Definition and Concept of the Circular Economy in a European Perspective.

Barros, M.V., Salvador, R., de Francisco, A.C. and Piekarski, C.M. 2020. Mapping of research lines on circular economy practices in agriculture: from waste to energy. Renew. Sustain. Energy Rev. 131: 109958. https://doi.org/10.1016/j.rser.2020.109958.

B´en´e, C., Prager, S.D., Achicanoy, H.A.E., Toro, P.A., Lamotte, L., Bonilla, C. et al. 2019. Global map and indicators of food system sustainability. Sci. Data 6: 1–15. https://doi.org/10.1038/s41597-019-0301-5. Beter Leven, 2021. Wat is Beter Leven? Beter Leven.

Billen, G., Lassaletta, L., Garnier, J., Le Noé, J., Aguilera, E. and Sanz-Cobeña, A. 2019. Opening to distant markets or local reconnection of agro-food systems? Environmental consequences at regional and global scales. Agroecosystem Diversity. Elsevier, pp. 391–413. https://doi.org/10.1016/B978-0-12-811050-8.00025-X.

Boqvist, S., Söderqvist, K. and Vågsholm, I. 2018. Food safety challenges and one health within Europe. Acta Vet. Scand. 60: 1. doi: 10.1186/s13028-017-0355-3.

Borsellino, V., Kaliji, S.A. and Schimmenti, E. 2020. COVID-19 drives consumer behaviour and agro-food markets towards healthier and more sustainable patterns. Sustain. Times 12: 1–26. https://doi.org/10.3390/su12208366.

Chaudhary, A., Gustafson, D.I. and Mathys, A. 2018. Multi-indicator sustainability assessment of global food systems. Nat. Commun. 9: 848. https://doi.org/10.1038/s41467-018-03308-7.

Cooper, R. and Chew, W.B. 1996. Control tomorrow's costs through today's designs. Harvard Business Review. https://hbr.org/1996/01/control-tomorrows-costs-through-todays-designs.

Corrado, S., Caldeira, C., Carmona-Garcia, G., Körner, I., Leip, A. and Sala, S. 2020. Unveiling the potential for an efficient use of nitrogen along the food supply and consumption chain. Glob. Food Sec. 25: 100368. https://doi.org/10.1016/j.gfs.2020.100368.

Chichaibelu, B.B., Bekchanov, M., von Braun, J. and Torero, M. 2021. The global cost of reaching a world without hunger: Investment costs and policy action opportunities. Food Policy [Internet].; 104(August):102151.

DG Enterprise and Industry and DG Energy European Commission. 2014. Ecodesign - your future. How ecodesign can help the environment by making products smarter.

EC [European Commission]. 2016. Circular economy in Europe, Developing the knowledge base, EEA Report no 2/2.

EC [European Commission]. 2020a. FOOD 2030, Pathways for action FOOD 2030, Research policy as a driver for sustainable, healthy and inclusive food systems.

EC [European Commission]. 2020b. RESILIENCE AND TRANSFORMATION. Report of the 5th SCAR Foresight Exercise Expert Group – Natural resources and food systems: Transitions towards a 'safe and just' operating space. Luxembourg: Publications Office of the European Union, 2020.

EC [European Commission], Directorate-General for Research and Innovation and Bioeconomy Directorate. 2016. European Research & Innovation for Food & Nutrition Security. Luxembourg: Publications Office of the European Union. Doi:10.2777/069319.

EEA [European Environment Agency]. 2020. Food security and environmental impacts. Retrieved from: https://www.eea.europa.eu/themes/agriculture/greening-agricultural-policy/food-security-and-environmental-impacts [Online Resource].

EFSA and ECDC [European Food Safety Authority and European Centre for disease Prevention and Control]. 2019. Scientific report on the European Union One Health 2018 Zoonoses Report. EFSA J. 17: 5926. doi: 10. 2903/j. efsa.2019.5926.

FoodDrinkEurope. 2016. Ingredients for a circular economy. http://agricultura.gencat.cat/web/.content/04-alimentacio/malbaratament-alimentari/enllacos-documents/fitxers-binaris/Food-Drink-Europe_Ingredients-for-a-circular-economy.pdf.

Fanzo, J., Haddad, L., McLaren, R., Marshall, Q., Davis, C., Herforth, A. et al. 2020. The food systems dashboard is a new tool to inform better food policy. Nat. Food 1: 243–246. https://doi.org/10.1038/s43016-020-0077-y.

FAO. Sustainable Food and Agriculture. Retrieved from: https://www.fao.org/sustainability/background/en/ [Online Resource].

FAO. 2002. Safe Food and Nutritious Diet for the Consumer. World Food Summit – five years later. FAO, Rome, Italy.

FAO. 2009. How to Feed the World in 2050. Synthesis Report of the High-Level Expert Forum. Rome, Italy.

FAO. 2011a. The state of the world's land and water resources for food and agriculture (SOLAW)—Managing systems at risk. Food and Agriculture Organization of the United Nations, Rome Italy and Earthscan, London.

FAO. 2011b. Global Food Losses and Food Waste: Extent, Causes and Prevention. Study Conducted for the International Congress SAVE FOOD! at Interpack2011 Düsseldorf. Germany. FAO, Rome, Italy.

FAO. 2012. State of the World's Fisheries and Aquaculture. FAO, Rome, Italy.

FAO. 2014. Food Wastage Footprint: full-cost Accounting. FAO, Rome, Italy. ISBN 978-92-5-107752-8.

FAO. 2015. The Impact of Disasters on Agriculture and Food Security. Rome, Italy. Available online at: http://www.fao.org/3/a-i5128e.pdf.

FAO, IFAD and WHP. 2015. The State of Food Insecurity in the World 2015. Meeting the 2015 international hunger targets: taking stock of uneven progress. FAO, Rome, Italy.

FAO, IFAD, UNICEF, WFP and WHO. 2021. The state of food security and nutrition in the world 2021. Transforming food systems for food security, improved nutrition and affordable healthy diets. for all. FAO, Rome. https://doi.org/10.4060/cb4474en.

Fei, S., Ni, J. and Santini, G. 2020. Local food systems and COVID-19: an insight from China. Resour. Conserv. Recycl. 162: 105022. doi:10.1016/j.resconrec.2020.105022.

Fésüs, G., Rillaers, A., Poelman, H. and Gáková, Z. 2008. Regions 2020. Demographic challenges for european regions. directorate general for regional policy. Commission of the European Communities.

Foley, J.A., Ramankutty, N., Brauman, K.A., Cassidy, E.S., Gerber, J.S., Johnston, M. et al. 2011. Solutions for a cultivated planet. Nature 478: 337–342. http://dx.doi.org/10.1038/nature10452.

Garcia-Garcia, G., Azanedo, L. and Rahimifard, S. 2020. Embedding sustainability analysis in new food product development. Trends Food Sci. Technol. 104743. https://doi.org/10.1016/j.tifs.2020.12.018.

Garcia, S.N., Osburn, B.I. and Jay-Russell, M.T. 2020. One health for food safety, food security, and sustainable food production. Front Sustain Food Syst. 4: 1–9.

Ghisellini, P., Zucaro, A., Viglia, S. and Ulgiati, S. 2014. Monitoring and evaluating the sustainability of Italian agricultural system. An emergy decomposition analysis. Ecol. Model. 271: 132–148. https://doi.org/10.1016/j.ecolmodel.2013.02.014.

Global Report: The Edelman Trust Barometer. 2020. Edelman. https://www.edelman.com/sites/g/files/aatuss191/files/2020-01/2020%20Edelman%20Trust%20Barometer%20Global%20Report.pdf.https://www.edelman.com/ trustbarometer.

Gomez, J.M. 2020. Sustainability and Food Safety: What's the Connection?. https://globalfoodsafetyresource.com/sustainability-and-food-safety-whats-the-connection/.

Hafner, G., Barabosz, J., Schneider, F., Lebersorger, S., Scherhaufer, S., Schuller, H., et al. 2012. Determination of discarded food and proposals for a minimization of food wastage in Germany - Abridged Version - University Stuttgart, Institute for Sanitary Engineering, Water Quality and Solid Waste Management (ISWA) https://www.researchgate.net/publication/262728113_Determination_of_discarded_food_and_proposals_for_a_minimization_of_food_wastage_in_Germany.

Harvard School of Public Health. 2022. Plate and the Planet. Retrieved from: https://www.hsph.harvard.edu/nutritionsource/sustainability/plate-and-planet/[Online Resource].

Härtel, I. and Yu, H. 2018. Food security and food safety law. Handbook of Agri-Food Law in China, Germany, European Union: Food Security, Food Safety, Sustainable Use of Resources in Agriculture. 57–126.

Headey, D., Heidkamp, R., Osendarp, S., Ruel, M., Scott, N., Black, R. et al. 2020. Impacts of COVID-19 on childhood malnutrition and nutrition-related mortality. The Lancet 396(10250): 519–521. https://doi.org/10.1016/S0140-6736(20)31647-0.

Helland, J. and Sörbö, G.M. 2014. Food Security and Social Conflict. CMI Report 2014:1. Bergen: Christian Michelssen Institute.

Hessing, A., Schneider, R.M.G., Gutierrez, A., Silverberg, R., Gutter, M.S. and Schneider, K.R. 2020. The Cost of Food Safety. EDIS, (1): 5–5.

HLPE [High Level Panel of Experts]. 2014. Food Losses and Waste in the Context of Sustainable Food Systems. A report by the High Level Panel of Experts on Food Security and Nutrition of the Committee on World Food Security. Hlpe Rep, 1–6, 65842315.

Ibn-Mohammed, T., Mustapha, K.B., Godsell, J., Adamu, Z., Babatunde, K.A., Akintade, D.D. et al. 2021. A critical review of the impacts of COVID-19 on the global economy and ecosystems and opportunities for circular economy strategies. Resour. Conserv. Recycl. 164: 105169. https://doi.org/10.1016/j.resconrec.2020.105169.

Kearney, A.T. 2019. How Will Cultured Meat and Meat Alternatives Disrupt the Agricultural and Food Industry? Dusseldorf; AT Kearney Studie zur Zukunft des Fleischmarkts bis 2040. Available online at: https://www.atkearney.com/ retail/article/?/a/how-will-cultured-meat-and-meat-alternatives-disrupt-theagricultural- and-food-industry (accessed July 5, 2019).

Kirchherr, J., Reike, D. and Hekkert, M. 2017a. Conceptualizing the circular economy: an analysis of 114 definitions. Resour. Conserv. Recycl. 127: 221–232. https://doi.org/ 10.1016/j.resconrec.2017.09.005.

Kirchherr, J., Reike, D. and Hekkert, M. 2017b. Resources, conservation & recycling conceptualizing the circular economy: an analysis of 114 definitions 127: 221–232. https://doi.org/10.1016/j.resconrec.2017.09.005.

Kotsanopoulos, K.V. and Arvanitoyannis, I.S. 2017. The role of auditing, food safety, and food quality standards in the food industry: A review. Comprehensive Reviews in Food Science and Food Safety 16(5): 760–775.

Lawrence, M.A., Pollard, C.M. and Weeramanthri, T.S. 2019. Positioning food standards programmes to protect public health: current performance, future opportunities and necessary reforms. Public Health Nutrition 22(5): 912–926.

Leibler, J.H., Dalton, K., Pekosz, A., Gray, G.C. and Silbergeld, E.K. 2017. Epizootics in industrial livestock production: preventable gaps in biosecurity and biocontainment. Zoonoses Public Health 64: 137–145. doi: 10.1111/zph.12292.

McAloone, T.C. and Bey, N. 2009. Environmental improvement through product development: A guide. http://mst.dk/media/mst/9225391/environmental_improvement_through_product_development.pdf.

National Resources Defense Council (NRDC). 2017. Wasted: How America is losing up to 40 percent of its food from farm to fork to landfill (2nd ed.). Retrieved from https://www.nrdc.org/sites/default/files/wasted-2017-report.pdf.

Pietrzyck, K., Jarzębowski, S. and Petersen, B. 2021. Exploring sustainable aspects regarding the food supply chain, agri-food quality standards, and global trade: An empirical study among experts from the european union and the united states. Energies. 14(18).

Ritchie, H. and Roser, M. 2020. Environmental impacts of food production. Published online at OurWorldInData.org. Retrieved from: 'https://ourworldindata.org/environmental-impacts-of-food' [Online Resource].

Rowan, N.J. and Galanakis, C.M. 2020. Unlocking challenges and opportunities presented by COVID-19 pandemic for cross-cutting disruption in agri-food and green deal innovations: quo Vadis? Sci. Total Environ. 748: 141362. https://doi.org/10.1016/j. Scitotenv.2020.141362.

Sánchez García, J.L., Beiro Pérez, I. and Díez Sanz, J.M. 2019. Hunger and sustainability. Economic research-Ekonomska istraživanja 32(1): 850–875.

Slow Food Foundation. 2015. Let's not eat our planet. (Romanian) retrieved from: https://www.slowfood.com/wp-content/uploads/2015/11/DOC-CLIMA_ROM.pdf [Online Resource].

Springmann, M., Wiebe, K., Mason-D'Croz, D., Sulser, T.B., Rayner, M. and Scarborough, P. 2018. Health and nutritional aspects of sustainable diet strategies and their association with environmental impacts: a global modelling analysis with country-level detail. Lancet Planet. Heal. 2: e451–e461. https://doi.org/10.1016/S2542-5196(18)30206-7.

Swinnen, J., Maertens, M. and Colen, L. 2015. The role of food standards in trade and development. Food Safety, Market Organization, Trade and Development. Springer, Cham., 133–149.

Swinnen, J. 2017. Some dynamic aspects of food standards. American Journal of Agricultural Economics 99(2): 321–338.

Tseng, M., Chiu, A.S.F., Chien, C. and Tan, R.R. 2019. Pathways and barriers to circularity in food systems. Resour. Conserv. Recycl. 143: 236–237. https://doi.org/10.1016/j. resconrec.2019.01.015.

UNFAO. 2018. Food wastage footprint & climate change [Press release]. Retrieved from http://www.fao.org/3/a-bb144e.pdf.

UNICEF, WHO and World Bank. 2020. Levels and trends in child malnutrition: Key findings of the 2020 edition of the Joint Child Malnutrition Estimates. UNICEF/WHO/World Bank Group. https://www.who.int/publications/i/item/jme-2020-edition.

van der Wiel, B.Z., Weijma, J., van Middelaar, C.E., Kleinke, M., Buisman, C.J.N. and Wichern, F. 2020. Restoring nutrient circularity: a review of nutrient stock and flow analyses of local agro-food-waste systems. Resour. Conserv. Recycl. 160: 104901. https://doi.org/10.1016/j.resconrec.2020.104901.

Vanham, D., Bouraoui, F., Leip, A., Grizzetti, B. and Bidoglio, G. 2015. Lost water and nitrogen resources due to EU consumer food waste. Environ. Res. Lett. 10: 84008. https://doi.org/10.1088/1748-9326/10/8/084008.

Vågsholm, I., Arzoomand, N.S. and Boqvist, S. 2020. Food Security, Safety, and Sustainability—Getting the Trade-Offs Right. Frontiers in Sustainable Food Systems 4: 16. doi:10.3389/fsufs.2020.00016.

Verger, Y., Petit, C., Barles, S., Billen, G., Garnier, J., Esculier, F. et al. 2018. A, N, P, C, and water flows metabolism study in a peri-urban territory in France: the case-study of the Saclay plateau. Resour. Conserv. Recycl. 137: 200–213. https://doi.org/ 10.1016/j.resconrec.2018.06.007.

Westhoek, H.; DG SANTE, European Commission. EU Farm to Fork Strategy: Promoting Sustainable Food Consumption and Facilitating the Shift to Healthy, Sustainable Diets. Presentation at Online-Webinar by Valumics: Putting Solutions on the Table. Successful Approaches and Interventions to Support more Sustainable Food Consumption Behaviors in the EU. 16 July 2020. Available online: https://valumics.eu/events/(accessed on 5 April 2021).

Willett, W., Rockström, J., Loken, B., Springmann, M., Lang, T., Vermeulen, S. et al. 2019. Food in the Anthropocene: the EAT-Lancet Commission on healthy diets from sustainable food systems. Lancet 393: 447–492. doi: 10.1016/S0140-6736(18)31788-4.

World Bank. 2018. Food-borne illnesses cost US$ 110 billion per year in low – and middle – income countries [Press release]. World Bank. Retrieved from: https://www.worldbank.org/en/news/pressrelease/2018/10/23/food-borne-illnesses-cost-us-110-billion-per-year-in-low-and-middle-income-countries.

WRI, 2018. Creating a Sustainable Food Future: A Menu of Solutions to Feed Nearly 10 Billion People by 2050. Washington D.C.

14

Consumers' Acceptance (or lack thereof) of Sustainable Food Innovations

Setting Foundations for Convergence Innovation

Ghina ElHaffar[a] and *Laurette Dubé*[a,*]

1. Introduction

The food industry and related agronomic sector cause massive pressure on agriculture land quality, pesticide over-use, habitat destruction and freshwater availability (Crist et al. 2017), contributing directly to the intensification of the climate crisis. Food production and distribution alone account for more than 50% of the various environmental impacts of the total consumption (European Commission Joint Research Centre 2006).[1] It follows that investing in environmental solutions to food choices can prove relevant and efficient in lessening human pressure on the planet. Indeed, the urgency of the environmental situation has led to a plethora of innovations aimed at lowering the footprint of food products across the value chain from farming to waste management, offering better choices, limiting food waste, reducing green gas emissions, and ultimately narrowing the overall environmental impact.

With the availability of multiple environmentally friendly alternatives, consumers now have a choice: they can choose to limit their carbon footprint by altering their consumption habits and patterns. For instance, choosing a vegetarian burger over a meat burger, an option that is readily available in most food restaurants, can reduce the meal's environmental impact by roughly 77% (Saget et al. 2021). Another meat alternative would be cultured meat which limits both the types of green gas emissions and the overall climate warming effects in the short run (Lynch and Pierrehumbert 2019). These food alternatives and other low-carbon food innovations continue to surface as the environmentally safer option, as their potential for commercialization is getting more

---

[a] Faculty of Management, Desautels Faculty of Management, McGill University, Canada.
* Corresponding author: laurette.dube@mcgill.ca
[1] As cited by Alsaffar (2016).

feasible (Bryant and Barnett 2018). However, despite the widespread sustainable food innovations, there are varying levels of their acceptance from consumers. For example, consumers still hesitate to include cultured meat in their diets, and only 11% are willing to choose this option over conventional or plant-based alternatives (Slade 2018). Another example is organic foods, which occupies only 5.8% of total food sales in the USA (Wunsch 2021), and no more than 3.2% in Canada (EDC 2020).

There is still considerable effort to be made. However, to make change possible, it is crucial to comprehend the challenges surrounding sustainable food innovation, i.e., their acceptability, and demand adoption. While governments and industries are pacing to integrate low-carbon technologies and processes in their supply chains and infrastructures, consumers are not fully on board. So, what is standing in the way of consumers embracing this ecological vogue? And how can we overcome this impasse? In this chapter, we explore these issues. We first present a typology of low-carbon food innovations based on the life cycle of products. Then, we examine the evidence on the attitude-behavior gap, bounded rationality, psychological distance and product attributes, and their implications for consumers' adoptions (or lack thereof) of sustainable food alternatives. Further, we discuss possible solutions that could help transition individuals to more environmentally friendly food choices. Finally, we seal this chapter with a discussion on the potential of blockchain technologies in gaining consumers' trust and facilitating the ecological behavioral shift.

2. A Typology of Low-carbon food Innovation Across the Value Chain

With the increasing size of the global population, food production will need to double or triple by 2100 to meet the coming demand while maintaining the planet's environmental equilibrium to limit earth's depletion (Crist et al. 2017). Enhancing food production, consumption, and waste management was pointed out among the diverse steps that humanity can take to avert severe environmental consequences and to transition to sustainability (Ripple et al. 2017). Several innovative solutions across the value chain ought to join these steps, such as urban farming, genetic crop enhancement, conventional meat alternatives, meal kits, dietary gamification, and food sharing applications. We present these low-carbon food innovations in a life-cycle-based typology, and we identify key acceptability challenges among consumers.

2.1 Phase I: Farming

Our typology identifies five life cycle phases: farming, processing, access and consumption, and waste management. To begin with, farming represents the first phase of a food's life cycle and can generate a substantial toll on the environment through greenhouse gas emissions, chemicals residues and soil depletion (Aneja et al. 2008). To counter these effects, farmers have resorted to several solutions such as organic farming, local and urban farming, and genetically enhanced crops. Although these solutions hold potential for the environment, several issues arise at the time of their commercialization to consumers. For instance, despite the growing interest in organic alternatives and their health benefits (Rana and Paul 2020), organic produce still have low market share and low customer demand relative to conventional food. The most recurrent findings in this regard reveal an existing intention-behavior gap from the side of the concerned consumers (Chekima et al. 2017, ElHaffar et al. 2020).

Another issue arises at the market: when consumers approach food labeled as sustainable, whether organic, local, or deriving from urban farming, they find it challenging to make sense of the food's environmental attribute. This intangibility issue is due to the distant nature of the ecological issues (Segev et al. 2015) and the individuals' lack of awareness towards environmental matters. As for the genetic modification of crops, and even though it has been approved by scientists and judged safe to consume, people are still hesitating to adopt food derived from these enhancements for the related unhealthiness and unnaturalness perceptions (Bryant and Barnett 2018).

Table 1. Sustainable food innovations across the value chain.

Life cycle		Sustainable food innovation	Positive environmental effects	Acceptability challenges	References
Farming		Urban farming	Reduces transportation distance	Intangibility	(Segev et al. 2015)
		Local food		Intangibility, intention action gap	(Segev et al. 2015), (Chekima et al. 2017, ElHaffar et al. 2020)
		Organic food	Reduces chemical pesticide use, and land depletion	Price, intangibility, intention action gap	(Segev et al. 2015), (Chekima et al. 2017, ElHaffar et al. 2020)
		Genetically enhanced crops	Enhances crop quality and limits crop disease, and crop waste	Health, unnaturalness	(Bryant and Barnett 2018)
Production – Processing		Insect based alternatives		Neophobia, fear of insects, taste	(Dagevos 2021) (Hartmann and Siegrist 2017a,b)
		Cultured meat	Reduces carbon footprint	Health, neophobia, taste, price	(Bryant and Barnett 2018, Verbeke et al. 2015)
		Vegetarian meat alternatives	Reduces carbon footprint	Taste	(Boukid 2021)
		Clean energy solution (solar thermal heat)	Reduces carbon footprint	Intangibility	
Utilization	Access-Distribution	Eco-packaging- less packaging		Perceived contamination	(Magnier and Crié 2015)
		Eco-labels	Communicates the ecological value	Greenwashing	(Nuttavuthisit and Thøgersen 2017)
		Meal kits	Limits food waste	Decision and habit formation	(Mu et al. 2019).
	Consumption	Food sharing apps	Limits food waste	Trust	(Falcone and Imbert 2017).
		Food pairing apps	Limits food waste	Trust	(Falcone and Imbert 2017).
		Dietary change gamification		Habit formation	
	Waste Management	Apps for preventing food waste		Habit formation Usability Trust	
		Food waste reduction nudge		Decision and habit formation	

2.2 Phase II: Production-Processing

At the production phase, agricultural and animal resources are processed, transformed, and packaged before being expedited to the markets. At this phase, sustainable energy can be used to run the machinery and food processing plants, such as solar thermal heat and clean energy solutions (Schnitzer et al. 2010). Still, companies stumble again upon the intangibility challenge as this information is hard to communicate and explicit to consumers at the commercialization phase. Another low-carbon innovation in the production phase would include meat substitutes, as the production of beef meat is known for its relatively large carbon footprint (i.e., beef produces around

129 Kg of CO_2 compared to 2 Kg for vegetable substitutes, See Nijdam et al. (2012) for a review). Insect-based meat, cultured meat and vegetable-based meat are an illustration of such alternatives.

But getting people to accept these alternatives is challenging. For example, consumers express disgust and repugnance when presented with edible insects, mainly because insects are associated with danger and diseases in Western cultures and less with food or health (Dagevos 2021). Likewise, cultured meat is less appealing to consumers than standard meat because of the perception of human intervention in producing this meat. Hence, its unnaturalness provokes concern about its healthiness and safety. Taste and taste expectations also play a crucial role in consumers' acceptance of meat alternatives, as consumers recurrently report that vegetable meat alternatives such as tofu aren't as tasty as meat (Boukid 2021). The same goes for cultured meat, as consumers expect it to taste, smell and look different than normal meat (Bryant and Barnett 2018, Hartmann and Siegrist 2017a).

2.3 Phase III: Utilization (Distribution)

The next phase is access and distribution, where the food product is distributed to retailers and merchandised for consumers. Here, sustainable innovations include eco-packaging, edible packaging, and eco-labeling. While eco-packaging offers a better alternative to conventional packaging in fast-moving food produce at the selling point, consumers perceive it as less appealing visually, less effective in protecting its content, and less hygienic (Magnier and Crié 2015). Edible packaging is still a lab-based innovation, which faces many acceptability challenges such as health, safety, and toxicological effects perceptions (Jeevahan et al. 2020). Add to that the perceived contamination effect that influences product desirability and purchase intention (Gupta and Coskun 2021). Ecolabeling, on the other hand, discloses its own acceptance challenges among consumers. Although the labeling system has been initially created as a medium to address consumers' distrust in the food system (Tonkin et al. 2015), research has revealed a back-firing consequence and evidence of distrust in the control system and authenticity of food sold as organic (Nuttavuthisit and Thøgersen 2017).

2.4 Phase III and IV: Utilization (Consumption) and Waste Management

The final two stages of food's life cycle are consumption and waste management, whose main environmental contribution is to limit food waste and excessive consumption. Innovation in these two phases addresses consumers' behavior change and often includes a technological component such as digital mobile applications for leftover sharing, and restaurant choice. These innovations however come with their own share of challenges. On one hand, the applications are not yet optimized for user experience and can be daunting to use (Mu et al. 2019). On the other hand, sharing food and leftovers in a consumer-to-consumer setting, without any seal or third-party guarantees, rises doubts about the safety and quality of the food in question (Falcone and Imbert 2017).

Furthermore, getting consumers to adopt digital apps requires an amount of effort and commitment from consumers and complicates their food choices. It implies changing their habits and regularly making deliberate choices, which is cognitively tiring. Overall, sustainable food choices become surrounded by a behavioral cost, which limits its attractiveness for consumers.

Having presented the different innovations and relative challenges encountered at each of the food's life cycle phases, we will next elaborate on each challenge, and discuss potential solutions to overcome them.

3. Consumer Acceptability Challenges for Sustainable Food Innovations

Consumer's acceptance of sustainable food innovations comes with many challenges. Not only is it difficult to convince consumers that the ecofriendly alternative is a better option for them and the planet, but also getting them to act upon this conviction presents a more serious challenge. Acceptance therefore addresses consumer's attitudes towards the sustainable food alternative, their intentions to buy it, and most importantly their actual behavior in adopting and integrating it into their food diets and habits. In the following section, we discuss these challenges, starting with the behavioral aspect of acceptance, in which we present the attitude-behavior gap, the bounded rationality and the role of habits. Later, we explore the attitudinal and intentional aspect of acceptance by tackling consumers' perceptions of the sustainable alternative; specifically, we present the intangibility and psychological distance as well as product attributes and perceived trust as potential challenges.

3.1 Challenge 1: The Attitude-behavior Green Gap

When it comes to saving the planet, consumers are good intentioned. If you ask them about the environmental cause, they don't hesitate to convey their concern and their willingness to act in sustainable ways. Nevertheless, when in the marketplace, these good intentions fail to translate into real behavior. This inconsistency between consumers' declared values, attitudes, and intentions on one hand, and their real purchase behavior on the other, is well documented in the literature as the green attitude-intention-behavior gap (ElHaffar et al. 2020). This gap applies to sustainable food products, as well as other product categories such as personal care products, electronics, and apparel, and it can be traced both at the acquisition phase (at the purchase), and at the disposal phase (recycling, upcycling, reusing).

To approach the attitude-behavior gap as a primordial challenge in promoting sustainable food innovations, a thorough understanding of this phenomenon is key. The gap entails that for some sustainable food innovations, the problem is beyond attitudinal acceptance of the innovation, but rather implicates a behavioral reluctance when it comes to purchasing and integrating the innovation in consumers' diets. Several reasons could explain this hesitancy in action.

To begin with, consumers face a difficult choice when it comes to switching to eco-friendly food alternatives. If the decision was solely based on their values, action would be guaranteed, but consumers often assess the amount of costs associated with a behavior before undertaking it (Delmas and Colgan 2018). Take for instance the choice of buying organic fruits. If this was solely based on the environmental benefits of the organic fruits (soil is protected from bioaccumulation of chemical substances, groundwater is protected from contamination), consumers would buy the organic alternative in a heartbeat. But this specific example comes with substantial costs. Consumers would probably pay a considerably higher price. Moreover, switching costs are encountered as consumers would have to visit specific selling points such as farmers' market, and thus more time and effort are invested in the purchase. This situation creates a conflict of interest for consumers between their altruistic values (the environment) and egoistic values (their own comfort), which resolves into favoring the choice with less costs and most personal gains (Delmas and Colgan 2018). Thus, consumers are not able to prioritize the environment over their own benefits in consumption choices (ElHaffar et al. 2020), especially when the costs associated with the choice are larger than the personal benefits (Delmas and Colgan 2018).

Another reason for aggravating the green gap can be found in the defense and coping mechanisms developed by consumers to protect their mental wellbeing. Indeed, when there is an inconsistency between their values and their actual behavior, consumers experience a moral discomfort and accumulated feelings of guilt and shame (ElHaffar et al. 2020). To alleviate these negative emotions, consumers tend to justify their unsustainable behavior through neutralization techniques (Atkinson and Kim 2015, Gruber and Schlegelmilch 2014). These techniques include, among others, the denial of responsibility and the defense of the necessity (Gruber and Schlegelmilch 2014). For instance, when consumers don't go the extra mile to purchase bulk instead of canned food, they would blame the inaccessibility of these options and their urgent need to buy and consume this type of food. This is a serious problem, as it can easily turn into patterns of justification, and ultimately turn into indifference about sustainable behavior, establishing and maintaining the inconsistent behavior over time (Gruber and Schlegelmilch 2014).

The green gap has other antecedents that we will discuss separately in this chapter, such as the role of habits, the negative green perceptions, and lack of trust in the green labels.

3.2 Challenge 2: Human Bounded Rationality: Heuristics and Habits in Food Related Decisions

It is estimated that consumers make over 200 food related choices per day (Wansink and Sobel 2007),[2] and if they are to deliberately decide on each of these choices, they would need considerable time and effort. Indeed, most of our food choices are governed by unconscious cognitive processes. These processes are the manifestation of bounded rationality: they allow consumers to make quick and effortless but also 'satisficing' choices (Moser 2016). We will elaborate next on two types of these processes: habits and heuristics.

To begin with, habits are one of the behavioral techniques that facilitate our everyday consumption choices. Habits are repetitive behaviors that individuals perform automatically, without deliberate thinking. Generally, buying and eating food is an unimportant low involvement choice; therefore, when a consumer is used to buying non-sustainable food products, they rarely put effort into changing that, and would find it difficult to abruptly switch these habits (Aertsens et al. 2009). Habits play a crucial role, not only in buying food, but also in the decision phase that precedes the buying behavior. It has been found that habits of unsustainable food consumption can limit the information search towards sustainable food alternatives (Aertsens et al. 2009). This situation contributes to the difficulty of promoting sustainable food innovations and places an additional burden on green companies.

Other techniques which help consumers navigate everyday food choices, while avoiding cognitive effort, are heuristics. When deliberate decisions do arise, such as the choice between natural meat and plant-based meat, consumers base their choice on limited information processing through rules of thumb or heuristics (Siegrist and Hartmann 2020). Heuristics are a practical type of decision-making techniques which bases the choice on limited information, allowing individuals to make a quick decision based on one or few important attributes (Moser 2016, Schulte-Mecklenbeck et al. 2013).

In accepting sustainable food innovations, three main heuristics emerge. First, the affect heuristic comes in play when people rely on the feelings they associate with certain food in evaluating its risks or benefits (Siegrist and Hartmann 2020). For instance, when offered insect-based meals, consumers could feel disgust, and danger, and these emotions would drive a rejection of the food alternative. Second, trust heuristic prevails when individuals evaluate the source of the information instead of the information itself, and decide based on the level of trust they assign to the identified source (Siegrist and Hartmann 2020). This is especially true in food alternatives which are credence goods, and for which the quality of the promised attributes cannot be verified by consumers, even after purchase. For example, business buyers of local food products rely majorly on trust and personal relationships in their choice of local suppliers (Roy et al. 2017). Trust heuristics also play a role in assessing the credibility of environmental claims and eco-friendly labels on food products from the end-consumer (Ricci et al. 2018). We will review the importance of trust later on in this chapter. Lastly, the natural is better heuristic presumes that all human intervention to food products makes it less natural, and consequently less healthy (Siegrist and Hartmann 2020). Hence, consumers associate natural food with better food and prefer to choose them over processed food. Especially when consumers don't have elaborate knowledge on food's natural and environmental values (Siegrist and Hartmann 2020).

Besides habits and heuristics, consumers sometimes choose food rationally, ultimately to maximize the utility derived from it, and other times, they choose it to experience new things as sort

---

[2] As cited by Adamowicz and Swait (2013).

of variety seeking behavior (Adamowicz and Swait 2013). The multiple strategies and occasions of food choice are an additional factor for consideration when designing marketing campaigns, and when targeting consumers of sustainable food innovations.

3.3 Challenge 3: Intangibility Issue

Psychological distance relates to the mental representation of objects and events in the minds of consumers. This distance can spread over four dimensions: time, space, social and hypothetical (Liberman et al. 2007). The more the things are psychologically distant, the more they are construed as abstract in the mind of consumers (Trope et al. 2007). Research has demonstrated that climate change for instance is a psychologically distant issue on many levels (Spence et al. 2012), and by extension, products that are positioned as environmentally friendly are perceived as distant from the self (Segev et al. 2015). This is one of the reasons why consumers are reluctant to implement sustainable consumption into their lives: things that are abstract in their minds, i.e., buying sustainable food, are linked to a more distant time in the future. The consumer therefore *postpones* switching to an indefinite time in the future.

Another issue in this regard is the intangibility of the sustainable attribute. Sustainable foods (organic, local, urban farming, meat alternatives) are part of the credence good markets which are characterized by a high level of information asymmetry where consumers can never really verify if the product actually meets the standards that are promised by the producers (Janssen and Hamm 2012). In many cases, even after the purchase of a product certified as organic or as cruelty free, consumers are kept in the dark about its credibility.

Add to that the effect of green washing on consumers' trust in the sustainable food alternatives. Indeed, with the rapid growth in the eco-friendly products' market, and the trendiness of the green attribute, there is an increase in misleading advertising, which falsely promotes their products as green (Delmas and Burbano 2011).[3] The mere existence of this type of misleading advertising creates green skepticism, and consumer starts doubting every green claim, even the trustworthy ones. When consumers experience green skepticism, they experience anger and frustration (Terra Choice 2010),[4] and their knowledge and concern about the environment decreases (Goh and Balaji 2016). Hence, the accumulation of these three factors (i.e., psychological distance of the green alternative, intangibility of sustainable credence goods, and green skepticism) aggravate the problem of the green gap, and results in consumers renouncing eco-friendly food alternatives.

3.4 Challenge 4: Product Attribute Elusiveness: Naturalness, Health Concerns, Neophobia, and Taste Perception

Food product attributes that contribute to or hinder planetary and human health remain elusive (Dubé et al. 2021). To attain a lower carbon footprint, some food innovations consist of modifying core product attributes such as chemical and biological compositions, which are negatively perceived by consumers and accentuate the risk level associated with consuming these foods. Siegrist and Hartmann (2020) explain this phenomenon through the 'Natural is better' heuristic, where consumers automatically associate human intervention in food production with unnaturalness, harm, and contagion. The unnaturalness is a recurrent argument for rejecting sustainable food alternatives such as cultured meat (Bryant and Barnett 2018, Verbeke et al. 2015) and genetically modified food (Siegrist and Hartmann 2020). Consumers who oppose low-carbon food innovations based on their unnaturalness either believe that consuming this food is dangerous for human health and the environment or consider this kind of technology unethical for its excessive interference with nature (Bryant and Barnett 2018).

---

[3] As cited by Urbański (2020).
[4] As cited by Urbański (2020).

Here, the healthiness concern is primordial as it relates to the nutritional content of the food and the safety consequences of its consumption. For instance, some consumers think that cultured meat is less healthy and nutritious than conventional meat (Bryant and Barnett 2018, Verbeke et al. 2015), or that it is linked to chronic diseases such as cancer (Laestadius and Caldwell 2015).[5] Despite the seriousness of these perceptions, research has shown that they are not detrimental in consumer acceptance (Bryant and Barnett 2018), and that they are subject to change, especially that they derive, not from established facts, but from skeptical anecdotes and uncertain rumors. It is expected that media could alter these perceptions by progressively unveiling the benefits associated with these food alternatives and pinpointing trustworthy organisms' guarantees regarding their safety (Verbeke et al. 2015).

Another matter that pertains to the unnaturalness argument is food neophobia where consumers exhibit a fear towards new foods, and which can be emphasized by the fear of the unknown and unfamiliar technologies used to produce these foods (Siegrist and Hartmann 2020, Verbeke et al. 2015). This is especially the case for cultured meat, genetically modified ingredients, and processed insect-based meals.

4. Behaviorally Informed Solutions

Although consumers' acceptance of novel sustainable foods is challenging for companies, governments and policy makers, behavioral research suggests a plethora of solutions that can weigh considerably in the problem's reconciliation. To begin with, highlighting the personal benefits that consumers derive from their sustainable green consumption has shown immense potential in alleviating the green gap. This approach entails intertwining environmental benefits with self-benefits when communicating and advertising the novel food alternative. It is known as green bundling and has been extensively reviewed and exemplified by Delmas and Colgan (2018) in their book 'the green bundle'. For instance, emphasizing the long-term and short-term health benefits of adopting a plant-based diet (vs. meat-based diet) while also communicating the comparative carbon food-print of each style gives consumers more reason to opt for sustainable meat alternatives.

Companies must also investigate and eliminate the behavioral costs or sludge that prevent consumers from adopting sustainable food alternatives. These behavioral costs include, but are not limited to, price premium, store availability, brand notoriety, and product usability. Consumers who are not familiar with food kits or food sharing apps can benefit greatly from awareness ads and 'how to' videos, breaking the usability barrier.

Awareness plays a crucial role in familiarizing consumers with the mere existence of sustainable food solutions, which represents a fundamental and primordial phase in behavioral change. In this realm, influencer marketing is a promising avenue (Johnstone and Lindh 2018), as it eases the commercialization and acceptance of sustainable food alternatives. It also implements social cues and helps alleviate the uncertainty and ambiguity surrounding the consumption of low-carbon food alternatives (White et al. 2019), especially for new and unfamiliar alternatives such as insect based-meals.

As for the intangibility issues, it is best addressed through communication strategies targeting changing perceptions and attitudes. When communicating the sustainable attribute, advertising should address local and proximal environmental impacts of the products/behaviors in a way that makes the rewards more tangible by individuals (White et al. 2019). Also, implementing virtual communities and gamification of the sustainable behavior into milestones could help minimize the abstractness of the ecological benefits (White et al. 2019).

Finally, and most importantly, companies' credibility is a primordial and pivotal pillar in regaining consumers' trust and encouraging them to undertake sustainable food choices. Trust influences behavior through its positive link to attitude towards the sustainable options, but also through its

---

[5] As cited by (Bryant and Barnett 2018).

influence on attenuating concerns related to such alternatives (Ricci et al. 2018). Companies should develop an authentic brand, build a solid consumer-brand relationship, and inspire transparency to boost consumers' trust. Like in the farmers' market, a short distribution circuit allows consumers to communicate directly with the producer, limit uncertainties, and strengthen transparency. Other strategies include introducing blockchain technologies to enact traceability and will be discussed further in detail in the following section.

5. Embedding Industry 4.0 into Society 5.0 for Ecosystem Transformation

The acceptability challenges that sustainable food innovations present are diverse, complex, and versatile, which entails that addressing them demands a convergent innovation approach (Dubé et al. 2014) rather than a simple 'technological fix' (Keogh et al. 2020). Interestingly, the vision of Society 5.0 marries both technological advances of Industry 4.0 as well as human-centered behavioral knowledge (Keogh et al. 2020), which would fit perfectly in the mix, help overcome the challenges, and generate value through both digital technologies and behavioral creativity. In the following, we will discuss the potential of behavioral change advances and industry 4.0 technologies in bridging the acceptability gap in a convergent innovation perspective focusing on critical challenges.

Industry 4.0 has given rise to new developmental orientations in the economy, and most importantly, it has emphasized the flexibility, decentralization, and individualization of product development and commercialization (Lasi et al. 2014). While human-centered behavioral interventions offer fertile soil for the resolution of several of the acceptability challenges, mainly the attitude-behavior gap and the bounded rationality, the nascent technological advances address the other challenges relating to consumer trust and product attributes.

For starters, the hyper connectivity between the different supply chain nexus, and the real-time surveillance of food conditions made possible by the integration of industry 4.0 technologies, play a pivotal role in harnessing and resurrection of trust in sustainable food products. The internet of things, for instance, coupled with sensors and drone technologies, allows for consistent monitoring of the weather conditions, health, and well-being of soil and crops (Keogh et al. 2020). Additionally, blockchain technology promotes food authentication and distribution transparency (Antonucci et al. 2019), and can restore sustainable foods' credibility, especially but not exclusively, the labeled and processed foods.

Furthermore, technologies could tackle the product attribute problems from naturalness to healthiness and taste perception. To illustrate, consider the potential of 3D printing in molding the internal structures of food products, their dimensions, shapes, appearances, and also taste and personalized formulas (Keogh et al. 2020). As such, vegan or insect-based meat would have an appealing appearance and taste, approximate to those of standard meat. Although this solution addresses the actual taste and not the taste perception, the appropriate behavioral interventions and communication plan would resolve the attribute perception's discrepancy. This interdependence between technological advances and behavioral knowledge will be discussed in what follows.

Table 2. Acceptability challenges and selected solutions from the convergent innovation approach.

	Farming	Production	Distribution/ Commercialization	Consumption	Waste Management
Attitude-Behavior Gap		3D printing	Green Bundling Cost/ Sludge analysis	Influencer Marketing	Predictive analytics
Bounded Rationality			Awareness Influencer Marketing Framing		Gamification
Intangibility and credibility	IoT Blockchain	IoT Blockchain	Framing virtual communities IoT Blockchain	Gamification Authentic branding IoT Blockchain	IoT Blockchain
Product attributes		3D printing	Framing		

6. The Interdependence of Technological and Behavioral Advances in a Convergent Innovation Approach

In a world struggling to meet its sustainability goals, including environmental, economic, and social justice, there is a dire need to 'innovate the way we innovate' and to place the human at the center of the century's transformation (Dubé et al. 2014). While the technological advances offer substantial ecological alternatives and solutions to the food sector, there is still considerable effort that needs to be made to overcome the human hesitancy to come on board this ecological transition. As we demonstrated earlier, technological advances can offer a quick fix to the food sectors' problems. However, this will not be possible without human-centered knowledge and advances in behavioral sciences. This interwoven approach is fundamental to bridging the gap between sustainable innovations and consumers' acceptability, as it meets the convergent innovation approach in which technological, organizational, social, financial, and institutional innovations complement each other (Dubé et al. 2014, Dube et al. 2020).

In the food sector, solutions to ecological crises as well as consumers' acceptability lie at the convergence point where we can address at the same time: what consumers want, what they need, what they can and want to pay for, what the actors in the industry are capable of delivering, and most importantly what the planet can offer sustainably and cost-effectively (Dube et al. 2020). Advanced behavioral analytics are needed to inform business decisions and social innovations to understand consumers' wants and needs. Issues such as the attitude-intention behavior gap can be resolved when we understand the convergence point of this behavioral bias in a context-specific and audience-centered situation. Bounded rationality and habits can be addressed with convenient human-centered knowledge on behavioral change, intertwined with digital applications and targeted advertising powered by big data and digital technologies. The credibility issue can be overcome by introducing blockchain technologies to the labeling process while accounting for consumers' literacy, user-friendliness, and avoiding information overload. Thus, resolving the challenges demands integrating the convergence point perspective into the structure and building blocks of the envisioned solutions.

Technological advances are pivotal in our evolution as a civilization, but without proper integration of human-centered behavioral knowledge, and most importantly, without the assimilation of the fundamental fact that resources of any type are not limitless as previously assumed, our present and future economic and social growth cannot be guaranteed. Taking the food sector as an example, we can readily observe that the technological advances that offered sustainable food alternatives were not enough to resolve the environmental problem. Instead, in this book chapter, we explored the acceptability challenges as human-centered problems preventing the ecological transition from materializing. We advocate addressing these challenges by adopting an integrated convergent innovation approach, marrying both technological advances and deep behavioral insights, aiming and hoping to meet both the well-being of individuals and that of our planet.

References

Adamowicz, W.L. and Swait, J.D. 2013. Are food choices really habitual? Integrating habits, variety-seeking and compensatory choice in a utility-maximizing framework. American Journal of Agricultural Economics 95(1): 17–41.

Aertsens, J., Verbeke, W., Mondelaers, K. and Van Huylenbroeck, G. 2009. Personal determinants of organic food consumption: a review. British Food Journal.

Alsaffar, A.A. 2016. Sustainable diets: The interaction between food industry, nutrition, health and the environment. Food Science and Technology International 22(2): 102–111.

Aneja, V.P., Schlesinger, W.H. and Erisman, J.W. 2008. Farming pollution. Nature Geoscience 1(7): 409–411.

Antonucci, F., Figorilli, S., Costa, C., Pallottino, F., Raso, L. and Menesatti, P. 2019. A review on blockchain applications in the agri-food sector. Journal of the Science of Food and Agriculture 99(14): 6129–6138.

Atkinson, L. and Kim, Y. 2015. "I drink it anyway and I know I shouldn't": Understanding green consumers' positive evaluations of norm-violating non-green products and misleading green advertising. Environmental Communication 9(1): 37–57.

Boukid, F. 2021. Plant-based meat analogues: From niche to mainstream. European Food Research and Technology 247(2): 297–308.

Bryant, C. and Barnett, J. 2018. Consumer acceptance of cultured meat: A systematic review. Meat Science 143: 8–17.

Chekima, B., Igau, A., Wafa, S.A.W.S.K. and Chekima, K. 2017. Narrowing the gap: Factors driving organic food consumption. Journal of Cleaner Production 166: 1438–1447.

Crist, E., Mora, C. and Engelman, R. 2017. The interaction of human population, food production and biodiversity protection. Science 356(6335): 260–264.

Dagevos, H. 2021. A literature review of consumer research on edible insects: Recent evidence and new vistas from 2019 studies. Journal of Insects as Food and Feed 7(3): 249–259.

Delmas, M.A. and Burbano, V.C. 2011. The drivers of greenwashing. California Management Review 54(1): 64–87.

Delmas, M.A. and Colgan, D. 2018. The green bundle: Pairing the market with the planet: Stanford University Press.

Dubé, L., Jha, S., Faber, A., Struben, J., London, T., Mohapatra, A., et al. 2014. Convergent Innovation for Sustainable Economic Growth and Affordable Universal Health Care: Innovating the way we Innovate.

Dube, L., Wolfert, S., Zimmerman, K., Yang, N., Diaz-Lopez, F., Arvanitis, R., et al. 2020. Convergence research and innovation digital backbone: Behavioral analytics, artificial intelligence and digital technologies as bridges between biological, social and agri-food systems. In How is Digitalization Affecting Agri-food? (pp. 111–125): Routledge.

Dubé, L., Soman, D. and Almeida, F. 2021. Precision retailing: building upon design thinking for societal-scale food convergence innovation and well-being. In Design Thinking for Food Well-Being (pp. 227–245): Springer.

EDC, E.D.C. 2020. The organic food market in Canada and its global influence. Retrieved from https://www.edc.ca/en/guide/canada-organic-report.html.

ElHaffar, G., Durif, F. and Dubé, L. 2020. Towards closing the attitude-intention-behavior gap in green consumption: a narrative review of the literature and an overview of future research directions. Journal of Cleaner Production, 122556.

Falcone, P.M. and Imbert, E. 2017. Bringing a sharing economy approach into the food sector: The potential of food sharing for reducing food waste. In Food waste reduction and valorisation (pp. 197–214): Springer.

Goh, S.K. and Balaji, M. 2016. Linking green skepticism to green purchase behavior. Journal of Cleaner Production 131: 629–638.

Gruber, V. and Schlegelmilch, B.B. 2014. How techniques of neutralization legitimize norm-and attitude-inconsistent consumer behavior. Journal of Business Ethics 121(1): 29–45.

Gupta, S. and Coskun, M. 2021. The influence of human crowding and store messiness on consumer purchase intention–the role of contamination and scarcity perceptions. Journal of Retailing and Consumer Services 61: 102511.

Hartmann, C. and Siegrist, M. 2017a. Consumer perception and behaviour regarding sustainable protein consumption: A systematic review. Trends in Food Science and Technology 61: 11–25.

Hartmann, C. and Siegrist, M. 2017b. Insects as food: Perception and acceptance. Findings from Current Research. Ernahrungs Umschau 64(3): 44–50.

Janssen, M. and Hamm, U. 2012. Product labelling in the market for organic food: Consumer preferences and willingness-to-pay for different organic certification logos. Food Quality and Preference 25(1): 9–22.

Jeevahan, J.J., Chandrasekaran, M., Venkatesan, S., Sriram, V., Joseph, G.B., Mageshwaran, G., et al. 2020. Scaling up difficulties and commercial aspects of edible films for food packaging: A review. Trends in Food Science and Technology 100: 210–222.

Johnstone, L. and Lindh, C. 2018. The sustainability-age dilemma: A theory of (un) planned behaviour via influencers. Journal of Consumer Behaviour 17(1): e127–e139.

Keogh, J.G., Dube, L., Rejeb, A., Hand, K.J., Khan, N. and Dean, K. 2020. The Future Food Chain: Digitization as an Enabler of Society 5.0. Building the Future of Food Safety Technology (1st Edition): London, Oxford, UK.

Laestadius, L.I. and Caldwell, M.A. 2015. Is the future of meat palatable? Perceptions of *in vitro* meat as evidenced by online news comments. Public Health Nutrition 18(13): 2457–2467.

Lasi, H., Fettke, P., Kemper, H.-G., Feld, T. and Hoffmann, M. 2014. Industry 4.0. Business and Information Systems Engineering 6(4): 239–242.

Liberman, N., Trope, Y. and Stephan, E. 2007. Psychological distance. Social psychology: Handbook of basic Principles 2: 353–383.

Lynch, J. and Pierrehumbert, R. 2019. Climate impacts of cultured meat and beef cattle. Frontiers in Sustainable Food Systems, 3: 5.

Magnier, L. and Crié, D. 2015. Communicating packaging eco-friendliness: An exploration of consumers' perceptions of eco-designed packaging. International Journal of Retail and Distribution Management.

Moser, A.K. 2016. Buying organic–decision-making heuristics and empirical evidence from Germany. Journal of Consumer Marketing.

Mu, W., Spaargaren, G. and Oude Lansink, A. 2019. Mobile apps for green food practices and the role for consumers: a case study on dining out practices with Chinese and Dutch young consumers. Sustainability 11(5): 1275.

Nijdam, D., Rood, T. and Westhoek, H. 2012. The price of protein: Review of land use and carbon footprints from life cycle assessments of animal food products and their substitutes. Food Policy 37(6): 760–770.

Nuttavuthisit, K. and Thøgersen, J. 2017. The importance of consumer trust for the emergence of a market for green products: The case of organic food. Journal of Business Ethics 140(2): 323–337.

Rana, J. and Paul, J. 2020. Health motive and the purchase of organic food: A meta-analytic review. International Journal of Consumer Studies 44(2): 162–171.

Ricci, E.C., Banterle, A. and Stranieri, S. 2018. Trust to go green: an exploration of consumer intentions for eco-friendly convenience food. Ecological Economics 148: 54–65.

Ripple, W.J., Wolf, C., Newsome, T.M., Galetti, M., Alamgir, M., Crist, E., et al. 2017. World scientists' warning to humanity: a second notice. BioScience 67(12): 1026–1028.

Roy, H., Hall, C.M. and Ballantine, P.W. 2017. Trust in local food networks: The role of trust among tourism stakeholders and their impacts in purchasing decisions. Journal of Destination Marketing and Management 6(4): 309–317.

Saget, S., Costa, M.P., Santos, C.S., Vasconcelos, M., Styles, D. and Williams, M. 2021. Comparative life cycle assessment of plant and beef-based patties, including carbon opportunity costs. Sustainable Production and Consumption 28: 936–952.

Schnitzer, H., Muster-Slawitsch, B. and Brunner, C. 2010. Low Carbon Solutions for the Food Industry.

Schulte-Mecklenbeck, M., Sohn, M., de Bellis, E., Martin, N. and Hertwig, R. 2013. A lack of appetite for information and computation. Simple heuristics in food choice. Appetite 71: 242–251.

Segev, S., Fernandes, J. and Wang, W. 2015. The effects of gain versus loss message framing and point of reference on consumer responses to green advertising. Journal of Current Issues and Research in Advertising 36(1): 35–51.

Siegrist, M. and Hartmann, C. 2020. Consumer acceptance of novel food technologies. Nature Food 1(6): 343–350.

Slade, P. 2018. If you build it, will they eat it? Consumer preferences for plant-based and cultured meat burgers. Appetite 125: 428–437.

Spence, A., Poortinga, W. and Pidgeon, N. 2012. The psychological distance of climate change. Risk Analysis: An International Journal 32(6): 957–972.

Tonkin, E., Wilson, A.M., Coveney, J., Webb, T. and Meyer, S.B. 2015. Trust in and through labelling–a systematic review and critique. British Food Journal.

Trope, Y., Liberman, N. and Wakslak, C. 2007. Construal levels and psychological distance: Effects on representation, prediction, evaluation and behavior. Journal of Consumer Psychology 17(2): 83–95.

Urbański, M. 2020. Are you environmentally conscious enough to differentiate between Greenwashed and sustainable items? A global consumers perspective. Sustainability 12(5): 1786.

Verbeke, W., Sans, P. and Van Loo, E.J. 2015. Challenges and prospects for consumer acceptance of cultured meat. Journal of Integrative Agriculture 14(2): 285–294.

White, K., Habib, R. and Hardisty, D.J. 2019. How to SHIFT consumer behaviors to be more sustainable: A literature review and guiding framework. Journal of Marketing 83(3): 22–49.

Wunsch, N.-G. 2021. Organic share of total food sales in the United States from 2008 to 2019. Retrieved from https://www.statista.com/statistics/244393/share-of-organic-sales-in-the-united-states/

15

Food Process Simulation and Techno-Economic Assessment in Sustainable Food Manufacturing

Alexandros Koulouris,[1,*] *Nikiforos Misailidis,*[2] *Avraam Roussos,*[2] *Jim Prentzas*[2] and *Demetri P. Petrides*[3]

1. Introduction

The main objective of process design is the determination of the type and quantity of resources (process equipment, materials, utilities, consumables, labor and capital) required to fulfill the designed plant operational needs at the desired throughput. Especially in food processing, design must also account for food safety and quality and comply with hygienic regulations. Process design activities can be undertaken for greenfield plants or the expansion/retrofit of existing ones. Process synthesis is a process design activity aiming at selecting the "optimal" flowsheet when multiple alternative flowsheets are possible. In food processing, the availability of alternative flowsheets for producing the same product may be limited due to the strict processing conditions and food sensitivity, safety and hygienic considerations (Maroulis and Saravacos 2003).

In the route to process or product commercialization, process design can be undertaken in various stages of varying detail and accuracy starting from the conceptual design (which aims at estimating the technological and economic feasibility of a new process/product idea), all the way to the actual plant construction.

Process simulation is an enabling technology that aims at predicting the process outputs for given flowsheet and process inputs (Foo et al. 2017). This is possible with the use of operation models that simulate the behavior of actual processes at given operating conditions. These mathematical models can be derived from first-principles (i.e., applying the laws of physics and chemistry that describe the underlying phenomena) or be represented by empirical equations that capture the process input-output dependencies.

---

[1] International Hellenic University, Dept. of Food Science and Technology, Thessaloniki, Greece.
[2] Intelligen Europe, Greece.
   Emails: nmisailidis@intelligen.com; aroussos@intelligen.com; jprentzas@intelligen.com
[3] Intelligen, Inc., USA.
   Email: dpetrides@intelligen.com
* Corresponding author: akoul@ihu.gr

Irrespective of the origin of the modeling equations, the solution of the process simulation model for a given flowsheet yields information about the material and energy balances (flow, composition, physical state and conditions of all material streams and states in the process), the type and required size of equipment, the capacity of supporting utility systems and the type and flow of wastes generated. If the process is first developed at a lab or pilot-plant level, process simulation facilitates process transfer to the scaled-up manufacturing level.

With the above information available, it is possible to perform an economic analysis study by estimating the required capital investment and operating costs, the expected revenues and the value of economic indices assessing the feasibility of the project. In the context of process synthesis and in an attempt to produce the optimal design solution, this process can be repeated for alternative flowsheets generated by varying the operating parameters or the structure of the flowsheet.

Process simulation can also be used for existing facilities for on-going optimization of the plant operation, identification and elimination of process bottlenecks, to facilitate decisions related to in-house manufacturing *versus* outsourcing and for production planning and scheduling.

Sensitivity analyses are greatly facilitated by process simulation tools as well. The objective of these studies is to evaluate the impact of critical process parameters on key performance indicators (KPIs) such as cycle times, plant throughput, and production cost.

Process simulation tools have been used for many decades in the chemical and petrochemical industry where small operational improvements translate to significant operating cost savings. Food industry, on the other hand, was slower in adopting this technology. However, low profit margins, short lead times, the ever-increasing palette of products, the demand for timely production and delivery due to the perishable nature of most foods are some of the factors that force the food industry into adopting more advanced digital technologies in the spirit of the Industry 4.0 initiative (Koulouris et al. 2021). In order to increase product safety and quality, minimize costs and assure timely delivery of orders, it is necessary to optimize the food process both at the design as well as at the operation level; process simulation can play a major role towards achieving these objectives.

The implementation of process simulation in the food industry depends on the type of processing and product characteristics. According to Maroulis and Saravacos (2008), food processing plants can be divided into three major groups: (1) Food Preservation Plants that utilize agricultural raw materials and implement preservation processes, (2) Food Ingredient Manufacturing Plants that use commodity agricultural products or by-products to produce various food ingredients, and, (3) Food Product Manufacturing Plants that produce a multitude of packaged food products ready for consumer use. By their nature, food ingredient manufacturing plants are much closer to traditional chemical processing utilizing many different unit operations and producing large quantities of bulk products; therefore, these types of industries are more apt to adopting process simulation tools. An important and ever-growing subset of the ingredient manufacturing plants category is composed of the food waste valorization industries that utilize food waste as inputs in order to produce a multitude of useful end products.

Food waste valorization is one of the most important areas in the food processing and production research. It is estimated that food production is responsible for about 26% of the global carbon dioxide equivalent emissions, corresponding to about 14 billion tons per year. This consists of the emissions of the entire supply chain including deforestation, agriculture, cultivation, farming and the related energy requirements, as well as food ingredients and food products manufacturing and the relevant wastewater treatment and disposal emission generated. Livestock and fisheries, crop production, land use and supply chains contribute 31%, 27%, 24% and 18% of the total food greenhouse gas emissions, respectively (Ritchie and Roser 2020).

In addition, a number of studies estimate that food waste is about one third to half of the global food production (Gustavsson et al. 2013, Stenmarck et al. 2016). This includes the edible and non-edible food waste in all the supply chain elements such as residues of agricultural production, post-harvest handling and storage, as well as food waste generated in the utilization elements such as food processing and packaging, distribution, retailers and consumption by the households. The developed countries typically waste more in the later stages of the supply chain such as in households and retailers, while the developing countries typically waste more in the early stages of agriculture and post-harvest handling and storage. The percentage of food waste is also highly variable with respect to the food category or commodity group (Gustavsson et al. 2013, Murugan et al. 2013). During the food processing stage, food waste is between 12–41% (Caldeira et al. 2017). Moates et al. (2016) provide a list of the top 20 food waste streams, which include spent grains and other organic waste streams from the production of alcoholic beverages, press cakes from vegetable oil processing and meat and dairy side streams, such as slaughter by-products and whey protein.

In order to reduce food waste in general, and during food processing in particular, broadly accepted waste management hierarchies such as the Lansink ladder can be applied to determine the order of the most favorable actions. This is frequently referred to as Reduce, Reuse, Recycle (3R). After recycling, the preferred action is to produce energy and/or to burn (incinerate) and then finally to dispose. This scheme, in fact, captures the efforts of food research and food processing industries. For many food processing industries, the focus is primarily on reducing the processing steps, the amount of solvents and other raw materials used as well as the energy requirements. In parallel, the yields and efficiencies of the main products are maximized and the by-products' streams are reduced. This is also achieved by extracting any valuable components from the by-products. Following this, potential markets are established for the by-products, such as animal feed. Finally, if all of the above options are explored and there are still considerable amounts of by-products and wastes, then the focus is on how to produce energy with these. There are many available options depending on the food process, the type of by-products and the chemical composition and moisture content of the relevant streams.

Particular food processing industries have different challenges and waste minimization approaches within the broader concept of food valorization. Cereal wet milling processes typically fractionate cereal grains into gluten, starch and low value fiber rich by-products used for animal feed. These processes primarily aim to operate at maximum yields, while maintaining quality specifications of the main products. In the corn wet milling, the process separates the germ, the bran, the gluten and starch of the grains in this particular sequence. In an effort to minimize water usage in the plant, fresh water is only added at the final step of the fractionation that washes the natural starch. From there, water is distributed back to the process to support the fractionation and wash the other products of the plant with a counter-current flow. Ultimately, the water is used to the initial corn processing step, which is the steeping of the grains. The water solution at this stage has effectively picked up all the solubles from the process and is sent to evaporation for concentration, producing the corn steep liquor (CSL). The CSL is then mixed with the solid bran particles to form the corn feed gluten, a by-product which is used as animal feed. Many other fermentation industries use CSL to enhance their fermentation performance, since it contains a number of soluble proteins and nutrients that facilitate microbial growth. This counter-current flow of water effectively minimizes the water usage in the plant (Reduce), and reuses the same solution in many steps, while specific markets absorb the low value by-product that is ultimately produced (Reuse and Recycle).

In some cases, similar dilute by-product wet streams of food processes may in fact be considered and treated as wastewater. Their potential value, if concentrated, dried and sold as low value by-product (e.g., for animal feed), might not even cover the associated energy cost. On the other hand, if these streams are not valorised they have an associated wastewater treatment cost. These streams are used for the production of energy, via anaerobic digestion to produce biogas, and then upgrade and burn the biogas to produce energy (electricity and steam). More concentrated low value streams with suitable chemical composition might be used for bioethanol production. So, there exist many options that have been adopted by different plants depending on the chemical composition and water content of these streams, the cost of water and wastewater treatment, the location of the plant and, ultimately, the availability of capital or subsidies for supporting waste-to-energy investments.

Other by-product streams might be obtained dry from the process. In this case, burning them directly to produce energy (e.g., steam) might be possible. For example, in the production of sunflower oil, the dry separated seed hulls are burned in order to produce energy to cover the process demand. Other non-edible lignocellulosic agricultural residues, such as wheat straw and corn cob, can be used as feedstocks for production of biofuels (ethanol or butanol), an indirect way to convert them to energy. Furthermore, non-edible woody or lignocellusic feedstocks can be biochemically or thermochemically converted (e.g., via pyrolysis or gasification) to many different forms of energy. Some innovative integration options of this type are summarized by project BRISK II (Duarte et al. 2020).

Food processing by-product streams are mixtures of many substances, some of which might have specific functionalities when used as food ingredients by themselves, and are, therefore, valuable. Extracting these molecules from the waste streams is the preferred food waste valorization method since it adds more value to a by-product stream than using it for producing energy (directly or indirectly), or even as animal feed. There are existing processes that partially address this valorization approach (a review of which is provided, among others, by Moates et al. (2016)). For example, citrus fruit peels or apple pomace, both of which are by-products of juice production, are utilized as the feedstock for the production of pectins. Examples of other waste valorization processes on existing by-product streams to produce a variety of products can be found in Mirabella et al. (2013), Dimou et al. (2019), Saini et al. (2019), Pires et al. (2021), Sharma et al. (2021) and many other research papers. Frequently, these novel processes are developed in a lab scale and then upscaled to industrial level using process simulation tools, in order to perform a Technological Economical Analysis (TEA) and evaluate the process economics and the feasibility of such a process. These include chitosan production from shrimp shell waste (Gómez-Ríos et al. 2017), arabinoxylans in the context of a wheat biorefinery aiming mainly at producing bioethanol (Misailidis et al. 2009), co-extraction of gelatin and lipids from fish skins (Karayannakidis et al. 2015), production of xylitol, lactic acid, activated carbon and phenolic acids from brewer's spent grain (Mussato et al. 2013), production of plasticizer, lactic acid and animal feed from food waste (Kwan et al. 2015, 2018), biodiesel and ethanol co-production from lipid-producing sugarcane (Huang et al. 2016), anthocyanin extraction and ethanol production from blue and purple corn (Somavat et al. 2018), mango processing waste integrated biorefinery (Arora et al. 2018), astaxanthin production from agro-industrial wastes (Dursun et al. 2020), integration of whey by-products in cheese-making processing (Gomez et al. 2020), and many others.

In most of the above references (and in many others not presented here), process simulation is used to assess the techno-economic feasibility of large scale production that encompasses the proposed technologies. Food waste may contain many valuable substances and the question which one of those is technically and economically feasible and environmentally meaningful to extract does not have an obvious answer. At the same time, there may exist multiple alternative technologies for processing the same waste leading to different products or uses and determining the optimal split of the waste throughput between the alternative processes is a frequent question that arises. For all these reasons, computer-aided process design is a valuable tool in assessing food waste valorization processes.

In this chapter, two food waste valorization cases are presented, simulated and analyzed: a pectin producing plant from citrus peels and an isobutanol producing plant from corn stover. These processes were simulated with the use of SuperPro Designer v12 by Intelligen, Inc. (www.intelligen.com). SuperPro Designer can be used to efficiently model any processing mode (batch, semi-batch or continuous) in a wide variety of industries including food and food waste valorization processing.

The two case studies represent different operation modes (continuous operation for the pectin case, batch for the isobutanol case), different processing approaches (chemical for pectin vs. bio-chemical for isobutanol) and generate products of different nature and use (food ingredient in the pectin case, bio-fuel in the isobutanol case).

2. Process Simulation Case Studies

2.1 Pectin from Citrus Peels

This case study demonstrates the development and analysis of a simulation model for the production of pectin from citrus peels. Pectin substances are complex mixtures of polysaccharide polymers found in most plants in varying concentrations. They behave like stabilizing gels and contribute to both the adhesion between the cells and to the mechanical strength of the cell wall. Typical molecular weights of pectins are within a range of 50–150 kDa. Commercial pectin is typically extracted either from citrus fruit peels or from apple pomace (Van Buren 1993).

Pectins have found many applications in the food industry mainly due to their rheological properties; they are used primarily as gelling agents, and secondarily as texturizers, emulsifiers, thickeners, and stabilizers. When pectin is dissolved in a hot aqueous mixture of sugars and then cooled down below the gelling temperature, a gel will form. The strength of the gel depends on parameters such as the type of pectin, the content and types of sugars, the pH of the solution, and the presence and concentration of bivalent salts, especially calcium. The basic commercial categorization of pectins depends on whether they are highly esterified (high methoxyl or HM pectins) or not (low methoxyl or LM pectins) (Hoefler 1991).

In the food industry, HM pectin is used as ingredient in jams and jellies, while LM pectins can be used in low calorie foods with less sugar. Pectins are also used as fat and/or sugar replacers in dairy products, desserts, soft drinks, and pharmaceuticals. For instance, pectins stabilize acidic protein drinks such as drinkable yogurt, and improve the mouth-feel and pulp stability in juices. Pectin also acts as a fat substitute in baked goods. In the pharmaceutical industry, it is used to reduce blood cholesterol levels and gastrointestinal disorders. Other applications of pectin include use in edible films, paper substitutes, foams and plasticizers.

The global market demand for pectin and other thickening agents is expanding. This trend is mainly due to the evolution of eating habits globally as people in developed and developing countries prefer processed foods and particular textures. The global pectin market in 2015 was estimated at about $964 million, and it is expected to grow at a 7.1% compound annual growth rate (CAGR) over the next nine years. The food and beverage industry is the largest pectin consumer in the USA (https://www.grandviewresearch.com/industry-analysis/pectin-market, accessed Nov. 2021).

The global production of all citrus fruits is more than 100 million tons per year. The juice industry uses more than 30% of the fruit production, from which it wastes about 50% as peels, seeds, segment membranes and generates about 10–25 million tons per year of waste. Citrus peels have a moisture content of about 80%. The remaining solid is composed mostly by fibers, which are primarily pectins but also hemicellulose, cellulose and lignin, soluble sugars, proteins, starch and others. The peels also contain about 0.5% of essential oils, of which about 90% is D-Limonene. The current industry valorization practises use a considerable part of available citrus peels to extract and produce pectins, and also essential oils, while the rest ends up as animal feed and disposed in landfills. Novel uses and processing of citrus peels include the extraction of cellulose, sugars, polyphenols, flavonoids and other antioxidant compounds via physical and chemical methods (Patsalou et al. 2020), or thermochemical processing for the production of bio-oil and biochar or syngas and hydrogen, via pyrolysis (Kwon et al. 2019, Abdelaal et al. 2020) and gasification (Chiodo et al. 2016).

The sections that follow describe and analyze a pectin production process from orange peels modeled in SuperPro Designer. The development of the model was based on data available in the patent and technical literature and the authors' experience. The modeled plant runs in continuous mode, processes 35000 kg/h of peels and produces around 645 kg/h of pectin.

Table 1 shows the assumed composition of the fed peels.

Table 1. Composition of the orange peels.

Component	(%w/w)
Ash	0.3
Cellulose	3.6
Fats	0.8
Flavonoids	0.4
Hemicellulose	1.4
Lignin	1
Pectin	4.2
Proteins	1.3
Soluble Sugars	7
Water	80

2.1.1 Process Description

The process is divided into six sections, i.e., 'Peel Preparation and Sugar Wash', 'Sugars Processing', 'Pectin Extraction', 'Pectin Purification', 'Pectin Precipitation' and 'IPA Recovery'. Sections in SuperPro Designer represent groups of processing tasks (called *unit procedures*) that serve a common processing objective. Because of its large size, the flowsheet is presented in different flow charts representing different sections. In the description that follows, the procedures are referred by their IDs (names) as shown in the relevant charts. Note that all procedures in SuperPro Designer are tagged by their ID, the hosting equipment ID and a textual description of the procedure. All these three tags are shown below every procedure icon in the flowcharts.

2.1.1.1 Peel Preparation and Sugar Wash

In this section (Fig. 1), the peels are ground and washed to remove all the soluble solids. The peels are fed to the plant through silo P-1, mixed with a water recycle stream in P-2 and wet ground in P-3. The slurry is then sent to a tank (P-4) where it is mixed with another recycled aqueous stream to allow soluble components to dissolve in the liquid phase. The slurry is then fed to the screw

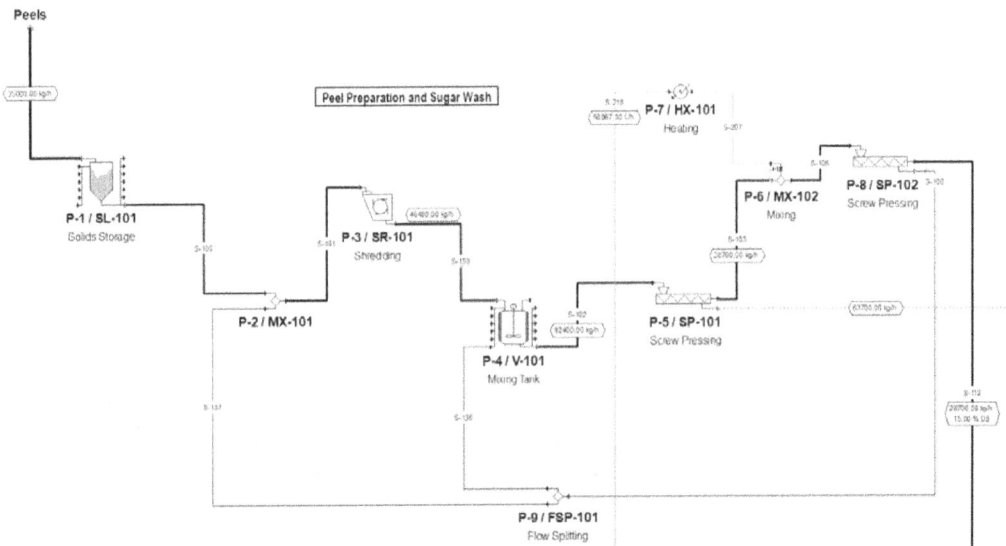

Figure 1. Flowchart of the Peel Preparation and Sugar Wash Section.

press (P-5) where the liquid phase is separated and sent to the sugar processing section. The cake from the press is mixed in P-6 with pre-heated (in P-7) recycled process water and sent to a second screw press (P-8). The liquid from the second press is recycled back via a splitter (P-9) to the mixing tank (P-4) and to the mixing procedure P-2. Note that there is a countercurrent pattern to the washing steps; fresh process water washes the cake in the final screw press, washing out most of the remaining solubles from the cake, and then the filtrate is sent back to the previous screw press. In reality, there exist more than two washing/pressing loops but the additional ones are omitted for the sake of simplicity.

2.1.1.2 Sugars Processing

The aqueous solution from the first screw press (P-5) containing most of the solubles is stored, then heated to 50°C and fed into an enzymatic treatment tank (P-12) as shown in Fig. 2. It is then pasteurized in P-13 (heated to 75°C and then cooled down to 50°C) and centrifuged (P-14). The centrifuge separates any remaining solids (sent back to pectin extraction) and sends the liquor to a tank (P-15) which acts as a balance tank for the evaporator (P-18). Before being fed to the evaporator, the liquor is preheated (in P-16) with the evaporator condensate and heated further with steam (in P-17). In a 4-effect evaporator (P-18), water is evaporated to reach a final dry solids (DS) concentration of 75%. The concentrated solution is then cooled down to 30°C in a cooler (P-19). This sugar solution is a by-product called Vinasse, and it is sold to the fermentation industries. The vapor from the final stage of the evaporator is condensed (in P-20), mixed (in P-22) with the condensates from the first 3 evaporator effects and used to preheat the feed and send to the water distribution system of the plant.

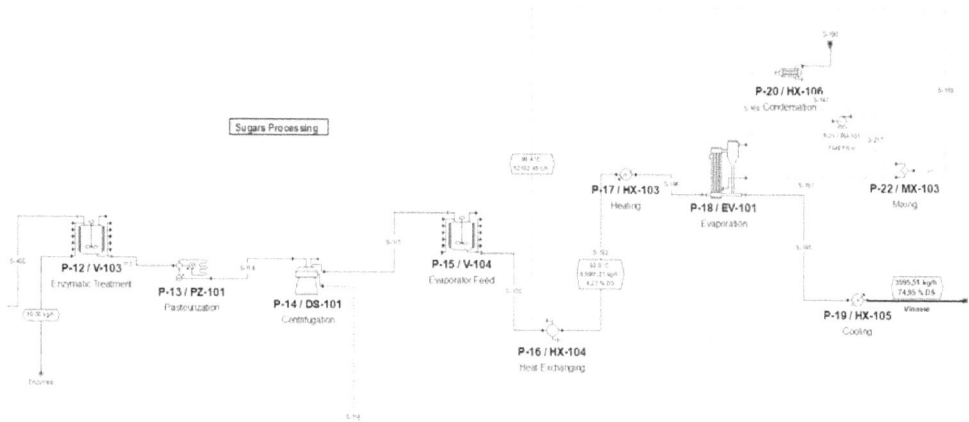

Figure 2. Flowchart of the Sugars Processing Section.

2.1.1.3 Pectin Extraction

The solids from the final screw press (P-8) that contain all the pectin are heated (in P-28) to 60°C and loaded into reactor P-29 (see Fig. 3). The reactor operates in batch mode and includes the following operations that are executed sequentially: charge of solids, charge of hot water, react, charge of sodium carbonate and transfer-out. Hot water is added to achieve a final water concentration of about 95%. Under the acidic conditions and high temperature in the reactor, pectins become soluble and are extracted from the solid to the liquid phase. However, a considerable quantity of pectin is also decomposed to various impurities. Similarly, other plant polymers such as cellulose and hemicellulose also decompose. These decompositions were modeled as reactions with the following mass stoichiometries:

1 Pectin → 1 PEC_W (40% conversion)

1 Pectin → 1 IMP (60% conversion)

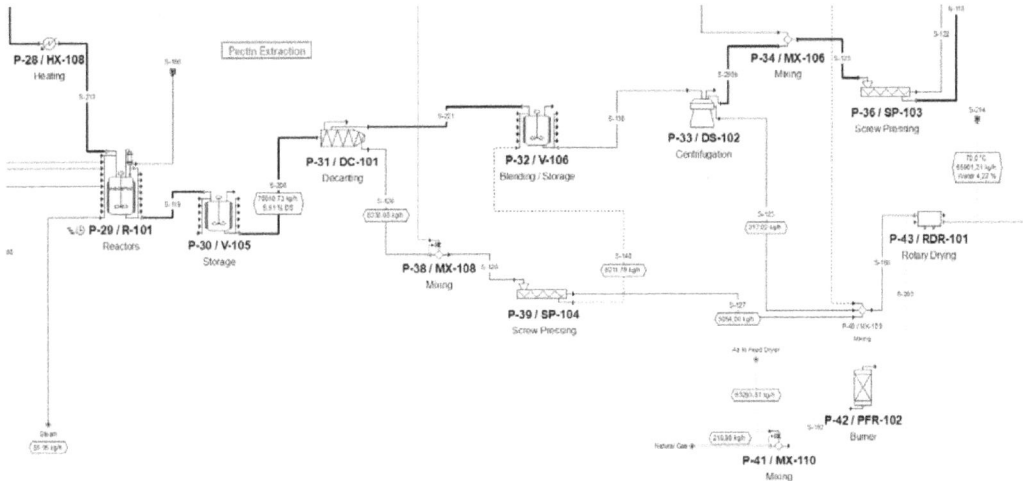

Figure 3. Flowchart of the Pectin Extraction Section.

198 Cellulose → 180 Glucose + 18 Water (10% conversion)

300 Hemicellulose → 82.5 Arabinose + 1.5 HMF + 36 Water + 180 Xylose (10% conversion)

PEC_W is the water solubilized pectin, while IMP and HMF (hydroxymethylfurfural) present impurities and furfurals, respectively. Pectin solubilization stops after sodium carbonate is added.

The solution is then transferred out to the buffer tank (P-30), which ensures constant flowrate to feed the downstream process. The slurry is sent to the decanter (P-31), where most of the solids are removed. The aqueous stream from the decanter is sent to another tank (P-32) which feeds a disc stack centrifuge (P-33) where most of the remaining solids are removed. The light stream from the centrifuge is then combined with filter aid (via P-34) in order to remove almost all remaining solids using a screw press (P-36). The filtered solution is then sent to another tank (P-37). Meanwhile, the solids collected from the decanter (P-31) are washed with water in order to recover any solubilized pectin from the solids stream. This is achieved by mixing water with the solids stream in a custom mixer (P-38), and then pressed in a screw press (P-39). The filtrate stream from the press is fed to the same tank (P-32) as the decanter's aqueous stream, while the cake is combined with the other suspended solids' streams and then sent to a rotary dryer (P-43). The dryer is direct fired and uses natural gas to heat the dryer inlet air (via the custom mixer P-41 and the burner P-42). The burner was modeled as a plug flow reactor (PFR), with two reactions to represent the combustion of the natural gas's methane and ethane. The reactions have the following molar stoichiometry:

1 Ethane + 3.5 Oxygen → 2 Carbon Dioxide + 3 Water (100% conversion)

1 Methane + 2 Oxygen → 1 Carbon Dioxide + 2 Water (100% conversion)

The dried solids which exit P-43 are sold as animal feed with a moisture content of about 10%.

2.1.1.4 Pectin Purification

In this section (Fig. 4), the filtrate from the screw press (P-36) containing the solubilized pectins is sent to a buffer tank (P-37), and then loaded onto a purification granular activated carbon (GAC) column (P-47). The column procedure has the following operations: loading, washing with NaOH and final washing with water.

During the load operation, the column activated carbon binds the impurities and the furfurals formed during extraction, and some of the flavonoids originating from the peels. The column is then washed with a sodium hydroxide solution (from mixer P-46) and with water (from the water distribution system, P-57). The output from P-47 is then loaded onto a mixed ion exchange column (P-51).

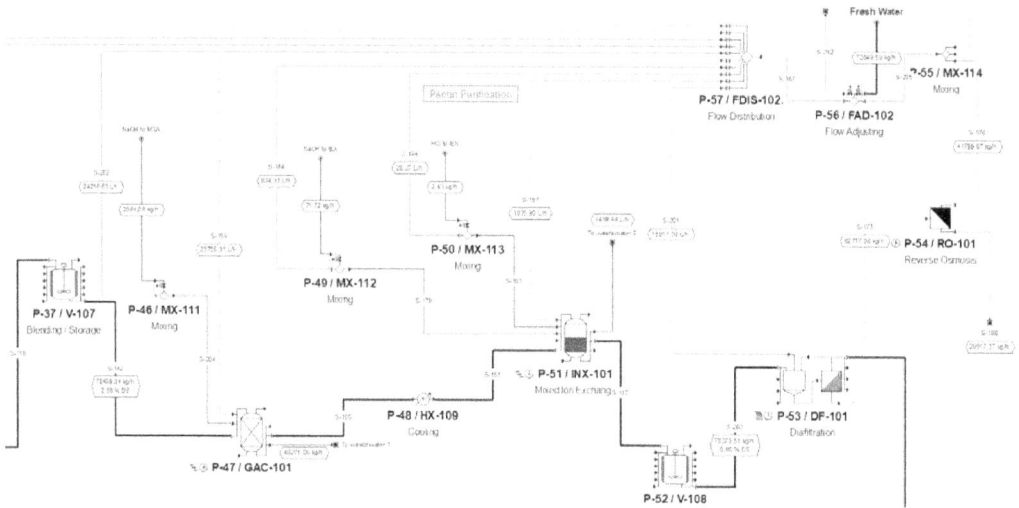

Figure 4. Flowchart of the Pectin Purification Section.

The following operations are executed sequentially in the column: load, wash with water, regenerate with HCl, regenerate with NaOH and wash with water. This column's resin binds the inlet stream's ions (ash, sodium carbonate, nitric acid) during its load operation; the ions are then removed during regeneration with sodium hydroxide and hydrochloric acid solutions which are prepared in custom mixers P-49 and P-50, respectively. The water needed for the column operations is supplied by distributor P-57.

The last purification step is a diafiltration (P-53). The membrane in this unit initially concentrates the pectin solution and then dia-filters the concentrate with water in order to wash out remaining small molecule impurities. The filtrate is then concentrated by a reverse osmosis membrane (P-54), in order to recover and reuse the water. This purified water is sent to a mixer (P-55) where it is combined with the condensate of the sugar solution evaporator (P-18) and then recycled back to the process (via flow adjusting unit P-56 and flow distributor P-57).

2.1.1.5 Pectin Precipitation

The purified pectin solution is sent to an evaporator feed tank (P-58) and through a preheater (P-59) to a four-stage evaporator (P-60) as shown in Fig. 5. The vapor from the last stage is condensed in a condenser (P-61) and pumped to a mixer (P-63). There, it is combined with the liquid condensate from P-60 and recycled back to the pectin extraction section for use in the reactors.

The purified and concentrated pectin solution is then mixed with a recycled aqueous solution of isopropanol (IPA) in P-64. The desired IPA concentration is adjusted using a custom mixer (P-65) and sent to the pectin precipitation reactor (P-66). In this plug flow reactor, the precipitation of the solubilized pectins was modeled as a reaction with the following stoichiometry:

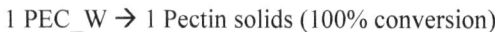

1 PEC_W → 1 Pectin solids (100% conversion)

The pectin precipitate is then separated in a decanter centrifuge (P-67). More IPA solution is then added to the pectin solids using a custom mixer (P-68). The solution is then pressed (in P-69) in order to concentrate and recover the pectin solids. The press uses additional IPA to wash the cake. The filtrate of the press, composed of aqueous IPA solution and trace amounts of washed-out impurities, is recycled back for use in pectin precipitation. Finally, the pectin solids are dried in a dryer (P-70) and then cooled (P-71). This is the main product of the process.

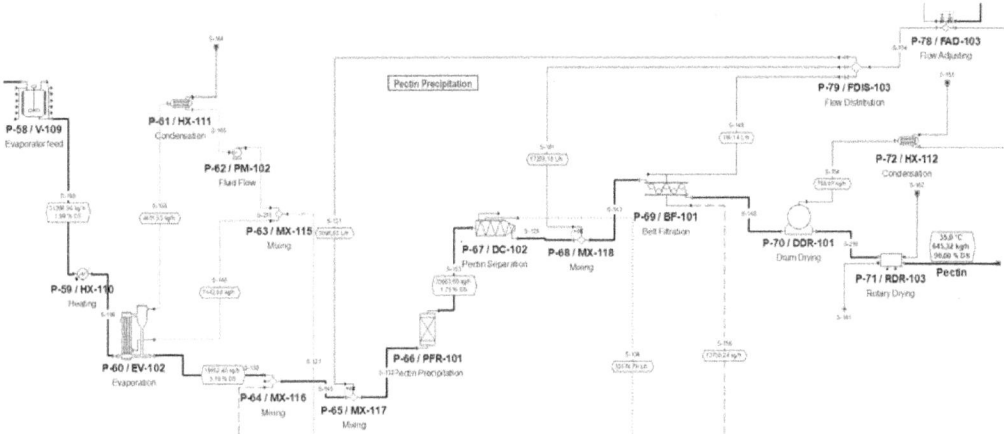

Figure 5. Flowchart of the Pectin Precipitation Section.

2.1.1.6 IPA Recovery

The vapors from the P-70 dryer (Fig. 6) are condensed in a condenser (P-72), combined with the filtrate of the pectin separation centrifuge, and sent to the distillation feed tank (P-73) in the IPA recovery section (Figure 6). The IPA solution is then preheated in a heat exchanger (P-74) and a heater (P-75) and fed into a distillation column (P-76). The purified aqueous IPA solution, which exits the top of the column with an IPA concentration of about 83% w/w, is used to preheat the distillation feed in P-74. Then it is recycled back to the process via a flow adjuster (P-78, which adds fresh "makeup" IPA), and a flow distributor (P-79) shown in Fig. 5. The stillage from the bottom of the distillation column, which contains all the washed-out impurities from the pectin separation centrifuge, is cooled down to 35°C using a cooler (P-77) and removed as waste.

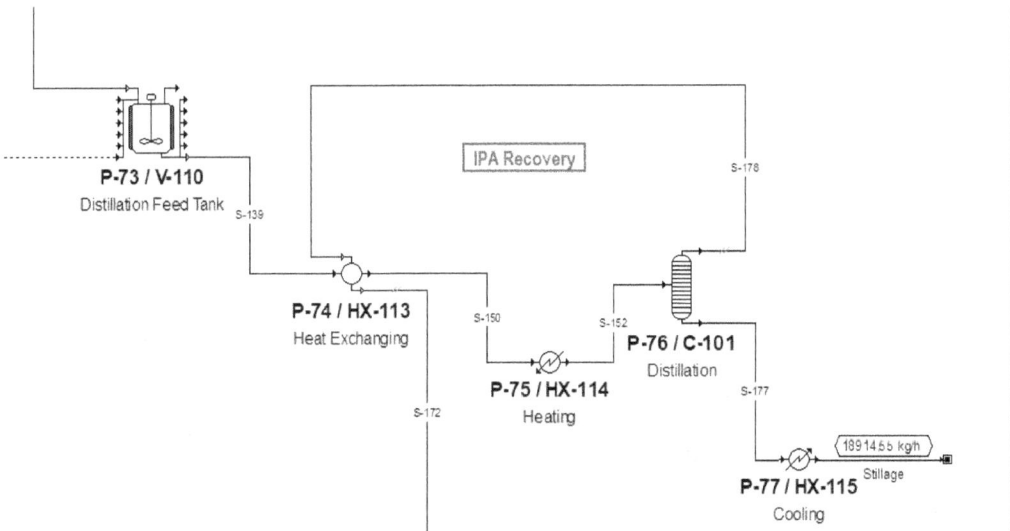

Figure 6. Flowchart of the IPA Recovery Section.

2.1.2 Simulation Results

Once the flowrate and composition of all input streams and the operating conditions of all processes are specified, SuperPro Designer solves the material and energy balance equations to calculate the state (chemical composition, flowrate, and condition) of all intermediate and output streams and of all internal process states, the required amounts of utilities and the required size of all equipment. The formulation of the equations is based on the mathematical models of all processes in the software library. The solution of the entire flowsheet is done using a sequential-modular strategy, i.e. procedures are solved sequentially following the material flow. If recycle loops exist, the tear-stream algorithm is implemented; using an iterative numerical solver, the calculations are repeated until the user-defined convergence criteria are satisfied.

In this case study, for an annual production of approximately 5111 metric tons (MT) of pectin, the calculated amounts of raw materials needed are shown in Table 2.

With respect to utility requirements as calculated by the energy balances, the plant operations are expected to require about 450,000 MT/yr of steam, 45,100,000 MT/yr of cooling water and 16,900,000 kWh/yr of electricity.

It should be noted that despite the recovery and recycling of water from many stages in the process (e.g., condensate streams, from the reverse osmosis units), the water requirement is still large (114.7 kg per kg of pectin produced). However, it would have been much larger (~ 300 kg per kg of pectin) if water recycling was not implemented. Similarly, the isopropanol requirement is kept low because of its recovery and reuse (0.02 kg of fresh IPA per kg of pectin produced instead of 22 kg in the absence of recycling). This recycling, however, comes at a price: because of the extra equipment needed (e.g., distillation column), the capital cost and utilities' requirement and cost are increased.

In this model, utility consumption is reduced somewhat by implementing heat integration with the use of several heat exchangers in the plant. Despite the use of these units, the process still requires around 88 kg of steam and 9 MT of cooling water for every kg of pectin produced. It would be possible to reduce the consumption of utilities (and their associated costs) further through more extensive energy integration. In general, optimizing a process with respect to energy and water integration is an important activity in process design with a potentially great impact in the economics and the environmental footprint of the process.

Table 2. Raw material requirements for the pectin process (MP: main product = pectin).

Material	kg/yr	kg/h	kg/kg MP
Enzyme	79,200	10.00	0.02
HCl	9,624	1.22	0.00
Isopropanol	99,688	12.59	0.02
Na$_2$CO$_3$	27,472	3.47	0.01
Natural Gas	1,670,956	210.98	0.33
Nitric Acid	256,565	32.39	0.05
Peels	277,200,000	35,000.00	54.24
Perlite	396,000	50.00	0.08
Sodium Hydroxide	8,525,071	1,076.40	1.67
Water	586,204,717	74,015.75	114.70

2.1.3 Economic Evaluation

As part of the simulation results, SuperPro Designer also calculates the size of all equipment in the process which are set at 'design mode' (i.e., their size is not set by the user). Table 3 presents the size (and estimated cost) of the most expensive equipment in the pectin process. The cost is estimated with built-in size-cost correlations or can be overwritten by the user with values obtained by equipment manufacturers and vendors.

The equipment cost is the basis for estimating the expected capital expenses (CAPEX) of the process. The cost of all resources used in the process (materials, utilities, consumables, labor etc.) is used for calculating the operating expenses (OPEX) of the process. Table 4 shows the main

Table 3. Size and cost of most expensive equipment for the pectin process.

Name	Description	Unit Cost ($)
EV-101	Multi-Effect Evaporator	4,531,000
	Evaporation Area = 431.54 m2	
SP-101	Screw Press	1,501,000
	Throughput = 92.40 MT/h	
SP-102	Screw Press	1,419,000
	Throughput = 86.10 MT/h	
SP-103	Screw Press	1,319,000
	Throughput = 78.57 MT/h	
EV-102	Multi-Effect Evaporator	1,236,000
	Evaporation Area = 88.24 m2	
DC-101	Decanter Centrifuge	1,035,000
	Throughput = 79723.39 L/h	
DC-102	Decanter Centrifuge	623,000
	Throughput = 40512.47 L/h	
RDR-101	Rotary Dryer	550,000
	Drying Area = 96.62 m2	
DS-102	Disk-Stack Centrifuge	523,000
	Throughput = 79.58 m3/h	
SB-102	Solids Bin	493,000
	Vessel Volume = 4.36 m3	
HX-104	Heat Exchanger	464,000
	Heat Exchange Area = 1002.36 m2	
DS-101	Disk-Stack Centrifuge	462,000
	Throughput = 64.59 m3/h	
GAC-101	GAC Adsorber (for Liquid Streams)	386,000
	Column Volume = 47.47 m3	
R-101	Stirred Reactor	375,000
	Vessel Volume = 196.36 m3	
DF-101	Diafilter	324,000
	Membrane Area = 795.51 m2	

economic values assumed for the calculation of OPEX as well as the selling price of pectin and by-products used for the estimation of revenues.

Using all above inputs, it is estimated that the plant requires a CAPEX of around $116 million while the annual OPEX is estimated at $52.9 million. Figure 7 shows the breakdown of the operating cost of its various components. It can be seen that fixed (facility-dependent) costs (e.g., depreciation and maintenance) are the major contributors to OPEX followed by the raw materials and utilities' cost.

Table 5 summarizes the main economic results and the values of the economic indices (ROI, PBT, IRR, NPV) regularly used to assess the economic feasibility of a process. Based on these indices, it can be concluded that the process is economically viable and seems attractive. However, it is worth noting that the operating cost per kg of pectin (10.35$/kg) exceeds its selling price (10$/kg); therefore, the economic viability should be attributed to the other assumed by-products of the

Table 4. Main economic inputs for the pectin process.

Peels cost ($/MT)	40
Isopropanol cost ($/kg)	0.6
Steam cost ($/MT)	12
Cooling water ($/MT)	0.050
Chilled water ($/MT)	0.400
Pectin selling price ($/kg)	10
Vinasse selling price ($/kg)	0.4
Animal Feed selling price ($/kg)	0.1

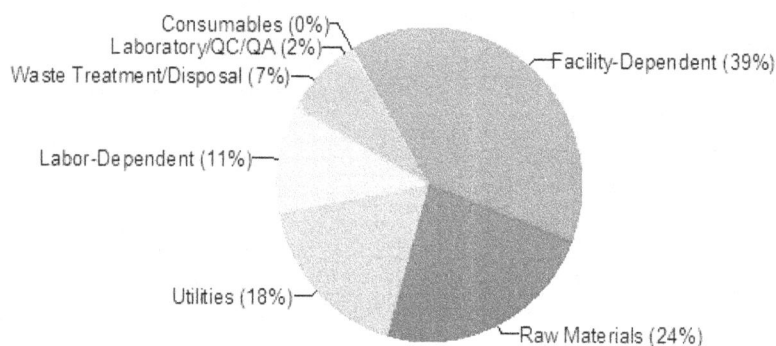

Figure 7. The operating cost breakdown for the pectin process.

Table 5. Main economic results and viability indices for the pectin process.

Total Capital Investment	115,756,000 $
Operating Cost	52,900,000 $/yr
Revenues	64,850,000 $/yr
Unit Production Cost	10.35 $/kg MP
Gross Margin	18.43 %
Return On Investment (ROI)	16.05 %
Payback Time (PBT)	6.23 years
IRR (After Taxes)	11.32 %
NPV (at 3.0% interest)	40,250,000 $

process (vinasse and animal feed). This is apparently a point of concern since fluctuations in the selling price of the by-products (as well as pectin itself) are expected to have a major impact on the economic bottom line. In these cases, a sensitivity or variability analysis should be performed to estimate the probability of profitable operation in the face of uncertainty in the product selling prices and other economic parameters that have significant effect on the economic results.

Process design modifications and additions should also be considered in the context of optimal process design. For example, the extraction and selling as by-products of flavonoids and citrus oils (such as d-limonene) contained in the peels may also be investigated using process simulation as a way to improve and robustify the process economics.

3. Isobutanol from Corn Stover

The production of biofuels from crop biomass has attracted enormous global attention in the last two decades, with bioethanol (produced from sugar or starch-based feedstocks) and biodiesel (produced from vegetable oils by transesterification processes or cracking) being the most eminent. While bioethanol is a well-established biofuel, there are several drawbacks associated with its use including a low energy content and an environmental regulation limit on the amount that can be blended with gasoline. On the other hand, biodiesel has been associated with extensive changes in land use which does not only have a negative effect on CO_2 emissions but also impacts food availability and prices. As a response to the above concerns, the EU resolved in 2016 to phase out first-generation biofuels by 2030.

As a result of the above limitations, the global interest has shifted to second-generation biofuels that can be produced from various types of non-edible biomass. One of the proposed alternatives is butanol. Butanol exhibits similar advantages as ethanol but possesses several superior characteristics including higher energy content and lower vapor pressure which facilitates blending with gasoline (Gevo 2011). As a result, U.S fuel specifications allow butanol to be blended in gasoline by up to 16% whereas the respective limit for ethanol is 10%.

Regarding the various candidate butanol isomers, isobutanol exhibits several advantages over n-butanol. Specifically, isobutanol has a higher octane number and is less toxic than n-butanol. Furthermore, given that n-butanol is typically produced along with acetone and ethanol in a process called ABE fermentation (Tao et al. 2014), isobutanol requires significantly less energy for its downstream processing. Isobutanol has a wide range of industrial applications: as a precursor, it is used in the production of various isobutyl esters, fine chemicals, and automotive paint cleaner additives, and as a solvent it is used in the paints and coatings industry. According to an Allied Market Research report (2021), the isobutanol market size was valued at $1.0 billion in 2020, and is projected to reach $1.9 billion by 2030, growing at a compound annual growth rate of 6.3% from 2021 to 2030.

The production cost of biobutanol is considerably affected by the feedstock price which accounts for as much as 60–65% of the total cost (Mansur et al. 2019). At present, biobutanol production mainly relies on sugar and starch-based feedstocks such as sugarcane, corn or wheat. While such feedstocks are globally available, they are also used as food supplies or main food ingredients for animals and humans. Therefore, alternative feedstocks with similar availability which are not food-related need to be utilized. A promising approach is the use of renewable lignocellulosic biomass from agricultural residues (corn stover, sugarcane, bagasse, wheat straw and rice bran), forest residues (sawdust, thinning rest) and energy crops (switchgrass and miscanthus). These raw materials are available in greater abundance and at a lower cost than the traditional edible feedstocks.

Renewable lignocellulosic biomass from agricultural and forest residues and energy crops is a promising alternative to the traditional sugar and starch-based feedstocks for the fermentative production of biobutanol. Besides the fact that it is not used as food ingredient, lignocellulosic biomass is available in greater abundance and at a lower cost. This is a very important aspect considering that the cost or raw materials accounts for 60–65% of the total production cost (Mansur et al. 2019). On the downside, lignocellulosic biomass requires an additional processing step to convert the cellulose and hemicellulose chains into fermentable monosaccharides that can be utilized by the biobutanol-producing microorganisms. This hydrolysis step involves either a chemical (acidic) or biochemical (enzymatic) agent that acts as a catalyst. Among the two alternatives, enzymatic hydrolysis is considered a more environmentally friendly option but, on the other hand, requires large amounts of enzymes (cellulases) which substantially increase the overall process cost (Kolesinska et al. 2019). The hydrolysis of cellulose generates glucose which can be utilized by a variety of microorganisms to produce biobutanol. On the other hand, the hydrolysis of hemicellulose results in a variety of sugars, mainly pentoses, which can be used by certain microorganisms, although with potentially lower conversion percentages. To improve the efficiency of the enzymatic hydrolysis, the lignocellulosic biomass is subjected to a pretreatment method that increases the surface area so that enzymes can penetrate more effectively. Common pretreatment methods include wet oxidation, liquid hot water, steam explosion, ammonia fiber explosion and supercritical CO_2 explosion (Shirkavand et al. 2016).

A known limitation of the fermentative production of biobutanol is product inhibition. Butanol is toxic to microorganisms even at concentrations as low as 13 g/L for n-butanol and 8 g/L for isobutanol (Mariano et al. 2011, Brynildsen and Liao 2009). To address this problem, the concentration of substrate during fermentation is typically limited to less than 60 g/L (Green 2011, Al-Shorgani et al. 2018). This practice results in increased water consumption and low butanol titers after fermentation which negatively affect both the energy requirement and the amount of wastewater generated in downstream processing. As it is clear from the above, the feasibility of the process depends, to a large extent, on the fermentation product titer (Stephanopoulos 2007).

There are two main approaches that attempt to overcome the toxicity limitation namely, genetic strain engineering and process modification. A number of researchers (e.g., Atsumi et al. 2018, Smith and Liao 2011, Shen et al. 2011, Jang et al. 2013) have attempted to improve the fermentation yield of biobutanol by using engineered toxicity-resistant microorganisms that are able to operate at increased product concentrations. With respect to process modifications, considerable improvement of the fermentation performance can be achieved by the implementation of integrated (*in situ*) product recovery during fermentation. The objective of this approach is to maintain butanol concentration in the fermentor below toxic levels, thereby enabling the use of denser substrate solutions and increased product titers, yields and productivities. Such integrated butanol recovery technologies include adsorption, liquid-liquid extraction, pervaporation, reverse osmosis, vacuum fermentation, vacuum stripping (flash fermentation) and gas stripping (Roussos et al. 2019).

This case study analyzes a process that produces cellulosic isobutanol from corn stover. The term "corn stover" encompasses whatever material is left in the field (leaves, stalks, cobs) after harvesting corn. Isobutanol can be produced by engineered *E. coli* and *Saccharomyces cerevisiae* strains (Ezeji et al. 2014). The technical challenges that must be overcome in order to render the production of isobutanol economically competitive can be summarized in two main objectives: the reduction of the cost of raw materials by the utilization of alternative feedstocks and the improvement of the fermentation yield, titer and productivity by genetic engineering and application of new process technologies (Cao and Sheng 2016, Haigh et al. 2018). The following simulated study can be regarded as a "what-if" analysis on a hypothetical isobutanol fermentation titer and can be used as a reference basis for the simulation and comparison of different scenarios based on more realistic or more optimistic data.

3.1 Process Description

This case study presents a conceptual design of a process producing isobutanol from corn stover using an engineered *E. coli* strain that can theoretically achieve an isobutanol titer of 65 g/L during fermentation with a vacuum stripping system. The operation mode is batch with a batch size of 247 metric tons (MT) of corn stover producing around 40.8 MT of isobutanol per batch or 40000 MT per year (assuming an annual operation time of 48 weeks). The process is divided into four sections: pretreatment and enzymatic hydrolysis, fermentation, purification and a utilities section producing steam and electricity from the lignin-rich residue of the pretreatment section. For brevity, the flowsheet of the entire process or for each section is not shown; instead, a schematic of the simplified process with the above sections is shown in Fig. 8.

3.1.1 Pretreatment/Enzymatic Hydrolysis

The assumed composition of corn stover is shown in Table 6. Corn stover is initially mixed with water up to 63 wt% in water content. Then, a 10% solution of sulfuric acid is added to achieve a sulfuric acid content of 0.3 wt%. The slurry is preheated before entering the steam explosion reactor. During this thermal pretreatment, it is assumed that 10% of cellulose is converted to hexoses and

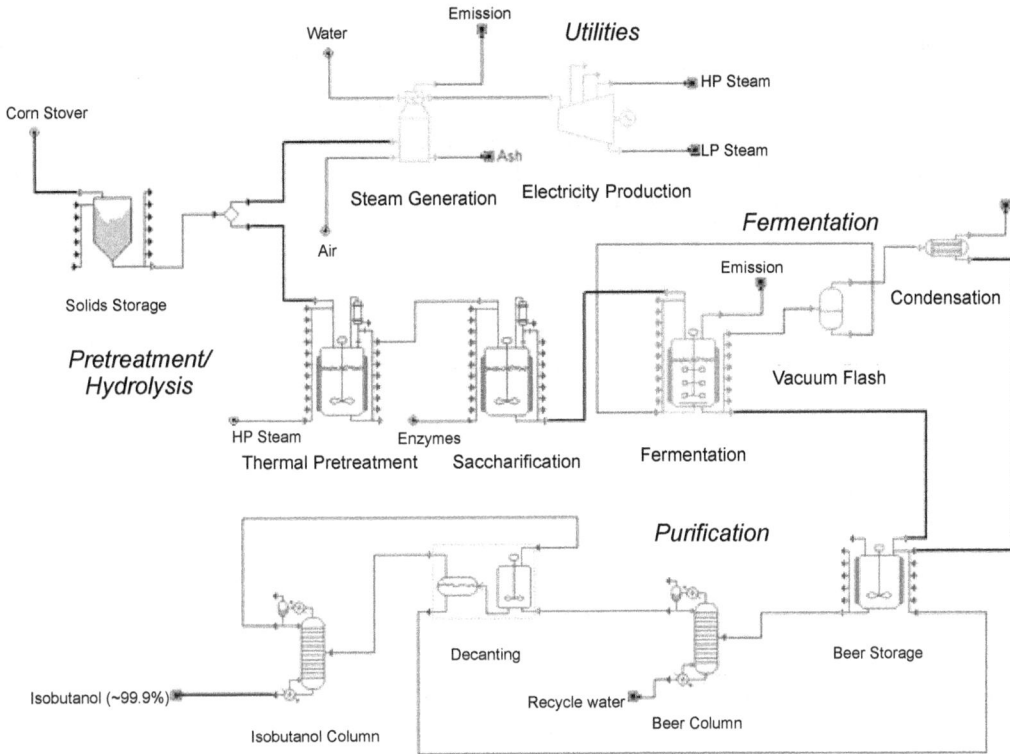

Figure 8. Schematic of a simplified flow diagram of the isobutanol process (Roussos et al. 2019).

Table 6. Composition of corn stover.

Component	(%w/w)
Ash	4.7
Cellulose	30
Hemicellulose	22
Lignin	16
Proteins	2.8
Soluble Sugars	4.5
Water	20

80% of hemicellulose is converted to pentoses. Subsequently, the slurry is neutralized with calcium hydroxide until a pH close to 5 is achieved. The mixture is then fed to the saccharification reactor where enzymatic hydrolysis takes place. The amount of enzymes fed is 20 mg per g of cellulose. The enzymatic hydrolysis is assumed to convert 90% of the remaining cellulose and hemicellulose to hexoses and pentoses, respectively. The hydrolysate, which contains 14 wt% of hexoses, 8 wt% of pentoses and 65 wt% of water, is filtered using a screw press and directed to fermentation. The lignin-rich filter cake is directed to the Utilities section for the generation of steam and electricity.

3.1.2 Fermentation

The fermentation media are first prepared in a blending tank operating in continuous mode with a residence time of 8   h, equal to the process cycle time (as explained below). Primary nutrients present in the fermentation media include hexoses and pentoses obtained from the hydrolysis of biomass (carbon sources) as well as diammonium phosphate ((NH_4)$_2$$HPO_4$) which is the phosphorus and nitrogen source. The media solution is distributed to a three-stage seed fermentation and one production fermentation stage. The fermentation time in each seed fermentation step is 48 h, and 72 h for the production fermentation. Ambient air is continuously filtered, compressed, cooled and distributed to the fermentors via a flow distribution procedure. Fermentation is modeled as a two-step process with the biomass formation first, followed by the isobutanol production stage. The mass stoichiometry of the biomass formation reactions in the fermentation steps is:

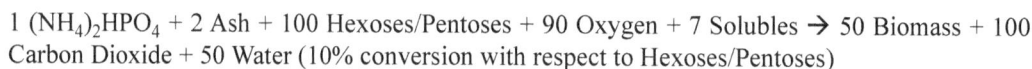

1 (NH_4)$_2$$HPO_4$ + 2 Ash + 100 Hexoses/Pentoses + 90 Oxygen + 7 Solubles → 50 Biomass + 100 Carbon Dioxide + 50 Water (10% conversion with respect to Hexoses/Pentoses)

The exothermic reaction enthalpy is assumed to be 3700 kcal/kg of oxygen consumed. In the production fermentation procedure, the pentoses and hexoses are also converted to isobutanol, according to the following molar stoichiometry:

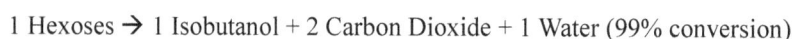

1 Pentoses → 1 Isobutanol + 1 Carbon Dioxide + 1.78 Water (95% conversion)

1 Hexoses → 1 Isobutanol + 2 Carbon Dioxide + 1 Water (99% conversion)

The conversion percentages refer to the amount of monosaccharides (pentoses or hexoses) left from the biomass formation step. The exothermic reaction enthalpy is assumed to be 250 kcal/kg of consumed monosaccharides.

The sequence of operations executed as part of each fermentation step is: SIP (Steam-In-Place) of the fermentor, Pull-In Substrate, Pull-In Water, Pull-In Inoculum, Ferment (biomass formation), Ferment (isobutanol formation), Transfer Out and CIP (Clean-In-Place). During fermentation, the temperature is maintained at 37°C with the use of cooling water. The aeration rate is 1 VVM (volume of air per volume of liquid per minute) in the seed fermentors and 0.5 VVM during the biomass production stage in the production fermentors. No air is supplied during the isobutanol production stage.

Because of the large batch size, a single batch requires 2 identical production fermentors of 416 m^3 each that run in parallel. In addition, in order to achieve the desired annual throughput (as explained below), 10 sets of 2 fermentors operate in Stagger Mode; in other words, a total of 20 production fermentors are employed in this process. The broth of each fermentor pair is then transferred to a blending tank which serves as a buffer storage between fermentation and purification.

The production fermentors are equipped with a vacuum stripping system that processes 78 MT/h of fermentation broth to maintain the isobutanol concentration in the fermentor below the toxicity limit of 22 g/L. The fermentation broth is circulated through a vacuum flash tank operating at very low pressure (~ 0.04 bar) where an isobutanol-rich vapor phase and an isobutanol-depleted liquid phase are generated. The liquid phase is recycled back to the fermentor. After the end of the fermentation process, the isobutanol titer in the tank that combines the final fermentation broth and the condensed vapor from the flash tank is around 65 g/L, and the overall fermentation yield is equal to 0.39 g of isobutanol/g of reducing sugars.

3.1.3 Purification

Isobutanol and water form an azeotrope when the alcohol fraction reaches 66.9 wt%. Nevertheless, isobutanol can be purified to a very high degree using a system of two distillation columns in combination with a decanter tank that exploits the low solubility of isobutanol in water to break the azeotrope (Vane 2008). In the first distillation column, the product is concentrated by removing approximately 95% of the water content. The concentrated top stream (with sub-azeotropic composition) is then allowed to settle in the decanter, where it forms two liquid phases: an aqueous phase with 8.6 wt% of isobutanol and an organic phase with 80.1 wt% isobutanol (i.e., the organic phase is above the azeotropic composition). Next, the organic phase goes to the second distillation column where isobutanol is recovered in the bottom stream with a very high purity (over 99.9 wt%). The top stream of the second column is recycled back to the decanter, and the aqueous phase of the decanter is recycled back to the first distillation column, to minimize product losses. In addition, the stream that feeds the first column is preheated by the wastewater stream that comes from the bottom of that column, in order to save energy.

3.1.4 Utilities

In the Utilities section, the lignin-rich residue from pretreatment and hydrolysis (19.7 MT/h) and untreated corn stover (1.0 MT/h) are used to produce high-pressure and low-pressure steam as well as a modest amount of electricity. The amount of untreated corn stover was adjusted so that the output of high- and low-pressure steam matches the requirements of the entire process. The electricity that is concomitantly produced covers a portion of the power requirements of the process and thus reduces the annual operating cost.

3.1.5 Process Scheduling and Debottlenecking

Because of the sequence of long fermentations, the process batch time is around 9 days. However, the availability of multiple fermentors that can operate in staggered mode (i.e., the fermentors alternate between batches) allows the implementation of an 8 h cycle time (i.e., a new batch starts every 8 hours). Figure 9 shows the process Gantt chart where the utilization of the seed and production fermentors over 20 consecutive batches (indicated by the different colors) is shown. The 8-hour cycle time allows the execution of 1026 batches per year and determines the annual throughput of about 40000 MT of isobutanol. Increase in the annual throughput is possible if the cycle time is further decreased by introducing additional seed and production fermentors. The downstream processing is continuous so the increased capacity by the use of additional fermentors can be absorbed by continuous equipment with larger throughput.

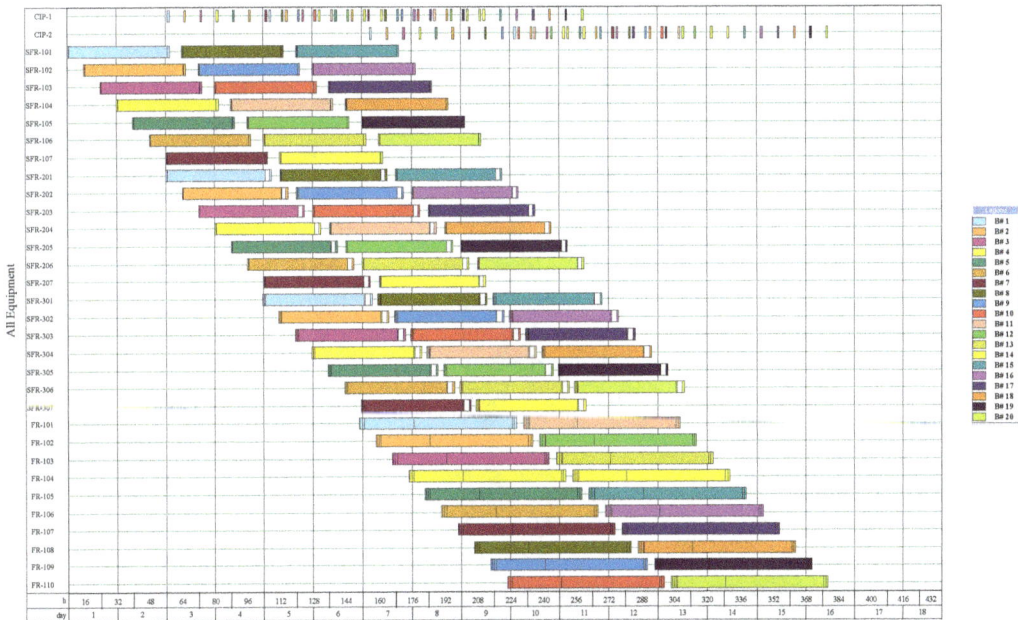

Figure 9. The equipment occupancy Gantt chart for the isobutanol process.

3.2 Simulation Results

The calculated amounts of raw materials needed for the above production throughput are shown in Table 7.

With respect to utility requirements as calculated by the energy balances, the plant is expected to require about 204,500 MT/yr of low-pressure steam, 12,700 MT/yr of high-pressure steam, 40,250,000 MT/yr of cooling water, 363,500 MT/yr of chilled water and 125,700 MWh/yr of electricity. However, the entire amount of steam (both low and high-pressure) and 14500 MW-h of electricity (~12% of total need) are produced in the co-generation plant by burning biomass (lignin-rich residue from the process and untreated corn stover). If the plant performance is to be optimized, the percentage of available corn stover that is used as fuel to the co-generation plant versus feed to the process should be one of the optimization design variables.

Based on the relative contribution of each section to the mass and energy consumption, the fermentation section is the hot spot with respect to utilities' consumption, the pretreatment/hydrolysis for materials' consumption and the purification for waste generation (Fig. 10). Therefore, any efforts in reducing any of the above operating costs should focus on the uses in the relevant sections.

Table 7. Raw material requirements for the entire process (MP: main product = isobutanol).

Material	kg/yr	kg/batch	kg/kg MP
(NH4)2HPO4	1,954	1.99	0.05
Air	1,485,068	1,513.83	37.13
Ca Hydroxide	1,493	1.52	0.04
Corn Stover	242,329	247.02	6.06
Hydrolases	1,266	1.29	0.03
NaOH (2%)	10,185	10.38	0.25
RO Water	224,741	229.09	5.62
Sulfuric Acid	1,568	1.60	0.04
Water	622,187	634.24	15.55

3.3 Economic Evaluation

Table 8 presents the economic assumptions on the cost of the main materials and utilities as well as the selling price of isobutanol.

The capital investment needed for the isobutanol plant with the calculated throughput is approximately $98 million (based on an equipment purchase cost of around $25 million). The annual operating cost (AOC) is estimated at $62 million or $60.9 million if the savings from the electricity produced in the Utilities section is taken into account. Likewise, the unit production cost is estimated at $1548/MT without considering the power savings, or $1522/MT with the savings.

Table 9 presents the main economic results assuming a selling price of isobutanol at $1800/MT. Based on these results, the isobutanol plant under the modeling assumptions seems like an economically attractive investment. However, this cost as well as the process throughput depends on many uncertain parameters: the % conversion of pentoses and hexoses to isobutanol, the enzyme loading ratio, the feedstock and enzyme costs etc. As demonstrated in Roussos et al. (2019), a variability analysis using a tool such as Monte Carlo simulation can reveal the extent to which the economic results are dependent on the above parameters and the certainty by which favorable economic results can be obtained in the face of uncertainty.

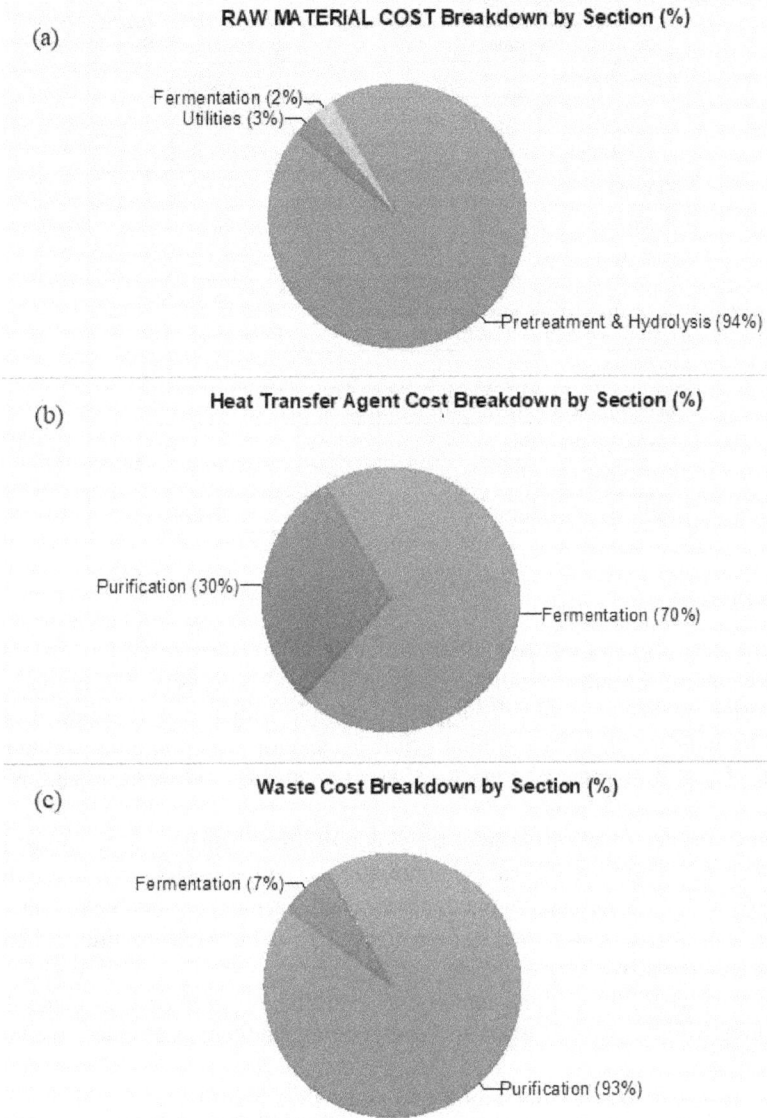

RAW MATERIAL COST Breakdown by Section (%)

(a)

Fermentation (2%)
Utilities (3%)
Pretreatment & Hydrolysis (94%)

Heat Transfer Agent Cost Breakdown by Section (%)

(b)

Purification (30%)
Fermentation (70%)

Waste Cost Breakdown by Section (%)

(c)

Fermentation (7%)
Purification (93%)

Figure 10. Operating cost breakdown per section for the isobutanol case (a) raw materials, (b) heat transfer agents, (c) waste treatment.

Table 8. Main economic inputs for the isobutanol process.

Corn stover cost ($/MT)	60
Enzyme cost ($/kg protein)	10
Electricity cost ($/kWh)	0.070
Cooling water ($/MT)	0.050
Chilled water ($/MT)	0.400
Isobutanol selling price ($/MT)	1800

Table 9. Main economic results and viability indices for the isobutanol process.

Total Capital Investment	98,212,000 $
Operating Cost	61,910,000 $/yr
Savings (due to Power Recycled)	1,019,000 $/yr
Net Operating Cost	60,891,00 $/yr
Revenues	72,001,000 $/yr
Batch Size	40.78 MT MP
Unit Production Cost	1,547.73 MT MP
Net Unit Production Cost	1,522.26 MT MP
Unit Production Revenue	1,800.00 MT MP
Gross Margin	15.43 %
Return On Investment (ROI)	13.80 %
Payback Time (PBT)	7.25 years
IRR (After Taxes)	9.89 %
NPV (at 3.0% Interest)	113,761,000 $

4. Conclusions

Food waste presents a global environmental issue; its valorization in the context of sustainable development is one of the most important areas in the food processing and production research. Current industrial practices do not fully exploit food valorization opportunities. As research and practical results indicate, there is a global effort to develop new technologies that will enable the extraction of useful ingredients from food waste or utilize food waste in an environmentally sound way.

Technology transfer of all these new technologies from bench-scale to industrial-scale can be greatly enhanced by the use of process simulation and other computer-aided tools. Designing an industrial-scale food valorization plant poses many challenging problems since there may exist multiple final product and process alternatives. The technical, economic and environmental assessment of all alternative designs can be facilitated by the use of a digital model that is able to predict the process yield, throughput, material and utilities requirements, estimate the important economic figures with respect to capital and operating expenditures and assess the environmental impact of the process.

In this context, two waste valorization processes were investigated in this chapter: a pectin producing process via extraction from citrus peels (an abundant by-product of the fruit juice industry) and an isobutanol producing process via bioconversion of corn stover (an agricultural residue). Both conceptual processes provide environmentally friendly processing solutions and, based on the economic evaluation results, also constitute attractive investments. Combined with optimization techniques and sensitivity/variability analyses, the simulated process models can yield optimal designs with the effect of uncertainties predicted and quantified.

References

Abdelaal, A., Pradhan, S., AlNouss, A., Tong, Y., Al-Ansari, T., McKay, G., et al. 2020. The impact of pyrolysis conditions on orange peel biochar physicochemical properties for sandy soil, Waste Management & Research: The Journal for a Sustainable Circular Economy 39: 7, https://doi.org/10.1177/0734242X20978456.

Allied Market Research. 2021. Isobutanol Market by Product and Application: Global Opportunity Analysis and Industry Forecast, 2021–2030.

Al-Shorgani, N.K.N., Shukor, H., Abdeshahian, P., Kalil, M.S., Yusoff, W.M.W. and Hamid, A.A. 2018. Enhanced butanol production by optimization of medium parameters using *Clostridium acetobutylicum* YM1. Saudi J. Biol. Sci. 25(7): 1308–1321. https://doi.org/10.1016/j.sjbs.2016.02.017.

Arora, A., Banerjee, J., Vijayaraghavan, R., MacFarlane, D. and Patti, A.F. 2018. Process design and techno-economic analysis of an integrated mango processing waste biorefinery. Ind. CropsProd. 116: 24–34.

Atsumi, S., Hanai, T. and Liao, J.C. 2018. Non-fermentative pathways for synthesis of branched-chain higher alcohols as biofuels. Nature 451(7174): 86–89. https://doi.org/10.1038/nature06450.

Brynildsen, M.P. and Liao, J.C. 2009. An integrated network approach identifies the isobutanol response network of *Escherichia coli*. Mol. Syst. Biol. 5(277): 1–13. https://doi.org/10.1038/msb.2009.34.

Caldeira, C., Corrado, S. and Sala, S. 2017. JRC Technical Report—Food Waste Accounting Methodologies, Challenges and Opportunities; Publications Office of the European Union: Luxembourg, DOI 10.2760/54845.

Cao, G. and Sheng, Y. 2016. Biobutanol production from lignocellulosic biomass: prospective and challenges. J. Bioremediation Biodegrad. 7(4).

Chiodo V., Maisano, S., Zafarana, G., Urbani, F., Galvagno, A., Prestipino, M., et al. 2016. Valorization of dry orange peel residues by gasification process, Conference Paper 24th European Biomass Conference and Exhibition.

Dimou, C., Karantonis, H.C., Skalkos, D. and Koutelidakis, E.A. 2019. Valorization of fruits by-products to unconventional sources of additives, oil, biomolecules and innovative functional foods. Current Pharmaceutical Technology 20(10), DOI: 10.2174/1389201020666190405181537.

Duarte, L.C., Carvalheiro, F., Lopes T., Gírio, C.O.F., Campo, I.D and Martinez, B.U. 2020. Development of system simulation tools for comprehensive modelling of biomass conversion and biorefinery, WP8 in EU Horizon 2020 funded Project: Biofuels Research Infrastructure for Sharing Knowledge II.

Dursun, D., Koulouris, A. and Dalgıç, A.C. 2020. Process simulation and techno economic analysis of astaxanthin production from agro-industrial wastes. Waste Biomass Valorization 11: 943–954.

Ezeji, T.C., Qureshi, N. and Ujor, V. 2014. Isobutanol Production from Bioenergy Crops. Elsevier, https://doi.org/10.1016/B978-0-444-59561-4.00007-3.

Foo, D.C.Y., Chemmangattuvalappil, N., Ng, D.K.S., Elyas, R., Chen, C.L., Elms, R.D., et al. 2017. Chemical Engineering Process Simulation, Elsevier.

Gevo. 2011. Isobutanol – A Renewable Solution for the Transportation Fuels Value Chain. http://www.etipbioenergy.eu/images/wp-isob-gevo.pdf.

Gomez, J.A., Sanchez, O.J. and Correa, L.F. 2020. Technoeconomic and environmental evaluation of cheesemaking waste valorization through process simulation using superpro designer,. Waste and Biomass Valorization 11(1), DOI: 10.1007/s12649-019-00833-4.

Gómez-Ríos, D., Barrera-Zapata, R. and Ríos-Estepa, R. 2017. Comparison of process technologies for chitosan productionfrom shrimp shell waste: a techno-economic approach usingAspen Plus®. Food Bioprod. Process. 103: 49–57.

Green, E.M. 2011. Fermentative production of butanol - the industrial perspective. Curr. Opin. Biotechnol. 22(3): 337–343. https://doi.org/10.1016/j.copbio.2011.02.004.

Gustavsson, J., Cederberg, C., Sonesson, U. and Emanuelsson, A. 2013. The methodology of the FAO study: Global Food Losses and Food Waste-extent, causes and prevention. SIK Report No 857, FAO 2011.

Haigh, K.F., Petersen, A.M., Gottumukkala, L., Mandegari, M., Naleli, K. and Görgens, J.F. 2018. Simulation and comparison of processes for biobutanol production from lignocellulose via ABE fermentation. Biofuels, Bioprod. Biorefining 12(6): 1023–1036.

Hoefler, A.C. 1991. Other pectin food products, Chapter 1. *In*: Walter, R.H. (eds.). The Chemistry and Technology of Pectin. Academic Press Inc. California, USA.

Huang, H., Long, S. and Singh, V. 2016. Techno-economic analysis of biodiesel and ethanol co-production from lipid-producing sugarcane. Biofuels Bioprod. Biorefin. 10(3): 299–315.

Jang, Y.S., Malaviya, A., Lee, J., Im, J.A., Lee, S.Y., Lee, J., et al. 2013. Metabolic engineering of *Clostridium acetobutylicum* for the enhanced production of isopropanol-butanol-ethanol fuel mixture. Biotechnol. Prog. 29(4): 1083–1088. https://doi.org/10.1002/btpr.1733.

Karayannakidis, P., Chatziantoniou, S. and Zotos, A. 2015. Co-extraction of gelatin and lipids from Yellowfin tuna (*Thunnus Albacares*) skins: Physicochemical characterization, process simulation and economic analysis. Journal of Food Processing and Preservation 39(6), DOI: 10.1111/jfpp.12484.

Kolesinska, B., Fraczyk, J., Binczarski, M., Modelska, M., Berlowska, J., Dziugan, P., et al. 2019. Butanol synthesis routes for biofuel production: Trends and perspectives. Materials (Basel) 12 (3): 350. https://doi.org/10.3390/ma12030350.

Koulouris, A., Misailidis, N. and Petrides, P. 2021. Applications of process and digital twin models forproduction simulation and scheduling in the manufacturing of food ingredients and products. Food and Bioproducts Processing 126: 317–333.

Kwan, T.H., Pleissner, D., Lau, K.Y., Venus, J., Pommeret, A. and Lin, C.S.K. 2015. Techno-economic analysis of a food waste valorization process via microalgae cultivation and co-production of plasticizer, lactic acid and animal feed from algal biomass and food waste. Bioresour. Technol. 198: 292–299.

Kwan, T.H., Hu, Y. and Lin, C.S.K. 2018. Techno-economic analysis of a food waste valorisation process for lactic acid, lactide and poly(lactic acid) production. J. Clean. Prod. 181: 72–87.

Kwon, D., Oh, J., Lam, S.S., Moon, D.H. and Kwon, E.E. 2019. Orange peel valorization by pyrolysis under the carbon dioxide environment Bioresource Technology 285, https://doi.org/10.1016/j.biortech.2019.121356.

Mansur, M.C., Rehmann, M.S. and Zohaib, M. 2019. ABE Fermentation of Sugar in Brazil, University of Pennsylvania. Available online: https://repository.upenn.edu/cbe_sdr/17/(accessed Mar 18, 2019).

Mariano, A.P., Qureshi, N., Filho, R.M. and Ezeji, T.C. 2011. Bioproduction of butanol in bioreactors: new insights from simultaneous *in situ* butanol recovery to eliminate product toxicity. Biotechnol. Bioeng. 108(8): 1757–1765. https://doi.org/10.1002/bit.23123.

Maroulis, Z.B. and Saravacos, G.D. 2003. Food Process Design, CRC Press.

Maroulis, Z.B. and Saravacos, G.D. 2007. Food Plant Economics, CRC Press.

Mirabella, N., Castellani, V. and Sala, S. 2014. Current options for the valorization of food manufacturing waste: a review. Journal of Cleaner Production 65: 28–41. http://dx.doi.org/10.1016/j.jclepro.2013.10.051.

Misailidis, N., Campbell, G.M., Du, C., Sadhukhan, J., Mustafa, M., Mateos-Salvador, F., et al. 2009. Evaluating the feasibility of commercial arabinoxylan production in the context of a wheat biorefinery principally producing ethanol: Part 2. process simulation and economic analysis. Chem. Eng. Res. Des. 87(9): 1239–1250.

Moates, G., Sweet, N., Bygrave, K. and Waldron, K. 2016. Top 20 Food Waste Streams, D6.9 within the EU funded REFRESH project.

Murugan, K., Chandrasekaran, S.V., Karthikeyan, P. and Al-Sohaibani, S. 2013. Current State of the art of food processing by-products, Chapter in Valorization of Food By-Products, edited by Chandrasekaran M., CRC Press Taylor & Francis Group, Florida USA.

Patsalou, M., Chrysargyris, A., Tzortzakis, N. and Koutinas, M. 2020. A biorefinery for conversion of citrus peel waste into essential oils, pectin, fertilizer and succinic acid via different fermentation strategies. Waste Management 113: 469–477, https://doi.org/10.1016/j.wasman.2020.06.020.

Pires, A.F., Marnotes, N.G., Rubio, O.D., Garcia, A.C. and Pereire, C.D. 2021. Dairy by-products: A review on the valorization of whey and second cheese whey. Foods 10(5), https://doi.org/10.3390/foods10051067.

Ritchie, H. and Roser, M. 2020. CO$_2$ and Greenhouse Gas Emissions, *Published online at OurWorldInData.org.* Retrieved from: https://ourworldindata.org/co2-and-other-greenhouse-gas-emissions.

Roussos, A., Misailidis, N., Koulouris, A., Zimbardi, F. and Petrides, D. 2019. A feasibility study of cellulosic isobutanol production—Process simulation and economic analysis. Processes 7(10): 667.

Saini, A., Panesar, P.S. and Bera, M.B. 2019. Valorization of fruits and vegetables waste through green extraction of bioactive compounds and their nanoemulsions-based delivery system. Bioresources and Bioprocessing 6: 26, https://doi.org/10.1186/s40643-019-0261-9.

Sharma, M., Usmani, Z., Gupta, V.K. and Bhat, R. 2021. Valorization of fruits and vegetable wastes and by-products to produce natural pigments, Critical Reviews in Biotechnology 41(4): 535–563, DOI: 10.1080/07388551.2021.1873240.

Shen, C.R., Lan, E.I., Dekishima, Y., Baez, A., Cho, K.M. and Liao, J.C. 2011. Driving forces enable high-titer anaerobic 1-butanol synthesis in *Escherichia coli*. Appl. Environ. Microbiol. 77(9): 2905–2915. https://doi.org/10.1128/aem.03034-10.

Shirkavand, E. Baroutian, S., Gapes, D.J. and Young, B.R. 2016. Combination of fungal and physicochemical processes for lignocellulosic biomass pretreatment - a review. Renew. Sustain. Energy Rev. 54: 217–234. https://doi.org/10.1016/j.rser.2015.10.003.

Smith, K.M. and Liao, J.C. 2011. An evolutionary strategy for isobutanol production strain development in *Escherichia coli*. Metab. Eng. 13(6): 674–681. https://doi.org/10.1016/j.ymben.2011.08.004.

Somavat, P., Kumar, D. and Singh, V. 2018. Techno-economic feasibility analysis of blue and purple corn processing for anthocyanin extraction and ethanol production using modified dry grind process. Ind. Crops Prod. 115: 78–87.

Stephanopoulos, G. 2007. Challenges in engineering microbes for biofuels production. Science 315(5813): 801–804. https://doi.org/10.1126/science.1139612.

Tao, L., Tan, E.C.D., McCormick, R., Zhang, M., Aden, A., He, X., et al. 2014. Techno-economic analysis and life-cycle assessment of cellulosic isobutanol and comparison with cellulosic ethanol and n-Butanol. Biofuels. Bioprod. Biorefining 8(1): 30–48.

Van Buren, J.P. 1993. Function of pectin in plant tissue structure and firmness, Chapter 3. *In*: R.H. Walter (eds.). The Chemistry and Technology of Pectin. Academic Press Inc. California, USA.

Vane, L.M. 2008. Separation technologies for the recovery and dehydration of alcohols from fermentation broths. Biofuels. Bioprod. Bioref., 553–558.

16

Bioeconomy of Sustainable Food Industries

Endang Chumaidiyah,* Wuryaningsih Dwi Sayekti and *Rita Zulbetti*

1. Introduction

World food demand shows an increasing trend—FAO (2018) estimates that food needs for developing countries will increase by 60% in 2030 and double by 2050. The inability to provide global food demand can lead to quite significant problems, namely the food crisis.

Meanwhile, a climate impact study shows that almost half of the world's food production has damaged the environment. The problems resulting from the food production include, among other, the damages to the balance of ecosystems, deforestation, and loss of biodiversity. Therefore, it is necessary to try a new approach of the agricultural system that is less harmful to the environment.

Bioeconomy becomes an approach to achieve transformation towards a more sustainable and equitable food system that seeks to provide healthy and nutritious food, as well as to create livelihood opportunities and reduce environmental impacts. It is in line with the Sustainable Development Goals (SDGs) proclaimed by the UN since 2015 with 17 goals that need to be achieved globally as an effort towards balance in all aspects of the world's people's lives as a whole including the social, economic, educational, health and environmental aspects in a sustainable manner.

2. What is Bioeconomy?

Bioeconomy is defined in various versions: according to the European Union, bioeconomy includes the production of renewable biological resources and their conversion into food, feed, bio-based products and bioenergy. It includes agriculture, forestry, fisheries, food, pulp and paper production, as well as parts of the chemical, biotechnology, and energy industries (European Commission 2012).

Finland, one of the leading countries in the application of bioeconomy, has its own definition. According to Finland, bioeconomy refers to an economy that relies on renewable natural resources to produce food, energy, products and services. This strategy aims to "reduce" the dependence on fossil natural resources, to prevent the loss of biodiversity and to create new economic growth and jobs in line with the principles of sustainable development (MEE 2014).

Industrial Engineering Faculty, Telkom University, Indonesia.
* Corresponding author: endangchumaidiyah@telkomuniversity.ac.id

In contrast, based on a report titled The Bioeconomy to 2030: Designing A Policy Agenda, OECD countries consider that the application of biotechnology in primary production, health and industry can lead to a new bioeconomy that involves three elements: the advanced knowledge of complex genes and cell processes, renewable biomass, and integration of cross-sectoral biotechnology applications (OECD 2006).

From several definitions of bioeconomy above, it can be concluded that bioeconomy means the sustainable extraction, exploitation, growth, and production of renewable resources from land and sea as well as environmentally friendly conversion into food, feed, fuel, fiber, chemicals, and materials, for being consumed and recycled sustainably (Sillanpää and Ncibi 2017).

According to Biernat (2019), bioeconomy is based on the utilization of renewable resources derived from biological materials to obtain benefits with new characteristics because of:

- Renewable natural resources,
- Resources with low or neutral greenhouse gas emissions,
- Resources that are used repeatedly (cascade) in the production process,
- Resources with high potential for useful properties according to the product's purposes (such as lower or non-existent toxicity, higher stability, higher and stronger durability, limited water consumption, etc.).

This new approach in the field of economics seeks to examine the application of innovation and its combination with industrial biotechnology applications. Activities such as research and innovation, coherent policy and setting bioeconomic strategies at the country and regional level as well as international and cross-sectoral cooperation are indispensable for bioeconomic development. Bioeconomic priorities seek to achieve the economic growth on the basis of traditional and emerging industries. The improvement will be embodied through the creation of new value chains based on the biological origin resources that will deliver the products with high added value to the market.

2.1 The Objectives of Bioeconomy

The objectives of bioeconomy development (e.g., in the European Union) are in line with the Sustainable Development Goals (SDGs), especially for the five objectives as shown in Fig. 1.

Ensuring food and nutrition security

The four pillars of food security from FAO are availability, access, utilization, and stability.

1. Availability: including the indicators regarding food resource supply, food production and storage.
2. Access: including the concept of physical and economic access to food.
3. Utilization: indicators for the sufficient food quality and quantity
4. Stability: indicators of the stability of the above components.

To ensure the food and nutrition security, the four components above must be available in such a way that none of the factors can adversely affect the others. For example, an individual may have adequate economic status but may be at risk of becoming food insecure if an event that reduces the availability of food resources or limits physical access to those resources occurs.

Managing natural resources' sustainability

The second objective refers to the conservation, protection and restoration of ecosystems, as well as the sustainable management of primary production systems, with the aim of maintaining the healthy and resilient ecosystems. The European Union's 2018 Bioeconomic Strategy states that managing natural resources sustainably is more important than ever in today's context of increasing the

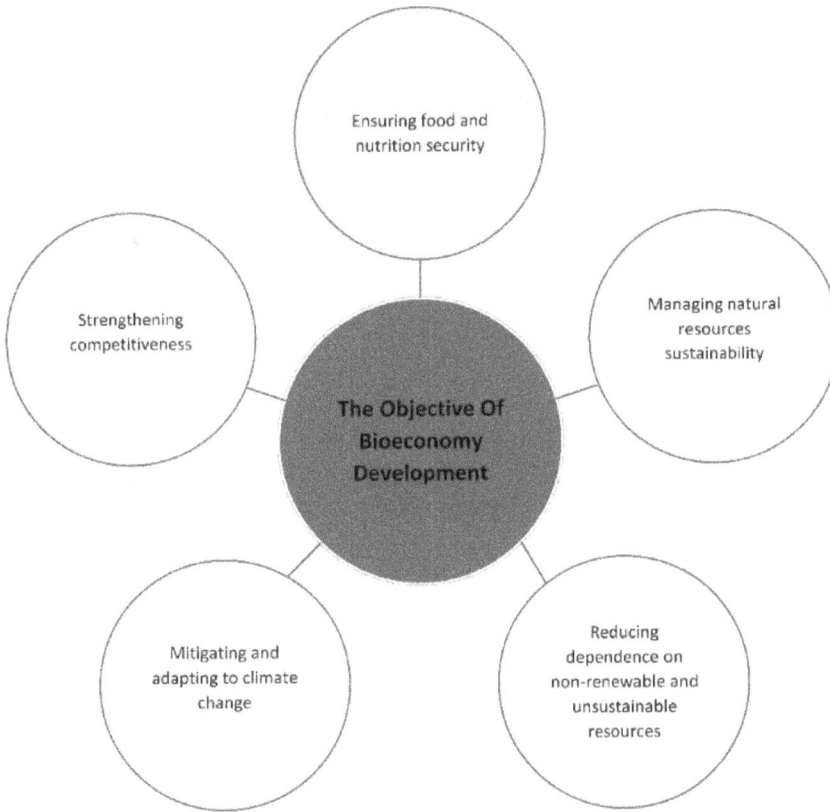

Figure 1. The Objectives of Bioeconomy Development. Adapted from (Giuntoli et al. 2020, Wohlfahrt et al. 2019).

environmental pressures and biodiversity loss. Furthermore, actions are needed to avoid ecosystem degradation, to protect natural capital, to restore, assess and enhance ecosystem functions, which in turn can improve food and water security, and contribute substantially to climate change adaptation and mitigation through "negative emissions" and carbon sinks.

Reducing dependence on non-renewable and unsustainable resources

The goal is achieved from various aspects of economic development, policy instruments, community governance, and a special focus on the rural area economy.

Mitigating and adapting to climate change

There are several key components that are indicators of the importance of mitigating and adapting to climate change in all value chains and in all sectors (Giuntoli et al. 2020).

1. Indicators related to climate change mitigation efforts, for example: carbon balancing due to bioeconomy sectors.
2. Indicators related to adaptation to climate change, in natural ecosystems through the specific management actions (for example: species/plant selection as the function of future environmental constraints).
3. Referring to resilience and adaptation in the built environment if it involves a biomass component (for example: urban trees, green roofs, etc.).

Strengthening competitiveness

Bioeconomy support strengthening competitiveness through sustainable production and consumption along the value chain, as well as the principles of a circular economy focuses on the resource efficiency, energy efficiency, waste reuse, and waste treatment.

2.2 Sectors in Bioeconomy

The European Commission lists the sectors that make up the entire bioeconomy. The sectors considered are directly related to the Nomenclature Statistique des activités économiques dans la Communauté européenne (NACE) system.

Bioeconomy industry can be broadly divided into three distinct types of economic activities (Kardung 2019):

1. Natural resource-based activities that directly utilize the biological resources (agriculture, forestry, fisheries) and provide biomass as the input to other industries.

2. Conventional activities to further process biomass from food, feed, tobacco, beverages, wood and wood products, textiles, apparel, leather, paper and pulp, furniture.

3. New activities to further process biomass and/or biomass residues from food, feed, tobacco, beverages, wood and wood products, textiles, apparel, leather, paper and pulp, furniture or to use processing residues from biorefineries, biofuels, chemical based bio, bio-based plastics, and biogas.

For a more detailed description of these sectors, see Table 1.

3. Food and Agriculture

Food is a basic human need, therefore its history is as long as the human life. At first, the human need for food was only met by taking the food around him. The increase in human population resulted in insufficient food supply in nature, and this is where agricultural activities began. Agriculture is defined as the activity of managing biological natural resources with the help of technology, capital, labor, and management to produce agricultural commodities that include plants and animals that are useful for meeting human needs. Formerly, agriculture only discussed how to produce plants and animals. The definition of agriculture is developing, as presented by Zocca et al. (2018) that agriculture includes a very broad sense, and refers to almost all the basic instincts of human life such as the ability to produce food to satisfy hunger and the sustainability of species.

To fulfill hunger and sustain species, agricultural activities focus on food commodities, although in the end also other commodities are needed by humans. In other words, it can be said that human life can continue if agricultural activities exist. The success of agriculture is indicated by the world freedom from hunger as stated as one of the goals of the SDGs, namely zero hunger.

At the beginning, hunger was simply stated as insufficiency to meet the energy needs for human life, but the concept of hunger later developed into the concept of food security. The concept of food security also continues to develop. Starting from only focusing on the side of food production and consumer purchasing power of available food, it then developed on the consumption side and quality of life. The more comprehensive concept of food security is in line with the agricultural objective presented by FAO in 2015 that agriculture does not only grow food crops but also grows healthy and nutritious food to create a well-nourished population. The realization of good resilience is not only providing sufficient food for the community but also how humans can obtain food for a healthy life. Globally, FAO in 2003 stated that food security consisted of four pillars, namely supply, accessibility, utilization, and stability. The linkage among the four pillars of food security can be seen in Fig. 2.

Table 1. Description of the Bioeconomy Sector.

Broad Sector Code	Aggregated Sector Code	Description
Primary agriculture (Agri)	Cereal	Cereals (paddy rice, wheat, barley, maize, other cereals)
	Veg	Vegetables (tomatoes, potatoes, other vegetables)
	Fruit	Fruits (grapes, other fruits)
	Oilseeds	Oilseeds (rape, sunflower, and soya seeds)
	OilPlant	Oil plants (olives, other oil plants)
	IndCrop	Industrial Crops (sugar beet, fiber plants, tobacco)
	Ocrop	Other crops (live plants, other crops)
	ExtLiveProd	Extensive livestock production (live bovine, sheep, goats, horses, asses, mules)
	IntLiveProd	Intensive livestock production (live swine, poultry)
	OliveProd	Other live animals and animal products
	RawMilk	Raw milk
	Fishing	Fishing
Food processing (Food)	AnFeed	Animal feed, fodder crops, biodiesel by-product oilcake
	RedMeat	Red meat (meat of bovine, meat of sheep, goats)
	WhMeat	White meat (meat of swine, poultry)
	VegOil	Vegetable oils
	Dairy	Dairy
	Rice	Rice, processed
	Sugar	Sugar, processed
	OliveOil	Olive oil
	Wine	Wine
	BevTob	Beverages and Tobacco
	OfoodProd	Other food products
Bio-mass supply (BioMass)	EnergyCrops	Energy crops
	Pellet	Pellets
	Forestry	Forestry, logging, and related service activities
Bioenergy (BioEne)	BioElectricity	Bioelectricity
	Biofuel1	Biofuel 1st generation (bioethanol, biodiesel)
	Biofuel2	Biofuel 2nd generation (biochemical and thermal technology biofuel)
Bio-industry (BioInd)	Wood	Wood products
	Textile	Textiles, wearing apparel and leather
	BioChem	Biochemicals
Nonbio-based activity (NonBio)	NatRes	Natural resources (coal mining, petroleum, and coal, raw minerals)
	Energy	Energy (electricity and gas)
	Manu	Manufactures
	Service	Services

Sources: Fuentes-Saguar et al. 2017

From Fig. 2, it can be seen that food supply for the population can be obtained from production and imports; a region imports food if the food production in the region is insufficient. The adequacy of food production in the region for its population is influenced by various production factors which include, among others, land, labor, capital and technology, as well as infrastructure.

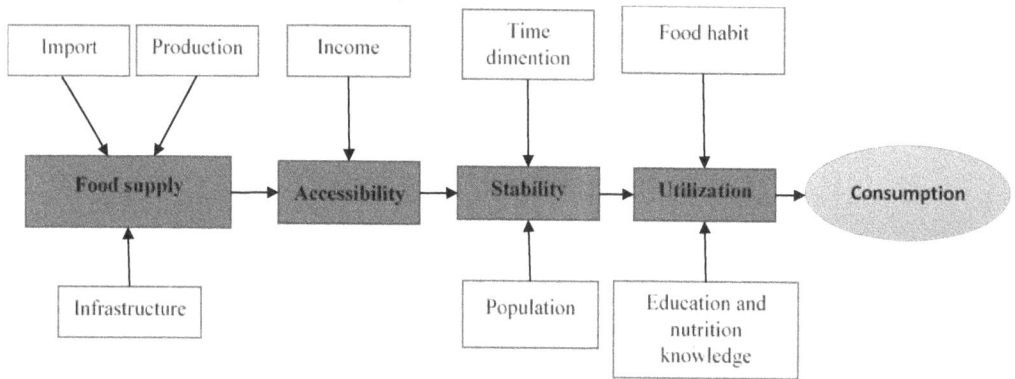

Figure 2. The Linkage of Four Food Security Pillar.

The adequacy of food supply in an area does not guarantee that all residents can access the food properly, as it is influenced by several factors. Accessibility to food can be divided into two, namely physical access and economic access. Physical access is related to the food distribution system-how food are distributed in the entire area. This is related to the transportation system in the area. Economic access is related to the purchasing power of the population, where purchasing power is determined by income and food prices.

The next pillar in the food security system is stability that indicates the ability of the system to maintain its optimum condition over time. In this case, it is related to time which is identical to change, the difference in time will have implications for changes in the system. Therefore, efforts to manage changes with respect to time will determine the assurance of whether the system will run well.

Accessibility to food that is maintained over time will determine the utilization of that food that is manifested in food consumption. Food utilization is influenced by individual factors that determine whether the food that is accessed is used properly. Some of these individual factors include eating habits, education, and knowledge of nutrition.

The aim of agriculture is to grow healthy and nutritious food to create a well-nourished population, then the extent to which this agricultural goal is achieved can be seen from the nutritional state of the population. Carolan (2022) states that based on a measure of calorie/energy hunger, it is known that in 2017 as many as 800 million people (1 in 9) people in the world suffer from malnutrition. The nutritional condition of the world's population is summarized from the FAO report. Richie et al. (2018) state that there are about one billion people suffering from protein deficiency, a third of children under five suffer from stunting, and more than two billion suffer from micronutrient deficiencies. However, there is a paradox as there are more than two billion people suffering from obesity.

The paradox of the nutritional status of the population as described above shows that the world will continue to face the nutritional problems. The problems of undernutrition will be faced by many low-income countries, while high-income countries face nutritional excess. This condition also occurs among the income groups of people in an area.

The OECD (2017) states that the increasing productivity of various agricultural commodities over time in developing countries will result in the increase of production, the decrease of prices, and stimulated consumption. However, the increase in food supply and income has only a small impact on food security, both globally and within ASEAN countries. This fact shows that efforts to achieve the good food security are not enough with the conventional agricultural development approaches.

The conventional agriculture is the agriculture of the past, which is in accordance with the description of most people that when they hear the word agriculture, it is synonymous with the conditions of low education and heavy physical work in cultivating land, but earning inadequate income. The conventional agriculture then changed globally, where food and agriculture are in a system called the agri-food system. The system includes the Input Sector, Production Sector, and Processing-Manufacturing Sector. Beierlein et al. (2008) mention that the Agri-food System is still working; in the current economic conditions, the system is becoming more interdependent than in the past, and it will continue into the future.

Because of food, energy, and financial crisis in 2008, there is a growing interest in the increasing use of bioenergy and biofuels to reduce pressure on demand for energy from fossils. The researchers further explored the shift towards bioenergy from fossil energy. The shift has wider implications for the transformation of the food system, that is from food-based economy to bioeconomy (Babu and Debnath 2019).

4. Bioeconomy and Food Security

The definition of bioeconomy has been elaborated in the previous section that there are several key words if the bioeconomy is associated with food security. Some of these keywords include: renewable resources, biological technology, conversion into food, feed, and products based on biological resources, and the scope is agriculture, forestry, and fisheries.

Besides having positive potentials for food security, bioeconomy is also known to have negative ones. Bioeconomy problems in food security occur if the use of biofuels causes an unstable food supply, especially if the price of agricultural products results in energy prices to fluctuate rapidly.

The link between bioeconomy and food security can be explained by integrating bioeconomy into the agri-food system and its relation to the pillars of food security in Fig. 3.

The integration of bioeconomy into the agri-food system occurs in two sectors within the system; the food production and the food processing and manufacture sector. The first sector includes the input sector because it is a component of the production sector. Examples of bioeconomy activities in the production sector are the applications of biotechnology and tissue culture in obtaining superior seeds and organic pesticides and the use of organic fertilizers. In addition to that, the hydroponic cultivation is also the application of bioeconomy principles in the food production sector.

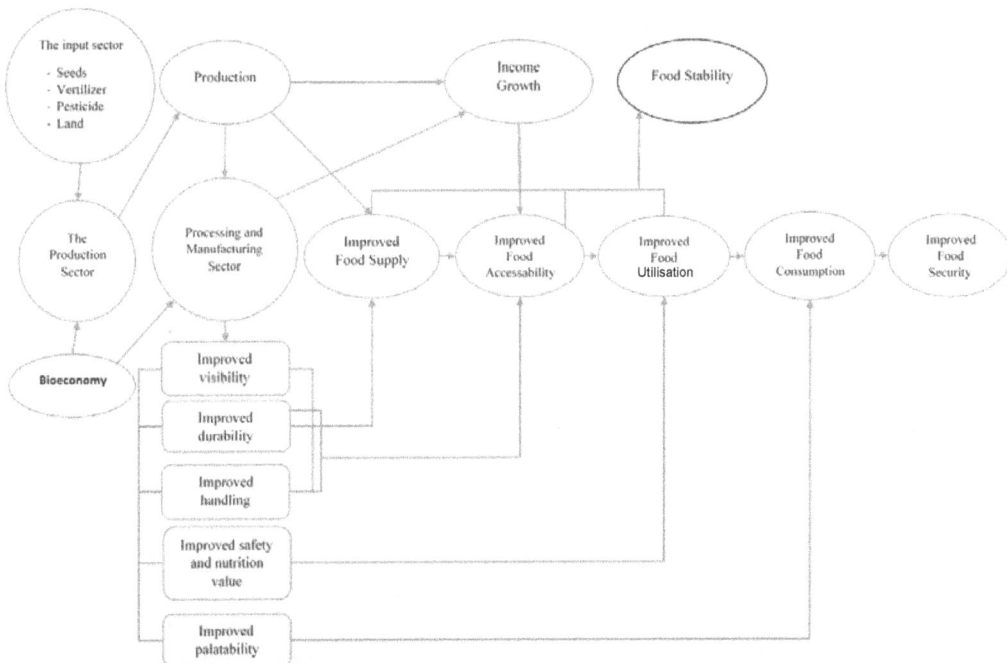

Figure 3. Integration of The Agri-Food System with The Pillars of Food Security.

The application of biological technology will improve the food production both in quantity and quality. The improved food production will directly increase the availability of food, both in the form of fresh food and processed food that results from processing fresh food into intermediate products and final products.

The use of biological technology in food processing, for example, is the use of microorganisms in the manufacture of cheese, yogurt, and the modified cassava flour-mocaf. Food processing will improve food variability, durability, handling, safety and nutrition value, and palatability. Processed food which are easier to handle, more various, and more durable will increase the food availability for the community. The availability of good food will increase people's physical accessibility to food.

In addition to increasing the availability of food in the community, the increase of food production and the improvement of food processing will improve the community's economy. The community economy improvement will increase the purchasing power, and in the end, increase people's economic access to food.

Food processing will improve safety and nutrition value. This condition improvement will provide opportunities for the increasing consumption of this food by individuals. They will choose foods that are considered safe and meet their nutritional needs.

People's food habits have evolved from time to time. This evolution will have implications for changing consumers' food preferences. They will choose the food that is considered adequate to their needs. Good food processing will be able to meet consumer preferences that results in better food consumption.

From the aforementioned description, it is clear that the application of bioeconomy principles in the agri-food system will have an impact on various pillars of food security in the food security system. The stable conditions of food supply, food accessibility, food utilization, and food consumption will have an impact on the food security status of the community.

5. Food Industries

The food industry is one of the sectors in the agri-food system. There are two types of companies in this sector: commodity processors and food manufactures. The food industry will provide food based on the consumers' preferences at the right time and place.

In general, industry is defined as an effort, process or activity of processing raw materials, either raw or semi-finished, so they become goods with better economic value and benefit the community. The term industry is identified with all human economic activities that manage raw goods or raw materials into intermediate goods and finished goods. Raw materials in this case can be various kinds, including food. The food industry is a form of business type in which there is a food production process which includes the selection of raw materials, food processing, food quality testing, packaging, and food distribution activities.

According to Beierlein et al. (2008), food processors take the agricultural products from producers (for example: milk, livestock, wheat) and process them into more acceptable forms by consumers or the food industry (such as pasteurized milk, packed meat, and wheat flour). The characteristics of commodities are generally still visible at the end of the processing. Food manufactures mix the raw (intermediate) agricultural products and process them together into products, for example bread, jam, and corned beef. The purpose of the food industry is to provide food according to the time, place, and form desired by consumers. Zocca et al. (2018) state that the broader purpose of the food industry besides consumers is also aimed at the community, that is to provide safe, healthy, nutritious, economically accessible food, and sustainable food production. Figure 4 illustrates the scope and objectives of the food industry.

Agricultural products in the form of plants and animals can be consumed in a natural state or need to be processed in the food industry so that they can be consumed or stored for a longer time. Consumers choose food based on a number of factors which are generally included in quality

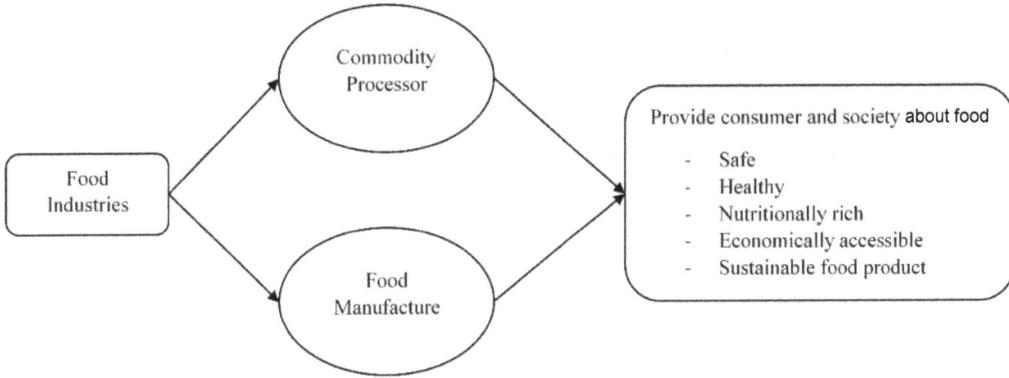

Figure 4. The Scope and the Objectives of the Food Industry. Adapted from Beierlein et al. (2008) and Zocca et al. (2018).

terminology (Tzia et al. 2016). Although the quality is the most essential in the food industry, it is difficult to define it due to the variation in consumers' food expectations. Food preference is determined by many factors including age, religion, culture, social, psychological, and health as well as the expectations of the look, texture, aroma, and flavor. The food processing industry in general is complex, diverse, and requires various techniques to meet consumers' increasingly complex demands. In line with it, the process innovation is compulsory in the food industry.

Food processing can be classified into the chemical, physical, and biological processings. Apart from this, there are also many big issues that affect food processing, including the use of genetic molecules with the use of GMOs (Genetical Molecular Organisms) and the use of animal cloning. The use of biological components in food processing, including the use of GMOs and animal cloning (although it is still controversial) is a bioeconomy implementation in food processing.

Schaschke (2018) says that the main process for producing food that meets biological standards and is acceptable in terms of quality for consumption include mechanical processes, heating, cooling, and fermentation. Fermentation is one of the bioeconomy processes. Several food processing processes are described in the following sections:

5.1 Mechanical processes

Some raw food materials require a mechanical process at the beginning of the process; for example, the process of reducing the size in the cutting process of apples or fruit containing tannins will result in a brown color on the material surface. Another mechanical treatment is filtration and centrifugation, which is used to separate liquid from solid. The last mechanical treatment is protecting packaging like cans, jars, and plastic sachets. They protect food from damage, organisms, dust, and mechanical damage.

5.2 Heating

Heating is intended to eradicate pathogenic and other destructive microorganisms. It is also intended to reduce humidity resulting in physical and chemical changes. Heat treatment and temperature regulation are carried out in various ways, including sterilization and pasteurization.

5.3 Mixtures and Emulsions

Most food is consumed in mixed forms. Mixing is essential to induce the desired reaction. Emulsification occurs when two liquids do not dissolve in each other producing unwanted droplets. To avoid it, an emulsifier is used.

5.4 *Novel Food Processing*

Novel Food is defined as food that has not received appreciation for consumption or has never been processed before. An example of this new food processing is the use of high pressure. Recently, consumers generally expect high quality food with minimal processing, free of additives, and highly nutritious. Fortunately, this high-pressure processing has unique effects that fulfill the aforementioned consumers' expectation.

6. Bioeconomy and Food Industries

The definition of bioeconomy was presented at the 2018 Global Bioeconomy Summit. Bioeconomy is the production, utilization and conservation of biological resources, including science, technology, and innovation, to provide information, products, processes and services in all economy sectors with the aim of achieving economic sustainability. Although definitions vary in the side of academics, government, and industry, in principle, bioeconomy is an effort to manage biological resources in a sustainable economy for the welfare of the community while preserving the natural environment.

The current world population reaches 7.8 billion in 2022 as reported by the United States Census Bureau. The population will keep increasing every year and this is a challenge, especially in terms of providing food which is the basic need of the world community. Additionally, the increasing population also requires wider land area for the settlement. This results in a decrease in agricultural land due to the function conversion to residential and industrial needs. Population growth, limited land, and the carrying capacity of the natural environment raise not only various problems but also ideas to overcome them and bioeconomy approach is one of them.

Bioeconomy or bio-based economy is related to the use of biological resources and their substitutes to produce energy, food, feed, fiber, and other production goods by applying biological processes (BMBF 2011). It includes various sustainable economy activities in the fields of agriculture, fisheries, forestry, and other biological resources by taking into account the environmental factors. The outputs of the various activities become sources of raw materials for the food industry that produces various kinds of food and beverage products as shown in Fig. 5.

Food and feed production is a top priority in formulating strategies and action plans for the use of natural resources efficiently and sustainably. Efforts to increase agricultural productivity, land management, logistics, and storage techniques can increase the efficiency of the food and feed supply chain. However, they are not enough. Efforts are also needed to utilize the agricultural waste or by-products that can be used as raw materials to produce other products (Trigo et al. 2021) so it leads to zero waste or environmentally friendly processes.

Agriculture, fisheries, livestock and forestry industries are generally located in rural areas where the nature supports these activities. In the production activity, the processing of biological resources involves the surrounding community who work in the field such as a large number of farmers or ranchers who form a community. Being farmers or ranchers are the livelihoods of people living in rural areas. Both are their source of income in meeting their economic needs.

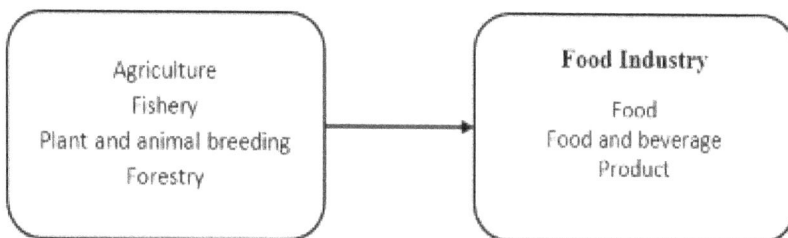

Figure 5. Material Supply Food Industry. Adapted from (Kristinsson and Jorundsdottir 2019, Robert et al. 2020, von Braun 2018, Wohlfahrt et al. 2019).

To illustrate, a dairy farming community that produces fresh milk in a rural area in West Java, Indonesia involves around 2,600 dairy farmers and manages around 4,000 cows. The farmer community forms a Cooperative to collect fresh milk produced by farmers using Milk Collection Point (MCP) technology to maintain the quality of fresh milk according to the standards set by the Dairy Processing Industry. As a manufacturing company in the field of milk processing, the Dairy Processing Industry produces various dairy products and their derivatives and the raw materials are supplied by dairy farmers. This Dairy Processing Industry is a food industry that supplies dairy products which are one of the basic needs of the community in meeting their nutritional needs.

The processing output of agriculture, fishery, plant and animal breeding, and forestry as the supplier of raw materials is needed by the food industry in producing various food and beverage products. In general, the fields of agriculture, fishery, plant and animal breeding, and forestry involve a large number of farmers, breeders, or others so it tends to be labor intensive, while the food industry is generally managed by several companies as a manufacturing industry with adequate capital and is managed more professionally.

7. Bioeconomy and Food Industries System

The concept of bioeconomy is a comprehensive approach in meeting the increasing world food demand sourced from the renewable resources by taking into account economic and social factors, as well as climate change.

Bioeconomy and food industries are interrelated to form a value chain system starting from natural resources to food and beverage products to meet community's basic consumption needs. The bioeconomy and food industry system as shown in Fig. 6 forms a sub-system that is interconnected with one another to form an integrated supply chain system by also taking economic, social, environment and SDGs factors into account.

Resources are production factors that are used as the facilities and infrastructures in the transformation process from inputs to outputs to create added value. The resources are needed in every transformation in the system, from the bioeconomy process, downstream logistics, food industry, and upstream logistics, before finally arriving at consumption.

The logistic system in the bioeconomy consists of downstream and upstream logistics (Kristinsson and Jorundsdottir 2019, Robert et al. 2020). The bioeconomy and food industry system involve a strong and efficient logistic system as a supply chain starting from raw materials, semi-finished materials, to finished products that are ready to be distributed to meet the community needs. The logistic system consists of downstream and upstream logistics, each forms a supply chain starting from distributors, retailers, agents, to local trading. The logistic system is a series of interrelated activities including planning, implementing, and supervising the process of moving goods.

In the end, all production results in the form of food and beverages will be distributed to meet people's consumption needs. This consumption forms a balance pattern between supply and demand as a price determinant. The increasing world population has resulted in an increase in demand for the food and beverage products as a basic need. The market mechanism will encourage producers with more efficient processing processes to have the opportunity to dominate the market because they can provide lower prices. People as consumers will choose the food and beverage products for their consumption based on the considerations between the offered quality and price.

The bioeconomy and food industry system is in line with the Sustainable Development Goals (SDGs) put forward by the United Nation which was agreed by 193 countries with the targets and indicators published in 2017. The SDGs contain 17 important points, namely: (1) No Poverty, (2) Zero Hunger, (3) Good Health and Well-being, (4) Quality Education, (5) Gender Equality, (6) Clean Water and Sanitation, (7) Affordable and Clean Energy, (8) Decent Work and Economic Growth, (9) Industry, Innovation and Infrastructure, (10) Reduced Inequality, (11) Sustainable Cities

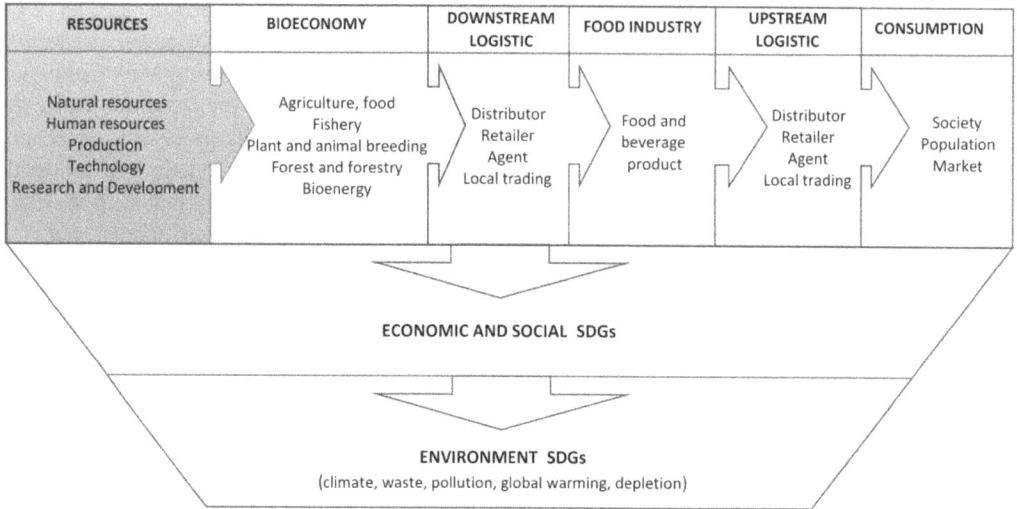

Figure 6. Bioeconomy and Food Industries System.

and Communities, (12) Responsible Consumption and Production, (13) Climate Action, (14) Life Below Water, (15) Life On Land, (16) Peace, Justice, and Strong Institutions, (17) Partnerships for the Goals. All the 17 points in the SDGs are related and aligned with the bioeconomy and industry system.

The transformation of the whole process in the system supports the quality of life activities of social community according to the SDGs (1, 2, 3, 4, 5, 10, 11, 16, 17), and has an impact on the community's economy according to the SDGs (8, 9, 12) especially in the increase of economic growth. In the bioeconomy and food industry system, the interaction among all processes in the system ultimately has an impact on the environmental factors such as the SDGs (6, 7, 13, 14, 15).

The bioeconomy and food industry system is an integrated system that is interrelated among its sub-systems consisting of natural resources, bioeconomy, downstream logistics, food industry, upstream logistics, consumption, economic and social, and environment. Each sub-system in the bioeconomy and food industry system is described in the following sub-chapters.

7.1 Natural Resources

Natural resources are everything coming from nature that can be used to meet the needs of human life. The existence of natural resources is not evenly distributed on Earth—some countries have abundant biological and non-biological resources (Venables and Anthony 2016). However, the wealth of natural resources is often not in line with the economic level of the countries; it happens when they are not able to process them using adequate technology.

Based on their nature, natural resources are classified into renewable and non-renewable natural resources. The renewable natural resources are natural resources that continue to exist as long as they are not overexploited, for example plants, animals, microorganisms, water, wind, and sunlight. Meanwhile, the non-renewable natural resources are natural resources whose formation takes thousands of years so that they are limited in number, for example oil, gold, iron, and various other mining materials.

The food industry is generally related to the further manufacturing processes of agricultural, fishery, animal husbandry products, which are classified as the renewable natural resources. To achieve maximum results, processing with modern technology is required to produce a high level of efficiency.

7.2 Human Resources

Human resources are individuals who work as movers, thinkers, and planners in carrying out organizational activities (Mathis and Jackson 2003). Human resources are the key determinants of business progress and success; therefore, adequate competencies are needed and they need to be trained and developed.

The definitions of human resources can be divided into two: micro and macro. The definition of human resources in a micro sense is individuals who work and become members of a company as employees, laborers, workers and so on. Meanwhile, the definition of macro human resources is the population of a country who have entered the age of the workforce, both those who have not worked yet and who have already worked.

Human resources in the micro sense generally refer to workers and employees who work formally for companies in the fields of agriculture, animal husbandry, and plantations. In downstream logistics, food industry, and upstream logistic companies, workers will be involved to carry out their business activities. Manpower deployment is carried out in accordance with the competencies required by the company which is carried out formally in accordance with the applicable labor laws.

Human resources in the macro sense are partly people of working age who are in the locations where natural resources exist and who have the opportunity to process natural resources in the vicinity. The process of natural resources was initially carried out conventionally and from generation to generation by the people who were in the location of these natural resources with limited competence. Even though they are a source of community livelihood, the process of natural resources is not optimal if the competencies are inadequate. It is one of the reasons why the location of the existence of natural resources is not in line with the income of people with low economic levels.

The increase of human resource competence, especially those around natural resources, is indispensable, so the processing of natural resources can be carried out efficiently and involves the surrounding community such as farmers, breeders, and others. The professions can even be passed on to the next generation in the processing of natural resources. It is a sustainable economic cycle that needs to be developed with a breakthrough in increasing human resource competencies.

7.3 Production and Technology

The production process in the food and beverage industry is currently moving towards the industry 4.0 with the concept of a smart industry where all elements are connected to electronics (Demir and Istanbullu 2020). The production process in the industrial revolution 4.0 is carried out with digital transformation to support more agile business processes with orientation to technological, human, and cultural aspects. The application of the industry 4.0 in the food industry has challenges in converting machines into automations and robotics where the production machines need to be integrated with an Artificial Intelligent (AI) system.

The utilization of modern technology in the natural resource management has been started since the industrial revolution era. Processing natural resources with technology is proven to be able to provide more results in a shorter time. In the fields of agriculture, fisheries, and plantations, the use of technology increases productivity so it has been able to meet the increasing demand for food up to the present day.

Smart Farming 4.0 is a technology-based intelligent farming method that uses several agricultural technologies including automatic watering, drone sprayers (drones for spraying pesticides and liquid fertilizers), surveillance drones for land mapping, as well as soil and weather sensors. The existence of installed soil and weather sensors on agricultural land will assist farmers in obtaining data about their crops. The obtained data from the sensors include air and soil humidity, temperature, soil pH, water content, rainfall, and wind speed. The data can be used to monitor the condition of their land. The precise use of the technology helps farmers improve the quality and quantity of their crops.

Internet of Things (IoT) in agriculture is in the forms of sensor technology for water use, sensors to detect pest attacks, as well as sensors to determine environmental emissions. With the application of such technology, the agricultural output can be increased rapidly and accurately. In addition, IoT can make production areas' monitoring easier since it is connected to smartphones. The use of IoT can make the precision farming and smart irrigation happen. The use of sensors applied in agricultural land allows farmers to obtain detailed information on topography, fertility levels, acidity levels, soil temperature, and can even measure the weather and predict the weather pattern.

Blockchain is a distributed ledger containing several blocks of detailed transaction records that are connected chronologically to form a chain network. The technology can be used in the food and agriculture sectors to record the state of agriculture, inventories and contracts in agriculture accurately. The blockchain technology can trace the origin of food thereby helping to create a trustworthy food supply chain and building trust between producers and consumers. The technology can also be used to store, process and track the transaction data from the suppliers of raw materials, suppliers of semi-finished materials, to product distribution through the distributors and retailers.

7.4 Research and Development

Research and Development (R&D) is a study to develop new products, processes, or improve existing products (Melissa 2017). This is a way to encourage the development of a broad bioeconomy, due to the development of innovations in products, processes, or systems. The results of Research and Development can be a solution to problems in the processing of natural resources related to the social and environmental impacts. Research and Development involves science, high-level scientific expertise and multidisciplinary to solve the complex problem with various perspective of knowledge to get the best result of the research.

Achieving the transition towards the sustainable, low-carbon, resource-efficient, and recycling bioeconomy requires support from the adequate science and technology. The involvement of scientists and researchers from various fields such as chemistry, biology, biochemistry, economics, agronomy, engineering, medicine, and social sciences is necessary for the success of a bioeconomy concept.

Research and Development activities support the innovations in all processes in the bioeconomy and food industries system, starting from the processing of resources, bioeconomy, downstream logistics, food industry, upstream logistics, to consumption. The produced findings and innovations can be in the forms of big concepts, structural changes, quality improvement, efficiency, and productivity in the whole processes in the bioeconomy and food industry system.

7.5 Economic and Social Impact

Various definitions of bioeconomy link the sustainability of processing the natural resources with economic, environmental and social perspectives; especially from an economic perspective, they are clearly defined by the words 'bio' and 'economy'. The transformation of the whole processes in the bioeconomy and food industry system has the main goal of increasing economic growth. This can be used as a way to increase the sustainable economic growth of a country and this has been studied in the United States, Finland, Germany, Netherland, United Kingdom, and China (Sillanpää and Ncibi 2017).

The implementation of bioeconomy in the food industry requires the investment costs and operational costs which will be calculated by the size of market demand and product prices (Wright and Brown 2007). The economic indicators of the invested capital will be seen from the calculation of the resulted ROI, the profit, net present value, and sensitivity analysis. There are 3 economic aspects related to the SDGs, namely Decent Work and Economic Growth (SDGs 8); Industry, Innovation and Infrastructure (SDGs 9); and Responsible Consumption and Production (SDGs 12).

Social factors are of particular concern in increasing industrial growth, including the food, agriculture, fisheries, and plantations industry. It appears as the reflection of corporate social responsibility on all stakeholders including employees, buyers, investors, government, community (Aguilera et al. 2007), and the environmental sustainability of future generations.

Socially sustainable industries are those that have limited negative impacts on society during production, utilization, recycling and final disposal. Success indicators are related to social dimensions including food security, energy security, employment, inequality, and social welfare (Fleurbaey 2015, Dempsey et al. 2011). Other social aspects also include the respect for land rights, and social acceptance.

In the bioeconomy and food industry system, the social factor is one of the main goals of all processes in the system. It is in line with the SDGs that has 9 points that are mostly related to the social dimension in the bioeconomy and food industry system. The social factors on the SDGs that have the most impacts (9 points) on the ecosystem of bioeconomy and food industry system are as follows: No Poverty (SDGs 1), Zero Hunger (SDGs 2), Good Health and Well-being (SDGs 3), Quality Education (SDGs 4), Gender Equality (SDGs 5), Reduced Inequality (SDGs 10); Sustainable Cities and Communities (SDGs 11); Peace, Justice, and Strong Institutions (SDGs 16); and Partnerships for the Goals (SDGs 17).

7.6 Environment

The environmental impact is a decrease in the quality of the human environment caused by human activities. The processing activities of natural resources including food, agriculture and animal husbandry are one of the causes of the environmental problem emergence. Therefore, it is necessary to make efforts to maintain environmental sustainability. In the bioeconomy and food industry system, the efforts to protect the environment are one of the considerations in the transformation process at the end of the system.

The environmental quality can be monitored from the air quality, pollution, climate, soil, water, solid waste and wastewater processing, gas emissions of greenhouse, biodiversity and wildlife conservation, energy, waste, global warming, and depletion (Ruiz et al. 2012, Tanzil and Beloff 2006). The production and transformation processes need to be carried out in an environmentally friendly manner so they do not cause a decline in the carrying capacity of the environment. It is in line with the SDGs; there are at least 5 points related to the bioeconomy and food industry system, namely Clean Water and Sanitation (SDGs 6), Affordable and Clean Energy (SDGs 7), Climate Action (SDGs 13), Life Below Water (SDGs 14), and Life On Land (SDGs 15).

8. What is Sustainability?

Sustainability is an ability to survive and adapt to changes. In the context of ecology, it means that biological systems are still able to sustain unlimited biodiversity and productivity. This is achieved by means of not utilizing and exploiting natural resources to damage the ecology or ecological balance in the area.

Earth provides food, energy, and materials for the survival of living. For decades, the dependence on fossil resources and derived fuels, chemicals, and materials has resulted in unsustainable economic models that have created complex situations around the world. The demand for raw materials continues to increase, and the depletion of most resources requires humans to think and operate in a holistic sustainable manner, especially by managing abundant fossil resources in an efficient and environmentally friendly manner, especially coal whose reserves are estimated to be available until the end of the 21st century (Mohr and Evans 2009).

Sustainability is basically the result of balancing human needs with available resources to meet the current and future needs without depleting them. Overall, sustainable development is a paradigm built to meet human needs economically, environmentally and socially. In this regard, bioeconomy, as an economic model based on the utilization and conversion of biomass to produce commodities, is the core of a global sustainable economic strategy.

9. Circular Economy

According to the European Union Parliament, the circular economy is a model of production and consumption which consists of sharing, leasing, reusing, repairing, refurbishing, and recycling existing materials and products. In this way, the product life cycle is extended (European Parliament 2015). In practice, the circular economy implies reducing waste to minimum. When a product is at the end of its use, the product can be reused as productively as possible so as to provide added value. The circular economy is a regenerative system that minimizes the uses of resources, waste, emissions, and excess energy by slowing, closing, and narrowing the energy and material cycles (Geissdoerfer et al. 2017).

The circular economy is important to maintain the availability of raw materials for industry in the midst of the increasing human population, where the raw materials available from nature are limited. In addition, the circular economy can also reduce environmental impacts due to the smart use of raw materials, is able to reduce carbon dioxide emissions, and benefits the economic growth (Hysa et al. 2020).

The bioeconomy has similarities with the circular economy. Some of the similarities are increase of resources with the higher eco-efficiency and lower Green House Gas (GHG) effects. In addition, they also reduce the demand for fossil carbon and lead to the valorisation of waste and side streams (Carus and Dummer 2018).

However, the circular economy is a broader cross-sectoral concept when compared to the bioeconomy. The circular economy strengthens the eco-efficiency processes and uses recycled carbon to reduce the use of additional fossil carbon. Bioeconomy replaces fossil carbon with bio-based carbon from biomass.

References

Aguilera, R.V., Rupp, D.E. and Williams, C.A. 2007. Putting the S back in corporate social responsibility: a multilevel theory of social change in organizations. Acad. Manag. Rev. 32: 836–63.

Allwood, J.M. 2014. Squaring the circular economy: the role of recycling within a hierarchy of material management strategies. In Handbook of recycling (pp. 445–447).

Babu, S.C. and Debnath, D. 2019. Bioenergy economy, food security, and development. Dalam Suresh Chandra Babu and Deepayan Debnath (eds). Biofuels, Bioenergy and Food Security. Elsevier Inc. United Kingdom.

Beierlein, J.G., Schneeberger, K.C. and dan Osburn, D.D. 2008. Principles of Agribusiness Management. Waveland Press Inc. United States of America.

Biernat, K. 2019. Introductory chapter: objectives and scope of bioeconomy. In Elements of Bioeconomic (pp. 1–12).

Carolan, M. 2022. The Sociology of Food and Agriculture. Third edition. Routledge 2 Park Square, Milton Park, Abingdon, Oxon.

Carus, M. and Dammer, L. 2018. The "Circular Bioeconomy" –Concepts, Opportunities and Limitations, Nova, Institut, Hürth (Germany).

Cichocka, D., Claxton, J., Economidis, I., H€ogel, J., Venturi, P. and Aguilar, A. 2011. European Union research and innovation perspectives on biotechnology. J. Biotechnol. 156: 382–91.

Demir, Y. and Istanbullu, F.D. 2020. The Effects of Industry 4.0 on the Food and Beverage Industry. Journal of Tourismology. DOI: 10.26650/jot.2020.6.1.0006.

Dempsey, N., Bramley, G., Power, S. and Brown, C. 2011. The social dimension of sustainable development: defining urban social sustainability. Sustain. Dev. 19: 289–300.

Dobrijevic, G., Boljanovic, J.D., Dokovic, F., Pejanovic, R., Skataric, G. and Damnjanovic, I. 2019. Bioeconomy-based food industry of serbia: the role of intellectual capital. Economics of Agricultur, Belgrade, pp. 51–62.

Errol, S. van Engelen. New World Technologies 2020 and Beyond. Business Expert Press, 2020.

European Commission. 2012. Commission adopts its Strategy for a sustainable bioeconomy to ensure smart green growth in Europe. Statistics, February, 1–5.

Eueopean Parliament. 2015. EuCircular economy: definition, importance and benefitso Title.

Federal Ministry of Education and Research (BMBF). 2011. Berlin: National Research Strategy BioEconomy 2030 – Our Route towards a biobased economy. p. 56. Available online at http://www.bmbf.de accessed March 10, 2022

Fleurbaey M. 2015. On sustainability and social welfare. J. Environ. Econ. Manag. 71: 34–53.

Fuentes-Saguar, Datricia, D., Mainar-Causape, Alfredo J. and Ferrari, Emanuele. 2017. The role of bioeconomy sectors and natural resources in EU Economies: A social accounting matrix-basec analysis approach.

Geissdoerfer, M., Savaget, P., Bocken, N.M.P. and Hultink, E.J. 2017. The Circular economy—A new sustainability paradigm? Journal of Cleaner Production, 143 (April 2018), 757–768. DOI: https://doi.org/10.1016/j.jclepro.2016.12.048.

Giuntoli, J., Robert, N., Ronzon, T., Sanchez Lopez, J., Follador, M., Girardi, I. et al. 2020. Building a monitoring system for the EU bioeconomy Progress Report 2019 (Issue January). DOI: https://doi.org/10.2760/717782.

Hysa, E., Kruja, A., Ur Rehman, N. and Laurenti, R. 2020. Circular Economy Innovation and Environmental, Sustainability Impact on Economic Growth: An Integrated Model for Sustainable Development, MPDI.

Kardung, M. 2019. Framework for Measuring The Size and Development of The Bioeconomy.

Kristinsson, H.G. and Jorundsdottir, H. olina. 2019. Food in the bioeconomy. Trends in Food Science & Technology, 84, 4–6.

Mathis, R.L and Jackson, J.H. 2003. Human Resource Management. Thomson.

MEE. 2014. The Finnish Bioeconomy Strategy. 31.

Melissa A Schilling. Strategic Management of Technologycal Inovation. Mc Graw Hill Education, Fifth Edition, 2017

Mohr, S.H. and Evans, G.M. 2009. Forecasting coal production until 2100. Fuel. 88: 2059–67.

OECD. 2006. OECD International Futures Programme The Bioeconomy to 2030: Designing a Policy Agenda, available at http://www.oecd.org/sti/emerging-tech/34823102. Accesssed March 8, 2022.

OECD. 2017. Building Food Security and Managing Risk in Southeast Asia. OECD Publishing, Paris.

Pemerintah Republik Indonesia. 2012. Undang-undang Republik Indonesia Nomor 18 tentang Pangan.

Pfau, S.F., Hagens, J.E., Dankbaar, B. and Smits, A.J.M. 2014. Visions of sustainability in bioeconomy research. Sustainability 6: 1222–49.

Ritchie, H., Reay, D.S. and Higgins, P. 2018. Beyond Calories: A Holistic Assessment of the Global Food System, Frontiers in Sustainable Food Systems 2, www.frontiersin.org/articles/10.3389/ fsufs.2018.00057/full.

Robert, N., Giuntoli, J., Araujo, R., Avraamides, M., Balzi, E., Barredo, J.I., et al. 2020. Development of a bioeconomy monitoring framework for the European Union: An integrative and collaborative approach. New Biotechnology 59: 10–19.

Ruiz-Mercado, G.J., Smith, R.L. and Gonzalez, M.A. 2012. Sustainability indicators for chemical processes: I. Taxonomy. Ind. Eng. Chem. Res. 51: 2309–28.

Schaidle, J.A., Moline, C.J. and Savage, P.E. 2011. Biorefinery sustainability assessment. Environ. Prog. Sustain. Energy 30:743–53.

Schaschke, C.J. 2018. Food Processing. Accessed at https://bookboon.com/ downloaded on March 2022.

Sillanpää, M. and Ncibi, C. 2017. A sustainable bioeconomy: The green industrial revolution. In A Sustainable Bioeconomy: The Green Industrial Revolution. DOI: https://doi.org/10.1007/978-3-319-55637-6.

Tanzil, D. and Beloff, B.R. 2006. Assessing impacts: overview on sustainability indicators and metrics. Environ. Qual. Manag. 15: 41–56.

Trigo, E., Chavarria, H., Pray, C., Smith, S.J., Torroba, A., Wesseler, J., et al. 2021. The Bioeconomy and Food System Transformation.

Tzia, C., Giannou, V., Lignou, S. and Lebesi, D. 2016. Raw materials of foods: handling and management dalam Da-Wen Sun (ed) Food Processing (Food Safety, Quality, and Manufacturing Process). Taylor and Francis Group. London.

Venables, Anthony J. (February 2016). Using natural resources for development: "Why Has It Proven So Difficult?" Journal of Economic Perspectives. 30(1): 161–184. DOI: 10.1257/jep.30.1.161.

Von Braun, J. 2018. Bioeconomy—the global trend and its implications for sustainability and food security. Global Food Security 19: 81–83.

Wohlfahrt, J., Ferchaud, F., Gabrielle, B., Godard, C., Kurek, B., Loyce, C., et al. 2019. Characteristics of bioeconomy systems and sustainability issues at the territorial scale. A review. Journal of Cleaner Production 232: 898–909. DOI: https://doi.org/10.1016/j.jclepro.2019.05.385.

Wright, M. and Brown, R. 2007. Comparative economics of biorefineries based on the biochemical and thermochemical platform. Biofuels Bioprod. Biorefin. 1: 49–56.

Zocca, R.O., Gaspar, P.D., Nunes, J. and de Andrade, L.P. 2018. Introduction to sustainable food production (Chapter 1). dalam Charis M. Galanakis (eds.). Sustainable Food Systems From Agriculture to Industry. Elsevier Inc. United Kingdom.

17

Smart Logistics for Sustainable Food Industries

Abderahman Rejeb,[1],* Alireza Abdollahi,[2] Karim Rejeb[3] and John G. Keogh[4]

1. Introduction

The food supply chain is vital to the global economy (Sufiyan et al. 2019) and is one of the pillars of human civilization (Aqeel-Ur-Rehman et al. 2014). Logistics is critical to the food supply chain's success (Fredriksson and Liljestrand 2015) and foods once deemed as exotic are available year round in many markets. It is critical to meet customer needs by supplying the appropriate quantity and quality of food products at a reasonable price and at the right time with little or no waste (Jagtap et al. 2021). Scholars have emphasized the need to transform the conventional food supply chain toward more sustainable solutions. Organizations have historically focused on agrifood economics and favorable business models in their decisions to source and import foods from all over the world. Due to mounting pressures, which include empirical evidence on the effect of climate change, and evolving regulations related to supply chain due diligence is an focused term used in the context of ESG. For example, the German Government far reaching Supply Chain Due Diligence Act that came into effect 1-Jan-2023 (Zhu et al. 2018).

The food chain must operate efficiently in order to feed an estimated 9.7 billion people by 2050 (UN 2019). Additionally, increased water and energy consumption, climate change impact, greenhouse gas emissions, increased use of pesticides and fertilizers, and other resources (e.g., energy, water, soil), as well as pollution caused by supply chain management functions such as logistics, manufacturing, distribution, and transportation, are some of the environmentally and socially detrimental consequences of continuing this trend (Agyabeng-Mensah et al. 2020, Friha et al. 2021, Trivellas et al. 2020). Additionally, there are debates regarding food safety, quality, security, and waste throughout the food logistics process, all of which call for increased traceability, transparency, and monitoring of the food logistics process (Barbosa 2021, Huan et al. 2021, Zheng et al. 2021). Stakeholder expectations are becoming more complex, necessitating the development

[1] Doctoral School of Regional Sciences and Business Administration, Széchenyi István University 9026 Győr, Hungary.
[2] Department of Business Administration, Faculty of Management, Kharazmi University, Tehran, Iran.
Email: abdollahi.alirez@gmail.com
[3] Faculty of Sciences of Bizerte, University of Carthage, Zarzouna, Bizerte 7021, Tunisia.
Email: karim.rejeb@fsb.ucar.tn
[4] Henley Business School, University of Reading, Greenlands, Henley-on-Thames, RG9 3AU, UK.
Email: john@shantalla.org
* Corresponding author: abderrahmen.rejeb@gmail.com

of more integrated and traceable systems (Jagtap et al. 2021). Finally, sustainability's components are inextricably linked in a complex series of overlapping and interacting ecosystems. Like the domino effect, each of these ecosystems have variables that may have a positive or negative impact and cascading consequences on adjoined ecosystems (Keogh and Unis 2020). For instance, as Van Der Vorst et al. (2009) proposed, consumer awareness of sustainability-related issues is increasing, directly impacting their purchasing decisions.

Incorporating cutting-edge technologies has been viewed as a promising pathway to overcome these obstacles. Furthermore, these debates spawned the concept of smart food logistics (Verdouw et al. 2013), which aims to transform various aspects of food logistics through the use of novel technologies. The term "smart logistics" is loosely defined, and there are currently no unifying conceptualizations. However, the term 'smart logistics' is frequently used to refer to a variety of logistical operations, such as transportation, warehousing, distribution, and customer service, that are more intelligently planned, managed, and regulated than traditional solutions (Ding et al. 2021). In this chapter, we examine the transformative role of Industry 4.0 technologies and pillars in smart logistics (Rüßmann et al. 2015), focusing on their impact on economic, environmental, and social sustainability. We devote a section to each technological enabler and conclude with a concise summary.

2. Technological Enablers for Smart Food Logistics

2.1 Robots

Robots and automation represent major elements of the Industry 4.0 vision (Bader and Rahimifard 2020) and are widely used across several industrial sectors. Robotics and automation are utilized to create devices and equipment that operate with little or no human intervention (Vallandingham et al. 2018). Robotic technology aspires to alleviate and eventually eliminate strenuous or unsafe human labor while contributing to productivity (Duckett et al. 2018). While robotics is the process of developing and using robots to perform certain jobs, automation is the use of different types of automated mechanized technology. Agri-food supply businesses have considerably benefited from incorporating robots in activities as various as monitoring (e.g., health monitoring of crops by drones), weed detection, precision agriculture, disaster management. In food logistics, robotics application encompasses several industrial operations. Originally, robots were primarily employed for packing and palletizing (Khan et al. 2018). Incorporating robots and automating food chain activities makes it possible to monitor, track and trace products throughout the supply chain. This is especially critical for perishable food products that need careful monitoring and handling conditions (Pal and Kant 2019). The significant advantage of robotics use in logistics is the optimization of products' handling and transportation. Robotic applications provide more efficient material movement and tracking throughout the logistics process.

From a sustainability perspective, robotics could be utilized to increase productivity, reduce costs, and boost production volume by reducing the time necessary to complete some tasks. The task accuracy associated with automation contributes to minimizing defects and waste in food logistics (Duong et al. 2020). Robotics and automation can also reduce human error and could help with consistent product quality and aid in adherence to production standards and regulatory compliance, thus leading to consumer satisfaction and more demands for the products. Similarly, automation could improve operational flexibility and help firms to adapt and meet changing consumer needs (Echelmeyer et al. 2008). Additionally, robots enable businesses to protect workers from dangerous working conditions. Typically, these environments require workers to load and unload heavy products, standing in production and assembly lines for long periods leading to physical strain, or utilizing equipment in unsanitary conditions. Robots could be used to retrieve, move, sort, package, and distribute items within and between warehouses and production lines (Karabegović et al. 2015). Furthermore, mobile or autonomous robots are equipped with sensors and warning systems that reduce the risk of collisions and enhance worker safety. The automation of such activities and systems may lead to the redistribution of employees and the enhancement of working environments and conditions (Duong et al. 2020, Echelmeyer et al. 2008). The improvements in traceability and monitoring capabilities enable food organizations to meet social and environmental expectations. From the environmental perspective, the efficient use of resources and waste reduction are considerable contributions of implementing robotics and automation.

2.2 Big Data Analytics

In recent years, big data analytics (BDA) approaches have arisen as an enabling tool for organizations to gain a competitive advantage (Fosso Wamba et al. 2015, 2018, Tan et al. 2015). BDA garnered considerable interest from academics, practitioners, and decision-makers (Dubey et al. 2016). Big data consists of diversified (variety), voluminous (volume) data that is generated at a high rate of speed (velocity). These attributes are often referred to as the 3Vs. It was later developed by adding value and veracity and became the 5V concept (Queiroz and Telles 2018). Big data advancements have the potential to provide several advantages to sustainable food logistics, such as improvements in forecasting, operational performance, inventory management, distribution, and transportation management. Food logistics generates a considerable quantity of data due to the sheer volume of food products exchanged across the value chain. These characteristics may be linked to their provenance (origin), ownership chain, physical conditions (weight, size), quality (form), environment-related circumstances (e.g., temperature, moisture, microbial activity), freshness, and position (Jagtap and Duong 2019).

From a sustainability perspective, BDA could usher in the development of data-driven food supply chains with actionable insights for management decision making. BDA could facilitate coordination among supply chain exchange partners and improve supply chain responsiveness to changes in market demand resulting in better firm performance (Yu et al. 2018). BDA can be utilized to enhance efficiency and effectiveness of organizations. For instance, BDA can aid in the optimization of truck delivery routes by analyzing data to identify or adjust delivery routes in response to real-time road accidents or traffic congestion, hence avoiding food waste brought about by delays in transit. This, in turn, aids in the optimization of traffic, and reduces energy and resource consumption as well as reducing greenhouse gas emissions and other pollutants. The data could also be utilized for new product and services development. Integrating with other cutting-edge technologies, the information retrieved from BDA could contribute significantly to scheduling by aggregating available resources, capacity, and demand in real-time, enabling supply chain actors to make more informed decisions and perform preventive actions which lead to increased food safety and food waste mitigation (Parvin et al. 2019). An area requiring further exploration is the optimization of trucking capacity and their utilization to reduce the volume of empty cargo trucks on return journeys. This phenomena is often referred to as 'empty miles'.

2.3 Digital Twin and Simulation

A digital twin could be defined as a virtual representation that simulates the seamless integration and behavior of cyber-physical systems. These digital tools simulate physical environments and may contribute to the lowering of manufacturing and supply chain lead times, resulting in a leaner, more adaptable, and intelligent manufacturing and supply chain environment (Abideen et al. 2021). Moreover, a digital twin is a digital/cyber rendition of various forms of data (e.g., operational data), processes and models which could include physical spaces, machinery or tools. It evolves in lockstep with the real system and is capable of predicting the outcome of adapted solutions in real-world systems (e.g., optimizing physical operations) (Boschert and Rosen 2016). Whereas a digital twin endures over time, technology-based simulation may be utilized in a 'one-off' manner to plan and construct exploratory models for effective decision-making, and to develop and simulate complex production systems' activities (Coelho et al. 2021).

From sustainability perspectives, simulation and digital twin technologies assist food businesses in identifying the costs and risks associated with their activities, as well as the hurdles to system adoption and their influence on operational performance, social and environmental environments (de Paula Ferreira et al. 2020). Since more and more food industry stakeholders are under intense pressure to cope with the highly dynamic business environment and maintain performance standards, novel processes and systems almost always contain several unknown activities that are difficult to foresee but also expensive and time-consuming to operate. The digital twin enables food businesses to simulate their operations in order to test and try new ways of working (Coelho et al. 2021). Now, simulation models may be used to assess and forecast resource and process-related changes under a variety of "what-if" scenarios. As a result, organizations are increasingly enjoying major benefits from the technologies, which enable the mapping and analysis of details related to operational performance, product and service innovation, and faster delivery (Ivanov and Dolgui 2020). The digital twin replicates real-world environments such as manufacturing lines and floors, which may be continually analyzed to discover areas for improvement. Food supply chain partners can forecast the future effect and result of any circumstance without incurring money or jeopardizing brand image. Similarly, modeling and simulation are used in food production and logistics systems to test and validate goods, processes, and system design, as well as to forecast system performance (Korth et al. 2018). Additionally, it facilitates decision-making, education, and training sessions, resulting in improving efficiency and human resource development (Negahban and Smith 2014). Furthermore, simulation could reduce truck traffic, minimize carbon emissions, congestion, and accidents (Hoffa-Dabrowska and Grzybowska 2020). Overall, by implementing various simulation methods, firms could mitigate economic, social, and environmental risks.

2.4 System Integration

The goal of system integration is to integrate disparate subsystems into a cohesive whole to provide overall functionality. Modularity in business activities through subsystems is often desired for ease of implementing technology developments in certain areas and the capacity to react swiftly to changing demands. However, in order to realize the full advantages of a distributed and modular system, essential aspects such as data integration, interoperability, resilience, stability, capacity, and scalability must be addressed.

From a sustainability perspective, system integration could result in efficiencies initiated from digitalization (e.g., business process and role automation), near real-time data acquisition and transmission, and paperwork minimization, which could also provide environmental-related benefits. Effective systems' integration benefits end-to-end supply chain traceability and enhances data and information transparency. Subsequently, data and information transparency is a powerful tool for supply chain stakeholders to monitor and positively impact food safety, food security and quality, and to address other social and environmental challenges. Further details will be discussed in next section, which describes the Internet of Things (IoT) as one of the key enablers of system integration.

2.5 Internet of Things (IoT)

IoT could be defined as an inter-connected network of numerous physical things or objects encompassing various technologies, actuators and sensors, as well as software that interchange data through the Internet (Gubbi et al. 2013). IoT is going to obliterate the boundary between the physical and virtual worlds. This is accomplished by making things (devices, machines, and objects, for example) intelligent and interconnecting them in a way that they can communicate and exchange information without human intervention (Al-Fuqaha et al. 2015, Atzori et al. 2010). IoT could alleviate global food logistics' complexity, already posing issues in terms of visibility, transparency, compliance, and ensuring product integrity.

From a sustainability perspective, IoT devices and IoT-enabled technologies could transform logistics massively. By making real-time data available, IoT improves process traceability and decision-making, allowing for more collaboration, and communication across food chain participants (Ahumada and Villalobos 2009). IoT sensors and other information and communication technologies (ICT) could be used to ensure the quality, security, and safety of the food (Zhong et al. 2017). Also, with IoT the agility of food chain could be improved. It is done by providing a flood of instant data that can be examined by powerful predictive data analytics, enabling food firms to discover supply chain gaps and implement preemptive remedies (Verdouw et al. 2016). Additionally, food businesses can use IoT in the supply chain to maintain the freshness of produce, ensure compliance with standards related to food safety, reduce wastage of the food, and promote its security (Ray et al. 2019).

IoT and IoT-enabled technologies such as Radio Frequency Identification (RFID), Wireless Sensor Networks (WSN), sensors, and could make transportation, warehouse management, and delivery process more efficient, convenient, transparent, and traceable. For transportation, RFID is frequently utilized in transportation to determine and trace products and vehicles. RFID can be used to obtain and trace data regarding logistics (Liu et al. 2019), freight data and containers' location (Zhang et al. 2014), and customer order information, such as the identity and quantity (Cheung et al. 2008). These data serve to optimize path planning, vehicle configuration (Liu et al. 2019), as well as routing strategies (Cheung et al. 2008). Other supplementary technologies, such as a geographic information system (GIS) and a global positioning system (GPS), are frequently used in conjunction with RFID to support route planning, navigation, and monitoring of transportation processes (Liu et al. 2019). IoT-enabled technology, particularly WSN, enables a higher control and security level during transit. WSN also allows remote temperature monitoring of product cores, thereby reducing food losses, ensuring food safety (Jedermann et al. 2014).

In a food warehouse, IoT can connect objects and operators. Connected goods such as products, shelves, trolley, and logistics operators can be traced and controlled through WSN embedded with environmental actuators, enabling decentralized decision support. RFID is frequently employed to locate and trace items for inventory management (Garrido-Hidalgo et al. 2019) and interior placement (Wu et al. 2019). Additionally, it is capable of collecting real-time warehouse data (Goudarzi et al. 2016). The combination of sensing technologies and WSN improve the security and transparency of the warehouse. Ambient sensors incorporated in WSN nodes control variations in environmental indicators in a warehouse, such as temperature and humidity (De Venuto and Mezzina 2018). Messages containing monitoring data may be sent to the central control station.

IoT technology enables monitoring and tracking the delivery process of food products, as well as making delivery data intuitive and exchangeable. A variety of IoT technologies include sensor and RFID technology, ad hoc networking and wireless technology, and embedded systems which allow delivery service providers to share information virtually and more efficiently (Jie et al. 2015). Increased delivery efficiency and resource optimization may be achieved by integrating physical logistics resources with the IoT (He et al. 2020). Different smart sensors, GPS tools, and RFID tags are implemented and linked to users' smart devices and servers on delivery vehicles, ensuring that sensing data is exchanged and shared (Sivamani et al. 2014). In addition to delivery trucks, a delivery person using a mobile terminal with a GPS device and an activated RFID tag may be monitored both indoors and outside (Lin et al. 2011). IoT technology adds to delivery dependability and failure reduction. RFID has traditionally been viewed as a means to locate and track the condition of food products and vehicles in order to improve delivery reliability, such as on-time delivery and accurate orders. With the implementation of RFID, losses of goods and incorrect destinations can be significantly reduced (Fu et al. 2015). By gathering, monitoring, and recording ambient environmental conditions throughout delivery, IoT-enabled technologies can ensure the quality and safety of commodities and perishable products. WSNs are capable of automatically capturing ambient temperature data and humidity data, which may be used to monitor food as it moves through the supply chain (Tsang et al. 2018). In addition, customers may provide comments and instructions to personnel by checking this information. Through the transportation of fresh fish and refrigerated vegetables in the cold chain, temperature sensors are often used (Ruiz-Garcia et al. 2010, Trebar et al. 2015), which require precise temperature control.

2.6 Cybersecurity

There is an abundance of data pertaining to logistics strategy and execution. The use of technology simplifies the process of storing and navigating this data, paving the way for more straightforward and quicker logistical planning. As a result, the days of paper-based monitoring are past, and supply chains are more dependent on computer-based intelligent management systems, making them susceptible to cyberattacks (Cooper 2015). Globally, targeted attacks have increased significantly, affecting many industries and sectors, including corporations, schools, hospitals, and government websites. When an industry adopts a new technology, cybersecurity becomes a concern. Businesses must safeguard the security of their systems to prepare for the negative repercussions of technology. Cybersecurity is critical for businesses that rely on computer-based systems (tablets, laptops, smartphones, GPS), RFID, wireless internet, sensors of any sort, automated guided vehicles (AGVs), or Artificial Intelligence (AI) (Cooper 2015). Cybersecurity issues in the food business are few; nevertheless, with the growth of Industry 4.0 and the rising use of technology, they must become a regular consideration (Culot et al. 2019). Due to the food industry's particular infrastructure, bioterrorism is a significant risk, endangering consumer welfare, natural resources, and the economy (Cooper 2015). Due to the huge number of parties engaged in the food industry, such as suppliers, manufacturers, and retailers, the food chain is regarded as even more susceptible and prone to cyberattacks. Cyber-Physical Systems are the consequence of connecting physical and virtual systems in organizations over the Internet. Using IoT in this context poses a number of issues since there are so many security unknowns. Because these new technologies gather data and manage the systems of the sector, they are vulnerable to intrusion and hacking. Guidance related to this field of Industry 4.0 is still scarce.

From a sustainability perspective, these are benefits for the firms that addressed cybersecurity beforehand as well as threats for those who postpone such a consideration. Cybersecurity acts as an enabler and infrastructure for implementing the discussed cutting-edge technologies. Hence, firms should secure their cyber and physical systems to flourish and take advantage of the opportunities that other technologies and their complementarities provide.

2.7 Cloud Computing

Cloud computing is an enabling technology that allows the faster and more efficient transfer, storage, processing, and sharing of supply chain information (Gnimpieba et al. 2015). The cloud is a data storage technology that utilizes numerous servers to store data. These servers are often located remotely from the data owner and are maintained by a hosting business responsible for ensuring the data is available and secure. Cloud logistics is a term that refers to the process of developing a logistics information sharing platform based on cloud computing's powerful communication, operation, and matching capabilities (Gong and Yang 2012, Holtkamp et al. 2010). Cloud manufacturing refers to the practice of utilizing the cloud in manufacturing businesses (Li et al. 2010). Cloud manufacturing enables users to access virtualized manufacturing resources, competencies, and capabilities via internet-based services (Siderska and Jadaan 2018).

From a sustainability perspective, cloud manufacturing has the potential to transform the manufacturing industry into one that is more service-oriented, collaborative, and innovative (Ren et al. 2017) by enabling the sharing of manufacturing services, including software, computational, and physical manufacturing resources that are available on-demand (Simeone et al. 2019). Hence, firms could cut costs based on the economics of scale of the cloud provider and integrate their systems with their supply chain members more effectively and efficiently, which may aid in achieving competitive advantage. Cloud computing provides various benefits to food chain trading partners, including parallel processing, data security, resource virtualization, and increased data storage capacity (Subudhi et al. 2019).

Cloud manufacturing is critical in developing sustainable industries thanks to the high automation, collaborative design, strong process resilience, and improved waste minimization, reuse, and recovery (Fisher et al. 2018). For instance, cloud technology could be utilized to access real-time data about traffic, resulting in distribution optimization and pollution mitigation (Chen 2020). Also, using cloud manufacturing and IoT, Qu et al. (2016) developed a control method for coordinating manufacturing logistics. This technology enables dynamic resource management and collection of real-time data. Furthermore, logistics scheduling issues, including cloud manufacturing, have been modeled by Zhou et al. (2020). In order to speed up the delivery of items to customers, they proposed a scheduling method. Employees from different departments, as well as whole companies, may easily access and examine data in real time. This improves interaction between the different food chain players and speeds up operations, particularly those requiring logistics based on customer demand. Zhou et al. (2020) developed a mathematical model for a logistics scheduling issue, including cloud manufacturing. They offered a scheduling strategy for reducing the time required for manufacturers to deliver goods to consumers. Also, data may be readily exchanged and viewed in real-time by employees from other functional departments and even firms. This enhances communication between the many agents in the food chain and accelerates procedures, especially those involving logistics dependent on client demand. Additionally, as the hosting business could do data storage and processing, the focal firm could reduce infrastructure and maintenance costs (Keogh et al. 2020). Although cloud computing is not free, users only pay for what they use (Benotmane et al. 2018). Alternatively, if the organization keeps its own data, it will likely have unused space on its servers and backup devices. Also, they could reduce energy use and cut expenses of energy bills. The businesses could enjoy better computing power compared to theirs and better host services such as apps improving their efficiency and backup systems and reducing their risks. As they outsource the process, they could concentrate on more strategic matters.

2.8 3D Printing

Although 3D printing or additive manufacturing has been around for some years, the technology is now gaining momentum. It is a new technological invention that has the potential to alter and revolutionize a wide variety of industries, including manufacturing, construction, aviation, education, and healthcare, as well as the food industry (Wieczorek 2017). 3D printing refers to methods that include joining or solidification of material(s) under computer control in order to construct a three-dimensional item. Through the successive accumulation of layers of materials, computer-aided software manages this approach, which creates three-dimensional objects (Jiang et al. 2019). Additionally, 3D printing enables the layer-by-layer printing of preset slices of intended and desired objects (Jiang et al. 2019). Numerous ways exist for 3D printing, each utilizing a different material (Portanguen et al. 2019). For example, binder jetting, material jetting, directed energy deposition, powder bed fusion/binding, vat photopolymerization, sheet lamination, and material extrusion are all techniques associated with 3D printing technology.

From a sustainability perspective, 3D printing establishes itself as a viable technique of preparing and presenting food in the modern day (Raji 2017). This, in turn, could improve consumer engagement and satisfaction and contribute to developing competitive advantage. 3D printing technology has been shown to be effective in reducing the complexity of the food supply chain by displacing old processing techniques (Chen 2016), and fabricating customized food (Sun et al. 2015). Implementing this technology will enhance the closeness of production operations to customers/end users, resulting in a reduction in transit volume and associated packing, distribution, and shipping costs. As a result, inventory management for food producers would become more cost-effective and simple, as they can make food products on demand. Additionally, less transportation within the food chain may result in less damage to the environment due to reduced traffic from shorter vehicle journey.

A benefit of reduced transportation is fewer automobile emissions, thus protecting the environment, especially compared to the minimal or non-existent material waste generated by 3D printing, since just the necessary raw materials are needed to manufacture food products. Utilizing 3D printing in product development and manufacturing opens up new possibilities for product design, such as the ability to customize product size and shape as well as interior structure and taste (Ricci et al. 2019). It is therefore possible to create and pre-manufacture meals that may be tailored to individual consumer demands and preferences using 3D printing (Portanguen et al. 2019). The unique use of 3D technology favors the trend toward personalized foods by making meals with complicated dimensions and improved sensations, altering nutritious content, and increasing customer pleasure (Godoi et al. 2016). Using this technology, food companies may design aesthetically attractive food packaging while reducing waste, food product management, and any contamination concerns that may arise as a result of the food they produce. The unique qualities of 3D printing, which encourage sustainable manufacturing, waste reduction, lower packaging, and improved worker safety, are the basis for integrating this technology with sustainable development objectives.

2.9 Augmented Reality (AR)

The concept of "augmented reality" (AR) refers to the use of digital resources overlaid on top of the current view of the real world (Longo et al. 2017). Virtual and actual objects must be precisely registered in three dimensions, as stated by Wu et al. (2013) in their discussion of AR systems that combine the real and virtual worlds. The hardware for AR includes a sensor that gathers real-time images of the surroundings and a display that shows both the real and virtual worlds simultaneously. So far, eyeglasses, smartphones, tablets, and even head-up displays have all been able to make use of AR. The implementation of AR in industrial settings is referred to as Industrial Augmented Reality (IAR). Even though IAR systems have increasingly been prevalent, this is a relatively young subject, with just a few industries benefiting from its use. To adopt AR successfully in the industry, both organizational and technology fit are essential determinants (Masood and Egger 2019).

From sustainability perspectives, IAR can assist in reducing inventory costs by streamlining warehouse management (Mourtzis et al. 2019). This may be accomplished by lowering the mistake rate, enhancing flexibility, boosting dependability, speeding up operations, increasing staff safety, and increasing operator engagement (Rejeb et al. 2021a). AR, according to Wang et al. (2019), contains a number of enjoyable properties that allow for the construction of participatory and hedonic encounters, permitting customers to enhance their feelings while purchasing or eating food items. Additional benefits of employing AR in smart food logistics involve greater visibility of food chain operations, contamination risk prevention, empowerment of food marketing, simplicity of food training, and improvement in food logistics (Luque et al. 2017).

Even though AR application in food logistics has not been widely investigated, the technology offers several opportunities in the food industry, including worker safety, maintenance, training, quality control, design and layout, communication, location, language translation, and product expiration. Packing and storing food in a warehouse is a common usage of heavy equipment in the food business. The IAR may alert the operator to potential hazards by providing real-time feedback on how the machine is doing (Mourtzis et al. 2018).

IAR can assist with the maintenance and repair of industrial equipment. IAR can display pertinent photos, films, or highlight specific places on the equipment, in addition to providing thorough sequential instructions to assist with machinery maintenance. In maintenance activities, IAR may employ videos and photos to teach workers how to operate equipment or follow a particular procedure. IAR can assist in identifying faults, damaged items, and products that do not meet quality criteria. For example, with the assistance of IAR, it is possible to spot missing information on a label or broken packaging. Furthermore, during the development of new logistics activity, IAR may assist in visualizing how the equipment will be installed in the production facility or warehouse, as well as the material flow and equipment use. This enables the detection of early problems and the optimization of logistical operations.

Beck et al. (2016) claim that AR technology has the potential to improve training methods by allowing for higher flexibility, greater control over culinary processes, and rapid learning capacities. In food production and processing facilities, the usage of AR headsets might offer assembly line employees all the information essential to assure appropriate food handling and packing. Clark and Crandall (2019) validated this capability, indicating that AR smart glasses were 50% more effective at educating food service personnel on-the-job than standard classroom or video training approaches.

IAR can connect team members from similar or various industrial sites. In this situation, communication is improved by displaying the visualization of the team member with whom workers are communicating, who may provide guidance on how to continue. In the same vein, IAR can support team meetings for brainstorming and debate in order to optimize operations. In addition, IAR can assist in determining the location of a certain product that has to be retrieved within the warehouse and directing personnel on how to locate or find it (Rejeb et al. 2021). Similarly, regions of the warehouse with accessible storage space for items might be discovered to expedite logistical processes. Additionally, IAR may be used to depict the locations of various tools, machines, and places inside an industrial food facility.

Food companies obtain ingredients from a diverse variety of nations through global food supply networks. Their labels may be written in multiple languages, which personnel may not understand. IAR may provide product label translations by enabling consumers to scan the labels in their native language and perform an automatic translation. Also, if expiry dates are maintained for each product, IAR can readily detect things that are about to expire and so have a higher priority for sale.

2.10 Blockchain Technology

Sensors deployed across the food production chain create a tremendous quantity of data. While these data may be viewed in real-time, they must be securely accessed. This need may be accomplished by utilizing blockchain technology. Conceptually, blockchain is a chronologically ordered chain of (data contained) blocks stored among decentralized or distributed nodes. In this distributed setting, transactions should be verified by the nodes, which eliminates the need for a central authority, forming a new form of trust regarded as digital trust (Rejeb et al. 2019, Treiblmaier 2018). Starting with bitcoin and cryptocurrencies, it is now creatively disrupting and transforming various industries and domains, including food logistics (Biswas and Gupta 2019, Rejeb et al. 2020). Since data can't be manipulated with blockchain technology, businesses are rushing to embrace it. In a word, blockchain is a secure and distributed ledger that authorized nodes can read and write, and the data is preserved and shared on a peer-to-peer system instead of a central system.

From sustainability perspectives, blockchain technology provides a number of benefits for smart food logistics, including enhanced transportation and shipping efficiency, transparency, shipment or product tracking, fewer concerns stemming from misplacement or theft of commodities, and speedier invoicing and payment processing. It can assist in mitigating the majority of logistics-related challenges, such as transportation, procurement, supply chain traceability, customs' coordination, and trade financing (Rejeb et al. 2020). The globalization of food supply chains has created various issues for food businesses in terms of securing and protecting food products and information sharing within the supply chain. Nevertheless, when businesses operate in a blockchain setting, they will be able to maintain transparency and visibility over food processes, instantly detect data provenance, and trace the physical route taken by the food product (Tan et al. 2018).

Importing and exporting food goods entails a great deal of paperwork. According to Maersk, a single container may require permission from up to thirty officials, comprising health, customs, and tax regulators. While containers may load in minutes, a missing piece of paper might hold them up for days at the destination port, consequently causing food loss. Occasionally, the cost of relocating and maintaining all of the documentation is comparable to the cost of the container transportation around the world. Maersk in a collaboration with IBM created a blockchain for food chain traceability and digitization of trade protocols. This technology enables the organization to monitor the movement and precise location of food products across the supply chain. Moreover, blockchain technology enables the safe interchange of data and the creation of an immutable ledger for this material. It can generate substantial savings by reducing delays and fraud (Dobrovnik et al. 2018).

Moreover, ZIM, an ocean shipping firm, effectively demonstrated how bills of lading are issued, transported, and received via their decentralized network through their pilot project using blockchain technology. By gathering and preserving data on the production of foods, their sourcing location, and the ways foods are processed, the blockchain system contributes significantly to supply chain transparency. The data recorded on the ledger is permanent and easily exchanged, providing food chain partners with more comprehensive traceability capabilities than ever before. Thus, besides establishing authenticity and legitimacy, this method has the potential to verify product's ethical sourcing and storage circumstances, and eliminate counterfeiting. Blockchain technology was used by Walmart to ensure the safety of meat from China and mangoes from Mexico (Kamath 2018). Additionally, blockchain technology can assist with invoices that include inaccurate data, resulting in disputes and process inefficiencies, which can occur as a result of missing or delayed delivery. By digitizing papers and real-time delivery data and generating smart contracts, blockchain technology can settle these issues. For instance, when blockchain-connected food goods are delivered, the time and condition of delivery are immediately validated. Food products may be certified to have been delivered in accordance with agreed-upon conditions and terms like intact packaging, temperature and so on, and payments made to the right entities can be validated to be accurate, thereby boosting efficiency and reliability.

Furthermore, retailers may monitor the shelf life of food goods within their stores and use extra protections and measures to maintain the authenticity of food (Galvez et al. 2018). In addition, the technology may be utilized to improve food traceability (Mattila et al. 2016) and the capacity to detect and recall dangerous items promptly. For instance, Walmart demonstrated great traceability and the capacity to cut the time necessary to track down and retrieve a product from seven days to few seconds according to Kamath (2018).

2.11 Artificial Intelligence (AI)

There are three types of AI in industry: descriptive AI, predictive AI, and prescriptive AI (Olsen and Tomlin 2020). It is possible to improve process design and control by using descriptive AI techniques on the data collected from sensor-enabled activities. In order for AI to be able to forecast and prescribe, it must be able to connect the dots between what is happening and why. A good example of predictive AI is condition-based maintenance, which relies on accurate predictions about the timing of future failures. Prescriptive AI generates ideas for operational action. Connecting food logistics needs with appropriate AI will be the first hurdle for researchers to tackle in this new field. Despite the fact that decision support systems are not new, current advances in machine learning are increasingly being used in all three AI categories to transition from model-based methods to the use of training data approaches and statistical techniques that depend on limited or no models.

From a sustainability perspective, AI could be leveraged to significantly impact various supply chain activities and processes, including supplier selection, warehouse status and priorities control, food waste reduction, and shipment delays. Through process automation and interoperable and intelligent integration of food production activities, AI has the ability to boost the efficiency of food supply chains (Keogh et al. 2020). AI adoption could aid in delegating decision-making authority, enabling quick adaptability to the changing corporate circumstances and agility (Schuh et al. 2017).

By utilizing real-time data produced by sensors, AI technologies and approaches can help improve the management of food chain activities. Machine learning, as an application of AI, is the process through which computers learn to understand and act independently like humans using data. Automated route optimization supported by machine learning may help the food industry cut down on logistical expenses, which has been a top priority in the last few years (Makkar et al. 2019). Machine learning can be used in a variety of ways in smart food logistics, including anticipating customer demands to increase retention and forecasting product demand to manage supply.

3. Conclusion

We examined the primary technological enablers for smart food logistics in this chapter, as well as their contributions to the development of a sustainable food industry. Modern technologies are gaining traction in the food chain due to their capacity to promote sustainability, improve food quality and safety, and reduce food chain inefficiencies. New technologies are expected to result in cost savings, operational efficiencies, and competitive advantages from an economic standpoint. Through the use of technology, food logistics' stakeholders will be able to mitigate the negative environmental impacts of food logistics. The process of digitization (from analogue) and the further digitalization (business process and role automation) using technology tools provides improved governance and controls that support cleaner production, waste reduction, and effective resource consumption. Smart food logistics promotes employee safety and their development by minimizing hazardous working conditions and supporting training. While smart food logistics holds the promise of a greener food chain and sustainable development, the food industry is yet to adopt and incorporate new technologies that enable complete automation, maximize production capacity, and allow stronger collaboration and information exchange between food chain stakeholders. A few additional obstacles must be overcome to ensure effective integration of modern technologies in smart food logistics. For example, organizations may struggle with prioritization of time and key resources to invest in exploring and determining the most appropriate and affordable technologies. Moreover, the integration of new technologies in conventional food supply chains necessitates the development of organizational skills and capacities to enhance food logistics and increase consumer trust in food products. Finally, future research should examine the current limitations of emerging technologies and devise ways to overcome them in order to establish smart and sustainable food logistics.

References

Abideen, A.Z., Sundram, V.P.K., Pyeman, J., Othman, A.K. and Sorooshian, S. 2021. Digital Twin Integrated Reinforced Learning in Supply Chain and Logistics. Logistics 5(4): 84.

Agyabeng-Mensah, Y., Ahenkorah, E., Afum, E., Dacosta, E. and Tian, Z. 2020. Green warehousing, logistics optimization, social values and ethics and economic performance: The role of supply chain sustainability. The International Journal of Logistics Management.

Ahumada, O. and Villalobos, J.R. 2009. Application of planning models in the agri-food supply chain: A review. European Journal of Operational Research 196(1): 1–20.

Al-Fuqaha, A., Guizani, M., Mohammadi, M., Aledhari, M. and Ayyash, M. 2015. Internet of things: A survey on enabling technologies, protocols and applications. IEEE Communications Surveys and Tutorials 17(4): 2347–2376.

Aqeel-Ur-Rehman, Abbasi, A.Z., Islam, N. and Shaikh, Z.A. 2014. A review of wireless sensors and networks' applications in agriculture. Computer Standards and Interfaces 36(2): 263–270. https://doi.org/10.1016/j.csi.2011.03.004.

Atzori, L., Iera, A. and Morabito, G. 2010. The internet of things: A survey. Computer Networks 54(15): 2787–2805.

Bader, F. and Rahimifard, S. 2020. A methodology for the selection of industrial robots in food handling. Innovative Food Science and Emerging Technologies 64: 102379.

Barbosa, M.W. 2021. Uncovering research streams on agri-food supply chain management: A bibliometric study. Global Food Security 28: 100517.

Beck, D.E., Crandall, P.G., O'Bryan, C.A. and Shabatura, J.C. 2016. Taking food safety to the next level—An augmented reality solution. Journal of Foodservice Business Research 19(4): 382–395.

Benotmane, Z., Belalem, G. and Neki, A. 2018. Towards a cloud computing in the service of green logistics. International Journal of Logistics Systems and Management 29(1): 37–61.

Biswas, B. and Gupta, R. 2019. Analysis of barriers to implement blockchain in industry and service sectors. Computers and Industrial Engineering 136: 225–241.

Boschert, S. and Rosen, R. 2016. Digital twin—The simulation aspect. In Mechatronic futures (pp. 59–74). Springer.

Chen, Y. 2020. Intelligent algorithms for cold chain logistics distribution optimization based on big data cloud computing analysis. Journal of Cloud Computing 9(1): 1–12.

Chen, Z. 2016. Research on the impact of 3D printing on the international supply chain. Advances in Materials Science and Engineering, 2016.

Cheung, B.K.-S., Choy, K., Li, C.-L., Shi, W. and Tang, J. 2008. Dynamic routing model and solution methods for fleet management with mobile technologies. International Journal of Production Economics 113(2): 694–705.

Clark, J. and Crandall, P.G. 2019. Educational affordances of google glass as a new instructional platform for foodservice training. Management and Education 13(1): 28–32.

Coelho, F., Relvas, S. and Barbosa-Póvoa, A. 2021. Simulation-based decision support tool for in-house logistics: The basis for a digital twin. Computers and Industrial Engineering 153: 107094.

Cooper, C. 2015. Cybersecurity in food and agriculture. Protecting Our Future, 2.

Culot, G., Fattori, F., Podrecca, M. and Sartor, M. 2019. Addressing industry 4.0 cybersecurity challenges. IEEE Engineering Management Review 47(3): 79–86.

de Paula Ferreira, W., Armellini, F. and De Santa-Eulalia, L.A. 2020. Simulation in industry 4.0: A state-of-the-art review. Computers and Industrial Engineering, 106868.

De Venuto, D. and Mezzina, G. 2018. Spatio-temporal optimization of perishable goods' shelf life by a pro-active WSN-based architecture. Sensors 18(7): 2126.

Ding, Y., Jin, M., Li, S. and Feng, D. 2021. Smart logistics based on the internet of things technology: An overview. International Journal of Logistics Research and Applications 24(4): 323–345.

Dobrovnik, M., Herold, D.M., Fürst, E. and Kummer, S. 2018. Blockchain for and in Logistics: What to Adopt and Where to Start. Logistics 2(3): 18.

Dubey, R., Gunasekaran, A., Childe, S.J., Wamba, S.F. and Papadopoulos, T. 2016. The impact of big data on world-class sustainable manufacturing. The International Journal of Advanced Manufacturing Technology 84(1–4): 631–645.

Duckett, T., Pearson, S., Blackmore, S., Grieve, B., Chen, W.-H., Cielniak, G., et al. 2018. Agricultural robotics: The future of robotic agriculture. ArXiv Preprint ArXiv:1806.06762.

Duong, L.N., Al-Fadhli, M., Jagtap, S., Bader, F., Martindale, W., Swainson, M., et al. 2020. A review of robotics and autonomous systems in the food industry: From the supply chains perspective. Trends in Food Science and Technology.

Echelmeyer, W., Kirchheim, A. and Wellbrock, E. 2008. Robotics-logistics: Challenges for Automation of Logistic Processes. 2099–2103.

Fisher, O., Watson, N., Porcu, L., Bacon, D., Rigley, M. and Gomes, R.L. 2018. Cloud manufacturing as a sustainable process manufacturing route. Journal of Manufacturing Systems 47: 53–68.

Fosso Wamba, S., Akter, S., Edwards, A., Chopin, G. and Gnanzou, D. 2015. How 'big data' can make big impact: Findings from a systematic review and a longitudinal case study. International Journal of Production Economics 165: 234–246. https://doi.org/10.1016/j.ijpe.2014.12.031.

Fosso Wamba, S., Gunasekaran, A., Dubey, R. and Ngai, E.W.T. 2018. Big data analytics in operations and supply chain management. Annals of Operations Research 270(1): 1–4. https://doi.org/10.1007/s10479-018-3024-7.

Fredriksson, A. and Liljestrand, K. 2015. Capturing food logistics: A literature review and research agenda. International Journal of Logistics Research and Applications 18(1): 16–34.

Friha, O., Ferrag, M.A., Shu, L., Maglaras, L.A. and Wang, X. 2021. Internet of things for the future of smart agriculture: A comprehensive survey of emerging technologies. IEEE CAA J. Autom. Sinica 8(4): 718–752.

Fu, H.-P., Chang, T.-H., Lin, A., Du, Z.-J. and Hsu, K.-Y. 2015. Key factors for the adoption of RFID in the logistics industry in Taiwan. The International Journal of Logistics Management.

Galvez, J.F., Mejuto, J.C. and Simal-Gandara, J. 2018. Future challenges on the use of blockchain for food traceability analysis. TrAC Trends in Analytical Chemistry 107: 222–232.

Garrido-Hidalgo, C., Olivares, T., Ramirez, F.J. and Roda-Sanchez, L. 2019. An end-to-end internet of things solution for reverse supply chain management in industry 4.0. Computers in Industry 112: 103127.

Gnimpieba, Z.D.R., Nait-Sidi-Moh, A., Durand, D. and Fortin, J. 2015. Using Internet of Things technologies for a collaborative supply chain: Application to tracking of pallets and containers. Procedia Computer Science 56: 550–557.

Godoi, F.C., Prakash, S. and Bhandari, B.R. 2016. 3d printing technologies applied for food design: Status and prospects. Journal of Food Engineering 179: 44–54.

Gong, X. and Yang, R. 2012. Explore the Application of "Cloud Computing" and "Cloud Logistics" to Logistics [J]. China Business and Market 10: 29–33.

Goudarzi, P., Malazi, H.T. and Ahmadi, M. 2016. Khorramshahr: A scalable peer to peer architecture for port warehouse management system. Journal of Network and Computer Applications 76: 49–59.

Gubbi, J., Buyya, R., Marusic, S. and Palaniswami, M. 2013. Internet of Things (IoT): A vision, architectural elements and future directions. Future Generation Computer Systems 29(7): 1645–1660.

He, Y., Zhou, F., Qi, M. and Wang, X. 2020. Joint distribution: Service paradigm, key technologies and its application in the context of Chinese express industry. International Journal of Logistics Research and Applications 23(3): 211–227.

Hoffa-Dabrowska, P. and Grzybowska, K. 2020. Simulation modeling of the sustainable supply chain. Sustainability 12(15): 6007.

Holtkamp, B., Steinbuss, S., Gsell, H., Loeffeler, T. and Springer, U. 2010. Towards a Logistics Cloud. 305–308.

Huan, M., Ding, Z., Li, S. and Zhang, C. 2021. Blockchain Consensus Mechanism Design for Food Safety Traceability. 620–627.

Ivanov, D. and Dolgui, A. 2020. A digital supply chain twin for managing the disruption risks and resilience in the era of Industry 4.0. Production Planning and Control, 1–14.

Jagtap, S. and Duong, L.N.K. 2019. Improving the new product development using big data: A case study of a food company. British Food Journal.

Jagtap, S., Bader, F., Garcia-Garcia, G., Trollman, H., Fadiji, T. and Salonitis, K. 2021. Food Logistics 4.0: Opportunities and Challenges. Logistics 5(1): 2.

Jedermann, R., Pötsch, T. and Lloyd, C. 2014. Communication techniques and challenges for wireless food quality monitoring. Philosophical Transactions of the Royal Society A: Mathematical, Physical and Engineering Sciences 372(2017): 20130304.

Jiang, H., Zheng, L., Zou, Y., Tong, Z., Han, S. and Wang, S. 2019. 3D food printing: Main components selection by considering rheological properties. Critical Reviews in Food Science and Nutrition 59(14): 2335–2347.

Jie, Y., Subramanian, N., Ning, K. and Edwards, D. 2015. Product delivery service provider selection and customer satisfaction in the era of internet of things: A Chinese e-retailers' perspective. International Journal of Production Economics 159: 104–116.

Kamath, R. 2018. Food traceability on blockchain: Walmart's pork and mango pilots with IBM. The Journal of the British Blockchain Association 1(1): 3712.

Karabegović, I., Karabegović, E., Mahmić, M. and Husak, E. 2015. The application of service robots for logistics in manufacturing processes. Advances in Production Engineering and Management, 10(4).

Keogh, J.G., Dube, L., Rejeb, A., Hand, K.J., Khan, N. and Dean, K. 2020. The future food chain: digitization as an enabler of society 5.0. Building the Future of Food Safety Technology (1st Edition): London, Oxford, UK.

Keogh, J.G. and Unis, C. 2020. Rethinking future food chains: systems thinking and the cascading consequences of system failures. In: Food Safety Magazine. USA: Food Safety Magazine.

Khan, Z.H., Khalid, A. and Iqbal, J. 2018. Towards realizing robotic potential in future intelligent food manufacturing systems. Innovative Food Science and Emerging Technologies 48: 11–24.

Korth, B., Schwede, C. and Zajac, M. 2018. Simulation-ready Digital Twin for Realtime Management of Logistics Systems. 4194–4201.

Li, B.-H., Zhang, L., Wang, S.-L., Tao, F., Cao, J. W., Jiang, X.D., et al. 2010. Cloud manufacturing: A new service-oriented networked manufacturing model. Computer Integrated Manufacturing Systems 16(1): 1–7.

Lin, C.-Y., Cheng, W.-T. and Wang, S.-C. 2011. An end-to-end logistics management application over heterogeneous location systems. Wireless Personal Communications 59(1): 5–16.

Liu, S., Zhang, Y., Liu, Y., Wang, L. and Wang, X.V. 2019. An 'Internet of Things' enabled dynamic optimization method for smart vehicles and logistics tasks. Journal of Cleaner Production 215: 806–820.

Longo, F., Nicoletti, L. and Padovano, A. 2017. Smart operators in industry 4.0: A human-centered approach to enhance operators' capabilities and competencies within the new smart factory context. Computers and Industrial Engineering 113: 144–159.

Luque, A., Peralta, M.E., De Las Heras, A. and Córdoba, A. 2017. State of the Industry 4.0 in the Andalusian food sector. Procedia Manufacturing 13: 1199–1205.

Makkar, S., Devi, G.N.R. and Solanki, V.K. 2019. Applications of machine learning techniques in supply chain optimization. 861–869.

Masood, T. and Egger, J. 2019. Augmented reality in support of Industry 4.0—Implementation challenges and success factors. Robotics and Computer-Integrated Manufacturing 58: 181–195.

Mattila, J., Seppälä, T. and Holmström, J. 2016. Product-centric information management: A Case Study of a Shared Platform With Blockchain Technology.

Mourtzis, D., Zogopoulos, V., Katagis, I. and Lagios, P. 2018. Augmented Reality based Visualization of CAM Instructions towards Industry 4.0 paradigm: A CNC Bending Machine case study. Procedia CIRP 70: 368–373.

Mourtzis, D., Samothrakis, V., Zogopoulos, V. and Vlachou, E. 2019. Warehouse design and operation using augmented reality technology: A papermaking industry case study. Procedia Cirp. 79: 574–579.

Negahban, A. and Smith, J.S. 2014. Simulation for manufacturing system design and operation: Literature review and analysis. Journal of Manufacturing Systems 33(2): 241–261.

Olsen, T.L. and Tomlin, B. 2020. Industry 4.0: Opportunities and challenges for operations management. Manufacturing and Service Operations Management 22(1): 113–122.

Pal, A. and Kant, K. 2019. Internet of perishable logistics: Building smart fresh food supply chain networks. IEEE Access 7: 17675–17695.

Parvin, S., Venkatraman, S., de Souza-Daw, T., Fahd, K., Jackson, J., Kaspi, S., et al. 2019. Smart Food Security System Using Iot and Big Data Analytics. 253–258.

Portanguen, S., Tournayre, P., Sicard, J., Astruc, T. and Mirade, P.-S. 2019. Toward the design of functional foods and biobased products by 3D printing: A review. Trends in Food Science and Technology 86: 188–198.

Qu, T., Lei, S., Wang, Z., Nie, D., Chen, X. and Huang, G.Q. 2016. IoT-based real-time production logistics synchronization system under smart cloud manufacturing. The International Journal of Advanced Manufacturing Technology 84(1–4): 147–164.

Queiroz, M.M. and Telles, R. 2018. Big data analytics in supply chain and logistics: An empirical approach. The International Journal of Logistics Management.

Raji, I.O. 2017. 3D Printing Technology—Applications, Benefits and Areas of Opportunity in Nigeria. Int. J. Adv. Acad. Res.: Sci. Technol. Eng. 3: 21–30.

Ray, P., Om Harsh, H., Daniel, A. and Ray, A. 2019. Incorporating block chain technology in food supply chain. International Journal of Management Studies 6(1): 5.

Rejeb, A., Keogh, J.G. and Treiblmaier, H. 2019. Leveraging the Internet of Things and blockchain technology in supply chain management. Future Internet 11(7): 161. https://doi.org/10.3390/fi11070161.

Rejeb, A., Keogh, J.G., Zailani, S., Treiblmaier, H. and Rejeb, K. 2020. Blockchain technology in the food industry: A Review of Potentials, Challenges and Future Research Directions. Logistics 4(4): 27. https://doi.org/10.3390/logistics4040027.

Rejeb, A., Keogh, J.G., Leong, G.K. and Treiblmaier, H. 2021a. Potentials and challenges of augmented reality smart glasses in logistics and supply chain management: A systematic literature review. International Journal of Production Research 59(12): 3747–3776. https://doi.org/10.1080/00207543.2021.1876942.

Rejeb, A., Keogh, J.G., Wamba, S.F. and Treiblmaier, H. 2021b. The potentials of augmented reality in supply chain management: A state-of-the-art review. Management Review Quarterly 71(4): 819–856. https://doi.org/10.1007/s11301-020-00201-w.

Ren, L., Zhang, L., Wang, L., Tao, F. and Chai, X. 2017. Cloud manufacturing: Key characteristics and applications. International Journal of Computer Integrated Manufacturing 30(6): 501–515.

Ricci, I., Derossi, A. and Severini, C. 2019. 3D printed food from fruits and vegetables. In Fundamentals of 3D food printing and applications (pp. 117–149). Elsevier.

Ruiz-Garcia, L., Barreiro, P., Robla, J.I. and Lunadei, L. 2010. Testing ZigBee motes for monitoring refrigerated vegetable transportation under real conditions. Sensors 10(5): 4968–4982.

Rüßmann, M., Lorenz, M., Gerbert, P., Waldner, M., Justus, J., Engel, P., et al. 2015. Industry 4.0: The future of productivity and growth in manufacturing industries. Boston Consulting Group 9(1): 54–89.

Schuh, G. anderl, R., Gausemeier, J., ten Hompel, M. and Wahlster, W. 2017. Industrie 4.0 maturity index: Managing the digital transformation of companies. Herbert Utz Verlag GmbH.

Siderska, J. and Jadaan, K.S. 2018. Cloud manufacturing: A service-oriented manufacturing paradigm. A review paper. Engineering Management in Production and Services, 10(1).

Simeone, A., Caggiano, A., Boun, L. and Deng, B. 2019. Intelligent cloud manufacturing platform for efficient resource sharing in smart manufacturing networks. Procedia CIRP 79: 233–238.

Sivamani, S., Kwak, K. and Cho, Y. 2014. A study on intelligent user-centric logistics service model using ontology. Journal of Applied Mathematics, 2014.

Subudhi, B.N., Rout, D.K. and Ghosh, A. 2019. Big data analytics for video surveillance. Multimedia Tools and Applications 78(18): 26129–26162.

Sufiyan, M., Haleem, A., Khan, S. and Khan, M.I. 2019. Evaluating food supply chain performance using hybrid fuzzy MCDM technique. Sustainable Production and Consumption 20: 40–57.

Sun, J., Peng, Z., Zhou, W., Fuh, J.Y., Hong, G.S. and Chiu, A. 2015. A review on 3D printing for customized food fabrication. Procedia Manufacturing 1: 308–319.

Tan, B., Yan, J., Chen, S. and Liu, X. 2018. The impact of blockchain on food supply chain: The case of walmart. 167–177.

Tan, K.H., Zhan, Y., Ji, G., Ye, F. and Chang, C. 2015. Harvesting big data to enhance supply chain innovation capabilities: An analytic infrastructure based on deduction graph. International Journal of Production Economics 165: 223–233.

Trebar, M., Lotrič, M. and Fonda, I. 2015. Use of RFID temperature monitoring to test and improve fish packing methods in styrofoam boxes. Journal of Food Engineering 159: 66–75.

Treiblmaier, H. 2018. The impact of the blockchain on the supply chain: A theory-based research framework and a call for action. Supply Chain Management: An International Journal.

Trivellas, P., Malindretos, G. and Reklitis, P. 2020. Implications of green logistics management on sustainable business and supply chain performance: evidence from a survey in the greek agri-food sector. Sustainability, 12(24): 10515.

Tsang, Y., Choy, K., Wu, C., Ho, G., Lam, H. and Tang, V. 2018. An intelligent model for assuring food quality in managing a multi-temperature food distribution centre. Food Control 90: 81–97.

UN. 2019. World population prospects 2019.

Vallandingham, L.R., Yu, Q., Sharma, N., Strandhagen, J.W. and Strandhagen, J.O. 2018. Grocery retail supply chain planning and control: Impact of consumer trends and enabling technologies. IFAC-PapersOnLine 51(11): 612–617.

Van Der Vorst, J.G., Tromp, S.-O. and Zee, D.-J. van der. 2009. Simulation modelling for food supply chain redesign; integrated decision making on product quality, sustainability and logistics. International Journal of Production Research 47(23): 6611–6631.

Verdouw, C., Sundmaeker, H., Meyer, F., Wolfert, J. and Verhoosel, J. 2013. Smart agri-food logistics: Requirements for the future internet. In Dynamics in logistics (pp. 247–257). Springer.

Verdouw, C.N., Wolfert, J., Beulens, A. and Rialland, A. 2016. Virtualization of food supply chains with the internet of things. Journal of Food Engineering 176: 128–136.

Wang, Q.J., Mielby, L.A., Thybo, A.K., Bertelsen, A.S., Kidmose, U., Spence, C., et al. 2019. Sweeter together? Assessing the combined influence of product-related and contextual factors on perceived sweetness of fruit beverages. Journal of Sensory Studies 34(3): e12492.

Wieczorek, A. 2017. Impact of 3D printing on logistics. Research in Logistics and Production, 7.

Wu, H., Tao, B., Gong, Z., Yin, Z. and Ding, H. 2019. A fast UHF RFID localization method using unwrapped phase-position model. IEEE Transactions on Automation Science and Engineering 16(4): 1698–1707.

Wu, H.-K., Lee, S.W.-Y., Chang, H.-Y. and Liang, J.-C. 2013. Current status, opportunities and challenges of augmented reality in education. Computers and Education 62: 41–49.

Yu, W., Chavez, R., Jacobs, M.A. and Feng, M. 2018. Data-driven supply chain capabilities and performance: A resource-based view. Transportation Research Part E: Logistics and Transportation Review 114: 371–385.

Zhang, R., Lu, J.-C. and Wang, D. 2014. Container drayage problem with flexible orders and its near real-time solution strategies. Transportation Research Part E: Logistics and Transportation Review 61: 235–251.

Zheng, M., Zhang, S., Zhang, Y. and Hu, B. 2021. Construct food safety traceability system for people's health under the Internet of Things and big data. IEEE Access 9: 70571–70583.

Zhong, R., Xu, X. and Wang, L. 2017. Food supply chain management: Systems, implementations and future research. Industrial Management and Data Systems.

Zhou, L., Zhang, L. and Fang, Y. 2020. Logistics service scheduling with manufacturing provider selection in cloud manufacturing. Robotics and Computer-Integrated Manufacturing 65: 101914.

Zhu, Z., Chu, F., Dolgui, A., Chu, C., Zhou, W. and Piramuthu, S. 2018. Recent advances and opportunities in sustainable food supply chain: A model-oriented review. International Journal of Production Research 56(17): 5700–5722.

18

Water Footprint of Food Systems

Shreeya Shrestha,[a] *Bhintuna Vaidya*[a] and *Anish Ghimire**

1. Introduction

The availability of freshwater resources for agricultural production is crucial to food security. Agricultural (food and nonfood) production consumes over 70% of freshwater globally each year. Climate change adds to the uncertainty by altering rainfall patterns and increasing the frequency of extreme weather patterns like floods and droughts (Schewe et al. 2014). Increased worldwide water use in agriculture has resulted from growing populations and rapidly shifting diets, particularly increased consumption of animal source foods (ASFs) (Vörösmarty et al. 2000). Knowing the influence of food production and population-level dietary patterns on water use is crucial for long-term water management, and understanding the impact of food production and population-level dietary patterns on water use is critical in this context (Harris et al. 2020, Latham 2000).

Globally, many people are transitioning to food consumption habits that include more animal products as part of a nutrition transition (Kearney 2010). Most parts of the world are witnessing economic growth, which leads to higher purchasing power, resulting in increasing demand for food as well as a shift in the food varieties consumed (Latham 2000). Changes in food consumption patterns have increased demand for animal sources, milk, and eggs in recent decades, and producing animal-sourced foods sustainably while delivering enough protein to fulfil the requirements of a balanced diet has proven to be a particularly significant challenge (Henchion et al. 2021). Protein intake in developed nations is often higher compared to developing countries, primarily due to the excessive consumption of ASFs. Meat and other animal products are often more expensive per capita consumption as it rises with the average per capita income (Gerbens-Leenes et al. 2010). If population increase continues in developing countries, especially when combined with economic growth, as is predicted in countries such as China and Brazil, demand for animal products is expected to rise (Gerbens-Leenes et al. 2013). When it comes to meat products, China has seen massive growth in consumption. The consumption of 7 million tonnes of meat in 1975 has increased by nearly 13% reaching 86.5 million tonnes by 2018 turning it into the world's greatest meat consumption as shown in Fig. 1 (Chapagain and Hoekstra 2013).

Mapping and quantifying national water footprints (WFs) has been an evolving field of study to measure the environmental impacts of anthropogenic activities (Mekonnen and Hoekstra 2011b).

Resource Recovery Research Group (Re3G), Department of Environmental Science and Engineering, Kathmandu University, Dhulikhel 45200, Nepal.

[a] S. Shrestha and B. Vaidya contributed equally to this work as first authors.

* Corresponding author: anishghimire@ku.edu.np

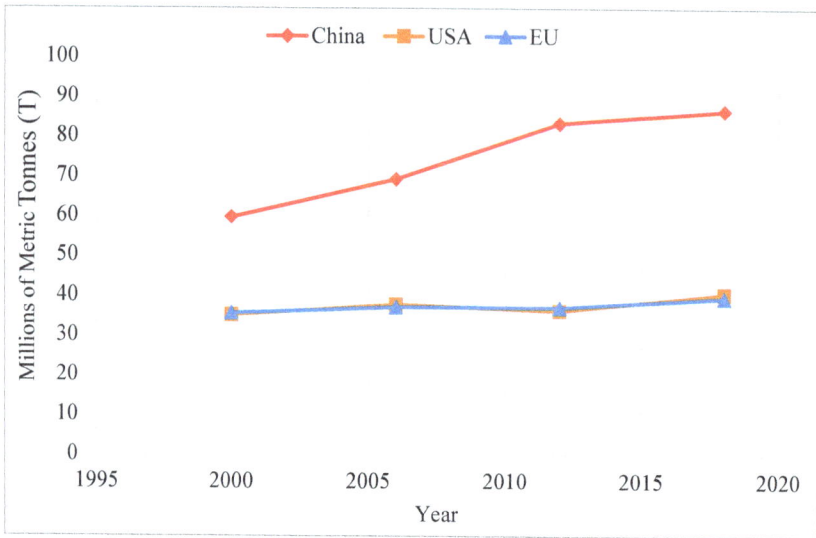

Figure 1. Growth of meat consumption in China, US and EU. Based on data from (Chapagain and Hoekstra 2013).

The notion of WF was proposed in 2002 to provide a consumption-based measure of water use in addition to the traditional production-sector-based measures of water use (Chapagain and Hoekstra 2013). Water consumption and pollution may be analyzed using the WF concept, which differentiates between a blue WF (consumption of surface and groundwater), a grey WF (amount of water required to dilute the pollution of surface or groundwater) and a green WF (consumption of rainfall). The WF related to agricultural production obtains the largest share of the total WF. Globally, agriculture contributes to 92% of the global fresh WF; 29% of the water in agriculture is directly or indirectly used for an animal (Mekonnen and Gerbens-Leenes 2020).

A high blue WF indicates that considerable amounts of irrigation freshwater are utilized during agricultural production, which can be significant in regions where surface and groundwater sources are being overexploited (Mekonnen and Hoekstra 2016). A WF with a high concentration of green or total (green + blue) could suggest low yields or inefficient water consumption in crops (Harris et al. 2020). A high blue and low green WF indicates that rainwater has been used unsustainably, which may lead to surface and groundwater overexploitation (Aleksandrowicz et al. 2016).

Several studies have demonstrated the significance of virtual water trade in available water conservation and scarcity management, and it will effectively reduce the threat of water crisis (Chapagain et al. 2006). This chapter aims to provide an overview of the water demand for various food products and diets, as well as the WF of food wastage. Finally, the chapter highlights the global WF status of food and dietary habits along with a review of different methods of WF calculations. It also gives an insight into the WF of food waste and global policies highlighting essential measures toward food sustainability, highlighting the contributions of major countries and crops to the unsustainable WF and various measures leading toward food sustainability.

2. Water Consumption Pattern and Virtual Water Trade Status

2.1 Water Consumption in Global Agriculture

Agricultural production reportedly accounts for 85% of global water usage (Shiklomanov and Rodda 2004), with this figure expected to quadruple by 2050 (Tilman et al. 2002). Around 90% of the growth in worldwide food production required by 2050 is expected to occur in developing countries, whose share of world food production is expected to rise by 2050—it will have risen to 74% (from 67% in 2007) (FAO 2017). By 2050, the irrigated area is predicted to increase by a factor of 1.9, while climate change is exacerbating water stress by altering water supply patterns in many parts of the world (Lobell et al. 2008). Furthermore, there is a global increase in the development of biological energy resources, which is accelerating agricultural production growth (Melillo et al. 2009). Water scarcity and land clearance are key environmental challenges as a result of these pressures around the world (Pfister et al. 2011).

The increase in agricultural productivity in developing nations will be especially beneficial to animal production, given their share of global output rising from 55% in 2005/2007 to 68% in 2050. Figure 2 shows predicted agricultural production growth by region, with South Asia nearly doubling and Sub-Saharan Africa nearly tripling.

It has been a critical requirement for improving agricultural production while preserving and strengthening natural resources such as water for farmers if global food supplies are to be increased on a sustainable basis. Smallholder farmers and their families will play a critical role in sustaining agricultural productivity growth. Farmers' fields had a lower water use efficiency than well-managed experimental sites, indicating that additional efforts are needed to transfer water-saving technologies to farmers. Water-saving agricultural and irrigation methods, such as deficit irrigation, low-pressure irrigation, subsurface drips, drip irrigation beneath plastic covers, furrow irrigation, rainfall harvesting, and conservation agriculture, will be extremely beneficial in such scenarios (Sharma 2020). Farming strategies that can take full advantage of natural rainfall and irrigation facilities are referred to as water-saving agriculture. It is preferable to optimize yield per unit of water rather than yield per unit of land when water is more limited than land. Northern China regions: Limited or deficit irrigation is becoming more prevalent in parts of North Africa, West Asia, and northern China (Oweis and Hachum 2009).

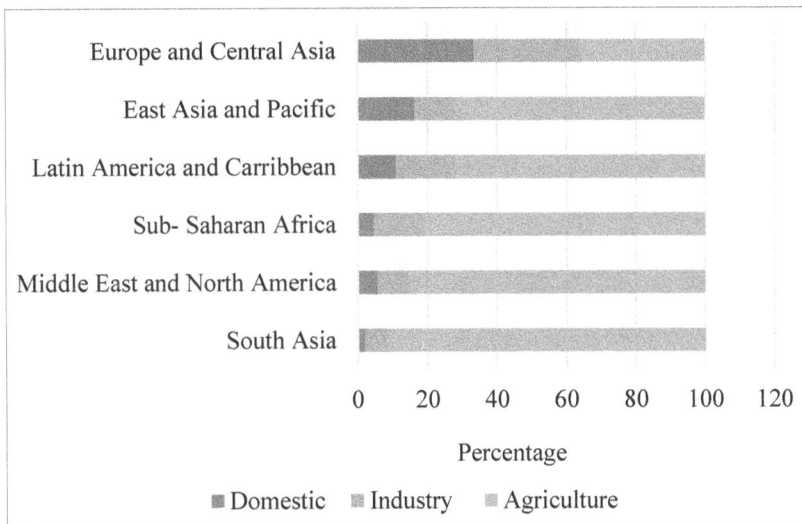

Figure 2. Share of freshwater withdrawal by sectors (%). Data Source: (Sharma 2020, Chouchane et al. 2015).

2.2 Global Status of Virtual Water Trade in the Food System

The notion of virtual, or embedded, water originated from a need to comprehend how water-stressed countries might provide food, clothes, and other water-intensive commodities to their people. Global economics in commodities has allowed states with scarce water resources to rely on the water resources of other countries to fulfil the needs of their citizens. While food and other commodities are traded on a global scale, its WF analyzes them in the form of virtual water (WFN 2021).

Over the previous two decades, the volume of virtual water commerce has more than quadrupled, and international trade accounts for more than a third of global water extraction (Chen and Chen 2013). The region's majority of countries rely heavily on water-intensive crops, mainly cereals and sugar, for their food. The limited water resources are used to irrigate high-value crops such as vegetables and fruits for domestic consumption and export to neighbouring countries (Mohammed and Darwish 2017).

For the years 1996 to 2005, the total international virtual water flow was 2320 Gm^3 per year (13% blue, 19% grey and 68% green). International trade in crops and crop-derived goods accounts for the majority (76%) of countries' virtual water flows. The global virtual water flows were 12% each due to trade in industrial items and animal products. Global virtual water flows connected with domestically manufactured goods were 1762 Gm^3/yr (Mohammed and Darwish 2017).

On a global scale, the blue and grey shares of the total WF of internationally traded items are slightly greater than those of locally consumed products. This implies that export goods are more directly related to water consumption due to surface and groundwater contamination than non-export ones. The total WF of internationally traded items has a green component of 68%, compared to 74% for total world production. The US (314 Gm^3/yr), China (143 Gm^3/yr), India (125 Gm^3/yr), Brazil (112 Gm^3/yr), Argentina (98 Gm^3/yr), Canada (91 Gm^3/yr), Australia (89 Gm^3/yr), Indonesia (72 Gm^3/yr), France (65 Gm^3/yr), and Germany (64 Gm^3/yr) are the major gross virtual water exporters, accounting for more than half of global virtual water export (Hoekstra and Mekonnen 2012).

The United States, India, Australia, Pakistan, Uzbekistan, Turkey, and China are the top blue virtual water exporters, accounting for 49 % of worldwide blue virtual water exports (Alcamo and Henrichs 2002). Given that inefficiencies and shortage rents are rarely included in water prices, particularly in agriculture, it is unrealistic to assume regional water scarcity patterns to be automatically compensated for by production and trade patterns. The United States (234 Gm^3/yr), Japan (127 Gm^3/yr), Germany (125 Gm^3/yr), China (121 Gm^3/yr), Italy (101 Gm^3/yr), Mexico (92 Gm^3/yr), France (78 Gm^3/yr), the United Kingdom (77 Gm^3/yr), and the Netherlands (71 Gm^3/yr) are the top gross virtual water importers (Hoekstra and Mekonnen 2012).

3. Methodologies for Evaluation of WF in the Food System

The methodologies were based on the quantity of water used while producing a product or service, along with the potential environmental impact. The acceptability of the results is determined by the method and quality of data used to calculate WF. Chouchane et al. (2015) revised the general WF technique by determining the scarcity of water in a particular country by comparing blue water consumption with the renewable blue water resources, resulting in a ratio that can be used to classify a country or society as water secure or stressed. Initially, Volumetric WF was used to estimate the volume of water used in rice production and consumption (Chapagain and Hoekstra 2010). The method can be used to analyze the trade impact of agricultural commodities because the calculation was done in a worldwide context. The water footprint of bioenergy crops can be calculated using the same method (Su et al. 2015). Results were presented in the form of water use efficiency in various climatic regions, making the method useful in determining which crops to produce in various areas for optimal water use efficiency (Su et al. 2015, Chouchane et al. 2015). The stressed-weighted approach of WF was created in Italy and is used to analyze the WF for wines and winery goods. This strategy can be used in the wine business, as well as other agricultural goods that are input-demanding, such as those that require a lot of fertilizers and pesticides. It delivers accurate greywater use due to its real data utilization, which is lacking in other approaches.

3.1 Calculation of WF Related to Animal Products

The WF of a live animal consists of two major components: indirect WF from the feed and direct WF from the drinking water and service water utilized (Chapagain and Hoekstra 2010). The WF of an animal is calculated as:

$$WF [a,c,s] = WF_{feed} [a,c,s] + WF_{drink}[a,c,s] + WF_{serv}[a,c,s] \qquad (1)$$

where, WF_{feed} [a,c,s], WF_{drink} [a,c,s], and WFs_{erv} [a,c,s] indicate WF of the animal category a, in nation c, in production systems s for feed, drinking water, and service-water consumption, respectively. The water used to clean the farm, bathe the animals, and conduct other environmental services is known as service water. The WF of an animal and its three components can be expressed in terms of $m^3/y/animal$ or terms of $m^3/animal$ when totaled across the animal's lifespan. It is crucial to analyse the WF of the animal at the end of its lifecycle for animals that supply their products after they have been slaughtered such as beef cattle, sheep, goats, pigs and broiler chickens. For dairy cattle and layer chickens, it is most straightforward to look at the animal's WF each year (averaged over its lifetime), as one can simply relate this annual WF of an animal to its average annual productivity (milk, eggs).

The WF of an animal's feed consumption consists of two parts: the WF of the various feed ingredients and the water needed to mix the feed:

$$WF_{feed}\ [a,c,s] = \frac{\sum_{p=1}^{n} \square \left(Feed[a,c,s,p] \times WF_{prod}^*[p] \right) + WF_{mixing}[a,c,s]}{pop^*[a,c,s]} \tag{2}$$

An animal category is linked to a specific country (referred to as "c") and production systems measured in tonnes per year. The expression "Feed [a,c,s,p]" denotes the yearly consumption of a particular feed ingredient (referred to as "p") by the animal category. The term "WFprod * [p]" signifies the water footprint of the feed ingredient p, measured in cubic meters per tonne. Moreover, "WF mixing [a,c,s]" represents the volume of water needed for mixing the feed for the animal category, Pop* [a,c,s] denotes the number of slaughtered animals per year or the number of milk or egg-producing animals in a year for the animal category.

The approach established by Mekonnen and Hoekstra (2010) can be used to calculate the by-products used by farm animals. The WFs of feed crops can be calculated using a crop water usage model that calculates crop WFs at a global spatial resolution. Grey WFs can be assessed by studying the amount of leaching and runoff of the nitrogen fertilizers as proposed by Mekonnen and Hoekstra (2011b) and Chapagain and Hoekstra (2010). As animal feed in a country consists of both domestic and imported products, the WF of animal feed in a country is calculated using the weighted average WF based on the relative amounts of domestic and imported products:

$$WF_{pop}^*[p] = \frac{p[p] \times WF_{prod}[p] + \sum_{ne} (T_i[n_e, p] \times WF_{prod}[n_e, p])}{P[p] + \sum_{ne} T_i[n_e, p]} \tag{3}$$

in which P[p] is the quantity of production of feed product p in a nation (tonne/y), T_i [n_e,p] refers to the quantity of feed product p imported from the exporting nation n_e (tonne/y), WF_{prod}[p] the WF of feed product p is produced in a nation ($m^3/tonne$) and WF_{prod} [ne,p] the WF of the exporting nation's n_e ($m^3/tonne$) feed product p. The WF of crop leftovers such as bran, chaff, straw and tops from sugar beet leaves has already been accounted for in the primary product, their WF has been set to zero.

The WF_{cons} (in m^3/yr), are calculated using the bottom-up strategy (based upon consumption data), stated as follows (Mekonnen and Hoekstra 2011a)

$$WF_{cons} = WF_{cons.direct} + WF_{cons.indirect\ (agricultural\ commodities)} + WF_{cons.indirect\ (industrial\ commodities)} \tag{4}$$

$$WF_{cons,indir\ (agricultural\ commodities)} = \sum_{\beta}(C(p) \times WF_{product[p]}^*) \tag{5}$$

where C[p] is the agricultural product p consumed within the EU28 (tonne/yr) and $WF^*_{prod}[p]$ is the average WF of this product (m³/tonne). The set of products under discussion includes the entire range of finished agricultural products.

3.2 Calculation of WF Related to Crops

The total blue and green water consumption in terms of depth (mm) can be computed using CROPWAT8.0's stated as crop water requirements (CWR) (Chapagain and Hoekstra 2013). The green water use (WUgreen, m³/ha) can be determined using the minimum CWR and effective rainfall (the amount of rainwater that will be available in the soil for crop growth). The computation of blue water use (WUblue, measured in cubic meters per hectare) involves multiplying the minimum irrigation water requirement and effective irrigation supply by the proportion of the total area of irrigated crops. Effective irrigation supply represents the stored soil moisture that is available for crop evaporation. The volume of fresh water needed to dilute leached pollutant loads to an acceptable level is referred to as greywater use (WUgrey, m³/ha). It is derived using Eq. (1) and is based on the rate of application of various chemicals like insecticides, pesticides, and fertilizers.

$$WFgrey = [(\alpha * AB)/(Cmax - Cmin)]/\gamma \qquad (6)$$

where, AB (kg/ha) represents the pollutant application rate per unit area; α is the rate of leaching; C_{max}(kg/m³) represents the maximum allowable concentration; C_{nat} (kg/m³) represents natural concentration, and Y(ton/ha) represents a yield of the crop. Due to a lack of local data, this analysis used nitrate-N as the typical pollutant, with a concentration of 10% (at a flat rate) as proposed by the previous study (Chapagain et al. 2006), C_{nat} as zero, and C_{max} as 11.5 mg/l under Nepal water quality guidelines. However, soil conditions (irrigation frequency, percolation rate, rainfall pattern, soil texture, and so on) along with fertilizer application methods (rate, timing, agronomical practices, and so on) directly impact the amount of nitrogen leached (Chapagain and Hoekstra 2010).

3.3 WF status Associated with Food Production

The overall volume of water used in agriculture is 7,980 Gm³/yr, with blue water accounting for one-third and green water accounting for the remaining two-thirds (soil water) (Mekonnen and Hoekstra 2011). Wheat consumes the most of this total volume, accounting for 1,087 Gm³ each year (70 % green, 19% blue, 11% grey). The lowest proportion of green water to the total consumptive WF among the principal crops are date palm (43%) and cotton (64%). Large volumes of water are required and polluted in the production of meat, especially for the production of animal feed (Chapagain and Hoekstra 2013). The amount of water utilized in the production of a crop is calculated using the total volume of the crop produced and the crop's water content (WF). In the case of animals, the model from Eq. (2) takes into account the water content of feed as well as the amount of water consumed over the lifetime of an individual. The internal (water used for products generated in that nation) and external (water used for products imported into that nation) water consumption of a nation can be calculated using these calculations. With global averages of 907 m³/capita per year for internal production and 160 m³/capita per year for external production, the WF for US agricultural items was expected to be 1192 m³/capita per year for internal production and 267 m³/capita per year for external production. WFs in developed countries are often larger than in poor countries. External WFs are high for countries that heavily rely on imported foods majorly several European countries, such as the United Kingdom and Italy.

Animal products require high water input among which pork and beef have the highest green WF of all animal products. Rice is the most water-intensive crop among plant foods, using over

21% of the total volume of water required by all field crops followed by wheat which is the second most water-intensive crop. Coffee, cocoa, and olive oil were among the most water-intensive crops, whereas fruits and vegetables have lower WFs (MacDonald and Reitmeier 2017).

The overall WF animal category and in-country c and production system s (tonne/y) of animal production (2,422 Gm^3/y) accounts for 29% of the WF of total of agricultural production (8,363 Gm^3/y) globally. It was then calculated as the total the global WF of crop production (7,404 Gm^3/y) (Mekonnen and Hoekstra 2011a), the direct WF of livestock (46 Gm^3/y), and the WF of grazing (913 Gm^3/y). While examining at total WF by animal group (Table 2), beef cattle contributes the most (33%) to the global WF of farm animal production, followed by dairy cattle (19%), pigs (19%), and broiler chickens (11%). Over 97% of the WF related to feeding comes from grazing and fodder crops in the grazing system, and the WF is predominantly green (94%). The green WF accounts for 87% and 82% of the total footprint in mixed and industrial production systems, respectively. The blue WF in the grazing system accounts for 3.6% of the overall WF, with drinking and service-water consumption accounting for around 33% of that (Mekonnen and Hoekstra 2011).

From 1996 to 2005, the global WF of crop production was 7404 Gm^3 per year (78% green, 12% blue, and 10% grey) (Mekonnen and Hoekstra 2011a). Wheat consumes the most of this total volume, accounting for 1087 Gm^3 every year (70% green, 19% blue, 11% grey). Rice (992 Gm^3/yr) and maize (770 Gm^3/yr) are the other crops with high total WF. Figure 3 depicts the proportion of the key crops to the worldwide WF associated with crop production. Rainfed crops use 4701 Gm^3/yr, whereas irrigated crops use 1070 Gm^3/yr, resulting in a global average green WF connected to agricultural production of 5771 Gm^3/yr. Green WF accounts for more than 80% of total consumptive WF (green and blue) in most crops. Among the principal crops, date palm (43%) and cotton have the lowest proportion of green water to total consumptive WF (64%).

Figure 3 depicts the fraction of green water in total evaporative (green plus blue) WF for the major crops. The world's average blue WF for crop productivity was 899 Gm^3 per year. Wheat (204 Gm^3 per year) and rice (202 Gm^3 per year) have large blue WF, accounting for 45% of worldwide blue WF. The grey WF for nitrogen fertilizer used in crop cultivation was 733 Gm^3 per year. Wheat (123 Gm^3/yr), maize (122 Gm^3/yr), and rice (111 Gm^3/yr) are the three crops with the highest grey WF, contributing to nearly 56% of global grey WF.

The global WF of crop production was 7404 billion m^3/yr from 1996 to 2005. The high percentage of green water (78%) emphasizes the necessity of rain (Mekonnen and Hoekstra 2011b). The proportion of blue water is lower (12%), but the spatial analysis shows that the locations with substantial blue WFs are frequently arid and semi-arid regions with considerable water scarcity. The grey WF has a small share (10%), but this is a cautious estimate because the study only looked at the needed absorption quantity for leached nitrogen fertilizers, excluding relative contaminants

Table 1. Caloric value and WF of respective primary crops.

S. N.	Primary crop category	Caloric value (kcal/kg)	Water Footprint (litre/kcal)
1	Sugar Crops	290	0.68
2	Vegetables	240	1.34
3	Roots and tubers	830	0.47
4	Fruits	460	2.10
5	Cereals	3200	0.51
6	Oil crops	2900	0.81
7	Pulses	3400	1.19
8	Spices	3000	2.35
9	Nuts	2500	3.63

Source: (Mekonnen and Hoekstra 2011a)

Table 2. The WF of crop production of selected countries.

S. N.	Country	WF of crop production (Gm³/yr)
1	India	1047
2	China	967
3	USA	828
4	Brazil	329
5	Russia	327

Source: (Mekonnen and Hoekstra 2011a)

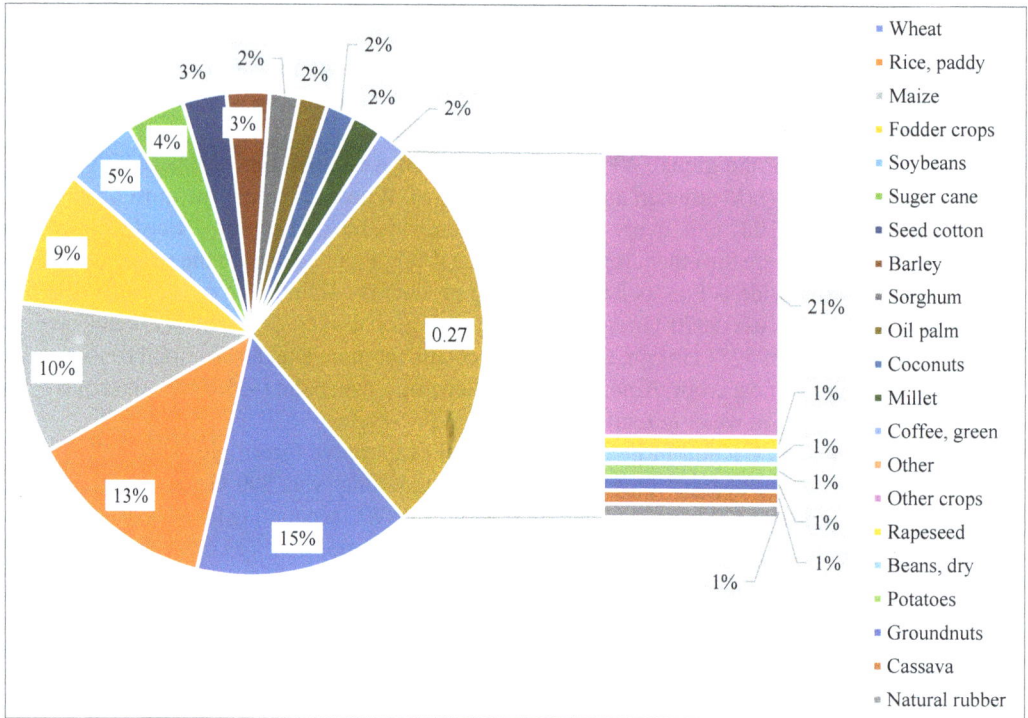

Figure 3. Contribution of different crops to the total WF of crop production (1996–2005) based on data from (Mekonnen and Hoekstra 2011).

like phosphorus and pesticides. According to the consistent studies carried out, green water has a considerable impact on global food output. Most countries, according to, have a potential of self-sufficiency on a green water-based and can produce their whole food requirement locally. There is a substantial possibility to increase water productivity in rain-fed agriculture by increasing production levels by up to fourfold within the existing water balance (Rockström 2003). This is a potential chance to boost water production in rain-fed agriculture by improving the water productivity omitting the need to introduce new blue water resources (Critchley et al. 1991). However, available precipitation in semi-arid and dry locations is extremely limited, making agricultural productivity without the use of blue water mostly impossible. If no blue water is used, global cereal output would be drastically reduced (Siebert and Doell 2010, Chapagain and Hoekstra 2013). As a result, in a world with finite freshwater supplies, to deal with the issue of increased water demand, a balanced green-blue water usage strategy would be required. Additional research is essential to identify the blue water availability by analysing the spatiotemporal variability, as well as the amount of blue water that may be utilised sustainably in a specific catchment without creating environmental damage.

To deal with the issue of increased water demand, a balanced green-blue water usage strategy would be required. More research is needed to identify the spatiotemporal variability of blue water availability, as well as the amount of blue water that may be utilized sustainably in a specific watershed without causing environmental damage.

3.4 Case Studies Associated with WF of Rice and Animal Products

Rice is a major crop for feeding the world's population, especially in Southern Asia and Africa. The construction of large irrigation systems is common to fulfill the water required in rice farming. Hence, rice is considered the most water-intensive crop on the planet. Rice production is divided into two categories: wetland and upland. Wetlands supply around 85% of the world's area where rice is harvested with irrigated wetland rice accounting for roughly 75% of total rice production (Chapagain and Hoekstra 2010).

Among the rice-producing countries, India consumes the largest amount of water for rice production followed by China, Indonesia, Bangladesh, Thailand, Myanmar, Vietnam, the Philippines, and Brazil as mentioned in Fig. 4. In comparison to India (239 m^3/cap/yr), Indonesia (299 m^3/cap/yr), China (134 m^3/cap/yr), and the United States (29 m^3/cap/yr), Thailand's per-capita WF for rice consumption is quite high (547 m^3/cap/yr). This difference is because some countries' diets contain more rice than others.

South Asian countries have high WF for rice production and consumption. In these countries, however, the majority of the WF is grounded in the monsoon, therefore the water scarcity is rather minor, contrary to popular opinion. The overall WF of rice uses almost similar quantities of green and blue water globally. When compared to the blue WF, the green WF (rain) has a lower production cost (evaporation of irrigation water from the field). The impact on the environment of rice production's blue WF is influenced by the location and when the water is used. It would take a detailed analysis to determine the location and timing of blue WFs in rice production that cause significant problems in the environment, but rice from the United States and Pakistan, which is majorly dependent on blue water, is likely to have greater impacts per unit of product than rice from the Vietnam. Comparing the local implications of different rice sources may be useful from a perspective of sustainable perspective for nations or areas where the import of rice import is maximum for their consumption (Chapagain and Hoekstra 2010).

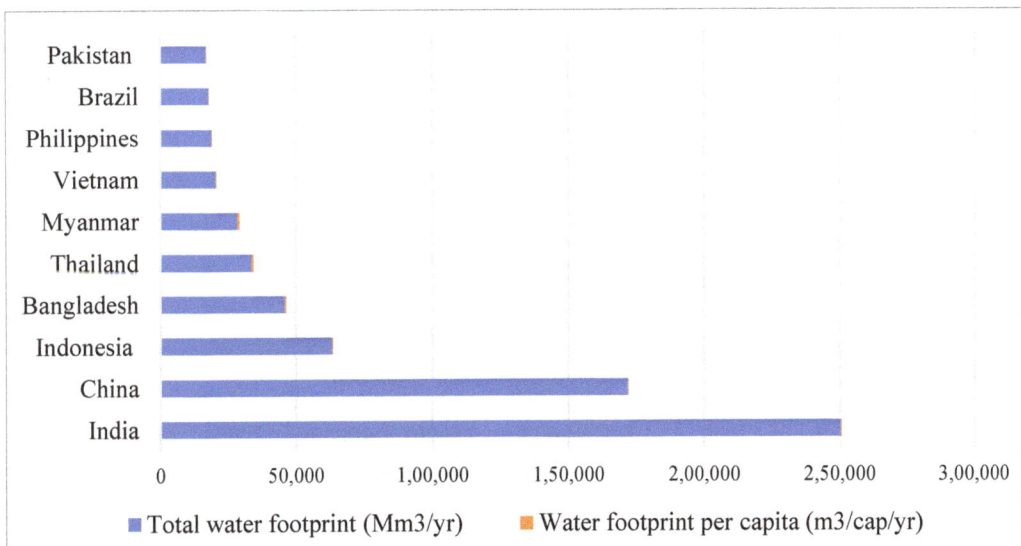

Figure 4. Top 10 countries with the largest WF related to rice consumption (Mm³/yr) (2000–2004) based on data from (Mekonnen and Hoekstra 2011b).

Global meat output nearly doubled between 1980 and 2004 and the graph is expected to rise with an estimation that meat production will double between 2000 and 2050 (Food and Agriculture Organization of the United Nations 2017). Although it is well known that products from animals use more water, little attention has been made to the livestock sector's overall impact on world freshwater demand (Chapagain and Hoekstra 2013). The majority of the water used in the supply chain for animal products is used in feed production.

Production of a kilogram of beef meat requires 8 times more feed (in dry matter) than the production of a kilogram of pig meat and 11 times more than chicken production. However, this isn't the only factor that can account for the differences. The composition of the feed is also crucial. Because concentrate feed accounts for a significant portion of the total, the WF of concentrate feed as a % of total feed is higher. Chickens are efficient in terms of total feed conversion efficiency, although their feed contains a significant portion of concentrates. This percentage is 73% (global average) for broiler chickens, but only 5% for beef cattle.

For the individual countries, the global average is not necessarily applicable. It can be demonstrated by the country average WFs for countries like China, India, the Netherlands, and the United States, as well as worldwide averages. The production system is not only responsible for the variation in the feed content but also by nation; as a result, the proportion of the various elements of the WF in different countries differs greatly in comparison to the worldwide average. In the

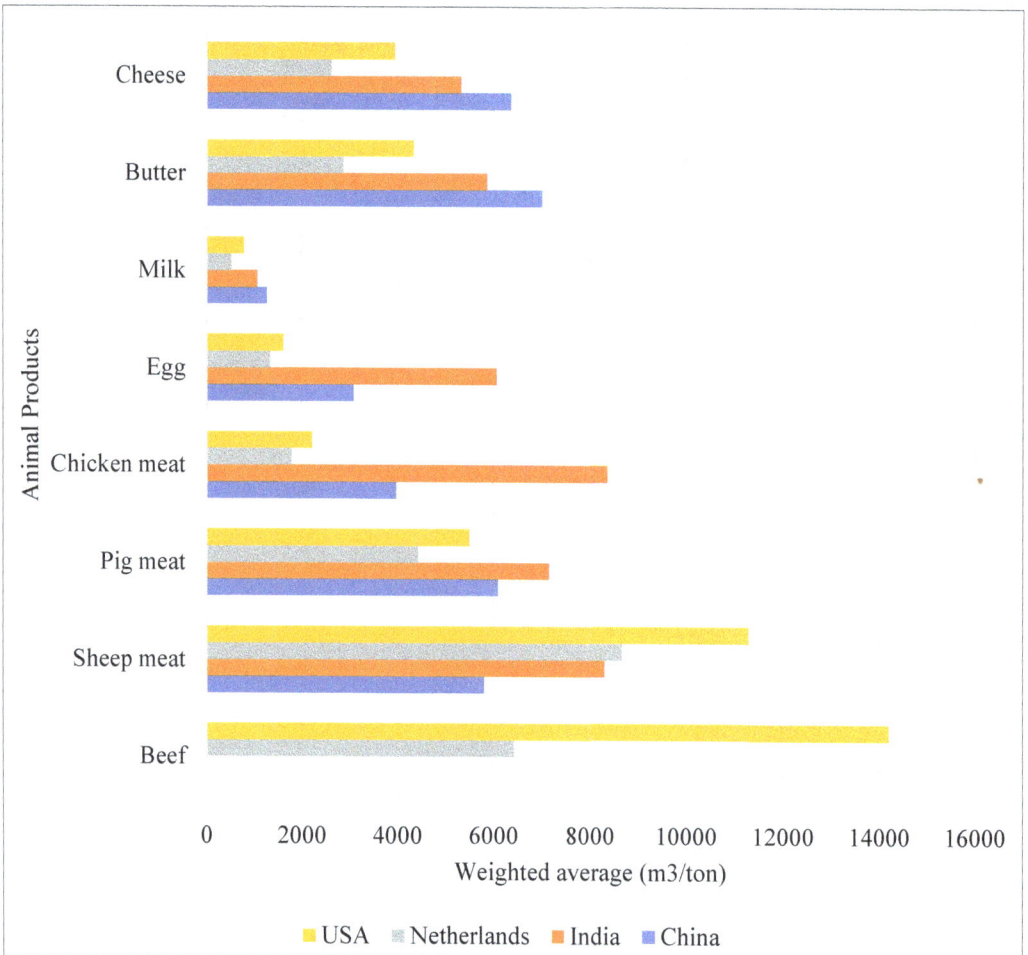

Figure 5. Animal Product consumption (m$_3$/tonne) in four representative countries. Based on data from: (Mekonnen and Hoekstra 2011a).

United States, cattle in the grazing system are fed with a proportionally large amount of irrigated and fertilized cereals, primarily corn, resulting in comparatively large blue and green WF of the US beef produced from the grazing system. Cattle in grazing as well as mixed systems in countries like China, and India are mostly supplemented with pasture and crop leftover, which have neither blue nor grey WFs, whereas the industrial systems of China and India produce beef with high blue and grey WF, which can be further explained due to the huge concentrates with high blue and grey WF in the Chinese and Indian industrial systems. Often, the WFs of feed ingredients vary by country, with variances in the total green, blue, and grey WFs of products from the animal.

Globally, the WF of the feed production is 2,376 Gm^3/yr, with crops accounting for 1,463 Gm^3/yr and grazing accounting for the remainder (Chapagain and Hoekstra 2010). The overall WF of feed crops is 20% of the total WF of global crop production, which is 7,404 Gm^3/yr (Mekonnen and Hoekstra 2011), whereas the global total blue WF of crop production of feed is 105 Gm^3/yr, which accounts for around 12% of total blue WF in crop production (Mekonnen and Hoekstra 2011a). This indicates feed accounts for about 12% of global groundwater and surface water usage for irrigation, rather than food, textile, or other crop products. Overall, WF of products from animals (2,422 Gm^3/yr) accounts for 29% of the WF in the agricultural sector (8,363 Gm^3/yr) globally. WF of farm animal production by animal category Table 3 depicts that the beef cattle contribute 33%, dairy cattle

Table 3. Animal Category WF and their Percentages.

Animal Category	Average annual WF of one animal (m^3/y/animal)	Annual WF of the animal category (Gm^3/yr)	% of total WF
Beef cattle	630	798	33
Dairy cattle	2,056	469	19
Pigs	520	458	19
Broiler chickens	26	255	11
Layer chickens	33	167	7
Sheep	68	71	3
Goats	32	24	1

Source: (Mekonnen and Hoekstra 2011a)

Table 4. Food items and their nutritional value, along with their WF based on data from (Mekonnen and Hoekstra 2011a).

Food items	Nutritional Content			WF per unit of nutritional value		
	Calorie (Kcal/kg)	Protein (g/kg)	Fat (g/kg)	Calorie (liter/kcal)	Protein (liter/g protein)	Fat (Liter/g fat)
Sugar crops	285	0.0	0.0	0.69	0.0	0.0
Vegetables	240	12	2.1	1.34	26	154
Starchy roots	827	13	1.7	0.47	31	226
Fruits	460	5.3	2.8	2.09	180	348
Cereals	3,208	80	15	0.51	21	112
Oil crops	2,908	146	209	0.81	16	11
Pulses	3,412	215	23	1.19	19	180
Nuts	2,500	65	193	3.63	139	47
Milk	560	33	31	1.82	31	33
Eggs	1,425	111	100	2.29	29	33
Chicken meat	1,440	127	100	3.00	34	43
Butter	7,692	0.0	872	0.72	0.0	6.4
Pig meat	2,786	105	259	2.15	57	23
Sheep/goat meat	2,059	139	163	4.25	63	54
Beef	1,513	138	101	10.19	112	153

contribute 19%, pigs 19%, poultry 19% and broiler chickens 11% to the global WF of farm animal production.

4. WF of Animal Versus Agriculture Products Per Nutritional Value Unit

According to the overall picture, animal goods have a higher WF per tonne of the product than agricultural items (Chapagain and Hoekstra 2010). From sugar crops (about 200 m³/tonne) and vegetables (300 m³/tonne) to pulses (4,000 m³/tonne) and nuts (9,000 m³/tonne), the worldwide average WF per tonne of crop increases, as seen in Table 3. From eggs (3,300 m³/tonne) to beef (15,400 m³/tonne) and milk (1,000 m³/tonne), the WF for animal products rises. Animal products also have a bigger WF than crop goods when measured in terms of calories. Beef has a 20-fold higher average WF per calorie than cereals and starchy roots. When observing the WF of protein, it is found that milk, eggs, and chicken meat have around 1.5 times the WF of pulses. Beef has a 6 times greater WF per gram of protein than pulses. When it comes to fat, we discovered that butter has a low WF per gram of fat, even lower than oil crops. In comparison to a vegetarian diet, meat-based diets have a larger WF. The study investigated the consequences of the findings by examining the diet of one industrialized nation, the United States, to see how food composition influences WF. Meat accounts for 37% of the average American's WF when it comes to food. The average American citizen's food-related WF will be reduced by 30% if all meat is replaced with the same amount of agricultural goods like pulses and almonds.

4.1 Dietary Factors and Dietary Pattern Analysis

Increased global water demand in agriculture has resulted from an increasing human population and rapidly shifting diets, including greater reliance on animal-sourced foods (ASFs). It has become increasingly vital to determine sustainable diets that promote health while decreasing environmental consequences, and understanding the influence of food production and population-level dietary behaviors on water usage is critical for long-term water management. An increasing number of studies indicate that lowering ASFs in the diet, particularly beef, poultry, and hog meat, is linked with fewer environmental effects and resource demands. Furthermore, reducing ASF content in diets often does not suggest reduced usage of water, specifically if ASF products are substituted with foods that are more reliant on irrigation, such as fruits and lentils.

4.2 Major Foods that Contribute to Dietary WF

Both commodities described in FAO FBSs and food ready for human consumption are referred to as food and food groups. ASFs, particularly meats, dominated total and green dietary WFs of "average" dietary patterns. Cereals placed second in terms of total and green dietary WFs. The primary contributors to blue WFs of "typical dietary patterns" were plant-based foods including cereals, nuts, and sugar. When people transition to better diets, the role of foods to dietary WF varies. Plant-based foods make up a considerable portion of total and green dietary WFs-based foods, which continue to dominate the blue WFs of healthy diets, with fruits playing an important role (Harris et al. 2020).

Table 5. The WF of two different diets in industrialized countries.

Items	Meat Diet			Vegetarian Diet		
	kcal/day³	L/kcal²	L/day	kcal/day³	L/kcal²	L/day
Animal origin	950	2.5	2,375	300	2.5	750
Vegetable origin	2,450	0.5	1,225	3,100	0.5	1,550
Total	3,400	-	3,600	3,400	-	2,300

Data Source: (Willett et al. 2019)

Several studies investigated the contribution of food categories to dietary WFs in "reduced ASF" or "no ASF" dietary habits (Harris et al. 2020). The meat was frequently removed first in "reduced ASF" diets, followed by other animal products. As a result, under the "reduced ASF" dietary patterns, the proportion of foods like milk to dietary WF rises relative to meat. Tea and coffee, for example, became important contributors to the total dietary WF when the function of food in the dietary WF for the "no ASF" pattern was examined; fruits and vegetables contributed 34% of the dietary blue WF for this pattern in the US (Harris et al. 2020).

4.3 Review of Dietary Habits and Water Consumption

According to Harris et al. (2020), lowering green WFs but not necessarily lower blue WFs will result from changing from current dietary habits to better ones (Harris et al. 2020). Reducing meat consumption would reduce green WFs and, in most circumstances, blue WFs when compared to "normal" dietary patterns. Switching from present "average" food patterns to diets with "no ASF" would reduce total WF by 25% and blue WF by 12%. The total WFs of dietary patterns with a "reduced ASF" were also lower than those with an "average" ASF. Dairy products had a lower WF than meat (Mekonnen and Hoekstra 2012) and because of the reduced ASF patterns, dairy products, oil crops, and lentils were frequently substituted with meat. Fruits, oils, and nuts, all of which are key components of a healthy diet, were considerable contributors to dietary blue WFs (Willett et al. 2019). The production of these crops, and thus the availability of healthy meals, could be affected by decreased surface water or groundwater availability, which could limit irrigation. Future studies should focus more on fruits, vegetables, and nuts, especially because additional productivity is necessary to meet global healthy eating guidelines.

5. WF of Food Waste

The Food and Agriculture Organization (FAO) of the United Nations (2017) estimates the food waste quantities for both edible and inedible foods (Food and Agriculture Organization of the United Nations 2017). Likewise, considering environmental implications about the whole product rather than just the edible portion, the great majority of studies include impact factors for the entire product rather than just the edible portion. As a result, the footprint estimates used food wastage volumes for "edible + non-edible portions". This also makes cross-component analysis easier. In 2007, global food waste was roughly 1.6 Gtonnes of "primary product equivalents". The edible portion of food waste amounts to 1.3 Gtonnes. The magnitude of these findings may be seen by comparing food waste volumes to total agricultural output quantities. It is worth emphasizing that the latter amount, which is estimated to be roughly 6 Gtonnes, includes agricultural production for reasons other than food (Chapagain and Hoekstra 2010).

In UK households, the total annual WF of food waste is 6262 million m³, of which 5368 million m³ is avoidable food waste and the remaining 894 million m³ is potentially avoidable wastage. The figures constitute 8% and 1% of the total food WF in the UK, respectively (DEFRA 2008). Overall, unwanted and potentially avoidable household food wastage in the UK has a WF of 284 litres per person per day in per capita figures. In the UK, the average daily home water use (i.e., tap water) is around 150 litres per person per day (DEFRA 2008). 243 litres per person per day (86%) of total home food waste is completely avoidable, while the rest is possibly avoidable. Imported foods account for a considerable portion (71%) of the unnecessary food waste in the UK. This is justified by the fact that unwanted waste (such as banana skins and bones) is an inevitable aspect of food consumption. Agricultural products (excluding textiles) account for around 60% of the UK's External WF (EWF), whereas domestic food waste accounts for 71%. It illustrates that imported food products use more water (m³ per tonne) than products made in the United Kingdom. Even though overall food waste is around 22% by weight, it is only 14% in terms of equivalent WF. This is due to the comparatively low water content of waste food per tonne of products. When it comes to wasted food, livestock goods (meat and dairy) have a higher internal WF than agricultural products.

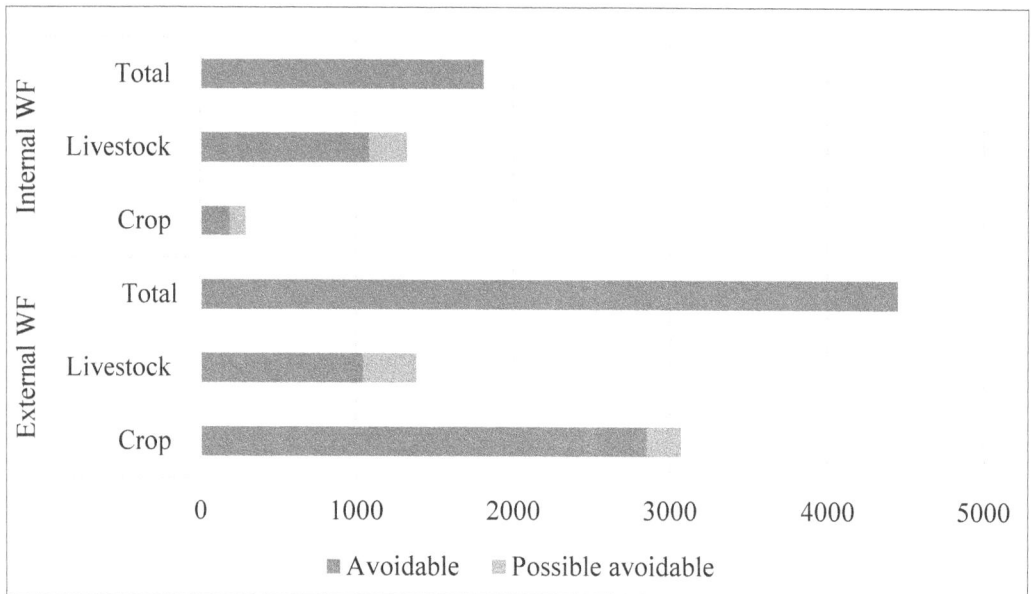

Figure 6. Internal and External WF of Household Food Waste of Crop and Livestock Products. Based on data from (Chapagain and James 2013).

In the case of imported goods, however, the agriculture product component has a larger EWF. Figure 6 shows the products that contribute the most to the WF of household food waste. The top two products in terms of WFs of household food waste are beef and cocoa products. According to WWF-UK research, they also score high in the UK for external WF across agricultural commodities. This should be emphasized that where feasible, an estimate of the composition has been supplied for sophisticated items including more than one element. However, when this is not feasible, the waste is assumed to be made up of the single element that contributes most significantly to the overall WF of the product (Chapagain and James 2013).

6. Policies and Measures for Sustainable WF

Addressing the demand for animal products by promoting a shift away from a meat-heavy diet will be an unavoidable component of government environmental policy (Mekonnen and Hoekstra 2012). Policies should neither compromise the essential rise in food security in less developed countries nor should the expansion of animal farming endanger the livelihoods of the rural poor (Mekonnen and Hoekstra 2012). There was some effort in the late 1990s to include virtual water considerations into policy frameworks. However, this was met with severe opposition by water managers and economists, who questioned the concept's maturity in the formation of water policy. While the official policy may have excluded deliberate decision-making on water comparative advantages, several countries had already done so, whether formally expressed or not. In the Middle East and North African region, countries with limited water resources were large importers of water-intensive foods. This seems to be significant for food safety and environmental sustainability since there is significant regional variability, which implies both solutions and threats (Harris et al. 2020). Dietary blue WFs, for example, were found to be unusually prevalent in Asia. Water shortage is an issue in this region because groundwater supplies are diminishing in certain locations, and climate change might alter typical rainfall and irrigation water supply patterns (Mekonnen and Hoekstra 2016). Changing food preferences in Asia may thus be insufficient to reduce local water usage unless combined with improved agricultural water management. Alternatively, the nutritional status might be improved by moving to more nutrient-dense and water-efficient crops. Cereals such as maize, millet, and sorghum, for example, have been demonstrated to be more productive than rice and wheat in India (Davis et al. 2018). Furthermore, food might also be imported from water-rich countries.

Promoting different water development strategies would need political will and cooperation from all organizations and stakeholders. Farmers and agricultural financiers are continuously attempting to cut costs and enhance output; however, this aim must therefore incorporate opportunity costs related to water, as well as a high return perspective that discourages people to take unsustainable development paths that will eventually require significant reinvestment.

7. Conclusion

Many governments are increasingly encouraging the analysis of the WF of various areas and industries to appreciate the economic reasons for the basins, WF and productivity evaluations, particularly on agricultural products. National and international water management authorities should examine WF parameters such as virtual water content, imported and exported virtual water quantities, and national and global water savings. It is projected to enhance the efficacy of water allocation for food consumption and transportation to improve planning and management, as well as contribute to the long-term sustainability of national and global water resources. In a nutshell, consistent data suggests that some dietary pattern rich in plant-based foods and limited in animal-based foods, along with low in total calories, is both healthier and linked with environmental sustainability in the future. The significant feasibility of dietary-focused food policies significantly contributes not only to improved global health but also enhanced environmental results and the possibility of food and water availability for upcoming generations.

References

Aleksandrowicz, L., Green, R., Joy, E.J.M., Smith, P. and Haines, A. 2016. The impacts of dietary change on greenhouse gas emissions, land use, water use, and health: a systematic review. PLOS ONE 11(11): e0165797. https://doi.org/10.1371/journal.pone.0165797.

Chapagain, A. and Hoekstra, A.Y. 2010. The blue, green and grey water footprint of rice from both a production and consumption perspective. https://research.utwente.nl/en/publications/the-blue-green-and-grey-water-footprint-of-rice-from-both-a-produ.

Chapagain, A.K., Hoekstra, A.Y. and Savenije, H.H.G. 2006. Water saving through international trade of agricultural products. Hydrology and Earth System Sciences 10(3): 455–468. https://doi.org/10.5194/hess-10-455-2006.

Chapagain, A. and James, K.D. 2013. Accounting for the Impact of Food Waste on Water Resources and Climate Change. https://doi.org/10.1016/B978-0-12-391921-2.00012-3

Chapagain, A.K. and Hoekstra, A.Y. 2013. Virtual water flows between nations in relation to trade in livestock and livestock products. Value of Water Research Report Series No. 13. UNESCO-IHE, Delft, the Netherlands. https://ihedelftrepository.contentdm.oclc.org/digital/collection/p21063coll3/id/10894/.

Chen, Z.-M. and Chen, G.Q. 2013. Virtual water accounting for the globalized world economy: National water footprint and international virtual water trade. Ecological Indicators. https://doi.org/10.1016/j.ecolind.2012.07.024.

Chouchane, H., Hoekstra, A., Krol, M. and Mekonnen, M. 2015. The water footprint of Tunisia from an economic perspective. Ecological Indicators, 52. https://doi.org/10.1016/j.ecolind.2014.12.015.

Critchley, W., Siegert, K., Chapman, C. and Finkel, M. 1991. Water harvesting. A manual for the design and construction of water harvesting schemes for plant production. Rome (Italy. http://www.fao.org/docrep/U3160E/U3160E00.htm.

Davis, K.F., Chiarelli, D.D., Rulli, M.C., Chhatre, A., Richter, B., Singh, D., et al. 2018. Alternative cereals can improve water use and nutrient supply in India. Science Advances 4(7): eaao1108. https://doi.org/10.1126/sciadv.aao1108.

DEFRA. 2008. Department for Environment, Food and Rural Affairs Departmental Report 2008 Cm 7399. 230.

Food and Agriculture Organization of the United Nations (Ed). 2017. The future of food and agriculture: Trends and challenges. Food and Agriculture Organization of the United Nations.

Gerbens-Leenes, P.W., Nonhebel, S. and Krol, M.S. 2010. Food consumption patterns and economic growth. Increasing affluence and the use of natural resources. Appetite 55(3): 597–608. https://doi.org/10.1016/j.appet.2010.09.013.

Gerbens-Leenes, W., Mekonnen, M. and Hoekstra, A. 2013. The water footprint of poultry, pork and beef: A comparative study in different countries and production systems. Water Resources and Industry s 1–2, 25–36. https://doi.org/10.1016/j.wri.2013.03.001.

Harris, F., Moss, C., Joy, E.J.M., Quinn, R., Scheelbeek, P.F.D., Dangour, A.D., et al. 2020. The water footprint of diets: a global systematic review and meta-analysis. Advances in Nutrition 11(2): 375–386. https://doi.org/10.1093/advances/nmz091.

Henchion, M., Moloney, A.P., Hyland, J., Zimmermann, J. and McCarthy, S. 2021. Review: Trends for meat, milk and egg consumption for the next decades and the role played by livestock systems in the global production of proteins. Animal 15: 100287. https://doi.org/10.1016/j.animal.2021.100287.

Hoekstra, A.Y. and Mekonnen, M.M. 2012. The water footprint of humanity. Proceedings of the National Academy of Sciences of the United States of America 109(9): 3232–3237.

Kearney, J. 2010. Food consumption trends and drivers. Philosophical Transactions of the Royal Society B: Biological Sciences 365(1554): 2793–2807. https://doi.org/10.1098/rstb.2010.0149.

Latham, J.R. 2000. There's enough food for everyone, but the poor can't afford to buy it. Nature 404(6775): 222. https://doi.org/10.1038/35005264.

Lobell, D.B., Burke, M.B., Tebaldi, C., Mastrandrea, M.D., Falcon, W.P. and Naylor, R.L. 2008. Prioritizing climate change adaptation needs for food security in 2030. Science (New York, N.Y.): 319(5863): 607–610. https://doi.org/10.1126/science.1152339.

MacDonald, R. and Reitmeier, C. 2017. Understanding Food Systems—1st Edition. https://www.elsevier.com/books/understanding-food-systems/macdonald/978-0-12-804445-2.

Mekonnen, M. and Hoekstra, A. 2010. The green, blue and grey water footprint of farm animals and animal products. American Journal of Hematology - AMER J HEMATOL.

Mekonnen, M. and Hoekstra, A. 2016. Four billion people facing severe water scarcity. Science Advances 2: e1500323–e1500323. https://doi.org/10.1126/sciadv.1500323.

Mekonnen, M. and Hoekstra, A.Y. 2011a. National water footprint accounts: The green, blue and grey water footprint of production and consumption. https://research.utwente.nl/en/publications/national-water-footprint-accounts-the-green-blue-and-grey-water-f.

Mekonnen, M.M. and Hoekstra, A.Y. 2011b. The green, blue and grey water footprint of crops and derived crop products. Hydrology and Earth System Sciences 15(5): 1577–1600. https://doi.org/10.5194/hess-15-1577-2011.

Mekonnen, M.M. and Hoekstra, A.Y. 2012. A global assessment of the water footprint of farm animal products. Ecosystems 15(3): 401–415. https://doi.org/10.1007/s10021-011-9517-8.

Mekonnen, M.M. and Gerbens-Leenes, W. 2020. The water footprint of global food production. Water 12(10): 2696. https://doi.org/10.3390/w12102696.

Hoekstra, A.Y. and Mekonnen, M.M. 2012. The water footprint of humanity. Proceedings of the National Academy of Sciences of the United States of America 109(9): 3232–3237.

Melillo, J.M., Reilly, J.M., Kicklighter, D.W., Gurgel, A.C., Cronin, T.W., Paltsev, S., et al. 2009. Indirect emissions from biofuels: How important? Science (New York, N.Y.): 326(5958): 1397–1399. https://doi.org/10.1126/science.1180251.

Mohammed, S. and Darwish, M. 2017. Water footprint and virtual water trade in Qatar. Desalination and Water Treatment 66: 117–132. https://doi.org/10.5004/dwt.2017.20221.

Oweis, T. and Hachum, A. 2009. Supplemental irrigation for Improved Rainfed Agriculture in WANA Region. In Rainfed Agriculture: Unlocking the Potential. https://doi.org/10.1079/9781845933890.0182.

Pfister, S., Bayer, P., Koehler, A. and Hellweg, S. 2011. Projected water consumption in future global agriculture: Scenarios and related impacts. The Science of the Total Environment 409(20): 4206–4216. https://doi.org/10.1016/j.scitotenv.2011.07.019.

Rockström, J. 2003. Water for food and nature in drought-prone tropics: Vapour shift in rain-fed agriculture. Philosophical Transactions of the Royal Society B: Biological Sciences 358(1440): 1997–2009. https://doi.org/10.1098/rstb.2003.1400.

Schewe, J., Heinke, J., Gerten, D., Haddeland, I., Arnell, N.W., Clark, D.B., et al. 2014. Multimodel assessment of water scarcity under climate change. Proceedings of the National Academy of Sciences of the United States of America 111(9): 3245–3250. https://doi.org/10.1073/pnas.1222460110.

Sharma, S. 2020. Nepal Medical Council Interim Guidance for Infection Prevention and Control When COVID-19 Is Suspected. 19.

Shiklomanov, I.A. and Rodda, J.C. 2004. World water resources at the beginning of the twenty-first century. Cambridge University Press. https://scholar.google.com/scholar_lookup?title=World+water+resources+at+the+beginning+of+the+twenty-first+centuryandauthor=Shiklomanov%2C+I.+A.andpublication_year=2004.

Siebert, S. and Doell, P. 2010. Quantifying blue and green virtual water contents in global crop production as well as potential production losses without irrigation. Journal of Hydrology 384: 198–217. https://doi.org/10.1016/j.jhydrol.2009.07.031.

Su, M.-H., Huang, C.-H., Li, W.-Y., Tso, C.-T. and Lur, H.-S. 2015. Water footprint analysis of bioethanol energy crops in Taiwan. Journal of Cleaner Production 88: 132–138. https://doi.org/10.1016/j.jclepro.2014.06.020.

Tilman, D., Cassman, K.G., Matson, P.A., Naylor, R. and Polasky, S. 2002. Agricultural sustainability and intensive production practices. Nature 418(6898): 671–677. https://doi.org/10.1038/nature01014.

Vörösmarty, C.J., Green, P., Salisbury, J. and Lammers, R.B. 2000. Global water resources: Vulnerability from climate change and population growth. Science (New York, N.Y.): 289(5477): 284–288. https://doi.org/10.1126/science.289.5477.284.

WFN. 2021. Water Footprint Network. https://waterfootprint.org/en/.

Willett, W., Rockström, J., Loken, B., Springmann, M., Lang, T., Vermeulen, S., et al. 2019. Food in the Anthropocene: The EAT-Lancet Commission on healthy diets from sustainable food systems. Lancet 393(10170): 447–492. https://doi.org/10.1016/S0140-6736(18)31788-4.

19

Carbon Footprint of the Food Supply Chain

Mariany Costa Deprá* and Eduardo Jacob-Lopes

1. Introduction

Food production plays a significant role in population survival. Likewise, agriculture is the cornerstone of humanity's existence and global development. However, in recent decades, agriculture and the food production chain have been plagued by two major contradictions: the first is the conflict between supply and demand for resources, and the second is the conflict between economic development and environmental protection (Yue and Guo 2021).

Given its importance, it is estimated that the intensification of agricultural and food production is a substantial source of greenhouse gas (GHG) emissions, generating around 13.7 billion metric tons of carbon dioxide equivalent (CO_2eq) per year. This represents approximately 29% of global GHG emissions (Marbach and Gaillac 2021). In addition, climate change is predicted to cause income losses of 10%–15% of global food production, resulting in an increase in food commodity prices of between 32–37% by 2050 (Zhang et al. 2021a). Therefore, immediate responses and actions to reduce emissions from the food supply chain are needed at all levels (El Geneidy et al. 2021).

As a starting point, efforts to slow global warming are already transforming food industries, business models, and governance practices (Heinonen et al. 2020). Based on broad international cooperation, many countries have started to address measures leading to the reduction of carbon emissions through the carbon footprint as an important strategic environmental management platform (Shi and Yin 2021).

The carbon footprint is an extension of the ecological footprint. Although a unified definition has not yet been established, the carbon footprint can be understood as a measure of the total unique amount of carbon dioxide emissions—directly and indirectly—caused by an anthropogenic activity throughout the life stages of a product (Wiedmann and Minx 2008). However, with the continuous advances in research methodologies, supported by life cycle assessment tools, the effective measurement of the carbon footprint began to be applied in the most diverse industrial segments.

The potential environmental impacts related to food production processes have become the stage for investigation. Proof of this is that many studies have tried to elucidate the carbon

Bioprocess Intensification Group, Federal University of Santa Maria, UFSM, Roraima Avenue 1000, 97105-900, Santa Maria, RS, Brazil.
Email: ejacoblopes@gmail.com
* Corresponding author: marianydepra@gmail.com

footprint of food supply chains (Vauterin et al. 2021, Zhang et al. 2021b). However, the steps in the operational chain—from production to final disposal—make the results of the analyses doubtful, as it is difficult to identify the specific sources and mechanisms of emissions due to their diffuse and complex nature.

In light of the above evidence, this chapter aims to contribute to an understanding of the gaps surrounding carbon footprint analysis and food supply chains. Here, we address the key carbon footprint factors in the food production process, through a selected review of relevant carbon footprint studies applied to the food industry. With the help of facilitators, carbon footprint reduction strategies were provided for the food supply chain. In the end, the chapter reports criteria beyond sustainability through the following research question: how do consumers view carbon footprint labels on food? To answer this question, we discuss the pro-environmental behavior of consumers regarding their food choices and their relationship with sustainability.

2. Key Factors of Carbon Footprint in the Food Supply Chains

Strengthening environmental responsibility, developing responsible production methods and proactive communication between consumers and decision-makers are essential objectives of key carbon footprint actors in food supply chains. However, to date, the main bottleneck for science is providing those involved with the information and tools they need to understand and influence key issues such as how to effectively determine the environmental profile, especially the carbon footprint. To this end, market regulators often follow the development and initiatives of actors in the processing chain, as guiding principles to indicate potential hot spots in the processes, as well as potential future improvements.

As is well known, the food supply chain is complex and nuanced as it progresses through each stage of the life-cycle. However, although there is this oscillation of procedures, it is possible to consider a basic processing pattern that is commonly applied to the most different food products (Table 1).

As can be seen in Table 1, considering the large segments of the food industry, its base activity comes from agricultural practices. These, in turn, are mainly responsible for the natural resources' use, such as demand and contamination of water resources, use, and soil transformation, beyond the crops and soil fertilizations. Consequently, these practices result in direct impacts on the environment, being mainly associated with environmental categories such as land use, greenhouse gases, nutrient depletion and demand for fossil resources and, therefore, considered dominant agents in the contribution of the carbon footprint.

Among the main food products affected by these categories are products of animal and vegetable origin, and the fertilization of crops and the soil, as well as the fortification of animal feed, are the main critical points associated with the impacts of eutrophication and acidification (terrestrial or aquatic). On the other hand, the steps considered downstream within the food industry differ according to the degree of processing of the raw material. For example, grain and cereal processing chains have the grain drying step as a unit operation with greater environmental impact, due to high energy demand (Renzulli et al. 2015). The same profile can be attributed to the fruit and vegetable processing sectors. This is because, in addition to considering the energy demand in relation to storage systems, often considering cooling chambers or controlled atmosphere, they also present the demand for water for sanitation as a relevant factor in processing (Martin-Gorriz et al. 2021).

On the other hand, the animal protein and meat products' sectors, as well as the dairy chain, present high-water consumption, release of effluents with a high content of organic components, and energy consumption as the main villains of the environment (Del Borghi et al. 2021). Therefore, the stages of cooling carcasses and milk, storage, and pasteurization are concentrated as the operations with the highest energy consumption.

Under these scenarios, it is important to emphasize that any activity that modifies or interferes with the environment plays an important role in terms of the carbon footprint. This is because

Table 1. An overview of the main sectors of the food industry, unit operations considered, and additional process attributes under carbon footprint assessment.

Food industry sector	Functional units	Main unit operations	Additional attributional processes	Higher environmental impact categories
Cereal and grain products	• 1 ha • 1 kg	Farm production → harvest → pre-drying → threshing → cleaning → selection → drying → storage → milling → packaging	Soil preparation and cultivation; External transportation; Fertilizer productions; Semi-products' production, if applicable;	Energy demand; Depletion of soil nutrient resources; Soil loss; Water demand; Land and water contamination; GHG emissions
Fruits and vegetables production	• 1 ha • 1 kg	Tillage→ manure application → planting → hoeing and furrowing → fertigation → pesticides application → harvesting → transporting → reception → sorting → cleaning → bleaching → peeling → packaging	Seed production; Fertilizer production; Auxiliary products as detergents for cleaning	Energy demand; Depletion of soil nutrient resources; Soil loss; Water demand; Land and water contamination; GHG emissions
Meat and animal products	• 1 kg animal (live weight) • 1 kg carcass yield • 1 kg protein	Animal production → pre-slaughter → bleeding → skimming → evisceration → inspection → carcass washing → cooling → storage → products meat processing	Feed production; Animal breeding (including enteric fermentation); Manure management; Final production (cure, milling and mixing, cooking, emulsion, pasteurization, fermentation, drying, dehydration, smoking); Wastewater treatment;	GHG emissions; Energy demand; Water consumption; Depletion of soil nutrient resources; Soil loss
Milk and dairy products	• 1L • 1 kg	Milk collection → reception → filtration → cooling → clarification → homogenization → standardization → pasteurization → filling → packaging	Feed production; Animal breeding (including enteric fermentation); Management; Final production (salting, drying, maturation, etc.); Wastewater treatment;	Energy demand; Water consumption; GHG emissions
Beverages	• 1L	Main feedstock → mixing → heating → filtration/decantation → fermentation/distillation → filling → storage	Fermentation/distillation; Clarification/filtration;	Energy demand; Water consumption
Food consumption	• 1 kg • 1 portion pack	Packages → transports → cooking → consumption patterns	Transportation of the product to an average customer or consumer Customer or consumer use of the product End-of-life processes of any wasted part of the product End-of-life processes of packaging waste	Energy demand; GHG emissions; Water consumption

even operations that do not directly involve carbon emissions, in their primary process, are also responsible for global carbonization. An easy-to-understand example is energy demand, which is the main factor in the food chain. Electric energy from coal processing is estimated to emit about 3.08×10^{-01} kg CO_2 eq/kWh. On the other hand, alternative sources of renewable energy, even if in smaller proportions, also present carbon emissions with approximate values of 1.05×10^{-01} and 2.50×10^{-02} kg CO_2 eq/kWh for photovoltaic panels and wind energy, respectively (Deprá et al. 2020).

Additionally, another activity very present in the farm stage is the soil fertilization process. Currently, nitrogen and phosphorus are estimated to be responsible for about 4.94 and 1.88 kg CO_2 eq/kg, respectively. Therefore, being able to map the entire food chain becomes an even more multifaceted process.

Therefore, when referring in numerical terms, the representation of the environmental impact associated with these practices can vary greatly depending on the choice of the functional unit, the limits of the system evaluated, as well as the additional attributes considered (Shakhbulatov et al. 2019). However, relative data from the largest meta-analysis of the carbon footprint applied to food, published to date by Poore and Nemecek (2018), show us even deeper insights into environmental performance. So, to help elucidate the main scientific question of the moment—where do the carbon footprints of our foods come from?—a graphical qualitative illustration of the percentage average carbon footprint at different stages is shown in Fig. 1.

Although the steps behind the food supply chain can vary considerably, it is possible to segment the chain into seven main stages: land-use change, farm, animal feed, processing stage, transport, retail, and packaging (Fig. 1). From this perspective, for most foods—and particularly the largest emitters—the carbon footprints majority result from land-use change (shown in green) and farm-stage processes (burnt yellow). As mentioned earlier, agricultural stage emissions include processes such as fertilizer applications (organics and/or synthetics), and enteric fermentation. Combined, land- and farm-use emissions account for more than 80% of the footprint of most food.

Taking a closer look at animal products, now considered the main villain of global warming, it is possible to observe that in terms of internship on the farm, carbon emissions for beef represent 74% of the total emitted, while its product alternative, tofu, have values less than 20%. However, when we evaluate the processing step, the scenario is reversed. This is because the carbon emissions of beef cattle represent magnitudes below 6%, while tofu presents values above 25%. Given these results, even if the global magnitude of the carbon footprint of tofu is lower than beef production,

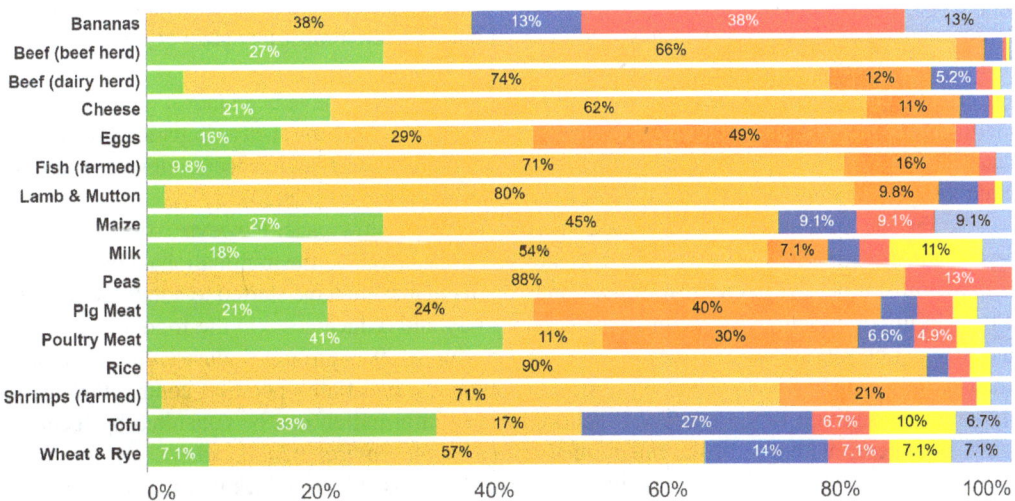

Figure 1. Carbon footprint estimates in percentual rates of main internationally traded foods. The data used refers to the study developed by Poore and Nemecek (2018). For absolute values and more details, see DOI:10.1126/science.aaq0216.

it is relevant that the industry assumes its role in terms of environmental impacts within the processing chain, and not just assign responsibility for agriculture. Therefore, although recent, strategies for the decarbonization of industrial activities are emerging in all sectors. Alternatives for energy and mass integration can help even further in improving the environmental profile of the food chain.

At the other end of the carbon footprint spectrum is food transport as a key factor. This stage of the supply chain represents, on average, 10% of total carbon emissions. With regard to beef, the proportion of GHGs that transport emits is even smaller, just 0.5% of total emissions. On the other hand, for the fruit and vegetable scenarios, such as bananas and peas, this step represents values between 13–38% of total emissions. According to studies developed by Aikins and Ramanathan (2020), the results suggest that different modes of transport can further impact the global chain. The explanation for this can be attributed to the origin of the fuel, or the choice of routes, whether road or sea. However, it is found that over long distances, heavier road vehicles can increase by up to 11 tons/CO_2 eq.

Evidently, these values presented are nothing less than a partial schematic representation of the environmental impact of a food product or industrial activity. Furthermore, it is important to emphasize that the main studies involving carbon footprint are used as an auxiliary method for the internal improvement of the product, in order to improve external communication, through labels, with environmentally conscious consumers. However, by mapping process hotspots, the carbon footprint makes it possible to identify potential carbon footprint enablers and, therefore, bring to the industrial community insights and insights into the main strategies for reducing the environmental burden of its products and processes.

3. Enablers of Food Supply Chain for Carbon Footprint Reduction

From a resource-based perspective, the practices associated with the food chain involve causally ambiguous and socially complex resources—which include all energy, material consumption, solid, liquid, and gaseous waste related to an ecosystem's internal and external processes, interconnected between various members of the supply chain, partners and consumers (Vargas et al. 2018). Therefore, companies need to develop strong coordination and collaboration among different stakeholders to achieve supply chain sustainability (Luthra et al. 2020).

Therefore, facilitator strategies are proposed in order to facilitate the adoption of sustainable supply chain management practices (Vargas et al. 2018). Similarly, food supply chain management is also applying these enablers. However, beyond the sustainable bias, this follow-up also needs to deal with ensures of food quality and safety. Therefore, selection criteria regarding storage facilities, routes, and means of transport, in addition to labor and operation must be used very efficiently, so that they do not affect the final product (Sharma et al. 2020).

To this end, several studies were carried out, and researchers have been investigating the main facilitators to increase sustainability in operations. Among them, criteria such as logistics, traceability, data sharing systems, and cyber networks, inventory management, and process integration are the main key points in improving environmental burdens (Parashar et al. 2020).

So far, the progress of applying these facilitators is at a slow pace. However, some promising results can already be seen. Proof of this are the new paths taken by the multinational company that manufactures food packaging, Tetra Park. Recently, its managers have started to adopt the Industry 4.0 as an important alternative to increase productivity and reduce costs in the food and beverage industry, without increasing extraction of natural resources. It is through Industry 4.0 that criteria such as the optimization of in- and out-of-plant logistics, real-time operation control throughout the plant, in addition to the accessible sharing of key information can be consulted by technical staff anywhere, without displacement charges (whether environmental, social, or financial). Consequently, with the help of the facilitators, Tetra Park was able to reduce its carbon footprint by around 12 million tons kg CO_2 eq by mid-2019. The future prospects are also promising, as they have committed to reducing it by up to 46% of its emissions by 2030, compared to the base year 2019 (Tetra Park Reports 2021).

In the same perspective, the respected Mars has also developed its own blueprint, known as the Generation Sustainable Plan. Through its mapping, it was possible to identify that the carbon footprint of the entire value chain was estimated at 33 million tons of carbon dioxide equivalent (MtCO$_2$ eq) in 2015. In view of the worrying data, definitions of targets for the reduction of emissions at 42% by 2025 and 100% (net zero) by 2040 have been established. However, to achieve them, enabling strategies such as the use of transparent and reliable GHG calculation and accounting methods will be essential. In practice, it is estimated that since 2015, there has already been a reduction of approximately –7.3% of GHG emissions, –16% of the water footprint, –8.4% of the demand for land use and transformation, in addition to a 21.7% increase in recycling and reuse of plastic materials (Mars 2021).

However, although some facilitators have already shown promise, the main bottleneck to be overcome in the food industry is still associated with waste. Today, it is estimated that ⅓ of all food produced in the world goes to waste. This equates to about 1.3 billion tons of fruits, vegetables, meats, dairy products, seafood, and grains that never leave the farm, are lost or deteriorate during distribution, or are wasted in their final stage of consumption (hotels, supermarkets, restaurants, residences). Indeed, food waste is not just a social or financial concern—it is also an environmental concern. When we waste food, we also waste all the energy and water needed to grow, harvest, transport and pack. Furthermore, in its deterioration and decomposition stage, its residue produces methane—a greenhouse gas even more potent than carbon dioxide (UNEP 2021). Thus, it is estimated that globally around 8–10% of all greenhouse gas emissions are attributed to food waste. Also, according to FAO data, the total carbon footprint of food waste is about 4.4 GtCO$_2$ eq per year (Scialabba 2015, FAO 2021). To circumvent this waste, reducing post-harvest handling and adopting improved technologies, planning efficient logistics, as well as conscientious purchasing and consumption, represent important strategies for the carbon footprint. Thus, despite the uncertainties, the magnitude of the carbon footprint requires that better use of food be applied, since the reductions in losses would have substantial positive effects on society's resources and, in particular, on climate change.

Given these perceptions, it is important that food and beverage companies start thinking about the opportunities and needs that drive the now. In fact, just a few years ago, industry 4.0 deployment strategies, the internet of things, sustainable plan developments, and supply chain logistics were not on the table for any food following. However, these enablers have shown prosperous results and are on a growing trendline to assist in consolidating the sustainable food industry. Therefore, it is necessary for decision-makers to feel that innovations should be placed on the market in a position to help them, and not just as a value-adding alternative.

4. Beyond Sustainability—How do Consumers see Carbon Footprint Labels on Food?

In recent decades, different scientific attempts have been made to try to develop a framework that would explain all the elements that could affect the pro-environmental behavior of consumers (Rondoni and Grasso 2021).

Eco-labelling, also known as sustainable labels, are properties attributed to products and processes in order to inform consumers about their environmental performance. Furthermore, they can communicate to consumers about certain measures considered by producers to neutralize or reduce the environmental burden (Canavari and Coderoni 2020). Therefore, environmentally sustainable labels, which are responsible for transmitting to the consumer the concepts related to all facets of sustainability, are the most common tools to support decision-making and the change in consumption patterns towards a sustainable model.

Under this concept, one of the main sustainable labeling models is the carbon footprint scheme. In short, carbon footprint labeling was developed in the UK around 2007 along with the progress of the environmental assessment tool, the life cycle assessment. Subsequently, the carbon footprint, such as product labeling, can be known as a measure of the total emission of carbon dioxide (and other greenhouse gases such as nitrous oxide and methane) caused by a given product over its life-cycle.

Since then, the most diverse environmental labels can be found on a wide range of products sold, including in the food segment (Fig. 2). The most basic examples consist of (i) color codes— which indicate whether a product has a high, medium, or low carbon footprint, usually represented by the colors red, yellow, and green, respectively; (ii) carbon reduction labels—which quantitatively indicate (through percentages or mass units) the reduction of the product's carbon footprint; and (iii) carbon-neutral labels—which indicate carbon offsets, that is, they present carbon reduction measures, their remaining emissions, as well as the purchase of carbon credits involved in the entire production process of the product (Birkenberg et al. 2021).

However, although the justification for disseminating environmental labeling as an attribute of value to products is necessary, so far, the standardization lack of labels, as well as the application lack in the food market, make it difficult for consumers to buy food certificates. In addition, criteria such as consumer preferences and how much they are willing to pay for sustainable products are still the stage for the discussion.

In order to elucidate this gap, several studies related to purchasing power, willingness to pay and carbon footprint labels have been published in academic literature. So far, what has been clarified is that there was a tendency for consumers to become more aware of the negative impacts of food production. However, although consumers have shown concern and understanding regarding these issues, this has not translated into greater use of sustainability labels (Leach et al. 2016).

In fact, consumers' priorities regarding food choices are driven by a series of interconnected factors, such as taste, prices, quality taxes, health and, finally, environmental burdens. This propensity to put environmental attributes as the last factor as a purchase criterion is often associated with higher prices of labeled products, credibility of labels, availability of labeled products, lack of understanding of what the labels represent, or other perceptions. In this sense, willingness to pay, that is, the maximum amount a consumer is willing to pay, becomes a crucial factor in finding the best price to sell a product, both for the seller and for the buyer. Therefore, reaching a middle ground between market vs. consumer must be done to make a—sustainable—sale.

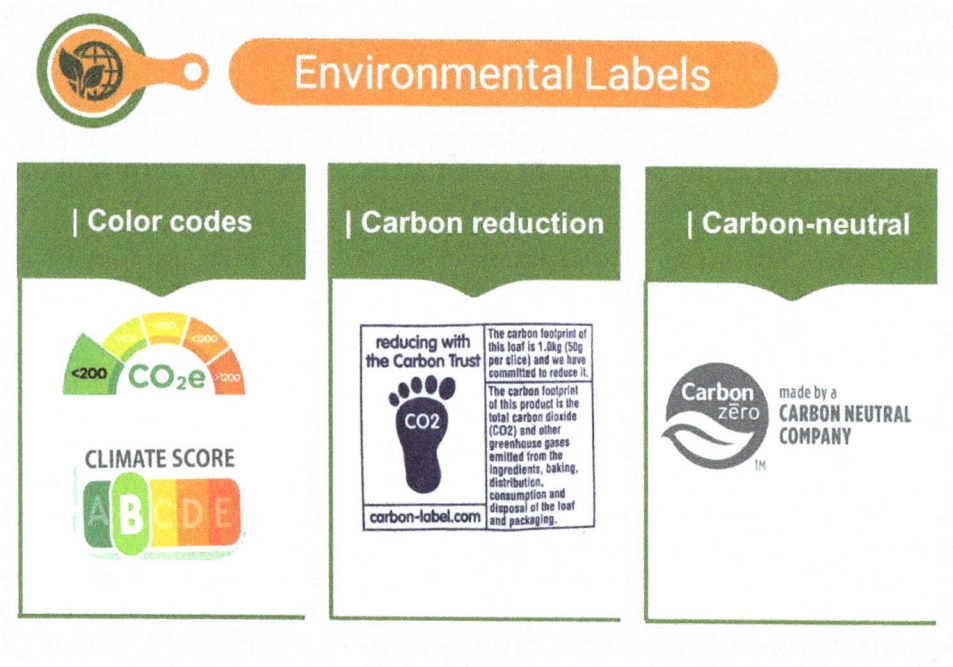

Figure 2. Examples of environmental labels used in food products.

To reinforce these assumptions, studies conducted by Li and Kallas (2021) concluded that the willingness to pay depended on the product, that is, the greater the share of a product in the family's monthly expenditure, the lower the willingness to pay for that product. In addition, other findings indicated that women and certain food categories, as well as specific regions, tend to value sustainable foods more, significantly influencing variations in the purchasing power of products. However, according to Canavari and Coderoni (2020), very few consumers look for carbon certificates in agri-food products, remaining an obscure gap that needs further investigation.

On the other hand, it is important to strengthen the recent progression of the recognition of sustainability as a fundamental factor for the protection of the ecosystem, by consumers, companies and governance. Undeniably, as there is greater awareness on the part of consumers, and measures taken by public and private companies to adopt, from outside standardized, carbon footprint labeling criteria and other sustainability indicators, there will be significant reductions in the carbon footprint of food baskets. Therefore, it is essential that manufacturers are advised to provide more information on this subject so that people can better understand its meaning. At the same time, policymakers should also aim to support measures to reduce the price of foods labeled with a carbon footprint in order to allow all consumers the resources to buy them (Rondoni and Grasso 2021). Through the application of these market strategies, it will then be likely that consumers will opt for more ecologically correct choices, increasing demand and sales for these products, and consequently, driving a low-carbon and, therefore, sustainable food market.

5. Conclusions

Managing the carbon footprint of food products throughout the supply chain is a key step in the effort to mitigate climate change. Therefore, using robust methodologies such as the carbon footprint to assess processes and products is essential, although there are some application limitations. In the near future, once the hotspots of the process are identified, the application of facilitating strategies may help in the journey towards a more carbon-constrained ecosystem. However, not only efforts on the side of agriculture, industry, and research will be crucial, but also educational policies and financial reinforcements will be essential to bring about the correct dissemination of strong marketing attributed to sustainable products, through educational labels about the carbon footprint of food products. Finally, it is necessary that everyone is willing to broaden their horizons so that we work collaboratively in order to create knowledge, build positive influences between farm-industry-product-society, reduce carbon emissions, and generate financial returns.

References

Aikins, E.F. and Ramanathan, U. 2020. Key factors of carbon footprint in the UK food supply chains: a new perspective of life cycle assessment. International Journal of Operations & Production Management. DOI: 10.1108/IJOPM-06-2019-0478.

Birkenberg, A., Narjes, M.E., Weinmann, B. and Birner, R. 2021. The potential of carbon neutral labeling to engage coffee consumers in climate change mitigation. Journal of Cleaner Production, 278: 123621. DOI: 10.1016/j.jclepro.2020.123621.

Canavari, M. and Coderoni, S. 2020. Consumer stated preferences for dairy products with carbon footprint labels in Italy. Agricultural and Food Economics 8(1): 1–16. DOI: 10.1186/s40100-019-0149-1.

Del Borghi, A., Parodi, S., Moreschi, L. and Gallo, M. 2021. Sustainable packaging: an evaluation of crates for food through a life cycle approach. The International Journal of Life Cycle Assessment 26(4): 753–766. DOI: 10.1007/s11367-020-01813-w.

Deprá, M.C., Dias, R.R., Severo, I.A., de Menezes, C.R., Zepka, L.Q. and Jacob-Lopes, E. 2020. Carbon dioxide capture and use in photobioreactors: The role of the carbon dioxide loads in the carbon footprint. Bioresource Technology 314: 123745. DOI: 10.1016/j.biortech.2020.123745.

El Geneidy, S., Baumeister, S., Govigli, V.M., Orfanidou, T. and Wallius, V. 2021. The carbon footprint of a knowledge organization and emission scenarios for a post-COVID-19 world. Environmental Impact Assessment Review, 91: 106645. DOI: 10.1016/j.eiar.2021.106645.

FAO. 2021. Food Outlook – Biannual Report on Global Food Markets. Food Outlook, November 2021. Rome. https://doi.org/10.4060/cb7491en.

Heinonen, J., Ottelin, J., Ala-Mantila, S., Wiedmann, T., Clarke, J. and Junnila, S. 2020. Spatial consumption-based carbon footprint assessments-A review of recent developments in the field. Journal of Cleaner Production 256: 120335. DOI: 10.1016/j.jclepro.2020.120335.

Leach, A.M., Emery, K.A., Gephart, J., Davis, K.F., Erisman, J.W., Leip, A., et al. 2016. Environmental impact food labels combining carbon, nitrogen, and water footprints. Food Policy 61: 213–223. DOI: 10.1016/j.foodpol.2016.03.006.

Li, S. and Kallas, Z. 2021. Meta-analysis of consumers' willingness to pay for sustainable food products. Appetite 105239. DOI: 10.1016/j.appet.2021.105239.

Luthra, S., Kumar, A., Zavadskas, E.K., Mangla, S.K. and Garza-Reyes, J.A. 2020. Industry 4.0 as an enabler of sustainability diffusion in supply chain: an analysis of influential strength of drivers in an emerging economy. International Journal of Production Research 58(5): 1505–1521. DOI: 10.1080/00207543.2019.1660828.

Marbach, S. and Gaillac, R. 2021. The carbon footprint of meat and dairy proteins: a practical perspective to guide low carbon footprint dietary choices. bioRxiv. DOI: 10.1101/2021.01.31.429047.

Mars (2021). Building a sustainable future. Accessed in August 8th, 2021 <https://www.mars.com/sustainability-plan>.

Martin-Gorriz, B., Martínez-Alvarez, V., Maestre-Valero, J.F. and Gallego-Elvira, B. 2021. Influence of the water source on the carbon footprint of irrigated agriculture: A regional study in South-Eastern Spain. Agronomy 11(2): 351. DOI: 10.3390/agronomy11020351.

Parashar, S., Sood, G. and Agrawal, N. 2020. Modelling the enablers of food supply chain for reduction in carbon footprint. Journal of Cleaner Production 275: 122932. DOI: 10.1016/j.jclepro.2020.122932.

Poore, J. and Nemecek, T. 2018. Reducing food's environmental impacts through producers and consumers. Science 360(6392): 987–992. DOI: 10.1126/science.aaq0216.

Renzulli, P.A., Bacenetti, J., Benedetto, G., Fusi, A., Ioppolo, G., Niero, M., et al. 2015. Life cycle assessment in the cereal and derived products sector. Life Cycle Assessment in the Agri-food Sector, 185–249. DOI: 10.1007/978-3-319-11940-3_4.

Rondoni, A. and Grasso, S. 2021. Consumers behaviour towards carbon footprint labels on food: A review of the literature and discussion of industry implications. Journal of Cleaner Production, 127031. DOI: 10.1016/j.jclepro.2021.127031.

Scialabba, N. 2015. Food wastage footprint & climate change. Food and Agriculture Organization of the United Nations. Retrieved June, 4, 2020.

Shakhbulatov, D., Arora, A., Dong, Z. and Rojas-Cessa, R. 2019. Blockchain implementation for analysis of carbon footprint across food supply chain. In 2019 IEEE International Conference on Blockchain (Blockchain) (pp. 546–551). IEEE. DOI 10.1109/Blockchain.2019.00079.

Sharma, J., Tyagi, M. and Bhardwaj, A. 2020. Parametric review of food supply chain performance implications under different aspects. Journal of Advances in Management Research. DOI:10.1108/JAMR-10-2019-0193.

Shi, S. and Yin, J. 2021. Global research on carbon footprint: A scientometric review. Environmental Impact Assessment Review 89: 106571. DOI: 10.1016/j.eiar.2021.106571.

Tetra Park Reports (2021). Tetra Park Sustainability Reports - Food. People. Planet. <https://www.tetrapak.com/content/dam/tetrapak/publicweb/gb/en/sustainability/documents/TP-Sustainability-Report-2021.pdf>.

UNEP - United Nations Environment Programme (2021). Food Waste Index Report 2021. Nairobi.

Vargas, J.R.C., Mantilla, C.E.M. and de Sousa Jabbour, A.B.L. 2018. Enablers of sustainable supply chain management and its effect on competitive advantage in the Colombian context. Resources, Conservation and Recycling 139: 237–250. DOI: 10.1016/j.resconrec.2018.08.018.

Vauterin, A., Steiner, B., Sillman, J. and Kahiluoto, H. 2021. The potential of insect protein to reduce food-based carbon footprints in Europe: The case of broiler meat production. Journal of Cleaner Production, 128799. DOI: 10.1016/j.jclepro.2021.128799.

Wiedmann, T. and Minx, J. 2008. A definition of 'carbon footprint'. Ecological Economics Research Trends 1: 1–11.

Yue, Q. and Guo, P. 2021. Managing agricultural water-energy-food-environment nexus considering water footprint and carbon footprint under uncertainty. Agricultural Water Management 252: 106899. DOI: 10.1016/j.agwat.2021.106899.

Zhang, F., Cai, Y., Tan, Q., Engel, B.A. and Wang, X. 2021b. An optimal modeling approach for reducing carbon footprint in agricultural water-energy-food nexus system. Journal of Cleaner Production 316: 128325. DOI: 10.1016/j.jclepro.2021.128325.

Zhang, L., Ruiz-Menjivar, J., Tong, Q., Zhang, J. and Yue, M. 2021a. Examining the carbon footprint of rice production and consumption in Hubei, China: A life cycle assessment and uncertainty analysis approach. Journal of Environmental Management 300: 113698. DOI: 10.1016/j.jenvman.2021.113698.

20

Environmental Sustainability Assessment of Food Waste Valorization Options

Ahmet Görgüç, Esra Gençdağ, Elif Ezgi Özdemir, Kardelen Demirci, Beyzanur Bayraktar and *Fatih Mehmet Yılmaz**

1. Introduction

The evaluation of food waste by converting into value-added products has been among major focus of attention in recent years, due to many advantages in economic, social and ecological terms. According to the Food and Agriculture Organization (FAO), food waste is defined as "losses in quality and quantity through the supply chain process occurring at the production, postharvest, and processing phases" (Sharma et al. 2020). Spoilage, damage, poor demand forecasting, overproduction, surplus stocks, inefficient management, exogenous factors (seasonal effects, supply and demand, etc.), contamination, and packaging issues are among the reasons of food waste formation (Lemaire and Limbourg 2019). FAO estimates that 1.3 billion tons of food is wasted annually around the world, accounting for one-third of all food produced (Sharma et al. 2020). The rapid increase in world population in recent years has resulted in a massive increase in food demand. By 2050, the world population is expected to exceed 9 billion people, necessitating a 60–70% increase in food production (Laso et al. 2016). Generally, wastage results in significant losses of other resources such as water, land, labor, and energy (Sharma et al. 2020). In order to avoid the environmental problems and to maintain the sustainability of the resources, waste streams must be properly managed. The consequences of food waste include not only ethical and economic concerns, but also environmental concerns that can jeopardize the food chain sustainability (Mosna et al. 2021). Food waste is estimated to contribute to greenhouse gas emissions by releasing approximately 3.3 billion tons of CO_2 into the atmosphere each year (Paritosh et al. 2017). According to a recent report of The World Bank, the current global average generation rate of municipal solid waste in urban areas is approximately 1.2 kg per person per day, and by 2025, this value is expected to increase to 1.42 kg/person/day (2.2 billion tons of waste per year on a global scale) (Storino et al. 2016). Solid waste is generally incinerated or dumped in an open area, which can result in serious health and environmental challenges. The processes associated with thermal recycling are central

Aydın Adnan Menderes University, Faculty of Engineering, Department of Food Engineering, Efeler-09010, Aydın, Turkey.
* Corresponding author: fatih.yilmaz@adu.edu.tr

approaches for waste management. All such procedures implement waste oxidation, which occurs as a result of incineration, gasification, or waste decomposition (e.g., thermal disposal). Thermal disposal can be used for a variety of waste types, including municipal waste, industrial waste, hazardous waste, and sewage sludge. For this reason, several countries restrict or prohibit disposal strategies such as incineration and landfilling. The ability to convert waste to a safe material, a significant reduction in the weight and volume of waste, and the ability to recover large amounts of heat are the main advantages of this process (Bujak 2015). However, the incineration of food waste with a high moisture content results in the release of dioxins, which can cause a number of environmental issues. Furthermore, incineration reduces the potential value of the substrate by impeding the recovery of nutrients and valuable chemical compounds from a destroyed substrate (Paritosh et al. 2017). Composting is a valuable way of waste treatment that contributes to reducing organic waste destined for landfill disposal or incineration, which can reduce costs and environmental impact due to collection, transportation, and treatment. However, the main problems with composting are the development of potential odors and the presence of insects and rodents (Storino et al. 2016). One of the issues for food and environmental scientists is to provide practical solutions for the disposal of food by-products and wastes, which actively contribute to pollution. Decantation, dissolved air flotation, de-emulsification, flocculation, adsorption, centrifugation, coagulation, oxidation technologies, bioreactors, ozonation, enzymatic treatments, and membrane technology are all used for the treatment of by-products or wastes (Castro-Muñoz and Ruby-Figueroa 2019). As a result, proper methods for managing food waste are required. Food waste prevention has gained prominence in the European political debate as a result of the recent Circular Economy (CE) package (EC 2015a). Indeed, the CE Action Plan (EC 2015b) is defined food waste as one of the "priority areas", or areas that should be carefully considered in order to strengthen the circularity of the European economy (Cristóbal et al. 2018). In line with these, current chapter aimed to provide a detailed overview of food waste valorization options, with particular emphasis on each major waste stream discarded during processing.

2. Fruit and Vegetable Waste and By-products

Fruit and vegetables are the most utilized and preferred commodities among consumers, and have a crucial role in human diet and lifestyle (Ganesh et al. 2022). On the other hand, considering the rate of food waste and by-products, fruits and vegetable wastes compose the largest resource by the ratio of 44% (Vilariño et al. 2017, Ganesh et al. 2022). Especially, juice processing industry has led to a huge amount of fruit and vegetable waste and by-products. The juice processing steps for various fruits and vegetables were given in Fig. 1 by presenting occurred waste types and amounts for each process.

Global production and waste rate of selected fruits and vegetables with major producing countries are illustrated in Table 1, and their composition in terms of waste/by-product streams are given in Table 2. Most of the fruits and vegetables include a considerable amount of flesh, peel/skin, and seed, which are disposed as wastes or by-products (Cheok et al. 2018). Recently, researchers have focused on the recovery of these wastes to overcome the adverse environmental effects, and to convert them into useful biomasses including health beneficial compounds (Ng et al. 2020). The applications of fruit and vegetable wastes include edible films and coatings (Torres-León et al. 2018a), enzyme processing technologies (Srivastava et al. 2021), microbiological media (Vieira et al. 2021), food additives (Hussain et al. 2020), green nanoparticles (Mahmoudi et al. 2020, Rambabu et al. 2021a) and the remedy for healthier human life (Chaouch and Benvenuti 2020). Besides, non-food usage of these wastes covers the production of biochar, adsorbent, biosorbent, carbon quantum dots, biogas, etc. (Ng et al. 2020).

The variety of potential biorefinery systems leading to the valorization of fruit and vegetable wastes and by-products is highly broad as seen in Fig. 2. Fruit and vegetable wastes are mostly known as having high moisture content and rich organic compounds with biodegradable properties

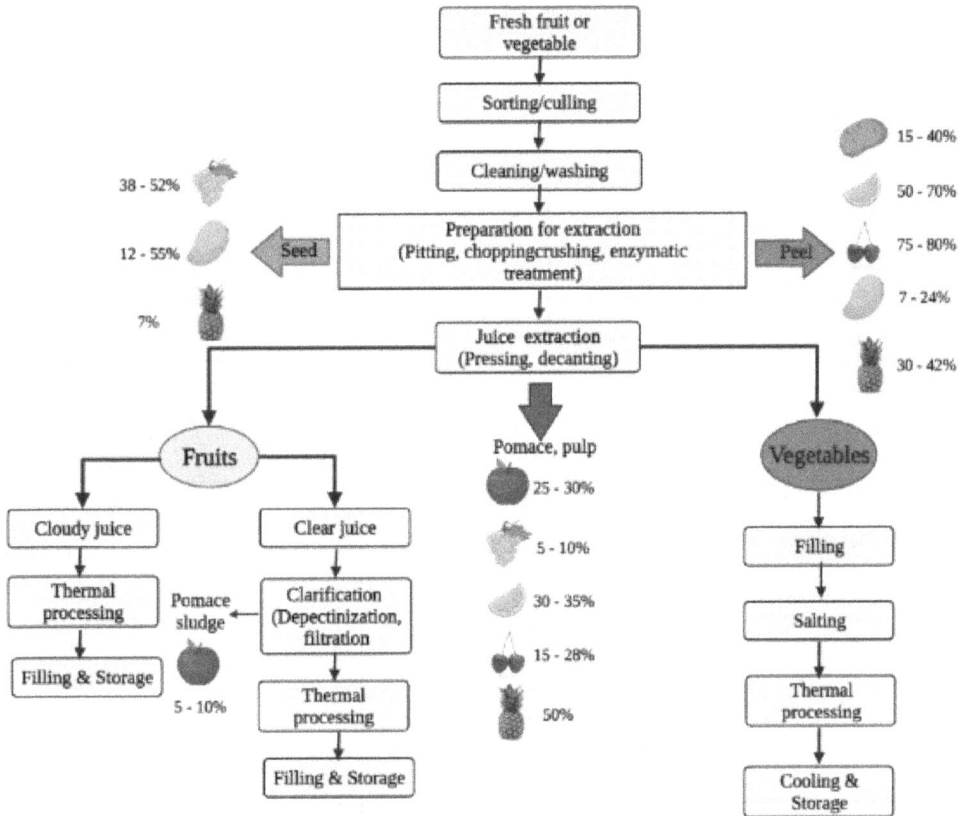

Figure 1. Juice processing steps and discarded wastes and by-products for selected fruits and vegetables.

including carbohydrates, lipids, and proteins. Additionally, poorly biodegradable lignocellulosic biopolymers (cellulose, hemicellulose, and lignin) are also found. These properties have adverse effects on conventional waste disposal systems due to high operating costs in the stages of transport, greenhouse gas emission and leachate discharge in landfills (Edwiges et al. 2018). Depending on the chemical composition of fruit and vegetable wastes, there are various valorization options such as extraction of bioactive compounds, enzyme production, and synthesis of bio-based products (bioplastics, biochar, biogas, biofuels, and degradable biofilms) (Esparza et al. 2020). The detailed composition for selected fruit and vegetable wastes discarded from different processing steps are given in Table 3.

2.1 Plum Waste and By-products

Plum (*Prunus domestica* L.) originates from Asia, and globally has a 12.6 million of production rate annually (Liaudanskas et al. 2020). During the past 25 years, plum harvesting has doubled worldwide, reaching 12.6 million tons in 2018 (Dołżyńska et al. 2020). Plum has various bioactive compounds such as dietary fibers, phenolic compounds and vitamins (Savic and Gajic 2020). These bioactive compounds have diverse benefits for human health, mainly prevention of cancer, diabetes, and obesity. The industrial processing of plum generates a huge amount of skin, pomace, seed/

Table 1. World production and waste rate of selected fruits and vegetables with major producing country.

Type of fruit/vegetable		Major producing country	Production rate (million tons/year)	Waste rate (million tons/year)	References
Plum (*Prunus domestica* L.)		Europe and Asia	12.6	0.12	Liaudanskas et al. (2020)
Banana (*Musa cavendish* L.)		India	115.7	36.0	Marenda et al. (2019), Barnossi et al. (2021)
Apple (*Malus* spp.)		China	86.1	17–21	Enniya et al. (2018), Liu et al. (2021a), Duan et al. (2021)
Pineapple (*Ananas comosus* L. Merrill)		Costa Rica	25.9	0.62	Casabar et al. (2019)
Grape (*Vitis* spp.)		China	73.3	11.1	Sun et al. (2020), Kovalcik et al. (2020)
Citrus	Tangerine/mandarin	China	21.2	15–25	Wikandari et al. (2014), Mahato et al. (2020), Rosas-Mendoza et al. (2020)
	Lemon/lime	Mexico	2.60		
	Grapefruit	China	4.80		
Sour cherry (*Prunus cerasus* L.)		Russia	1.50	0.60	Kasapoğlu et al. (2021), Almasi et al. (2021), Sezer et al. (2021)
Date (*Phoenix dactylifera* L.)		Egypt	8.17	4.50	Aslam et al. (2020), Oladzad et al. (2021)
Avocado (*Persea americana* Mill.)		Mexico	6.41	1.20	Castillo-Llamosas et al. (2021), Merino et al. (2021)
Mango (*Mangifera indica* L.)		India	55.0	0.21	Krishna et al. (2020), Manhongo et al. (2021b)
Potato (*Solanum tuberosum* L.)		China and India	383.0	4.80	George et al. (2017), Javed et al. (2019)
Onion (*Allium cepa* L.)		European Union	66.0	0.45	Campone et al. (2018), Mourtzinos et al. (2018), Celano et al. (2021)

kernel, and leaves as waste and by-product. Plum seed represents approximately 5% of the total weight of the fruit. Also, the plum seed is mainly composed of proteins, phenolic compounds, and oil. Plum seed oil can be an alternative for the production of biofuel, cosmetics, and pharmaceuticals thanks to high content of vitamin E. Also, plum seed is a great source of enzymes such as peptidase and angiotensin-converting enzyme (González-García et al. 2018). It can also be an alternative for a cheap source of bioactive peptides, bioadsorbents and activated carbon (Kostić et al. 2016). Vladić et al. (2020) investigated the plum seed as an alternative source of edible oil with respect to its fatty acids and tocopherols. They observed that the plum seed oil has a desirable fatty acid composition with a high content of oleic acid, linoleic acid, and low content of saturated fatty

Table 2. Percent of flesh, peel/skin, and seed in fruits and vegetables.

Type of fruit/vegetable	Peel/skin (%)	Seed/pit/ kernel (%)	Press cake (%)	Crown (%)	References
Banana (*Musa cavendish*)	30–40		60–70		Barnossi et al. (2021)
Apple (*Malus* sp.)	20–25	2–4	25–30		Dhillon et al. (2013), Chand and Pakade, (2015)
Grape (*Vitis* spp.)	10–20	38–52	5–10		Beres et al. (2017), Bordiga et al. (2019), Kim et al. (2019a)
Citrus fruits	50–70	< 10	30–35		Manhongo et al. (2021a)
Sour cherry (*Prunus cerasus* L.)	75–80	15	15–28		Yılmaz et al. (2019), Yılmaz et al. (2020), Kazempour-Samak et al. (2021)
Date (*Phoenix dactylifera* L.)	16.7	11–18	17–28		Guo et al. (2003), Oladzad et al. (2020), Ahmed et al. (2021)
Avocado (*Persea americana* Mill.)	12.1	13–16	15		Colombo and Papetti (2019), Manhongo et al. (2021a)
Mango (*Mangifera indica* L.)	7–24	12–55			Manhongo et al. (2021a)
Pineapple (*Ananas comosus* L. Merrill)	30–42	7	50	13	Roda and Lambri (2019), Banerjee et al. (2018)
Potato (*Solanum tuberosum* L.)	15–40		42–54		Torres and Domínguez (2020), Kot et al. (2020)
Onion (*Allium cepa* L.)	37				Cho et al. (2021)

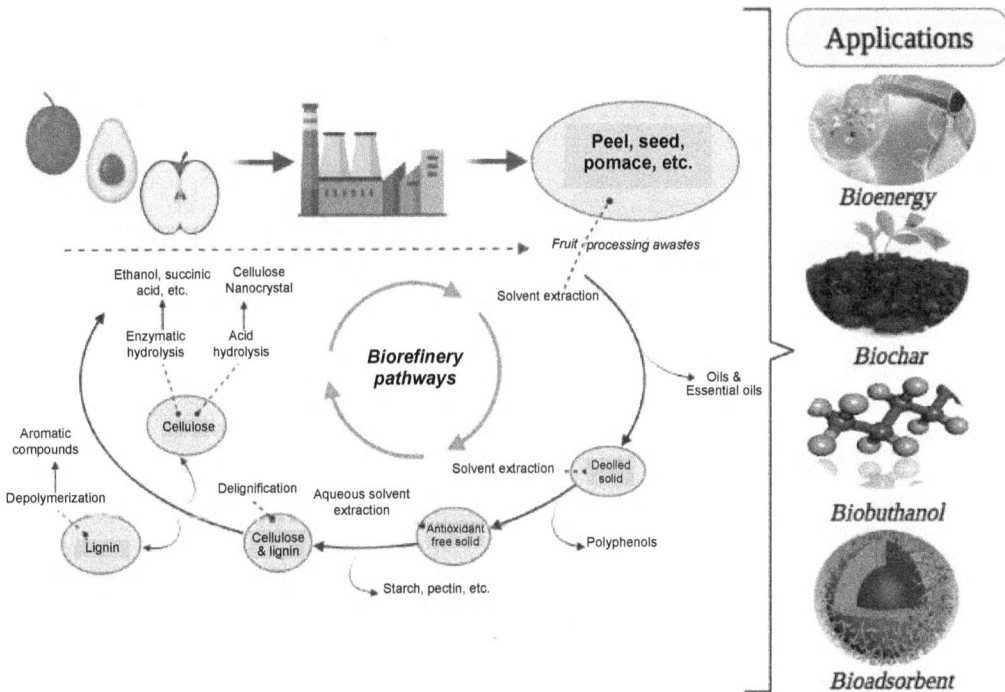

Figure 2. Biorefinery pathways for evaluating fruit and vegetable wastes and by-products.

Table 3. Composition for selected fruits and vegetables wastes discarded from different processing steps.

Type of fruit/vegetable	Waste/by-product	Composition (%)									References
		Moisture	Ash	Total lipid	Protein	Carbohydrate	Dietary fiber	Cellulose	Hemicellulose	Lignin	
Avocado (*Persea americana* Mill.)	Seed/Stone	56.0	3.70	3.70	4.90	69.1	19.5				Permal et al. (2020), Garcia-Vargas et al. (2020), Sánchez-Quezada et al. (2021), Alkaltham et al. (2021)
	Peel/Skin	69.1	6.90	3.60	8.10	1.20	81.4	12.1–27.6	11.5–25.3	4.40–35.3	
	Pulp	78.2	9.30	7.00	12.8	0.60	72.1	6.50–40.9	3.00–47.9	1.80–15.8	
	Wastewater	88.3	17.9	53.8	10.3						
Pineapple (*Ananas comosus* L. Merrill)	Peel/Skin	92.2	10.6					19.8	11.7		Selani et al. (2014), Roda and Lambri (2019)
	Pulp		0.20		4.60		45.2	14.3	22.1	2.30	
	Crown		0.40		4.20			29.6	23.2	4.50	
Mango (*Mangifera indica* L.)	Seed/Stone	2.30	5.70	5.40	7.20	81.9	22.1	14.1	2.80	2.00	Torres-León et al. (2019), Sánchez-Camargo et al. (2019), Patiño-Rodríguez et al. (2020)
	Peel/Skin	3.60		1.60	6.30	52.3	35.6	23.5			
	Pulp	4.20	2.60	1.20	3.90	41.1	11.8	6.50			
Banana (*Musa cavendish*)	Peel/Skin	78.4	13.4	11.6	7.90	59.5	7.70				Osma et al. (2007), Izidoro et al. (2008)
	Pulp	89.1	0.40	0.10	0.30	10.2	<0.10				
Apple (*Malus* sp.)	Seed/Stone	26.6	2.00	3.50	8.80	83.4	2.40				Akpabio et al. (2012)
	Peel/Skin	47.9	3.20	8.90	6.70	79.4	1.80				
	Pulp	32.7	3.30	10.0	4.70	79.0	3.00				

		Moisture	Ash	Total lipid	Protein	Carbohydrate	Dietary fiber	Cellulose	Hemicellulose	Lignin	
Grape (*Vitis* spp.)	Seed/Stone	81.7	3.20	11.6–19.6	11.0–13.0	73.0	> 75.0	40.0–49.0			Mironeasa et al. (2010), Mendes et al. (2013), Beres et al. (2017), Gowman et al. (2019)
	Peel/Skin	26.4	7.80		18.8	12.3		20.8	12.5		
	Pulp		7.50	1.10–5.10	5.40–13.8	<0.20	17.3–58.0	17.5	6.90	51.7	
Citrus	Seed/Stone	27.0–36.5	4.60–5.60		3.90–9.60		5.00–8.50				Anwar et al. (2008), Sharma et al. (2017)
	Peel/Skin	75.3	3.30	2.20	10.2		57.0				
	Pulp	85.7	6.50	4.90	8.60		7.30				
Date (*Phoenix dactylifera* L.)	Seed/Stone	5.80	5.10	10.2–12.7	5.20–5.60	81.0–83.1		59.8	9.50		Demirbaş, (2017), Majzoobi et al. (2019), Pal et al. (2022)
	Pulp	13.4	2.60	4.90	6.40	79.1	11.7				
Potato (*Solanum tuberosum* L.)	Peel	86.5	8.50–7.90	2.6–8.5	13.9–17.9	39.9–56.2	13.0–25.9				Chintagunta et al. (2016), Torres and Dominguez (2020), Kot et al. (2020), Suárez et al. (2020)
	Pulp		4.00		4.00	37.0	7.00	17.0	14.0		
	Leaves	87.7	0.70–4.90	0.2–0.7	0.2–1.5	4.40–7.40	23.9–25.4				
	Wastewater		0.20–2.30	0.20–0.50	15.1–19.2	1.50–21.1	0.20–1.30				
Onion (*Allium cepa* L.)	Peel	8.00	0.40	15.1	0.90	88.6	0.20				Yahaya et al. (2010), Kumar et al. (2022), Shaikhiev et al. (2022)
	Bulb	88.6–92.8	0.60		0.90–1.60	5.20–9.00	13.6				

acids. Accordingly, plum seed has a potential for the production of oil for food and pharmaceutical industries. More recently, the production of gallic acid using tannin-rich biowaste (black plum seed) was investigated by Saeed et al. (2020). This research provided a solution for the disposal of industrial processing waste of plum, and also ensured the production of cost-effective and high yield gallic acid at industrial scale.

2.2 Date Waste and By-products

Date is a traditional and economically valuable fruit, having an important role in human diet especially in the Middle East and North Africa (Ahmed et al. 2021). Dates are consumed freshly or processed into a broad range of food products, including date bars, syrup, juice concentrate, jam, candy, pickles and wine. Processing of dates can provide higher durability and lower transfer costs. Generally, processing steps of dates involve fumigation, washing, drying, sorting, grading, and packaging. Large amounts of waste and by-products are generated during these processing steps. Date seeds (11–18%) and press cake (17–28%) are the major processing wastes of date fruit (Siddiq et al. 2013). Seeds are commonly discarded during de-pitting step, while date press cake is discarded from the date syrup or juice industries after extraction. Besides, date press cake is a considerable waste which is mostly disposed of in open lands and drains. Many studies have revealed that date seeds contain carbon-rich macromolecules, including lignin, cellulose, hemicelluloses, and proteins (Oladzad et al. 2021, Rambabu et al. 2021b). Also, studies have reported some fatty acid contents as 41.1%, 21.4%, and 11.3% for oleic acid, linoleic acid, and palmitic acid, respectively. Additionally, it was predicted that 0.47 to 0.90 million tons of date seeds that contain 9% of oil was processed all over the world (Mrabet et al. 2020, Ahmed et al. 2021). Thus, date seeds are potential resources for several areas, i.e., feedstock for bio-oil (Oladipupo Kareem et al. 2021), biochar (Salem et al. 2021), bioethanol (Bouaziz et al. 2020), biohydrogen (Rambabu et al. 2021b), and activated carbon production (Heidarinejad et al. 2018). Date press cake also has a huge amount of dietary fiber, phenolic compounds, and antioxidant activity that could be considered as a source of natural antioxidants. Considering the production rate and the nutritional value of these wastes, valorization is gaining importance in the production of a wide range of products. Heidarinejad et al. (2018) investigated the conversion of date press cake to KOH-activated carbon. They observed high surface area (2633 m^2/g) and microporous texture in activated carbon prepared from date press cake. The usage of these valuable wastes not only address their problems related to waste management but also make new functionality to industries for sustainability.

2.3 Apple Waste and By-products

Apple wastes are collectively known as apple pomace, which contain the skin, pulp and seeds derived from the processing of concentrated apple juice, jam, and sweets (Dhillon et al. 2013). It is predicted that 20–40% of the total production of apple is processed annually, which generates 6 to 12 million metric tons of apple pomace during the processing of apple juice (Awasthi et al. 2021). The apple pomace is mainly composed of dietary fibers (cellulose, hemicellulose, lignin, gum, and pectin), carbohydrates, phenolics, proteins, minerals and vitamins. Pectin and oligosaccharides, *p*-coumaric acid, *p*-coumaroyl-quinic, caffeic, ferulic and chlorogenic acids, procyanidin B_2, kaempferol, rhamnetin, cyanidin-3-*O*-galactoside, phlorizin, and oleanolic acid are the major bioactive compounds found in apple pomace (Duan et al. 2021). Considering the dimension of apple wastes, especially apple pomace-based bio-refineries (biofuel, biochemical, biopolymer, etc.) could have an alternative for environmental, economic, and social areas through waste management and sustainability. Furthermore, apple pomace is one of significant sources of commercial pectin production. However, there is an increasing interest in utilizing apple pomace as a functional ingredient in food products such as bread, cookies and crackers, dairy products, etc. Heraldy et al. (2018) successfully utilized the wastes discarded from apple juice processing as a potential and low-price biosorbent for Pb(II) removal from wastewater. Another sustainable biomaterial production was studied by Hijosa-Valsero et al. (2017). They investigated the potential of apple pomace for biobutanol production. It is reported that the biobutanol production can be performed from lignocellulose-rich apple pomace (21% cellulose, 15% hemicellulose, and 18% lignin) without a detoxification step.

2.4 Pineapple Waste and By-products

Pineapple (*Ananas comosus* L. Merrill) has the highest production rate of all tropical fruits worldwide. The main producer and exporter of pineapple is Costa Rica, with a producing rate of approximately 25.9 million tons annually (Casabar et al. 2019). As the global demand and consumption rate of pineapple is rising, the processing waste of pineapple is also increasing. Industrialization of this fruit generates significant amounts of residues (nearly 25%), which consist mainly of the peel, core and crown of the pineapple (Conesa et al. 2016). Considering the different waste parts, peels represent the largest part of waste production with the rate of 30–42%, followed by core (10%) (Roda and Lambri 2019). Pineapple wastes in the form of crown, peel, stem, and core contain a wide range of bioactive compounds in each fragment. Namely, there are bromelain, ferulic acid, and vitamin A and C in pineapple peels, ascorbic acid and proteolytic enzymes in the core and stem, and citric acid in leaves (Hikal et al. 2021). Industrial processing wastes of pineapple are also high-quality feedstocks with a significant amount of lignin (1.5%), cellulose (14%) and hemicellulose (20.2%) for bioproducts' (bioethanol, bromelain, biofuel, bioenergy, etc.) production (Casabar et al. 2020, McCance et al. 2021). Pineapple wastes were evaluated by Anbesaw (2021) as a potential source for bromelain enzyme that can be used for different food and non-food applications. Authors used produced bromelain to recover silver from X-ray films with low-cost and in an efficient way. Pineapple wastes can also help to minimize the production cost of a huge number of polyester films. In another study on biomaterial production, bacterial cellulose and its nanocrystalline form is produced by using pineapple peels as the culture medium (Anwar et al. 2021). Similarly, Gil and Maupoey (2018) focused on the valorization of industrial processing waste of pineapple (core and peel) for bioethanol and bromelain production.

2.5 Citrus Waste and By-products

Citrus fruits are among the most popular and highly consumed fruits in the world, because of their nutritional values and health beneficial properties. Citrus fruits mainly cover sweet orange, mandarin, grapefruit, lime, and lemon (Satari and Karimi 2018). The majority of citrus fruits are utilized in citrus processing industries, causing approximately 50–60% processing wastes by weight. Citrus waste mainly comprises peel, pulp, membrane residue and seed parts (Mahato et al. 2020). Therefore, the valorization alternatives of citrus processing wastes are gaining importance to develop a sustainable bio-economy, and to reduce adverse effects on air, soil and water. Innovative approaches are required rather than traditional valorization methods such as pectin, animal feed, organic fertilizer and compost production. For instance, the bioconversion processes of these wastes have come to the fore considering the potential economic gains (Zema et al. 2018).

2.6 Banana Waste and By-products

Banana is the second most produced fruit, representing 16% of the total fruit production in the world with the production rate of 115.7 million metric tons. The peel (30–40%) is the major part of banana wastes with 34.7–46.3 million metric tons producing rate (Barnossi et al. 2021). Banana organic wastes can be mainly valorized directly as animal feed, bioadsorbent and biofertilizer, or for the functional product developments. The banana peel contains high amounts of bioactive compounds compared to the flesh part. Furthermore, the antioxidant capacity of the peel is higher than the fruit itself. Dopamine and catecholamine are known as the major antioxidants found in banana peel (Vu et al. 2020). Banana peel is an excellent bioresource as an adsorbent material with superior lignocellulosic biomass content (Lapo et al. 2020). Different usage areas of banana wastes were also investigated by researchers. More recently, Atchudan et al. (2021) investigated the potential of sustainable synthesis of carbon quantum dots from banana peels. They detected the banana peel waste-derived carbon quantum dots had no photobleaching under UV-light irradiation for a long period of time, indicating high photostability.

2.7 Cherry Waste and By-products

The rapid increase in the production rate of cherries and cherry products has caused the formation of significant amounts of cherry waste, especially from industrial processes (Bagheri et al. 2021). The most common and important species of cherry are *Prunus avium* L. known as sweet cherry and *Prunus cerasus* L. known as sour cherry (Yılmaz et al. 2019). Most of sweet cherries are consumed fresh, while sour cherries (approximately 85%) are generally consumed after processing into food products such as fruit juice, jam, puree, concentrate, wine, frozen, dried or canned fruit, marmalade, and jelly (Çavuşoğlu et al. 2021). Recently, many studies revealed that cherry by-products are plentiful sources of bioactive compounds such as phenolic acids and flavonoids. These compounds have a high antioxidant capacity and other health promoting properties (anti-inflammatory, anticancer, anti-diabetic activity, etc.). Additionally, these compounds are mainly taking part in cherry skin, contributing sensory and organoleptic properties of fruits. Maurício et al. (2018) investigated the recovery potential of cherry pomace discarded from liquor industry. Authors revealed that cherry pomace can be converted into extracts having eco-friendly and health promoting properties that can be used as functional ingredients in food, nutraceutical and cosmetic industries.

2.8 Grape Waste and By-products

Three-quarter of the total grape production is utilized into wine-making which is globally around 27 billion liters on annual basis. Grape pomace is mostly discarded from wine industry (nearly 1200 tons per year), and consists of skin, stem, stalks, pulp, and seeds that is equivalent to 25% of total grape weight (Beres et al. 2017). Phenolic compounds such as anthocyanins, hydroxybenzoic and hydroxycinnamic acids, flavonols, and flavan-3-ols are abundantly present in grape pomace. Additionally, malvidin, petunidine, cyanidin, peonidin and delphinidine are located in grape skin which are responsible for characteristic red color. Currently, grape seed oil is a commercialized product used in food and cosmetics formulations. Studies indicated that grape by-products also have potential for further valorization attempts thanks to their unique bioactive components (Bordiga et al. 2019).

2.9 Mango Waste and By-products

Mango is the second most-produced tropical fruit in the world, exceeding 55 million tons production rate in last two years (Manhongo et al. 2021b). A considerable quantity of processing wastes (350–600 kg/ton) of mango mainly consists of peel and seed parts (Banerjee et al. 2018). Especially, mango peels can be used in the formulation of biodegradable packaging materials thanks to high polysaccharide content. Accordingly, Torres-Leon et al. (2018b) observed that the biofilms prepared with mango peel showed good barrier properties and high water vapor permeability.

2.10 Avocado Waste and By-products

The avocado processing industries such as guacamole, avocado pulp and oil produce large amounts of waste and by-products, mostly the pulp, seeds and skin (Castillo-Llamosas et al. 2021). Nevertheless, these by-products are a significant source of extractable bioactive molecules, which could be used in the formulation of functional foods or cosmetic products. Particularly, pulp waste of avocado (nearly 15% of the fruit) is mainly discarded from the extraction process of avocado oil. A 500 kg of wastewater is discharged to produce 80 kg of oil from 1000 kg of avocado fruit (Permal

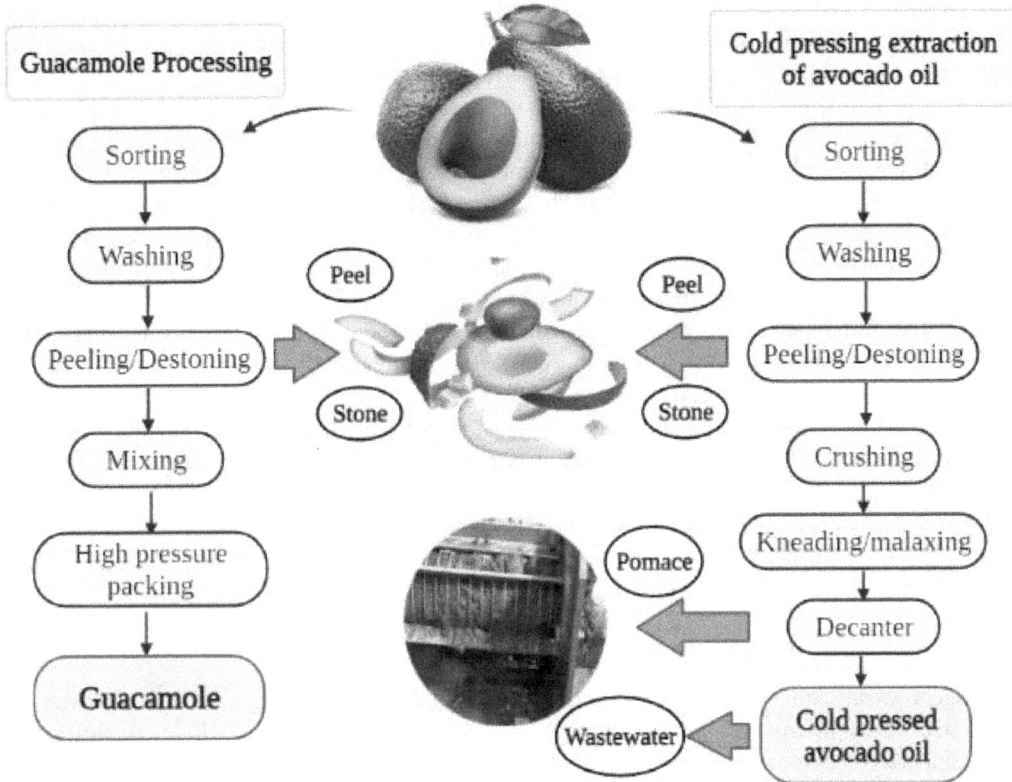

Figure 3. Wastes and by-products generated from cold pressing extraction of avocado oil and production of guacamole.

et al. 2020). Waste and by-products generated from cold press avocado oil extraction and guacamole production (Permal et al. 2020, García-Vargas et al. 2020) were illustrated in Fig. 3.

From an economic point of view, it is reported that the avocado oil market size reached 484.6 million USD in 2020 (Solarte-Toro et al. 2021). Permal et al. (2020) investigated the valorization potential of wastes (peel, stone, pomace, and wastewater) discarded from commercial cold pressed avocado oil process. They found that wastewater had significantly higher lipid content compared to skin, seed and pomace. Therefore, authors used wastewater powder in pork sausage formulation, and observed the inhibition of lipid oxidation. Avocado seeds contain 30% starch, making the seeds an alternative natural starch source. Additionally, avocado seed starch can be considered as a new source of raw material for the production of bioethanol. In a recent study, the production of industrial-grade starch from avocado seeds have been investigated (Tesfaye et al. 2021). Authors found that avocado seeds are technically and economically feasible for industrial-grade starch with 56% yield. Also, the selling potential of industrial-grade avocado seed starch of $0.8375/kg indicated the feasibility. Sánchez et al. (2021) produced biodegradable films from avocado seed and peels. They easily extracted starch and cellulose having high solubility, low water absorption and low swelling power, which affects the formation of biodegradable films. Similarly, Jimenez et al. (2021) stated that the biofilms' production from avocado wastes was feasible thanks to high biopolymer contents such as starch and pectin.

2.11 Potato Waste and By-products

Potato processing industry generates huge amount of wastes (accounts for 12–20% of the raw material), mainly composed of peel, flesh layers, and leaves (Chen et al. 2020). Additionally, starch processing industry generates large amount of wastewater containing gelling agents and various bioactive compounds (Torres and Domínguez 2020). Potato chips are the most commonly processed food product, resulting in potato peel (15 to 40%) as by-product (Gebrechristos and Chen 2018). Potato peels contain assorted phenolic compounds such as gallic acid, caffeic acid, ferulic acid, vanillic acid, chlorogenic acid, *p*-hydroxy benzoic acid, gentisic acid, and *p*-coumaric acid. Potato peels also contain glycoalkaloids, namely solanine and chaconine. These compounds can be used in pharmaceutical industry with their anti-inflammatory, antibiotic or anti-allergenic properties at certain doses. However, glycoalkaloids must be removed from potato peels before food applications since they can have toxic effects at doses over 200 mg/kg. Another important waste stream discarded after harvesting of potato tubers is leaves. They contain high level of carotenoids, particularly lutein (14.7–442.3 mg/g) which has anticancer, anticarcinogen and antiaging properties (Krishna et al. 2018). It has been reported that potato processing industry produces nearly 7 m³ wastewater during processing of 1 ton of potatoes (Kot et al. 2020). Potato industrial waste management is mainly composed of prevention strategy, usage in value-added food formulation, feeding, production of bioproducts (biogas, biobutanol, biochar, bioplastic, hydrochar, etc.), composting, and landfilling (Maroušek et al. 2020). For example, Chen et al. (2018) investigated the hydrothermal carbonization of potato waste, and observed that the aqueous phase significantly increased the hydrochar yield, while Weber et al. (2020) managed to produce bioethanol and distilled beverage to valorize potato wastes. The findings indicated that the utilization of potato wastes could bring up new opportunities for food sector with minimal greenhouse emissions from renewable resources.

2.12 Onion Waste and By-products

The major waste and by-products discarded from industrial processing of onion are skin, the outer two fleshy leaves and the top and bottom bulbs (Campone et al. 2018). Specifically, the major bioactive compounds found in onion skin are quercetin and quercetin glycosides having antimicrobial, antioxidant, cancer prevention, and anti-inflammatory effects. Pereira et al. (2017) studied on the valorization of onion juice waste for the production of *Pleurotus sajor-caju* fruiting bodies and pectinases. They found a feasible substrate for *P. sajor-caju* cultivation and obtained pectinase, an interesting industrial characteristic of onion waste. There are many research studies regarding the extraction of quercetin from onion wastes (Kim et al. 2019b, Santiago et al. 2020, Wianowska et al. 2021). It is reported that 400 mg quercetin/kg can be produced from onion wastes (Črnivec et al. 2021). Similarly, quercetin and derivatives found in onion leaf were evaluated as natural food colorant by Mourtzinos et al. (2018). The onion leaf extract was also tested as a food colorant in yogurt matrix. The findings indicated that it is possible to use onion waste extracts as a natural pigment alternative to synthetic food colorants. More recently, Cho et al. (2021) developed an alternative feed source for honey bee by bioconversion of onion waste into biosugar. Biomedical and pharmaceutical applications have been also evaluated as a different valorization strategy for onion peel. As an example, Kumar et al. (2022) highlighted that onion peel extracts can exhibit health promoting effects such as antimicrobial, anticancer (breast and liver cancer cells), antidiabetic, hypercholesterolemia and obesity, supported by *in vitro* and *in vivo* studies.

3. Cereal Waste and By-products

Cereals such as corn (*Zea mays* L.), wheat (*Triticum aestivum*), rice (*Oryza sativa*), sorghum (*Sorghum vulgare*), oat (*Avena sativa*), rye (*Secale cereale*), and barley (*Hordeum vulgare*) are among the basic food sources of the agro-food industry, which have been harvested since ancient times (Osorio et al. 2021). Due to the changes in eating habits, cereal products such as bread, pasta, biscuits, and semolina have become highly preferred product groups in recent years, comprising up to 20% of the daily diet (Belc et al. 2019, Gutierrez 2019). According to FAO, worldwide cereal production increased from 2.49 to 2.91 billion tons between 2009 and 2018, with a growing rate of 16.87%. Maize (corn) is the most commonly produced cereal with 1.15 billion tons, followed by wheat (766 million tons) and rice (755 million tons). These three cereals account for almost 90%

of the worldwide cereal production. However, the production and transportation of cereals cause enormous solid waste formation such as seeds, bran, germ, herm and waste water streams including condensates, washing water and other solutions (Shehu et al. 2019). Overall, almost 19% of the production is lost during post-harvest and distribution stages (Ganesh et al. 2021).

3.1 Waste Fractions Formed during Processing

The increase in global food production and accordingly food waste has increased the relevant studies on various applications and potential evaluation of discarded food wastes. Grain waste has a potential use as a source of fibrous biomass due to its high carbohydrate and fiber content. Reducing cereal waste may provide practical advantages such as redefining the cereal production chain, increasing the efficiency and improving the economic gain. Bioconversion processes also have environmental advantages such as reducing emissions (in terms of carbon and chemical emissions) and ecological footprints (Grippo et al. 2019). At the same time, these processes enable raw materials to be used for the production of biodiesel, paper, animal feed, ethanol, and therefore contribute to the environmental sustainability (Gutierrez 2019). Due to these reasons, valorization attempts in academic and industrial fields are increasing day by day with the aim of obtaining high value-added products from cereal wastes.

According to their physical properties, specific operations are required for the processing of cereal grains such as the size reduction, dehulling, grinding, milling and sieving (Osorio et al. 2021). During the processing of cereals, some streams such as husk, bran, hull, cob and straw may be released as by-products (Fig. 4). On average, 9.6 million tons of maize cob, 120 million tons of rice husk, 730 million tons of rice straw, 529 million tons of wheat straw and 51.3 million tons of barley straw are generated globally (Deshwal et al. 2021).

Corn (*Zea mays* L.), which is the most produced grain in the world, has become the primary food in many countries (Waheed et al. 2020a). 64% of the corn production is utilized for the animal feed industry, 19% for human consumption and 9% for the food processing industry, while nearly 5% of the produced corn is disposed as waste (Kim and Dale 2004). Waste products such as corn cobs, leaves, stems, shell and roots are generated starting from harvesting to the post-industrial distribution of corn (Fig. 5). The ratio of waste streams to the mass of corn crops is reported as 18.1% for stem, 12.9% for leaves, 8.8% for corn cobs and 18.8% for root and shell (Kaewdiew et al. 2019). These waste products of corn are utilized in the production of animal feed, bioethanol and biodegradable packaging materials (Deshwal et al. 2021). For instance, the processing of corn flour includes processes such as degermination, drying, aspiration, and milling, and the germ formed during dry or wet milling can be used as a by-product for corn oil production after pressing (Papageorgiou and Skendi 2018).

Wheat (*Triticum aestivum* L.) is one of the most important cereal products used as a staple food in human nutrition all over the world. Bread wheat (*T. aestivum*) and durum wheat (*T. durum*) are among the most well-known wheat varieties (Karakaş et al. 2021). The waste streams formed during wheat processing are illustrated in Fig. 6. In general, the wheat kernel consists of 81–84% of endosperm, 14–16% of bran and 2.5–3.8% of germ (Zou et al. 2018). The conversion of wheat into flour includes the industrial production steps of cleaning, conditioning, breaking, sieving, reducing and milling (Papageorgiou and Skendi 2018). During these production steps, some waste/by-products such as wheat bran, straw, germ and endosperm are formed as 23–27% of the milling output (Prueckler et al. 2014). Among them, wheat germ and bran are outshined others as being extremely nutritious by-products. It has been reported that the 15–20% of wheat is released as bran, and 0.2–0.5% as wheat germ during the processing of wheat flour (Demir and Tarı 2014, Zou et al. 2018).

In the production of 1 kilogram of pasta, 1.978 kilograms of waste and losses occur throughout the entire life cycle of pasta, while 16.6% of this amount constitutes the edible part. Here, the biggest portion of wastes occurs during harvest with a rate of 68.9%, followed by 17.2% at milling stage,

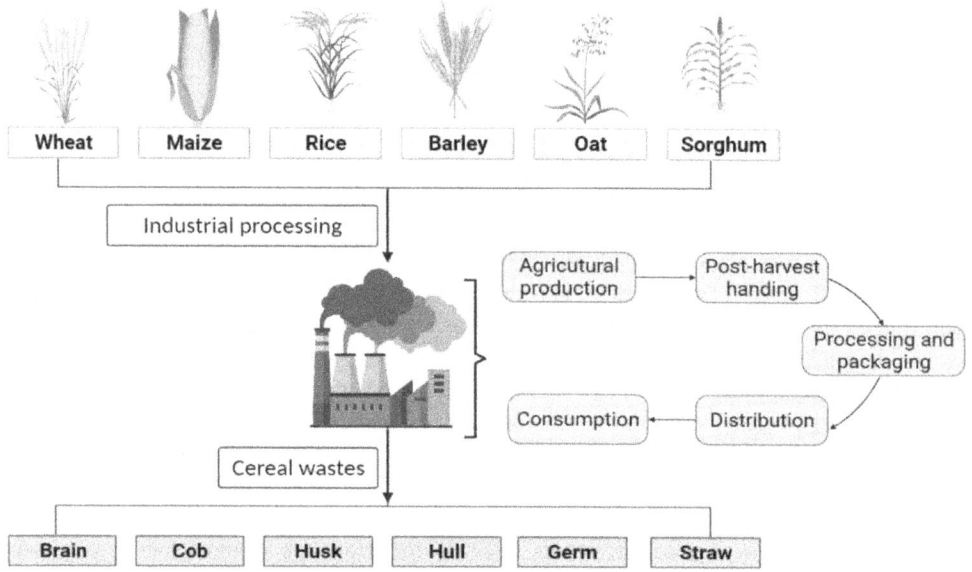

Figure 4. Wastes and by-products from the grain processing industry.

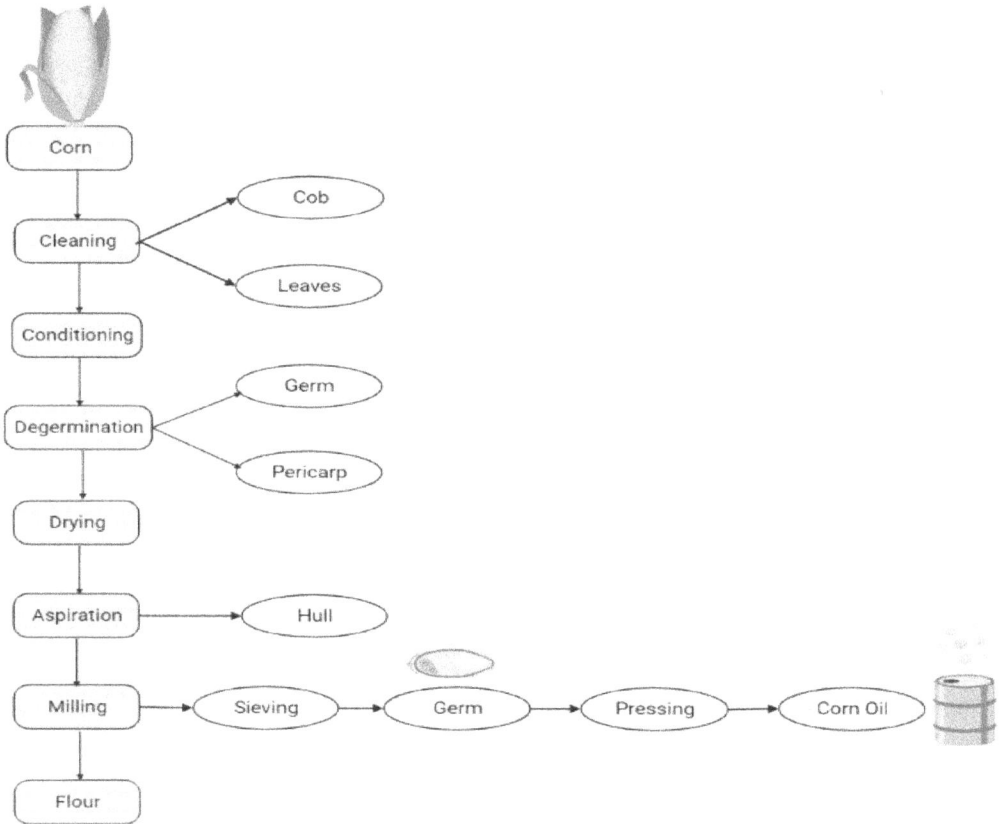

Figure 5. Flow chart of corn flour and corn oil production (Mangia 2010, Vanara et al. 2018).

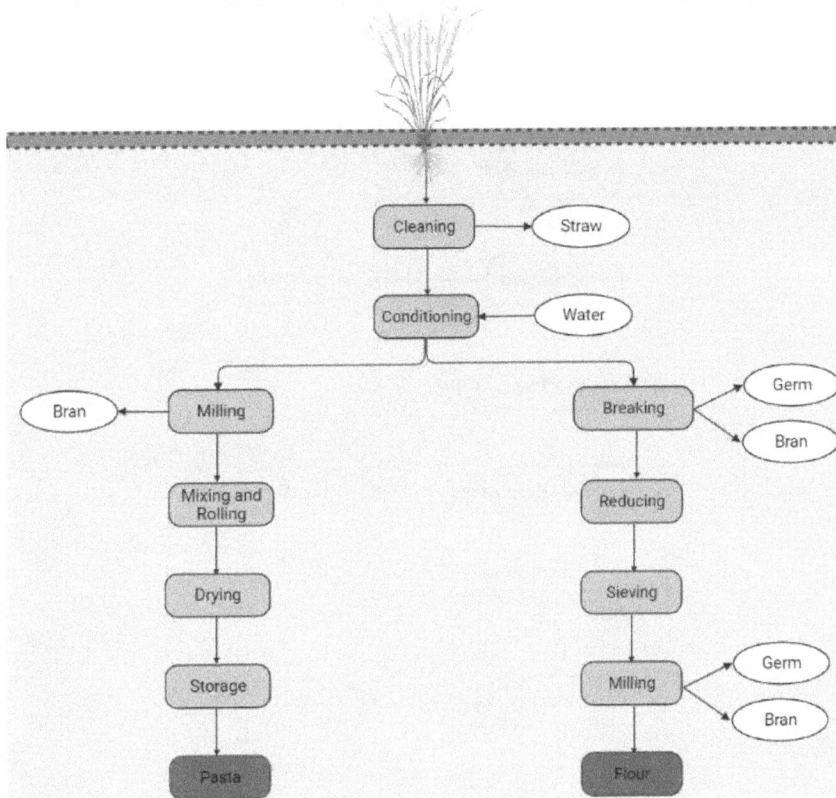

Figure 6. Waste fractions formed during flour and pasta production (Papageorgiou and Skendi 2018, Ruini et al. 2013).

and 12.61% during consumption. Straw, which is not suitable for production processes due to its physiological characteristics, is reused for energy recovery, fertilizer and animal feed. In addition, 17% of the loss in the milling phase is bran, which is a by-product of wheat, and can be used as an alternative for semolina production (Principato et al. 2019).

Rice (*Oryza sativa* L.) is one of the most important grain products, which is a significant source in the diet of almost a quarter of the world population (Alexandri et al. 2020). The process steps of milled rice, known as white rice, includes harvesting, cleaning, husking, milling, polishing and grading (Fig. 7). During these steps, some waste products such as husk, bran, germ and straw may be formed (Alexandri et al. 2020). The husk is the outer layer of the grain and is generally not considered edible, and is therefore removed during the de-husking process. Rice bran is a waste product discarded in milling process, and is obtained from the outer layer of brown rice. On average, by-product fraction of rice is 20% rice husk, 2% germ and 8% bran. In addition, 0.4–1.4 kg of rice straw is released for every one kilogram of milled rice (Alexandri et al. 2020, Papageorgiou and Skendi 2018).

Cereal brans such as from wheat, corn and rice are rich sources of bioactive phenolic compounds (especially ferulic and coumaric acid). The amount of phenolic acids in bran is reported to be 10–20 times higher than in the endosperm. Ferulic acid constitutes 60–70% of the phenolic acids found in grains, and is mostly found in the cell walls of the aleurone and pericarp layers (Hromádková et al. 2008, Zhao et al. 2009). Zhao et al. (2009) showed that the intake of corn bran increases the antioxidant levels in blood plasma, while the intake of wheat bran and corn bran increases phenolic antioxidant level in urine. Peanparkdee et al. (2019) reported that rice bran consists of many bioactive compounds such as phenolic acids (protocatechuic acid, vanillic acid, *p*-coumaric acid, ferulic acid, sinapic acid), anthocyanins (cyanidin 3-glucoside, peonidin 3-glucoside) and flavonoids

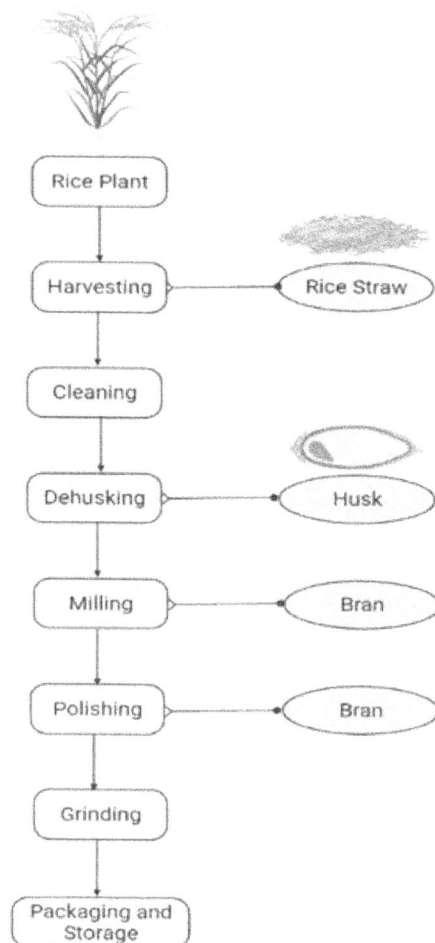

Figure 7. Flow chart of dry milled rice (Son et al. 2011, Papageorgiou and Skendi 2018, Alexandri et al. 2020).

(rutin, myricetin, quercetin-3-glucoside). Highly bioactive compounds obtained from rice bran have been utilized as food enrichment agents, colorants and preservatives. However, the use of rice bran is limited due to the rapid rancidity caused by the high amounts of unsaturated fatty acids, and strong hydrolytic and oxidative enzymes (such as lipase and lipoxygenase). For this reason, researchers suggested that fresh rice bran should be stabilized immediately after the milling step, and studies on the stabilization methods of rice bran have increased in recent years (Wu et al. 2020). In addition, phenolic compounds in rice bran are mostly present in bound form. Therefore, there has been a great interest to find an effective method for the release of bound phenolic compounds (Călinoiu and Vodnar 2020, Liu et al. 2017, Wanyo et al. 2014). Enzyme application is one of the promising techniques to increase the bioavailability of bound phenolics. Liu et al. (2017) showed that large amounts of phenolic compounds are released as free and soluble conjugate forms after enzyme hydrolysis. As a result, cereal brans can be utilized in beverage and other food industry branches to produce phenolic-, antioxidant- and flavonoid-rich products.

The losses of grain throughout the entire food chain arise from all processes, from harvest, postharvest and industrial processing, to the consumption steps. Globally, more than two billion consumers do not have direct access to the municipal solid waste collection; therefore, they dispose their wastes by burying, open burning or discharging into an open ground or waterways (Reyna-Bensusan et al. 2019). Unconsciously applied disposal methods cause negative effects on many environmental factors such as energy, soil and water resources (Skendi et al. 2020). The burning of solid waste in open areas causes a significant amount of carbon emission. This black carbon emission is a short-lived climate pollutant that contributes to climate change by absorbing solar radiation. It has also been reported that black carbon creates a global warming problem 5,000 times more than CO_2, and has negative effects on human health. Reyna-Bensusan et al. (2019) showed that black carbon emissions from open burning of wastes have more adverse effects on climate change compared to CH_4 emissions (Reyna-Bensusan et al. 2019). For instance, burning a ton of straw causes a loss of 5.5 kg of nitrogen, 2.3 kg of phosphorus, 25 kg of potassium and 1.2 kg of sulfur. In addition, the burning of waste also causes greenhouse gases that cause global warming, loss of biodiversity and reduced productivity of agricultural land (Bhuvaneshwari et al. 2019). In this context, studies on the evaluation of food waste also aim to improve air quality and respiratory health by preventing open burning. For example, the EU Waste Framework Directive (2008) of the European Parliament and the Council has introduced the 3 R (reduce, reuse, recycle) policy with the aim of reducing the waste source, ensuring its reuse and recycling. The studies carried out and the guidelines created highlighted the critical importance of transforming food wastes into value-added products, and contributed to a clean, reliable and sustainable environment (Fritsch et al. 2017, Skendi et al. 2020).

Table 4 summarizes the nutritional composition of different cereal wastes. Cellulose, the main component of waste fractions, is a renewable biopolymer found in abundance in the plant cell wall (Xu et al. 2018). Lignocellulosic materials are one of the most frequently used natural resources due to their biodegradability, renewability, low density, resistance to strength and non-toxicity (Xiao et al. 2019). Cellulose is usually produced from wood pulp, but this production requires high costs compared to renewable materials, so the focus of researchers on cellulose production from agricultural residues has been increased (Fritsch et al. 2017). Agricultural residues such as wheat

Table 4. The composition of cereal wastes.

Sample	Protein (%)	Fat (%)	Cellulose (%)	Hemicellulose (%)	Lignin (%)	Ash (%)	References
Wheat bran	17.4	3.71	31.1	34.3	16.3	4.98	Ye et al. (2021), Xiao et al. (2019)
Wheat straw	2.29	2.21	44.8	33.1	8.46	5.68	Liu et al. (2021b)
Rice husk	2.30	0.30	33.8	17.1	21.5	16.5	Collazo-Bigliardi et al. (2018), Bakari and Yusuf (2018)
Rice straw	21.6	10.4	28.0	55.0	11.0	6.00	Banik and Nandi (2004), Syaftika and Matsumura (2018)
Dried corn cob powder	2.86	0.61	45.0	33.1	13.8	3.10	Paynor et al. (2016), Louis and Venkatachalam (2020)
Corn stover	5.36	2.16	32.8	31.1	10.1	4.08	Wang et al. (2020)
Corn husk	8.93	9.44	46.2	33.8	3.92		Vitez et al. (2020), Herlina et al. (2018)
Oat hull	5.53	0.67	31.2	28.7	18.1	0.83	Debiagi et al. (2021)
Oat husk	3.32	2.60–5.20	40.1	25.1	26.1	8.70	Bulkan et al. (2021), Kouřimská et al (2021), de Oliveira et al. (2017)
Sorghum stalk	5.31	1.17	48.5	28.0	22.1	0.77	Jonathan et al. (2012), Wang et al. (2021)
Sorghum bagasse	5.40	10.4	40.4	35.5	3.90	0.20	Almodares et al. (2009), Dogaris et al. (2009)
Barley husk	4.00	4.00	39.0	12.0	22.0		Bledzki et al. (2010)

straw, rice straw, corn husk, soybean and corn stove have attracted great attention as alternative materials in cellulose production with their high cellulose content and low unit costs. Producing cellulose with the recovery of waste products provides benefits in terms of economic gain and environmental sustainability, as well as its renewability feature; therefore, the interest in innovative technologies for the evaluation of food waste streams is increasing (Xu et al. 2018). For the production of cellulose fiber, agricultural wastes are first subjected to cutting, milling and grinding processes in order to increase the surface area. The ground sample is then chemically treated using toluene/ethanol or petroleum ether to remove lipids, waxes and pigments. Next, alkaline treatment is carried out using sodium hydroxide, potassium hydroxide and organic solvents to dissolve some of the hemicellulose and lignin. Another important step is the bleaching process, in which the color of grain waste is bleached using sodium chloride or hydrogen peroxide bleaching agents. In this part, the remaining lignin, hemicellulose, pectin and wax substances are separated (Ahmad Khorairi et al. 2021). Grain wastes such as wheat straw, corn straw and oat hull are generally the most used raw materials for cellulose production (Fritsch et al. 2017).

Agricultural wastes include a wide range of raw materials suitable for industrial use. One of the most important components of grain wastes are dietary fibers such as cellulose, hemicellulose, lignin, pectin and digestive-resistant starches. β-glucans and arabinoxylans, which have positive health benefits, are the most well-known water-soluble dietary fibers found primarily in the outer layer and endosperm cell walls of all cereals. Arabinoxylans (AX) belong to the hemicellulose group and contain copolymers such as two pentose sugars, arabinose and xylose. Arabinoxylans are linear chains of D-xylopyranosyl units linked by β-(1→4) bonds to which α-L-arabinofuranosyl units are substituted at O(2), O(3) or both positions. Solubility is the most important feature of AX polysaccharides found in cereal wastes (Skendi et al. 2020, Skendi and Biliaderis 2016). AXs have positive effects on many health conditions such as reducing constipation, lowering LDL cholesterol, reducing blood glucose level and reducing the risk of tumors. Since AXs are dietary fibers that resist dissolution, they are not digested along the gastrointestinal tract (Rosicka-Kaczmarek et al. 2016). Therefore, xylo- and arabino-xylooligosaccharides produced from arabinoxylan have been evaluated as prebiotics using green processing techniques to minimize the environmental impact of waste products. Walton et al. (2012) gave 10 g/day arabinoxylan-oligosaccharides (AXOS) enriched bread to 20 healthy individuals for 21 days, and there was an increase in the number of probiotic *Bifidobacterium* and *Lactobacilli* in the gut microbiota of individuals fed with AX-breads compared to those of white bread. In a study by François et al. (2012), 60 healthy people consumed 0.3 and 10 g/day wheat bran extracts enriched with arabinoxylan oligosaccharides for 3 weeks. While daily intake of 10 g wheat bran extract increased fecal *Bifidobacteria* levels and short chain fatty acid concentration in fecal concentrations, it decreased fecal pH and frequency of constipation. Water-extractable AX is known not only for its emulsifier, prebiotic and antioxidant properties, but also for its gelling capacities with an oxidizing agent. Arabinoxylans can gel through oxidative cross-linking of ferulic acids by some chemical or enzymatic free radical-generating agents. The strength, pore size and crosslinking density of the gels formed by water-extractable AX depend on the ferulic acid content. AX gels have the potential to form strong and stable gels in foods as texturing agents. Kale et al. (2013) prepared gels by cross-linking ferulic acids from corn bran arabinoxylans by alkaline extraction method, and showed that the ferulic acid content and gel elasticity of AX decreased with the increase in extreme extraction conditions. In the maize processing industry, the stage in which maize kernels are cooked in an alkaline solution, usually lime water, is called nixtamalization. Although the supernatant (nejayote) formed in this process is disposed of as waste, some studies have reported that nejayote contains more than 60% non-starch polysaccharides. Niño-Medina et al. (2009) extracted water-soluble feruloylated AX from nejayote formed during tortilla making, and found that nejayote arabinoxylans had a high content ferulic acid content to support the formation of laccase-derived covalent gels. This study is promising for the use of nejayote arabinoxylans as texturizing agent with a commercial advantage over other gums in the food industry.

β-glucans are dietary fibers found in high amounts in grains, particularly oat and barley. Cereal β-glucans are found 3–7% and 3–11% in oat and barley, respectively, but in lower amounts in rye (up to 2.5% in the flour) and wheat (up to 2.5% in the bran). Cereal β-glucans are linear polysaccharides formed by the linking of D-glucopyranosyl units with β-(1→4) and β (1→3) bonds. Their ladder-shaped structures mainly consist of two parts: Cellotriosyl (58–72%) and cellotetraosyl (20–34%) units. The molecular weights of cereal β-glucans vary depending on their botanical origin, genotype and agronomic factors. High concentration and molecular weight β-glucans exhibit better viscosity properties. It has been reported that β-glucan lowers the glycemic index, blood sugar and LDL cholesterol levels (Skendi et al. 2020, Henrion et al. 2019). Gamel et al. (2014) investigated viscosity and solubility of β-glucan extracted from oat bran by enzymatic extraction with amylase, protease and lipase. The authors concluded that most of the final solution viscosity is due to β-glucan; however, high molecular weight β-glucan increased the final viscosity more because the separation of β-glucan from the food matrix was facilitated by the addition of enzymes. β-glucan has an important place in the food processing industry with its emulsifier, thickener and water holding capacities, as well as having desired health effects such as preventing the formation of diabetes, relieving constipation and reducing the risk of cancer (Liu et al. 2021a). Grain β-glucans are promising for use as emulsifiers in the food industry due to their rheological properties as they increase the viscosity of aqueous solutions and maintain emulsion stability. Karp et al. (2019) showed that cereal β-glucans exhibited a high degree of emulsion stability (80%) at the end of 14 days of storage and partially caused visible viscosity increase. Yu et al. (2021) reported that water-insoluble β-glucans extracted from oat bran exhibit a more fibrous layer and greater swelling power than water-soluble β-glucans. As a result of their animal experiments, authors also observed that water-insoluble β-glucans caused a reduction in body weight and an improvement in blood lipid levels in high-fat diet fed mice. In addition, Cheng et al. (2017) investigated the effects of β-glucan extracted from oat bran on glucose homeostasis in mice fed on high-fat diet, and found improved insulin sensitivity, body weight, and glucose homeostasis in mice fed with both oat bran and extracted β-glucan.

Protein is a very important part of the daily human diet, as well as the improving sensorial and functional properties of foods. Many cereal wastes, especially wheat and rice bran, contain significant amounts of proteins. However, protein macromolecules found in grains are mostly bound to the lignocellulosic components. This case creates a barrier to protein extraction so that the protein is unable to interact efficiently with enzymes or solvents (Leal et al. 2021). For this reason, many researches have been carried out to increase the efficacy of protein recovery by different methods such as ultrasound (Phongthai et al. 2017), microwave (Bedin et al. 2020, Hayta et al. 2021), extrusion (Leal et al. 2021, Roye et al. 2020) and high pressure (Zhu et al. 2017), in order to obtain protein fractions cereal wastes. Initial studies on the recovery of plant-based proteins have been mainly focused on rice bran as the raw material. Rice bran is a by-product that is released during milling and is very rich in terms of nutrients, especially protein (11–13%). Therefore, non-allergic rice bran is seen as a potential protein source. In addition to its nutritional properties, the protein obtained from rice bran can also be used in the food processing industry as a foaming agent and emulsifier (Hu et al. 2019). Rice bran oil is a vegetable oil obtained by processing rice bran and is frequently consumed in countries such as Japan, India, Korea, China and Indonesia. Rice bran oil contains high amounts of unsaturated fatty acids with 45% oleic acid and 33% linoleic acid content. Rice bran oil contains significant amounts of antioxidant compounds such as tocopherols and γ-oryzanol. The oil can be used as an anti-inflammatory and antioxidant agent to reduce the low-density lipoprotein and cholesterol concentrations in blood stream, and for the treatment of cardiovascular disease and atherosclerosis (Liu et al. 2019). Especially, ferulic acid esters in γ-oryzanol have been found to possess strong radical scavenging properties (Fraterrigo Garofalo et al. 2021). Considering all these features, it can be interpreted that many cereal wastes, especially bran, contain significant amounts of protein and other bioactive compounds with numerous potential evaluation areas. Table 5 summarizes some example studies on the valorization of cereal wastes.

Table 5. Example studies on the valorization of cereal wastes.

Source	Treatment	Final product	Results	Application	References
Rice husk	Alkaline treatment (2% NaOH) Bleaching treatment (2% H_2O_2)	Cellulose nanocrystals	Cellulose nanocrystals with particle size of 15–20 nm exhibited good thermal stability.	Reinforcement agent in nanocomposites	Fathi et al. (2018)
Maize stalk	Alkaline treatment (1.5% NaOH) Bleaching treatment (1.5% $NaClO_2$ + 1.5% KOH) Acid treatment (60% H_2SO_4)	Cellulose nanowhiskers	The crystallinity index of maize stalk increased after chemical and mechanical treatments.	Reinforcement in biopolymers	Motaung and Mtibe (2015)
Corn stove	Alkaline treatment (2% NaOH) Bleaching treatment ($NaClO_2$ + NaOH) Acid treatment (55% H_2SO_4)	Cellulose nanocrystals	While the crystallinity of the corn stove before bleaching was 25.3%, it increased to over 43% after acid hydrolysis.	Strengthening agents	Costa et al. (2015)
Wheat, maize, waxy maize starches	Homogenization Vacuum filtration Inoculation	Bacterial cellulose fibrils	Addition of bacterial cellulose fibrils to wheat, maize and waxy maize starch reduced the enthalpy of gelatinization.	Gelatinization	Díaz-Calderón et al. (2018)
Wheat straw	Chemical and thermal treatment (H_2SO_4) Enzymatic hydrolysis (Cellulase, beta-glucosidase and xylanase) Inoculation (*Gluconacetobacter xylinus*)	Biocellulose nanofiber	The enzymatically pre-treated wheat straw resulted in the highest bacterial cellulose of 10.6 g/L.	Production of biocellulose nanofiber	Al-Abdallah and Dahman 2013
Wheat bran	Ultrasound-assisted pretreatment Thermal treatment Acid and alkaline hydrolysis	Polyphenol	The combination of deep eutectic solvent with thermal treatment and ultrasound pre-treatment showed in the highest total polyphenol yield (94.62 mg ferulic acid eqv./g dry matter).	Wheat bran polyphenol recovery	Cherif et al. (2020)
Rice bran	Enzymatic pretreatments Extrusion pretreatments Combined pretreatments	Protein	The combination of pretreatment raised the protein yield (69.6%) and positively influenced the functional properties.	Protein recovery	Leal et al. (2021)
Rice bran	Hydrothermal treatment Lyophilization	Protein	Hydrothermal phosphorylation with sodium trimetaphosphate increased the solubility and emulsifying activity of rice bran protein.	Emulsifier	Hu et al. (2019)
Colored rice bran	Cold-press extraction (CPE) Solvent extraction (SE) Supercritical CO_2 extraction (SC-CO_2)	Rice bran oil	The rice bran oil obtained from the bran of black rice extracted by SC-CO_2 showed the highest antioxidant activity.	Antioxidant	Mingyai et al. (2017)

Source	Method	Product	Description	Application	Reference
Rice bran	Isolation of Arabinoxylan Fecal sample collection Fermentation	Arabinoxylans	Rice bran AX increased the number of *Collinsella, Blautia* and *Bifidobacterium* and decreased the number of *Sutterella, Bilophila* and *Parabacteroides* in human fecal samples.	Prebiotics	Gu et al. (2021)
Wheat straw	Alkaline extraction Enzymatic hydrolysis	Xylo Oligosaccharides (XOS)	XOS hydrolyzate produced from wheat straw was found to be a suitable carbon source for probiotic *Lactobacillus brevis* strain.	Prebiotics	Faryar et al. (2015)
Corn bran	Alkaline extraction	Arabinoxylans	Increasing harshness of alkali treatment caused a decrease in average ferulic acid content of AXs and elasticity of gels.	Gelling	Kale et al. (2013)
Dehulled barley bran	Ultrasonic extraction (UE) Hot water extraction (HWE) Microwave extraction (ME) Microwave-assisted ultrasonic extraction (MUE)	β-glucan	β-glucan extracted with MUE showed stronger emulsification stability, while the one extracted with UE showed stronger foaming ability.	Emulsifier Foaming capability	Liu et al. (2021)

3.2 Sustainable Valorization Options

Recycling of waste provides a significant amount of economic gain as well as environmental sustainability. The economic gain from the recycling of waste may increase even more with the factors such as recycling and reuse of waste products, minimizing losses, and increasing the utilizing efficiency of resources (Belc et al. 2019). It has been reported that the cereals generate $101,751 worth of waste products during agricultural production, $159,766 in post-harvest processing and storage, $96,608 in processing and packaging, and $52,146 in consumption only in South Africa (Nahman and de Lange 2013). Bread is the most important one as a lost grain product that causes disposal in natural resources such as water, soil and energy used during production and transportation; in addition to the loss of raw materials, while wasted bread causes an economic loss of 400 million Euros in the Netherlands, the situation is even worse in developed countries. For example, in the UK, 20 million slices of bread are wasted every day, resulting in 584,000 tons of CO_2 equivalent emissions (Narisetty et al. 2021). Straw is one of the most common grain wastes left in the field after harvest. In India, farmers dispose of 16% of their crop waste by burning, 60% of which is rice straw. The open burning of rice straw causes 7,300 kg of CO_2-equivalent greenhouse gas emissions per hectare, soil pollution and a decrease in biodiversity. The economic expenditure per hectare for bran disposal was reported as 281$, and the environmental loss as 46$. Moreover, while the product value of bioethanol from straw is 926$/ha, the fossil fuel saving would be 65$/ha (Bhattacharyya et al. 2021).

The use of cereals in animal feed ingredients is very costly, but the use of wastes from cereals in feed formulations provides an alternative both in terms of environmental sustainability and in terms of producing low-cost animal feeds. The inclusion of different levels of rice bran waste in the diet of sheep significantly reduced the feed cost per kg body weight gain. While the feed cost of sheep fed without adding any rice bran waste was 5.1$ per kg live weight gain, consuming feeds with 30% rice bran added reduced feed costs to 1.5$ per kg live weight gain (Muhammad et al. 2008). In wheat processing factories, significant amounts of water are used during the washing of wheat and this water is discarded as waste. Since the resulting wastewater has high biological oxygen requirements and high levels of dissolved solids, nutrients and minerals, more studies are required on recycling attempts. Balcı and Bayram (2015) used filter, centrifuge, cleaning column and UV processes during the wheat washing phase. As a result, the water recovery rate was 53%, and the biological oxygen demand, total solids, brix, conductivity and turbidity of the wastewater were reduced. Cooking in bulgur production is a process that requires high levels of energy and water output. In another study, filtration, centrifugation, filtering column and ultraviolet light were used for the bulgur cooking process. The authors found that the energy requirement for heating the water used in the cooking process was decreased by 57% (Balcı and Bayram 2020).

Today, consumer interest in functional food products has begun to increase for a healthy lifestyle. The interest in the consumption of dietary fiber and phytochemicals has started to attract attention especially in recent years due to their health beneficial effects such as the prevention of chronic and degenerative diseases. The World Health Organization (WHO) and FAO recommend a daily intake of 25–30 grams of dietary fiber for humans. Therefore, researchers and developers in the bakery industry has begun to create new sources of dietary fiber for new products, formulations and processes. Considering all this, studies on the development of fiber-enriched grain-based products are increasing day by day. For these reasons, as a result of the knowledge of the nutritional content of grain products and their wastes, cereal wastes have been used as new and promising raw materials in the production of functional food products.

Microfluidized corn bran was used to develop high-fiber wheat bread with a new strategy. Production of high fiber bread was successfully achieved by replacing 20% of flour with microfluidized corn bran. In addition, bread developed with flour containing corn bran had a smaller specific loaf volume and improved textural properties such as firmness, springiness and cohesiveness, compared to the white bread (de Erive et al. 2020). Concentrated and dephytinized wheat and rice bran exhibited higher dietary fiber content in the product, and concentrated brans decreased the springiness, cohesiveness, and resilience of the bread, while significantly increasing its hardness (Özkaya et al. 2018). In addition to the textural properties, the nutritional content of value-added bread is also very important and the functional compounds of the raw material to be used is considered in the search for new cereal-based sources in functional bread production. One study showed that replacing 20% of wheat flour with defatted rice bran provides maximum protein, ash, fiber, K, Ca, and Mg content in bread, but the addition of 5% defatted rice bran has created the highest quality bread for commercialization, based on all chemical and sensory analyses (Ajmal et al. 2006). It has been reported that sourdough fermentation improves some sensory properties of wheat bran and stabilize nutritional properties of the wheat germ. Wheat breads developed using fermented milling by-products (wheat germ and bran) showed better protein digestibility, nutritional indices and starch hydrolysis rate, as well as improved biochemical, functional and sensory properties of the bread compared to the sole use of wheat flour. At the same time, obtained fortified bread is characterized by a higher dietary fiber content and a lower glycemic index value (Pontonio et al. 2017). Mikušová et al. (2013) enriched wheat and rye flour breads with extruded wheat bran (10%), cereal β-D-glucan hydrogel (12.5%) and a lactobacilli starter culture. The consumption of functional bread resulted in lower glucose levels in healthy men compared to control-group bread after 120 min. In order to increase the soluble fiber content of cereals and improve the quality characteristics of cooked foods, some processes such as extrusion, fermentation and enzymatic hydrolysis can be applied to cereal bran. For example, glucose oxidase and hexose oxidase enzymes were used to improve the rheological properties of flour and dough enriched with wheat and corn bran. The addition of 15 mg/kg glucose oxidase and 30 mg/kg hexose oxidase was the most effective doses in improving the bread and dough properties (Gül et al. 2009).

Pasta is one of the staple foods which is very popular in the current lifestyle all over the world, thanks to its long shelf life, easy portability and preparation, high carbohydrate content and low glycemic index. Pasta, which is a universal food, can be used to provide appropriate dietary fiber support as a result of the increase in health awareness and consumer demands. For this reason, the production of functional and value-added pasta products with cereal wastes has gained momentum. Since cereal brans such as wheat, rice, barley and oat bran have high dietary fiber and protein content, they are used in pasta production to increase the functional properties of durum wheat semolina. Depending on the type and concentration of grain bran, adding fiber-based sources to pasta has been reported to improve the color, brightness, water absorbance, and volume expansion values (Kaur et al. 2012). In a study, enriched pasta production was carried out by replacing 40% and 50% of semolina with treated wheat bran. Preparation of pasta with 40% steam heat-treated wheat bran offered approximately 5.2 times more dietary fiber, while the samples containing processed bran had lower cooking loss (Sudha et al. 2011).

Turkish noodles are one of the traditional cereal products in Turkey that are obtained by mixing the ingredients, sheeting the dough, cutting the dough and drying the noodles (Tuncel et al. 2017). The addition of infrared stabilized rice bran increased the nutritional constituents of Turkish noodles such as crude fat, protein, dietary fiber, vitamins and minerals, while decreasing the textural property parameters such as adhesiveness, cohesiveness, springiness, and gumminess (Tuncel et al. 2017). In other study, noodle and pasta formulations were enriched with dephytinized cereal brans (20%) of rice, rye, wheat and oat to enhance the nutritional quality. Authors demonstrated that the rice bran improved the calcium, phosphorus, manganese and magnesium levels, while rye bran improved potassium, and wheat bran improved zinc, magnesium and iron contents in pasta and noodle products (Levent et al. 2020).

High fat intake has adverse effects on human health such as cardiovascular diseases, obesity, diabetes and myocardial diseases due to the high cholesterol and saturated fatty acid contents. Therefore, consumers mostly prefer low-fat foods in the modern society (Marvizadeh and Akbari 2019). In this direction, the use of fat replacers has begun to increase to reduce the amount of high fat in foods. Since carbohydrate-based components resemble the functional and sensorial properties of fats, they can be used as fat replacers. In addition, most of the cereal-based products have the ability to mimic some certain fat properties. In some studies, rice bran in hamburgers (Marvizadeh and Akbari 2019), wheat and oat bran in cookies (Milićević et al. 2020), rye bran in meatballs (Yılmaz 2004) and oat hull in burger (Summo et al. 2020) are among the substances with approved fat replacement ability. According to Marvizadeh and Akbari (2019), adding 4% rice bran to hamburger caused a 25% decrease in the fat content. Researchers have proven that rice bran can be used as a fat replacer in hamburgers without loss of quality in the final product. In another study, total trans fatty acid content of meatballs treated with rye bran was found to be lower than the control group (Yılmaz 2004). Beef with low fat and high beta-glucan contents was obtained using a fat replacer from an oat hull-based ingredient. Beef with 100% oat hull as a fat replacer presented very low final fat content after cooking (3.48%) and met the daily intake of beta-glucan (2.96 g) with one serving size (Summo et al. 2020).

Tarhana is a cereal-based fermented food product. Although the recipe for tarhana varies from region to region, the main ingredients are yogurt and flour. After adding salt, yeast and various spices and vegetables, tarhana is fermented for 1–7 days before consumption. Lactic acid, ethanol, carbon dioxide and other fermentation products that provide the specific taste, smell and aroma of tarhana are formed during fermentation (Aktaş and Akın 2020, Çelik et al. 2010). The use of cereal wastes may be beneficial to produce high value-added and functional tarhana. Corn and rice brans, which are dietary fiber sources, were used in tarhana formulation in order to increase the nutritional content. The 15% rice bran substitute showed the highest ash, protein, phenolic content, oil and antioxidant capacity, while the 15% corn bran substitute showed the highest cellulose content (Aktaş and Akın 2020). Wheat bran and germ were successfully added to the tarhana formulation, and an increase was observed in crude protein, mineral and total phenolic contents in final product (Bilgiçli et al. 2006). Çelik et al. (2010) also reported that the crude fiber content of tarhana increased almost 7 times by substituting 40% wheat flour in tarhana formulation. Considering all the studies mentioned above, it has been confirmed that food products enriched with grain waste and by-products show superior nutritional, functional and rheological properties. However, more studies are required in order to eliminate some restrictions during the production of functional food from cereal wastes and by-products.

In addition to the use of cereal wastes and by-products in the food industry, they can also be evaluated in different industries including pharmaceutical, fertilizer, cosmetics, chemicals, biofuel and packaging (Skendi et al. 2020). Considering the fuel demands, it will possibly become necessary in the future to obtain bioethanol from starch-rich grains, and agricultural and forestry residues. Among biofuels, ethanol has many advantages such as low unit cost, being environmentally friendly and having high hybrid efficiency. Since cereal brans are rich in both starch and cellulose/hemicellulose contents, they significantly increase the ethanol production yield (Kumar et al. 2020, Palmarola-Adrados et al. 2005). Cereal straw is considered as a safe raw material in the production of biofuels such as bioethanol, biobutanol, biohydrogen and biogas (Ghosh et al. 2017).

Packaging materials obtained from petroleum-based plastics are a concern due to the depletion of non-renewable resources as well as not being destroyed in nature. Therefore, it is promising to use biopolymers obtained from renewable resources such as agricultural wastes as raw materials in packaging production. For instance, polylactic acid (PLA) packaging films can be produced by lactic acid bacteria obtained from cereal-based by-products. PLA creates lower CO_2 emissions and environmental footprints compared to non-renewable polymers. Since CO_2 will be absorbed during the cultivation of cereal products for PLA production, the use of these films creates less greenhouse gas emissions than others (Skendi et al. 2020).

By-products of cereals are also used in the cosmetic industry thanks to their valuable bioactive components. It is thought that the production rate of cosmetic products will increase by 53% until 2023 due to the intense consumption of new generations (Osorio et al. 2021). In line with some studies, it has been confirmed that arabinoxylans and β-glucans obtained from cereal wastes can be used as wrinkle reducers, moisturizers and natural antioxidants in cosmetic applications (Arzami et al. 2021). In addition, it has been reported that the rice bran oil is used in cosmetics to improve skin defects caused by melanin, and as a sunscreen in cosmetic formulations due to its high antioxidant activity (Lerma-García et al. 2009).

Studies on the antiviral, antibacterial, antifungal and antioxidant effects of cereal-based waste and by-products have also gained momentum in recent years. It has been reported that wheatgrass has anticarcinogenic, diuretic and anti-inflammatory effects while the maize bran gum is used as a bioactive substance carrier in the human intestine (Sielicka-Różyńska and Gwiazdowska 2020, Skendi et al. 2020). Cereal wastes and by-products are also used in the treatment of acute and chronic diseases, thanks to the bioactive compounds they contain. Corn cob is known for its high antioxidant properties and high phenolic, glycoside and anthocyanin contents (Doan et al. 2020). In addition, the parts of the cereals left in the field after harvest can be used as animal feed due to their low-cost applicability and rapidly changing nutritional habit of animals. Quality crop residues of cereals are used in the feed of animals such as pigs, ducks and cattle, as they provide high energy and protein (Ajila et al. 2012). Especially wheat, barley and oat are cereal wastes that are frequently used in animal feed compositions.

4. Oilseed Waste and By-products

More than 200 oilseed crops are grown worldwide, which can be divided into two main groups: edible oil sources used for direct consumption, and inedible oil sources used in the production of oleo-chemicals. The most important oilseeds in the food processing industry are soybean (*Glycine max*), rapeseed (*Brassica napus* L.), sesame (*Sesamum indicum* L.), sunflower (*Helianthus annuus* L.), coconut (*Cocos nucifera* L.) and olive (*Olea europaea* L.) (El-Hamidi and Zaher 2018, Kotecka-Majchrzak et al. 2020). According to USDA (2021), major oilseed production in worldwide increased from 580 million in 2019/2020 to 603 million tons in 2020/2021 season with a growing rate of 3.9%, soybeans being the primary type of oilseed with 366 million tons of production. Rapeseed is the second most produced oilseed with a production of 73 million tons, while peanut ranks third with 50 million tons (Fig. 8). Refined edible oils from oilseeds are widely used in the manufacture of margarine, confectionery, baked goods and meat products. On the other hand, unrefined cold pressed oils are considered a pro-health ingredient to enhance the flavor of salad, sauce and pastes (Kotecka-Majchrzak et al. 2020). It has been stated that the oilseed wastes produced during processing constitute 20% of total wastes generated worldwide (Otles et al. 2015). Figure 9 shows percentage rates of evaluation steps of oilseeds in 2020 season.

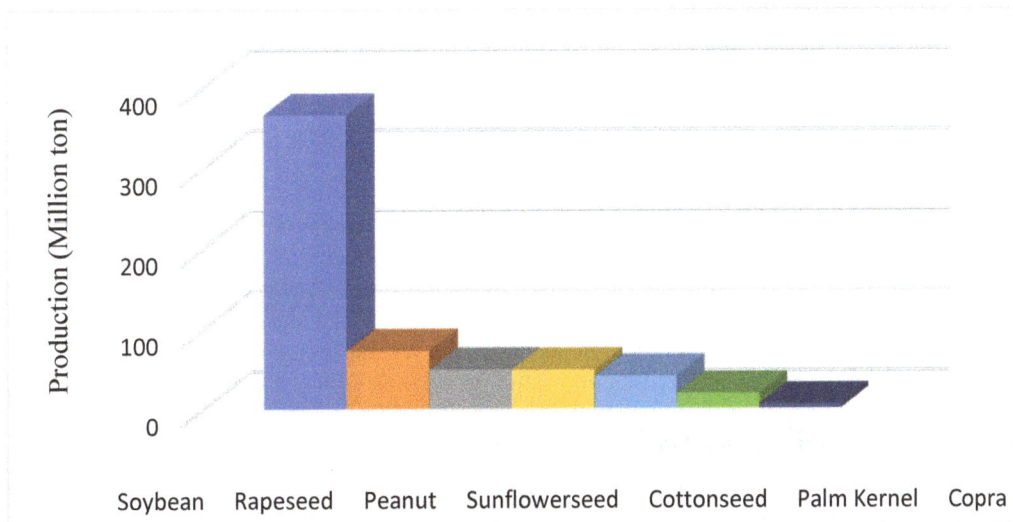

Figure 8. Production of major oilseeds all around the world in 2020/2021 season (USDA 2021).

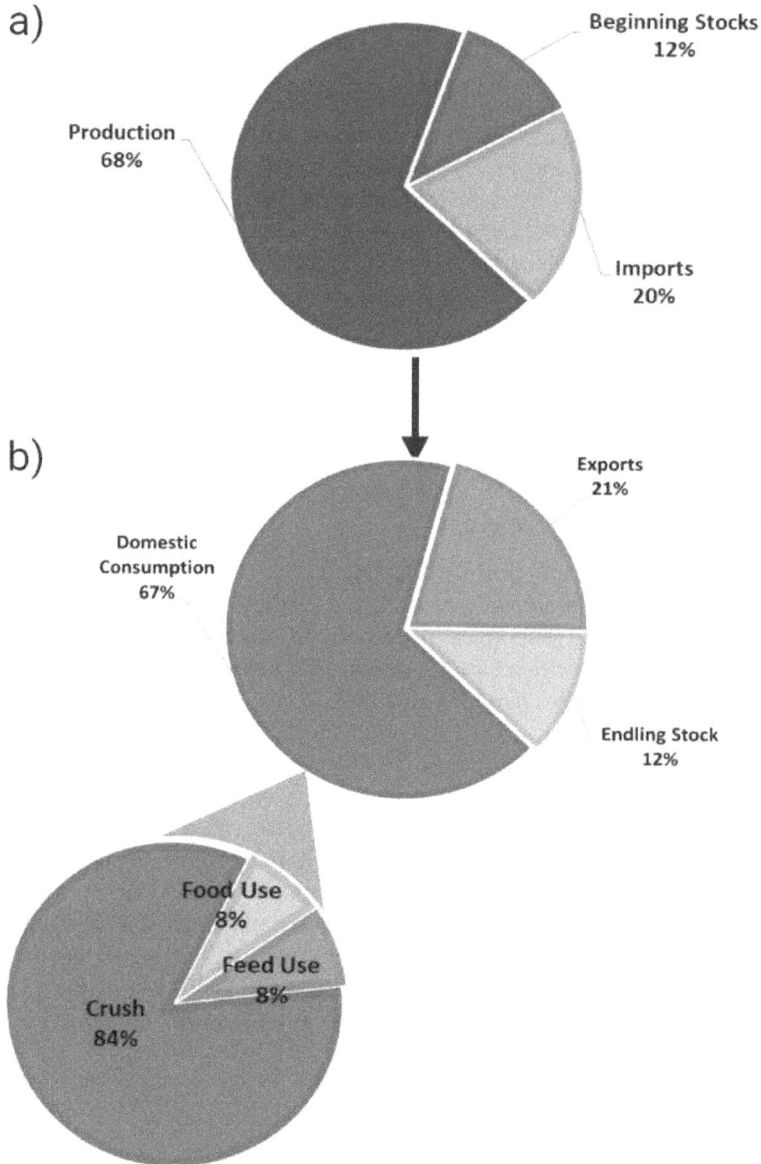

Figure 9. (a) Oilseeds obtainment percentages (b) Oilseeds evaluation percentages in 2020 season (USDA 2021).

Oilseed pulps, shells, leaves and seeds discarded in the food industry are of great interest worldwide since these streams contain high levels of phytochemicals. For this reason, considering the principle of zero waste, recovery studies have been carried out with the aim of increasing the energy savings and finding potential usage areas. In addition, it has become a trend to improve the quality of seed oils' waste and by-products, which are rich in phenolic compounds, carotenoids, tocopherols, pigments and vitamins, and to use them as a raw material source in the production of new functional products (Şahin and Elhussein 2018).

4.1 Waste Fractions Formed during Processing

Soybean is the most produced oilseed in the world. Waste and by-product fractions formed during soybean processing are illustrated in Fig. 10. Soybean meal, hull and straw are the main by-products

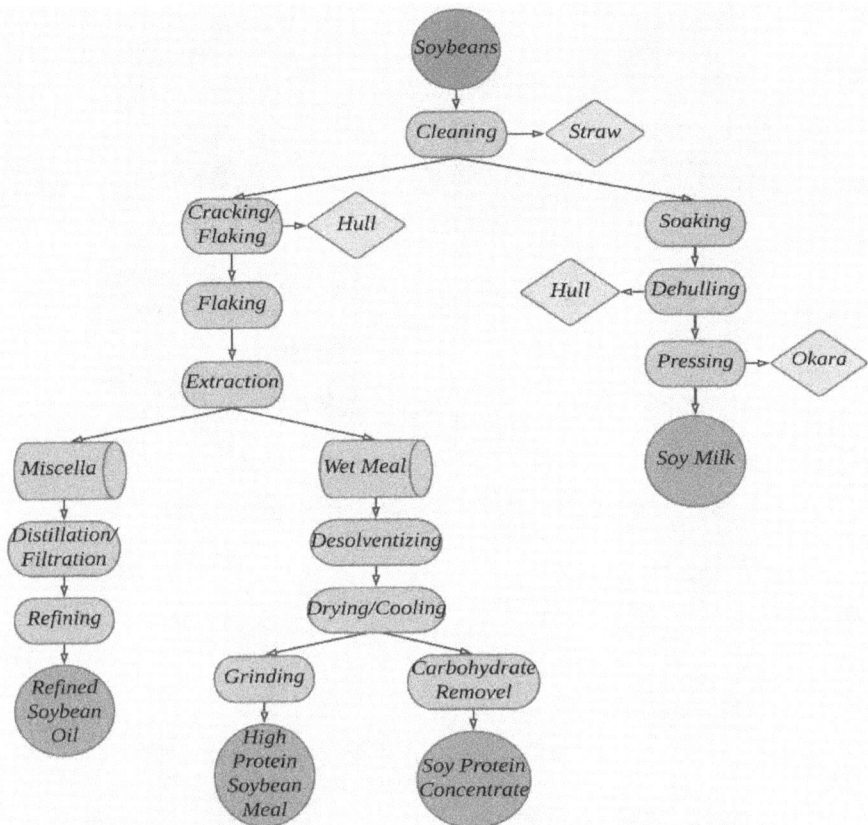

Figure 10. Flow chart of soybean products (refined soybean oil, soybean meal, flour, soy milk) (Dourado et al. 2011, Vishkarma et al. 2015).

of the soybean processing industry. Approximately 20 million tons of hull waste is generated annually which represents 8–10% of the whole grain (Cassales et al. 2011). The soy straw remaining after the soybean harvest is released as a by-product, approximately as much as the soybean equivalent produced each year. Although soybean straw is a cheap, abundant and renewable resource, its applications outside the animal feed industry are still limited (Reddy and Yang 2009). Soybean meal or cake obtained after oil extraction is used in feed production for livestock and fish (Dalgaard et al. 2008). Soy protein concentrate, extracted and toasted soybean meal, full fat soybean meal or low-oligosaccharide soybean meal are some products obtained from soybean processing by-products (Chou et al. 2004). While protein and oil components are considered valuable during soybean processing, other waste materials such as soy husks and straws are of less interest as low-cost raw materials (Cassales et al. 2011). The by-product of soybean milk and tofu production is called okara. In the soy milk industry, 1.1 kg of okara is released during the processing of each kilogram of soybean milk. High moisture content of okara makes it susceptible to decay, which poses a potential problem to be dumped into the environment. Therefore, finding alternative techniques for the use of okara in foods will not only eliminate the environmental pollution problem, but also transform this by-product into new value-added products (Redondo-Cuenca et al. 2008, O'Toole 1999).

Rapeseed (canola) is the most produced oilseed after soybean. According to a USDA (2021) report, worldwide rapeseed production increased by 26% between 2008 and 2020, due to the increase in yield and the expansion of cultivation areas (Bouchet et al. 2016, USDA 2021). Waste and by-product fractions formed during rapeseed oil processing are illustrated in Fig. 11. Rapeseed meal is produced as a by-product during rapeseed oil extraction. Although it is frequently used as

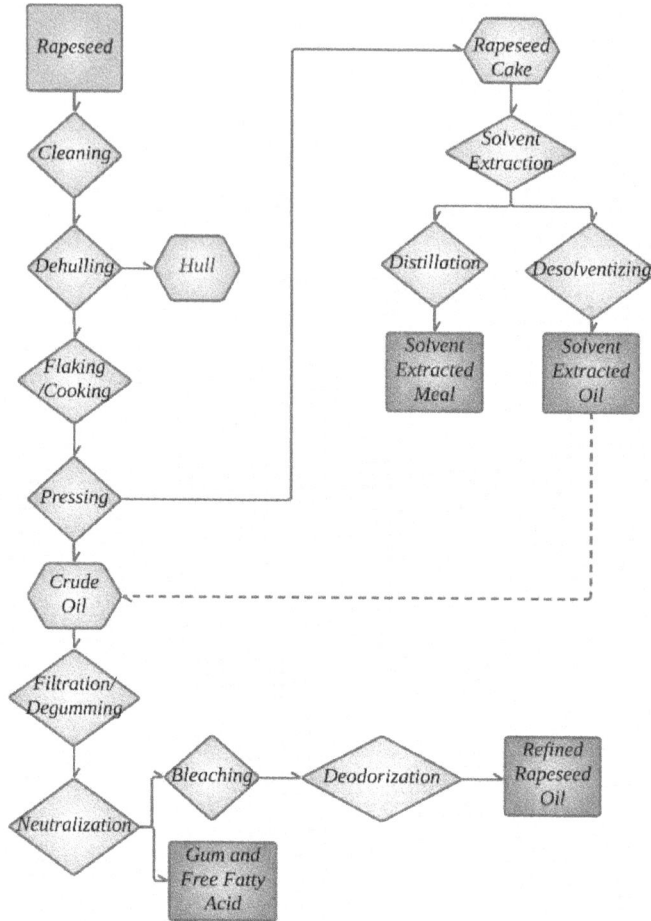

Figure 11. Flowchart of refined rapeseed oil production (Mosenthin et al. 2016, Gaber et al. 2018).

animal feed and organic fertilizer today, studies on its use in the food and biodiesel industry have been accelerated (Szydłowska-Czernia et al. 2010). It has been reported that 41 million tons of rapeseed meal was produced in the world only in 2020/2021 season (USDA 2021). In the rapeseed processing industry, the kernel and hull of whole seeds are separated to improve the quality of rapeseed meal, and to minimize the loss of solvent in the extraction meal (Matthäus 1998). In addition, hulls comprise 17–18% of the total rapeseed (Carré et al. 2016). The hulls are generally used in the animal feed industry after mixing with the extraction pulp, but the evaluation of these valuable by-products in other industries may provide further economic benefits (Lammerskötter et al. 2017).

Sunflower is one of the most consumed oilseeds in the world (Şahin and Elhussein 2018). During the production of refined oil from sunflower seeds, some by-products such as shell and meal are formed (Fig. 12). It has been reported that an average of 1 million tons of waste is generated annually during the processing of sunflower seeds (Mustafayev and Smychagin 2018), and approximately 50% of the sunflower seed weight is disposed of as a solid lignocellulosic waste (Casoni et al. 2019). Sunflower seed hull is a low-cost and abundant residue constituting 18–20% of the sunflower seed (Figlas et al. 2016). Sunflower seed hull is also rich in protein, carbohydrate, lipid, phenolic acids and anthocyanins (Hassannejad and Nouri 2018). Sunflower seed meal is also released as a by-product during oil extraction. In the 2020/2021 season, 19 million tons of sunflower seed oil was produced while 20 million tons of sunflower seed meal was released (USDA 2021).

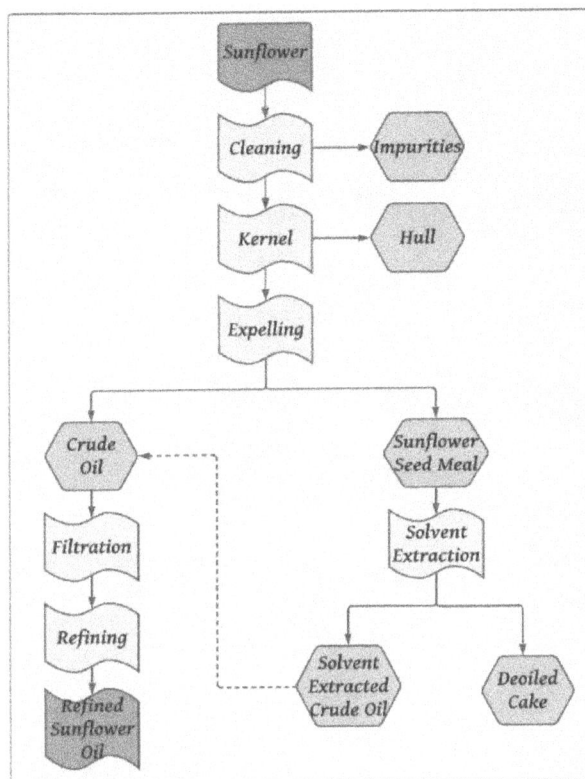

Figure 12. Flowchart of refined sunflower oil production (Le Clef and Kemper 2015).

Palm is another important oilseed for the oil processing industry. During palm oil production, many by-products such as kernel, nut, fiber, shell, empty fruit bunches and pressed cake are formed (Fig. 13). Palm oil and palm kernel oil are obtained from mesocarp and kernel parts, respectively. Approximately 20–22% crude palm oil and 5.5–6% crude palm kernel oil is produced from each fresh fruit bunch. In addition, 23–25% empty fruit bunches, 13–15% fiber and 6–6.5% palm shells are disposed of in the palm oil industry. It has been reported that 10 million tons of palm meal is released during oil production (USDA 2021). Other by-products e.g. fiber and husk can also be recycled for different purposes such as the energy source (Yan et al. 2019).

Although the nutrient content of oilseeds varies depending on oilseed genotype, soil type, climate, cultivation area, agricultural practices and processing conditions, they are quite rich in carbohydrate, protein, oil, fiber, vitamin, mineral and phytochemical contents (Table 6). The nutritional contents can be increased by using oilseed by-products as a raw material in the design of new value-added products with appropriate evaluation techniques. For instance, the disposal of oil cakes to the environment has started to decrease with the realization of their high phytochemical content since after oil extraction, some antioxidants such as phenolic acids, lignans or flavonoids remain in the cake. Although cakes from oilseed crops are traditionally used in the animal feed and fertilizer industry, the techniques for evaluating them as a source of processed ingredients, substrate and antioxidants have gained momentum in recent years (Ancuta and Sonia 2020).

Soybean cake is a by-product of soybean oil processing and is rich in isoflavones and other functional components such as saponins and phenolic compounds. Soybean cake contains four different isoflavone groups, i.e. aglycones, glucosides, malonylglucosides and acetylglucosides, and these isoflavones have been found to have different antioxidant activities. Kao and Chien (2008) reported that the malonylglucoside, glucoside, acetylglucoside and aglycone contents of soybean meal after solvent extraction were 2.41, 2.18, 0.26 and 0.16 mg/g, respectively. Isoflavones provide

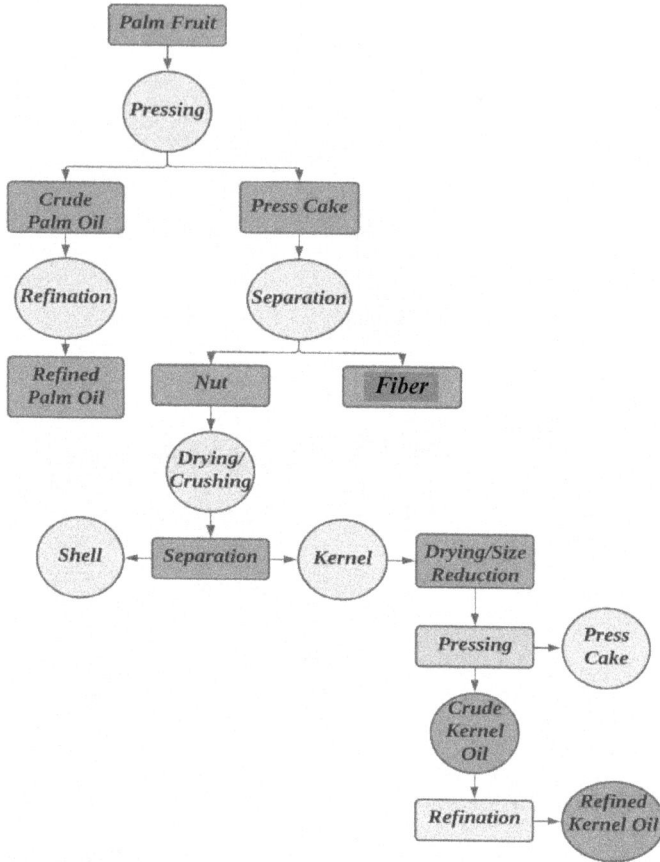

Figure 13. Flowchart of refined palm oil and palm kernel oil production (Obibuzor et al. 2012, Eckey 1954).

an advantage in the food industry thanks to their anti-inflammatory properties for the development of functional products with health benefits (Kao et al. 2007). Soybean cake has a protein content of approximately 50%, enough to make up for the protein deficiency of developing countries with high production quantity (Odebode et al. 2018). Soybean cake also contains polyunsaturated fatty acid (PUFA)/saturated fatty acid (SFA) ratio higher than 0.4 which is recommended by WHO/FAO for a balanced nutrition (Ivanov et al. 2010). Soybean meal is a by-product obtained from dehulled soybeans which is frequently used in diet formulations due to its high protein content and balanced amino acid composition. The growing area, soil type, climatic conditions and processing methods affect the nutrient content of soybean. For instance, the crude protein concentration of soybean meal grown in Brazil (49.3%) and India (49.5%) has been reported to be higher than those of in China (45.1%), Argentina (46.7%) or the USA (47.3%) (Lagos and Stein 2017). Soybean meal is the most widely used protein source in the animal feed processing industry, but despite its high production, studies have begun to find new alternative sources due to high demand and market scarcity. With the realization of the high protein content of rapeseed meal, it also has become a competitive raw material in animal feed production (Hao et al. 2020). In addition, the phenolic content of rapeseed meal is about 5 times higher than that of soybean meal (Szydłowska-Czerniak et al. 2010).

Rapeseed cake is a biomass waste released during cold pressing in rapeseed oil production. With the cold pressing method, the oil rate is reduced from 40–45% to 14–15%, but the oil remaining in the rapeseed cake can be recovered by solvent extraction (Smets et al. 2011). Cake obtained from rapeseed is considered as animal feed with its high protein and triglyceride contents. However, with the discovery of high erucic acid and glucosinolate contents, studies have begun on application

Table 6. Composition of oilseed waste and by-products.

Sample	Protein (%)	Fat (%)	Cellulose (%)	Hemicellulose (%)	Lignin (%)	Ash (%)	References
Soybean hull	14.9	3.01	28.6	20.0	13.1	0.20	Qing et al. (2017), Wang et al. (2009)
Soybean straw	6.59	2.07	34.1	16.1	21.6	5.16	Wan et al. (2011), Zen et al. (2020)
Soybean meal	53.5	1.40	5.78	5.88	0.90	6.40	Wang et al. (2006), Manzocchi et al. (2020), Alves et al. (2016)
Soybean cake	47.0	2.20	24.1	18.1	2.10	6.50	Maina et al. (2019)
Milled okara	34.0	3.70	64.3	7.87	1.14	0.80	Yang et al. (2021), Li et al. (2018)
Rapeseed hull	14.2	8.37	32.0	5.70	31.2	5.66	Carré et al. (2016), Naczk and Shahidi (1990), Mitaru et al. (1984)
Rapeseed meal	37.4	1.90	15.1	5.50	18.5	7.70	Kalaydzhiev et al. (2020), Pińkowska et al. (2013)
Rapeseed cake	30.0	12.2	9.00	8.00	4.00	4.90	Leming and Lember (2005), Smets et al. (2011)
Sunflower seed hull	4.00	5.17	39.0	18.0	20.0	2.10	Casoni et al. (2019), Cancalon (1971)
Sunflower seed cake	31.4	1.70	21.7	9.20	9.50	6.60	Ayrılmış et al. (2013)
Palm kernel shell			6.92	26.2	53.9	8.68	Okoroigwe et al. (2014)
Palm kernel cake	17.4	12.7	7.04	39.4	20.0	4.02	Chaipet et al. (2020)
Palm kernel fibre	14.8	9.00	52.0	14.0	31.0	6.30	Jabar (2016), Ho and Ofomaja (2006)

areas in other industries (Smets et al. 2013). Rapeseed meal is a valuable protein source with high bioactive contents such as tocopherols, B vitamins, choline, calcium, zinc, magnesium and copper (Szydłowska-Czerniak et al. 2010). The most abundant phenolic compounds in rapeseed meal are tannins and sinapic acids, accounting for 80% of the total phenolic content. Depending on the plant variety and the processing method of the oil, the total phenolic content of rapeseed meal may vary between 6.40 and 18.4 mg/g dry matter (Zago et al. 2015). The high levels of anti-nutrient compounds have limited the use of rapeseed meal, but its use as animal feed, fertilizer and a functional additive with balanced amino acid content has been extensively explored (Szydłowska-Czerniak et al. 2010).

Sunflower seed cake is a by-product of the sunflower oil industry. It is less used in the cattle feed due to its lower nutritional content than rapeseed cake and soybean cake. Sunflower seed cake contains approximately 1.5–2.5% oil, and this oil can be used as a lubricant to speed up processes in the extruder (Ayrışık et al. 2013). Sunflower seed cake proteins have been reported to exhibit antioxidant, anti-tumor, antibacterial and anti-hypertension properties. Considering their nutritional potential, researchers have started to work on the use of sunflower cake in health promoting functional foods (Chauhan 2021). Along with cottonseed meal and canola meal, sunflower seed meal has been also used as an alternative to soybean meal (Liu et al. 2015).

Palm kernel cake (PKC) is a by-product of oil extraction from palm fruit. Palm kernel cake is rich in saturated fatty acids, especially lauric acid, followed by myristic acid and palmitic acid (Ribeiro et al. 2018). In addition, PKC has a high fiber content (16–18%) and a high calcium/phosphorus ratio. In terms of amino acid content, glutamic acid (3.15%) is the most dominant one, followed by arginine (2.18%) and leucine (1.11%) (Sue 2004). PKC is a very suitable raw material for bioethanol production with its high polysaccharide and fermentable hexose sugar contents (Cerveró et al. 2010). The high fiber content of palm kernel cake also makes it an attractive ingredient for the food industry. Palm kernel fiber (PKF) is an agricultural waste produced during the pressing phase of palm oil production. The main components of PKF are carbohydrates such as lignin and cellulose, fibers and residual fats. There are also lipids and carboxylic acids on the PKF surface after extraction. Due to the polarity of the functional groups, the sorption may occur on the fiber surface. Therefore, PKF can be used to remove lead ions from aqueous solutions (Ofomaja et al. 2005, Ho and Ofomaja 2006).

In addition to meal and cake, agricultural waste materials such as hull, straw and shell are formed during oil extraction from oilseeds. Oilseeds are mostly coated with husk and shell layers. These parts are separated from oilseeds before processing, as they restrict the extraction of oil. Generally, shells and husks are widely used in the feed and fuel industry, but some oilseed shells and husks have high nutritional contents, so there are many areas of use in different industries. It has been reported that oilseed hulls and skins possess important antioxidant activities with the effect of tannin and proanthocyanidin compounds (Schmidt and Pokorný 2005). It was also reported that sunflower seed shell has high antioxidant capacity due to caffeic and chlorogenic acids (De Leonardis et al. 2005). For a comparison of antioxidative potential of different oilseed by-products, Amarowicz et al. (2000) reported that the crude tannins extracted from canola hull showed significantly higher antioxidant activity than rapeseed hulls.

4.2 Sustainable Valorization Options

In the food industry, high amounts of biodegradable wastes with high biochemical oxygen demand and chemical oxygen demand are formed. Some legal obligations have been created by governments for the evaluation and treatment of this high amount of waste. Applications such as incineration, fermentation, composting and landfill are the main methods used in the evaluation of agricultural wastes (Otles et al. 2015). Incineration takes place by oxidizing the combustible materials, and is more suitable for food wastes with a relatively low water content such as grains and oilseeds. Although the incineration process seems to be an easy method to dispose of waste, it raises concerns such as the negative effects of its emissions on the environment and yielding high costs (Despoudi et al. 2021). For example, burning sunflower hulls without the use of appropriate filtration system releases a large amount of dust into the atmosphere and increases the amount of CO emissions (Demir et al. 2005). The burning of agricultural waste straw causes air pollution and creates a worrying situation on human health, while burning leaves the fields vulnerable to wind and water erosion (Burton 2007). Besides incineration, agricultural wastes are also disposed of by landfill processes such as aerobic, anaerobic and semi-aerobic. However, disposal of wastes by anaerobic biodegradation method leads to the formation of methane gas as more than 45% of the total landfill gases. Methane gases are greenhouse gases that are much more powerful than carbon dioxide in terms of causing global warming. In 2004–2005, greenhouse gas emissions from the waste sector totaled 1.4 million metric tons of CO_2 equivalent. In addition, the concentration of methane gas has doubled in the past 150 years (Omar and Rohani 2015). Considering all these factors, the disposal of agricultural wastes with inappropriate methods creates many environmental and economic disadvantages. For this reason, studies on the evaluation, processing and use of wastes in the development of new functional products have gained momentum.

The use of plant proteins as a substitute for animal proteins has become more important considering the increasing world population, ecological and economical factors. The incorporation of plant proteins to foods improves expected functional properties such as whipping capacity, viscosity, emulsification, water/oil holding capacities, and organoleptic factors such as color, odor and taste. Meals resulting from the oil extraction of soybean, rapeseed, sunflower seed and palm kernel have high protein content. Many studies have evaluated the use of protein obtained from oilseed by-products as a substitute for human nutrition and animal feed as promising protein sources.

Preventing the disposal of oilseed meals will both reduce the environmental damage, and enable to use a cheap and protein-rich raw material source for new studies (Moure et al. 2006). Among oilseeds, the use of rapeseed as a protein source has been limited due to its anti-nutritional compounds such as glucosinolates and phytates. Glucosinolates become harmful when rapeseed seeds are crushed and the enzymes are released. For this reason, studies have focused on reducing the effect of rapeseed glucosinolates. Many studies have subjected rapeseed proteins to processes such as hot-pressing, cold-pressing and solvent-extraction for human consumption. In one study, the protein recovery of cold-pressed rapeseed cake (45%) was higher than that of hot-pressed rapeseed meal (26%) and solvent-extracted rapeseed meal (5%). In addition, cold-pressed rapeseed cake proteins exhibited better emulsification properties (Östbring et al. 2020). Rapeseed meal has higher protein yield than soybean meal, and has a potential to be used as a nutritional protein source due to its well-balanced amino acid composition. Tan et al. (2011) obtained canola proteins by Osborne method based on water, 5% NaCl, 0.1 M NaOH and 70% ethanol treatments, and standard alkaline extraction method. The authors showed that the Osborne method provides higher protein yield and recovery rate of water-soluble proteins than the alkaline method, and suggested that the canola meal proteins obtained by the Osborne method have good properties to be included in the human diet (Tan et al. 2011). Srilatha and Krishnakumari (2003) showed that dehulled sunflower meal offers higher protein and fiber content compared to whole seed meal and partially dehulled seed meal. Ermiş and Karasu (2020) stated that it is possible to use sunflower seed cake protein powder industrially in the formulations of vegan, confectionery and meat products.

In recent years, with the emergence of a healthy diet trend, consumers prefer natural antioxidants instead of synthetic ones. In this case, researchers are directed to find a new source of natural antioxidants such as some oilseed wastes rich in antioxidant compounds (Uluata and Özdemir 2017). Antioxidant compounds in oilseed meals and cakes are mostly in the form of lignans, phenolic acids and flavonoids (Terpinc et al. 2012). Solvent based, high pressure, microwave and supercritical fluid methods have been used to extract antioxidants from oilseed waste fractions. For instance, canola seed cake extracted with methanol:acetone:water solution showed the highest total phenolic (2104.67 mg GAE/100 g) and flavonoid contents (37.79 mg LUE/100 g), followed by hemp and lax seed cakes (Teh et al. 2014). Antioxidant compounds obtained from the by-products of oilseeds can be evaluated in the formulations of beverages, energy bars and bakery products (Ancuta and Sonia 2020, Terpinc et al. 2012). Many studies reported the presence of different antioxidant compounds in oilseed by-products: chlorogenic acid in sunflower seed cake (Zardo et al. 2019); ferulic acid, *p*-coumaric acid and protocatechuic acid in sesame bran (Görgüç et al. 2020); sinapic acid and flavonoid glycosides derivatives in canola seed cake (Zardo et al. 2020); isoflavone and gallic acid in soybean cake (Kao and Chen 2006). The oilseed meals rich in phenolic compounds also exhibit strong antioxidant activity. Some oil cakes show higher antioxidant activity than synthetic antioxidant butylated hydroxy toluene, which provides many benefits in terms of consumer health and economic recovery (Terpinc et al. 2012).

Various enzymes can be produced for industrial applications using solid state fermentation (SSF) from oilseed cakes and bagasses. These oilseed by-products are suitable raw materials that can be used in enzyme applications because they are rich in carbon and nitrogen content, and are inexpensive components (Ancuta and Sonia 2020). Many authors have worked on obtaining enzymes such as lipase (Parihar 2012), protease (Thakur et al. 2015), cellulases (Gupta et al. 2018), phytase (Singh and Satyanarayana 2006) and *L*-asparaginase (Satya et al. 2014) from oilseed cake substrates. Lipases are hydrolytic enzymes that catalyze esterification, interesterification and transesterification reactions in anhydrous environments. Lipases are widely used in food applications such as dairy and bakery products, as well as in the detergent, cosmetics, pharmaceutical and leather industries. Obtaining enzymes from agricultural wastes by applying SSF in a solvent-free system provides higher end product concentration and product stability, while avoiding solvent costs and providing economic improvement (Parihar 2012, Oliveira et al. 2017). Parihar (2012) obtained lipase enzyme from flaxseed and mustard oil cakes by SSF by bacterial strain of *Pseudomonas aeruginosa*.

Oliveira et al. (2017) found that the combination of palm kernel oil cake and sesame oil cake led to higher lipase production rate compared to their single use as substrate. The lipase enzyme has been also successfully produced from canola seed oil cake (Amin and Bhatti 2014), coconut kernel cake (Venkatesagowda et al. 2015) and peanut oil cake (Annamalai et al. 2011) under SSF by *Penicillium fellutanum, Lasiodiplodia theobromae,* and *Bacillus licheniformis* strains, respectively. Proteases are enzymes that play an important role in the hydrolysis of protein groups, and have wide application areas in many industries such as food, leather and detergent. Although proteases are derived from plant, animal and microbial sources, obtaining proteases from oilseed by-products has been promising with the increasing interest in natural components (Thakur et al. 2015). Rajmalwar and Dabholkar (2009) produced protease from soybean, groundnut, sesame, linseed, mustard and cotton oil cakes by *Aspergillus* strains using SSF method, and reported that soybean oilseed cake resulted in maximum enzyme activity, followed by sesame oil seed cake after 72 h of incubation.

Oilseed waste and by-products can also be used as an energy source. For example, biodiesel is produced from rapeseed meal and rapeseed straw, while the rapeseed meal provides 35% efficiency in coal-fired power plant, 30% electricity efficiency in combined heat and power plant, and 60% thermal efficiency. Rapeseed meal reduced the total energy requirement of a small-scale biodiesel production by 180% and the total energy requirement of heat and electricity production by 216% (Stephenson et al. 2008). The energy value of straw obtained from 1 hectare of rapeseed has an energy value equivalent to 65.7–81.2 gigajoules (Jankowski and Budzyński 2003).

Different literature studies evaluated oilseed waste products in the formulation and development of bread, biscuits, snacks, desserts and dairy products (Ancuta and Sonia 2020). For example, white bread enriched with walnut oil cake showed higher bread yield, phenolic acid content and antioxidant capacity (Pycia et al. 2020). Koneva et al. (2018) analyzed the rheological properties of the dough using different proportions of flour obtained from flaxseed cake (7.5%, 10% and, 12.5%), and reported that as the level of flaxseed flour was increased, the water absorption capacity was increased, and the viscosity, amylolytic activity and retrogradation properties were decreased. Authors stated that it is possible to enrich the bread and preserve the flavor parameters under industrial production conditions. Lu et al. (2013) replaced wheat flour with different levels (10%, 15%, 25%) of okara (a by-product of soy milk or tofu) for noodle, steamed bread, and bread making. They stated that the glycemic index of the obtained products was lower than the control group, and observed that hypoglycemic effect was due to the dietary fibers of okara. Biscuits are one of the most popular bakery products with their long shelf life, ready-to-eat properties and affordable costs. Biscuits enriched with oilseed wastes rich in dietary fiber can be used as functional foods. For example, replacing wheat flour with peanut and soybean oil cakes in biscuits offered better nutritional content in terms of protein, dietary fiber and polyphenols (Behera et al. 2013). Similar results were obtained when defatted sunflower, soybean and flaxseed flours were used in cookie formulation (Bhise et al. 2014). In addition, the authors reported that cookies prepared with combinations of sorghum-soy flour and wheat-soy flour were acceptable by school children. Consuming 28 grams of these two-combination biscuits provides 50% of the protein needs for children aged 3 to 10 years. For this reason, functional products such as the cookies prepared with defatted soy flour can be used as a protein source to eliminate the protein deficiency of children (Serrem et al. 2011).

In addition to bakery products, Kumar et al. (2011) added 4% soybean hull flour in chicken nuggets formulation, and reported that enriched chicken nuggets offered higher protein, ash and crude fiber contents. In another study, different concentrations of flaxseed oil cakes were used to make a new kefir-like beverage, and lactic acid bacteria and yeasts were found to grow well in flaxseed oil cakes without any requirement of additional substrate. In line with the results obtained, it is seen that kefir-like beverages enriched with oilseed cakes are a good source of antioxidants, and ideal for the consumption for vegans and lactose intolerant consumers (Łopusiewicz et al. 2019). In addition, okara drink co-cultured with *Lactobacillus paracasei* and *Lindnera saturnus* contains high amounts of free amino acids, isoflavone aglycones and probiotics, indicating that co-cultured okara can be used as a new functional probiotic beverage (Vong and Liu 2019). As a result, it is quite important to research new biotechnological process alternatives and usage areas of oilseed waste and by-products in order to obtain high value-added products in environmental aspect.

5. Meat and Fish Waste and By-products

5.1 Meat Waste and By-products

Meat is a type of animal flesh that contains high biological value protein, essential amino acids, micronutrients, and vitamins that are required for a healthy lifestyle (Saeed et al. 2020). The term "meat" does not refer to a single or even a few animal species, but rather to a wide range of species ranging from poultry to pigs, cattle, sheep, goats, wild game, and thousands of fish species. The meat processing industry generates a large amount of waste. Therefore, economic and social benefits of recycling were recognized by meat, poultry, and fish processors over a century ago and, many organizations and foundations worked together to develop novel technologies for utilizing co-products (Castro-Muñoz and Ruby-Figueroa 2019). In terms of food waste composition, fruit and vegetables cover 85% of the total waste produced; on the other hand, as far as environmental impact, they only contribute 46% of the total carbon footprint. However, the meat fractions show the opposite trend- meat accounts for only 3.5% of food waste, despite generating 29% of the carbon footprint (Mosna et al. 2021). As a result, when compared to all other food waste streams, meat waste has the greatest overall environmental impact in terms of greenhouse gas emissions. Furthermore, meat waste causes significant economic losses for the meat industry, affecting both global hunger and food security (Long and Mohan 2021). In modern technology, the production of meat waste has a significant environmental impact, resulting in increased levels of air, water and soil pollution. This leads to undesired outcomes for humans and other living organisms by fostering the growth of harmful microorganisms, which contributes to the spread of diseases (Thanikachalam and Ganesh 2020).

The meat industry waste consists primarily of waste streams from slaughterhouses, fishing, and poultry, the main ingredient of which is organic matter. The main characteristics of these wastes are that they are easily decomposed (due to high microbial activitiy), have high water content (70–95%) (which will increase the cost of transport), are easily oxidized (due to high fat content), and undergo changes due to enzyme activity (most of the enzymes are still active) (Shen et al. 2019). Recovering animal waste and by-products can help to solve a number of problems. On a larger scale, the valorization of animal by-products may result in a reduction in carbon footprint because non-recovered by-products must be incinerated. Economically, the valorization of animal by-products would enable food companies to benefit, and improve their circular economy (Bechaux et al. 2019).

It is possible to classify animal by-products as agricultural by-products (meat meal, bone meal, fertilizer, etc.), industrial by-products (gelatin, adhesive, casings, etc.) and pharmaceutical by-products (insulin, pepsin, biochemicals, hormones, etc.). Proper use of these by-products in the meat industry is needed to control pollution, provide better returns, provide highly nutritious animal feed, establish secondary rural industries, create jobs, and provide crop improvement (Irshad and Sharma 2015). Animal by-products and wastes can be characterized as edible or inedible, depending on whether or not they are used as food. Customs, religion, consumer acceptance, reputation, product availability, economics, and hygiene all play a role in determining edible food—what is edible in one region may be considered inedible in another. For instance, the use of blood for edible purposes is prohibited in Islam, despite the fact that it is widely used in the preparation of sausages in many western countries (Irshad and Sharma 2015). The European Commission issued regulations (EC) 1069/2009 and EC 142/2011, outlining health rules concerning animal by-products and their derivatives that are not intended for human consumption (Holloway and Wu 2019). Additionally, The European law EC no. 1774/2002 classified these by-products into three categories. Products in the first class are responsive to the transmission of contagious diseases to humans and/or animals while by-products in the second class are prohibited even for use in animal feeds. These products are derived from dead animals found outside slaughterhouses or from animals with drug residues.

The last class consists of by-products that have been approved for use in animal feed. In other words, they are all products derived from healthy animals slaughtered in slaughterhouses and certified fit for human consumption (Bechaux et al. 2019). The yield of edible by-products from animals varies greatly depending on species, sex, live weight, fatness, and collection methods (Castro-Muñoz and Ruby-Figueroa 2019). Different processes in the red meat industry, such as slaughtering, cutting, and processing generate different by-products (fat, leather, red and white offal, blood, bone, meat trimmings, hide, fatty tissues, horns, hoofs, feet, skull, etc.) that can be valorized (Holloway and Wu 2019). Biologically, the majority of non-carcass material is edible if the product is properly cleaned, handled, and processed. Red viscera, also known as "variety meat", include the liver, heart, kidney, tongue, and other products that are frequently used as edible co-products. Blood and trimmings would also be included in white offal (intestines and stomach).

Hides, skins, pelts, hair, feathers, hoofs, horns, feet, heads, bones, toenails, blood, organs, glands, intestines, fatty tissues, and shells, on the other hand, are inedible by-products. It has been indicated that tissues other than carcass meat and intestinal contents account for approximately 50% of the live weight of cattle, 42% of pigs, 28% broilers, and 24% of turkey (Castro-Muñoz and Ruby-Figueroa 2019). Meat and meat processing products generally contain high biological value protein and essential amino acids. Meat by-products, in addition to proteins, may be a good source of minerals (Ca, Fe, Mg, Cu, or Se) or vitamins (niacin, vitamin B_{12}, folate or vitamin C). Also, heme iron, which is primarily found in meat and their by-products, is more absorbable than non-heme iron (Mullen et al. 2017). It is possible to summarize the value-added and bioactive compounds obtained from the meat industry as fertilizer, animal feed, blood meal, meat and bone meal, feather meal, lactic acid, and probiotics (Sharma et al. 2020). Nevertheless, protein induced waste has gained popularity in the food processing industry for the production of bioactive peptides from animal by-products. Bioactive peptides are made up of amino acids that range in size from 2 to 30 units, and have specific sequences that show potential as functional food components (Saeed et al. 2020). Blood forms a major part of the animal weight (2.4–8.0%). Hemoglobin is the most abundant compound in animal blood, which is a valuable by-product. It can be hydrolyzed to produce bioactive peptides, and its hydrolysates possess antibacterial, antioxidant, angiotensin-converting enzyme (ACE) inhibitory (antihypertensive), and opioid activities, with antimicrobial peptides receiving the biggest interest (Shen et al. 2019). Saeed et al. (2020) investigated the anti-hypertension and antioxidant activity of meat by-product hydrolysates derived by enzymatic hydrolysis. According to the authors, liver and kidney hydrolysates had higher hydrolysis rates. In addition, the ACE inhibitory activity of liver hydrolysate was increased significantly after hydrolysis, while the kidney hydrolysates showed higher ABTS radicals scavenging activity than liver and heart. Bioactive peptide production is critical in the development of future functional foods as a potential solution for processors to address economic and environmental issues (Saeed et al. 2020). Bone in the animal body generally accounts for 20–30% of total body weight, and animal bones are rich in a variety of minerals, such as zinc, phosphorus, calcium and iron (Shen et al. 2019). Animal wastes from the meat industry also contain significant amounts of insoluble structural proteins such as anti-aging collagen, elastin, and keratin, which are major constituents of bones, organs, and hard tissues. These by-products are mostly high in protein content, and can be utilized for the discovery of novel functional ingredients (Brandelli et al. 2015). Furthermore, they contain chondroitin, phospholipids and phosphoproteins several times more than fresh meat, which is essential to the human brain (Shen et al. 2019). As a specific example of the utilization of meat by-products, Long and Mohan (2021) produced spray dried beef tongue powder (retained both protein and fat content) as a value-added and multipurpose food ingredient. Protein hydrolysates are also used in aquaculture as nutrients. The palatability of meat by-product protein hydrolysates in pet food can be increased by cleaving hydrophobic amino acids from peptides which cause bitterness. Also, keratin-containing by-products, such as hair, hoof, feather, and the outer layer of skin, can be used in pet food formulations after enzyme hydrolysis with keratinase (Holloway and Wu 2019). Another by-product obtained from animal bones is bone meal. Micro-bone meal is made up of bone granules with an average diameter of 100–500 μm which is obtained after crushing animal bone by micro-crushing technology. Because bone meal has a high surface area, it shows good adsorption and fluidity after crushing, making the nutrients in animal bone more easily absorbed. Therefore, bone meal can be an excellent source of valuable minerals, particularly calcium. For instance, ossein is a type of natural flavor compound obtained by the extraction of water-soluble substances from fresh bones, and is a promising dietary supplement affluent in amino acids, peptides, nucleic acids, sugars, and inorganic salts (Shen et al. 2019).

The extracellular matrix, which makes up the majority of the volume of animal tissue, is comprised of a variety of proteins and polysaccharides that help to maintain and protect the cells and tissues (Brandelli et al. 2015). Among them, collagen is the most abundant protein (30% of whole-body protein content) in many meat industry by-products. It is, in fact, the primary component of skin, hide, bones, and cartilages. Collagen has a low nutritional value because it lacks essential amino acids, but it is an excellent source of bioactive peptides. For instance, protein hydrolysates derived from collagen can be used to treat osteoarthritis by accumulating in joint cartilage (Toldrá et al. 2016). Purified collagen can be used in regenerative medicine and cosmetics, such as collagen injections for improving appearance, body lotions, and mascaras. Collagen is also used in casings, supplements, films, pharmaceuticals, as a precursor to biodegradable materials, for tissue engineering, and in 3D printing. The inner corium layer of bovine hide is rich in collagen and contains approximately 30% protein. Therefore, converting this waste into a high-value final product, such as collagen, will benefit both the environment and the leather processors (Noorzai et al. 2020). Collagen extracted from hide and skin may be also used as an emulsifier in meat products due to oil binding properties (Shen et al. 2019). Collagen-rich by-products can be heat-denatured and extracted, resulting in the formation of gelatin. Because of its gel-forming ability, gelatin is widely used in the food industry, but it is also used as a clarifying agent, stabilizer, protective coating material gelling (Toldrá et al. 2016), and ingredient for cooking (Zarubin et al. 2021). Gelatin derived from animal skins can be used to improve elasticity, consistency, and stability in foods, and it is widely added to desserts to form a jelly structure. Gelatin is also frequently used as a stabilizer in ice cream and other sweets. Because gelatin inhibits the formation of ice crystals and the recrystallization of lactose during storage, it is used as a protective colloid in ice cream, yogurt, and cream patties (Shen et al. 2019). In the chemical industry, collagen and gelatin are used as ingredients in surfactants, paints, varnishes, adhesives, antifreeze, cleaners, and polishes (Toldrá et al. 2016).

Rendering of perishable animal wastes provides a possible alternative method of eradicating the environmental problem while also generating revenue. Rendering industries produce meat and bone meal, hydrolyzed feather meal, blood meal, fish meal, and animal fats (Sharma et al. 2020). The yield of edible by-products including blood and organs in cattle averages 12%, in sheep 14%, and, if pork rinds are included, 14% for hogs. Also, most of the non-carcass material is biologically edible if cleaned, handled, and processed properly (Irshad and Sharma 2015). Blood can be used as an emulsifier, stabilizer, clarifier, coloring additive and nutritional component in foods, as well as a color generator by extracting hemoglobin. For instance, animal blood is used to make blood sausages, blood clots, blood pudding, biscuits and bread. However, the use of blood in the meat processing industry may cause an undesirable appearance due to the dark color of hemoglobin (Shen et al. 2019), but can be used as a color enhancer for sausages. Enzymatic hydrolysis of hemoglobin can produce a large amount of a heme iron polypeptide that improves iron absorption (Holloway and Wu 2019). Nevertheless, plasma can be an alternative to hemoglobin due to its functional and colorless nature (Toldrá et al. 2016, Shen et al. 2019).

Some animal fat by-products are used as raw materials to obtain a mixture of fatty acid methyl esters and glycerol. However, animal fat production mostly requires degumming process to remove protein and phosphoacylglycerols. Triacylglycerols (triglycerides) can also be converted into a mixture of iso-, and normal paraffin via heterogeneous catalytic hydrogenation to produce biogas (Toldra et al. 2016). Biomass-derived bio-fuel is biodegradable, non-toxic, sustainable, and can be used as an alternative to fossil fuels with low greenhouse gas emissions such as CO_2, CO, SO_2, HC, and NO_2 (Hossain et al. 2015). Biodiesel consists of mono-alkyl esters of long chain fatty acids derived from oil or fats, but the use of vegetable oil may raise the cost of biodiesel, prompting the use of animal fats as an interesting alternative. For instance, biodiesel produced from animal fat has been reported to reduce the fossil CO_2 emissions by 80%, a better rate compared to that of soybean oil (30%) (Toldrá-Reig et al. 2020).

Thermochemical processing of meat and bone meal, such as pyrolysis, coal combustion, or gasification, is used to produce high phosphate ashes useful as fertilizers, and the gas emissions from these processes are used to generate energy. Animal by-product incineration produces excellent mineral fertilizers while also providing an efficient source of heat (Holloway and Wu 2019). Conventional inorganic fertilizers contribute to increased atmospheric methane emission; however, organic fertilizers are considered as a good alternative to improve crop yield along with reduction in methane emission (Sharma et al. 2020).

The continuous growth of poultry industry results in increased amounts of solid waste (feathers, viscera, bones, etc.) generated by both production and processing facilities. These materials are currently rendered into meat and bone meal, feather meal, blood meal, and fat/oils (Brandelli et al. 2015). In contrast to the meat components, feathers, hoof, horn, and hair are processed separately to create the meal (Hicks and Verbeek 2016). Heads, gizzards, and blood have traditionally been used in meal production, and besides feet, skin, intestines, and glands can also be a source of poultry fat (Brandelli et al. 2015). The abdominal and gizzard fat that remain inside the poultry carcass, on the other hand, are a good source of fat due to their high content of monounsaturated fatty acids and vitamin A (Peña-Saldarriaga et al. 2020). Feather is a by-product of poultry processing and account for approximately 8% of live weight containing approximately 90% protein. Feather meal has a high energy content due to its lipid content, which ranges from 1.8% to 12%, and can be converted to biodiesel (Seidavi et al. 2019). Furthermore, feather hydrolysates could be converted into methane gas and fuel pellets for heating, as well as to produce biohydrogen (Brandelli et al. 2015). Keratin, the main protein found in feathers, is nearly insoluble. To improve digestibility, feathers are rendered in a batch or continuous hydrolyzer, or inoculated with a feather-degrading bacterium (*Bacillus licheniformis*), which excretes a powerful keratinase enzyme capable of hydrolyzing collagen, elastin, and feather keratin (Hicks and Verbeek 2016). Poultry hides can be used to produce elastin, which is commonly used in the manufacture of functional foods due to its antioxidant properties (Shen et al. 2019). Dry pet foods typically contain high amounts of protein and fat, and are low in ash content. Among animal proteins, poultry proteins (e.g., mechanically separated chicken) are widely used in both dry and wet pet foods (Donadelli et al. 2018). Traditionally, the protein sources have been provided from animal by-products in the meat industry during rendering process. Additionally, thermal rendering has been shown to be effective in destroying protozoa, bacteria, and viruses, including RNA viruses like avian influenza (Seidavi et al. 2019). Keratin-rich materials, such as nails and beaks, can be degraded by keratinolytic microorganisms, and are frequently used in the production of animal feed in conjunction with viscera or blood (Brandelli et al. 2015).

5.2 Fish Waste and By-products

Seafood losses are related to fishing, post-catch, and to the processing, distribution and consumption of fish. The processing stage, in particular, accounts for 5% of fish losses due to the generation of by-products that are edible for human consumption (Laso et al. 2016). A typical fish-processing operation includes stunning, grading, slime removal, de-heading, washing, scaling, gutting, fin cutting, meat bone separation, and steak and fillet preparation (Hicks and Verbeek 2016). The fish-processing industry produces by-products approximately 57% (*w/w*) of the total wild catch, which is not directly utilized for consumption (Lapeña et al. 2018). Also, various by-products are produced during the fish production process, accounting for more than 30% of the fish weight and, in some cases, up to 70% (depending on fish type or processing). Among these by-products, muscle pieces, viscera, thorns, heads, skin, fins, and scales are the most highlighted ones among others (Martí-Quijal et al. 2020). The disposal or non-valuation of these by-products will pollute the environment, and may even endanger the ecological health. Water soluble minerals, peptides, collagen, gelatin, enzymes, oligosaccharides, and fatty acids are all present in large quantities in the aquatic by-products (Shen et al. 2019). Therefore, all such resources can be used to valorize and produce a variety of high-valued goods such as bioplastics, lactic acid, or preservative compounds, which are very useful in food, pharmaceutical, nutraceuticals and cosmetic industries (Dave and Routray 2018, Martí-Quijal et al. 2020). For instance, in a study, antioxidant compounds were produced through the fermentation of fish by-products by bacteria isolated from sea bass viscera (Martí-Quijal et al. 2020). As a result, authors concluded that the lactic acid bacteria isolated from sea bass had an important proteolytic capacity, and were able to synthesize phenolic acids with strong antioxidant capacity.

There is an increasing worldwide awareness and interest in the valuing of available resources, particularly the recycling of nutrients and by-products, to support a circular economy in aquaculture. Therefore, The European Commission promoted "Blue Growth" in 2012, a green economy term focused on maritime and coastal sectors (EC 2012) (Campanati et al. 2021). To recover the bioactive components from aquatic wastes, three major techniques, namely chemical/physical, enzymatic and microbial methods, are used. The enzymatic method is regarded as one of the best techniques for all types of fish wastes among the three (Ashraf et al. 2020). The crude protein content of fish by-products varies from 8% to 35%. Several methods for recovering protein and peptides from fish by-products have been developed, including acid or alkaline hydrolysis, autolysis, and enzymatic hydrolysis (Zamora-Sillero et al. 2018). As the valuable components, fish collagen and fish gelatin (partially hydrolyzed collagen) can be extracted from heads (9.2–33.0%), fins (0.8–8.0%), skin (2.0–12.6%), scales (0.8–6.0%), bones (9.0–19.0%), and dead fish bodies (Zarubin et al. 2021). In a study, collagen was extracted from the skin of two species of teleosts and two species of chondrychtyes using pepsin, with yields ranging from 14.2% to 61.2% (Blanco et al. 2017). The main components of fish scales are hydroxyapatite (30%), and protein (70%) consisting primarily of collagen and keratin (Shen et al. 2019). While gelatin is used as a cooking gelling agent, fish collagen can be used for a variety of purposes, including tissue engineering, drug delivery, cell differentiation in cell culture, skin care, and as an amino acid source. However, due to the intense fish odor, fish gelatin is mostly suitable for cooking fish meals (Zarubin et al. 2021). Fish proteins have advantageous properties for bioplastic preparation, such as the network formation, plasticity, elasticity, and gas barrier properties. Araújo et al. (2018a) aimed to produce and characterize bioplastics prepared from myofibrillar proteins found in gilded catfish (*Brachyplatystoma rousseauxii*) waste. Consequently, the authors produced a homogeneous, transparent, strong and flexible bioplastic having low solubility and water vapor permeability.

Fish viscera is a potential source of bioactive materials from fish processing, and serving as a rich source of various digestive enzymes. Pepsin, trypsin, chymotrypsin, and elastase are the primary enzymes found in fish viscera, and proteases are the major group of naturally available enzymes in fish and aquatic invertebrates (Atef and Ojagh 2017). Fish offal is rich in proteases and also contains oil, which is abundant of eicosapentaenoic (EPA) and docosahexaenoic acids (DHA). Depending on the season and type of fish, fish oil contains 15–30% of these unsaturated fatty acids (Shen et al. 2019). EPA and DHA have biological effects on lipoproteins, blood pressure, cardiac function, colon cancer, metal health disorders, endothelial function, vascular reactivity, and cardiac electrophysiology, in addition to potent antiplatelet and anti-inflammatory effects. Moreover, fish oil is a good source of dietary vitamin D which is known to have a wide range of health effects (Atef and Ojagh 2017).

The bone fraction contains high amounts of minerals (such as calcium), phosphorus, and collagen proteins (Syazili et al. 2021). Abbey et al. (2017) produced fish powder from edible by-product fractions of fish processing plants. As a result, authors discovered that all fish by-products had high levels of iron (trimmings had 16.58 mg/100 g, while tuna frames and gills had 16.82 and 19.54 mg/100 g, respectively). Furthermore, zinc levels in tuna trimmings ranged from 0.41 mg/100 g to 1.88 mg/100 g in tuna gills (Abbey et al. 2017). Bones and offal from processed fish are wet or dry rendered to produce fish meal, a brown protein powder. Fish meal can be used in a variety of animal feeds, most notably pet food, where the fishy odor and flavors are beneficial. Moreover, fish meal has significantly higher calcium, phosphorus, and selenium content than other by-product meals, making them an ideal nitrogen and phosphorus-rich fertilizer for the selenium deficient soils (Hicks and Verbeek 2016).

6. Dairy and Egg Waste and By-products

Egg and dairy products have an important role in daily diet by providing the energy and nutrients needed by the human body. Especially, milk and dairy products are valuable nutritional sources of calcium and protein (Kart and Demircan 2014). While the types of dairy products vary from country to country depending on cultural and social factors, the amount of production and consumption varies according to consumer demand, population and nutritional habits. Dairy products can be categorized as cheese, butter, cream, ice-cream and yogurt. According to 2019 data, 883 million tons of milk was produced worldwide, 715 million tons of which was cow milk. European countries meet 26% of the total milk production with approximately 232 million tons (FAOSTAT 2019). According to 2018 data, approximately 35 million tons of dairy products were produced in the world, where cheese is the most produced dairy product with a total production rate of 11.7 million tons (FAOSTAT 2018). Egg is the most consumed protein source after meat. Apart from direct consumption, products obtained from eggs include whole egg powder, egg yolk powder, egg white powder, liquid egg white and liquid egg yolk. Egg is also widely used in the production of various ready-to-eat foods and pastry products in powder or pasteurized form (Ghosh et al. 2016). Additionally, whole egg powder has become popular among sportsmen due to its high protein content. In order to meet the increasing demands, egg production has also increased worldwide since 1990 (35 million tons), and reached 83 million tons in 2019. China is the leading country in egg production with an annual production rate of 28.8 million tons (FAOSTAT 2019).

The increase in dairy products and egg production directly affects the amount of waste and by-products generated during processing. Waste streams generated in the production of milk and dairy products vary depending on the product being processed: Buttermilk is a by-product of butter production while whey is discarded in cheese production (Yıldırım and Güzeler 2013). These by-products having valuable nutrients generated in the dairy industry can be utilized in value-added products or as raw material in the animal feed industries. In addition, more than 6 liters of waste water is generated for each 1 liter of milk processed (Ahmad et al. 2019). Although the eggshell is edible, it is mostly considered as waste since it is not generally consumed (Réhault-Godbert et al. 2019). Considering that a large egg weighs 60 g, 5.70 g is the eggshell, 37.8 grams is egg white and 16.5 grams is egg yolk (Yüceer and Caner 2019). Therefore, eggshell makes up about 9–12% of the egg weight, indicating that out of the 83 million tons of eggs produced in 2019, approximately 8.5 million tons of eggshell waste was produced (Waheed et al. 2020b). The solid wastes, which are directly released to the environment, may damage natural resources by causing environmental pollution. Besides, the wastewater generated in dairy production can pollute drinking waters due to improper waste management. However, studies and applications have revealed that dairy and egg by-products can be utilized in different applications instead of disposal practices (Abdulrahman et al. 2014). In order to protect the environment and ensure sustainability, food wastes should be disposed of properly or recovered for related industries. In particular, converting dairy and egg wastes, which constitute more than 26% of total food waste in the world, should become a priority for the food industry (Ganesh et al. 2021, Abdulrahman et al. 2014).

6.1 Waste Fractions Formed during Processing

According to FAO, approximately 1.3 billion tons of food waste is produced every year. These losses occur during post-harvest, processing, storage, distribution and also consumption. Milk and dairy losses occur during industrial treatments such as pasteurization and processing of milk into mainly cheese, butter and yogurt. The processing steps of the dairy production and formed by-product fractions are given in Fig. 14. Besides that, a high amount of wastewater is formed by contaminating the water with milk residues during production.

Dairy effluents contain dissolved lactose and proteins, fats, and food additives. Biochemical oxygen demand (BOD) and chemical oxygen demand (COD) are significant parameters that determine the chemical properties of effluents, and COD value is approximately 1.5 times the BOD value while BOD ranges from 0.8 to 2.5 kg/ton (Ahmad et al. 2019). Lactose, which forms 70–72% of the total solids, is the main component that causes an increase in BOD and COD values (Papademas and Kotsaki 2019). Especially, whey contains a large amount of lactose and 1 kg lactose is equal to 1.13 kg of COD (Ahmad et al. 2019). During cheese production, about 115 million tons of whey is produced each year, and 54 million tons of it is directly discharged into the drains (Zandona et al. 2021). There are two types of whey obtained by separating the coagulated casein using enzymes: sweet whey (minimum pH 6.3) and acid whey (pH 4.6) which contains 46.0–52.0 and 44.0–47.0 g/L lactose, respectively (Yıldırım and Güzeler 2013, Papademas and Kotsaki 2019).

The nutritional composition of selected dairy and dairy by-products is presented in Table 7. Buttermilk is the liquid phase formed after separation from the butter during churning stage. A 166 kg of buttermilk is formed for 100 kg of butter production. Buttermilk, which contains most of the water-soluble components and is similar in terms of skim milk composition, contains 5.1% lactose, 3.5% protein, 0.8% ash and 0.1% fat (Yıldırım and Güzeler 2013). Protein, lactose, minerals and fat is transferred into buttermilk as a result of the breakage of fat globule membrane by churning process. The milk fat globule membrane contains 4.43% phospholipid, 13.4% casein, 20.1% serum proteins and 21.6% membrane proteins in dry matter. Protein and milk phospholipids are the main

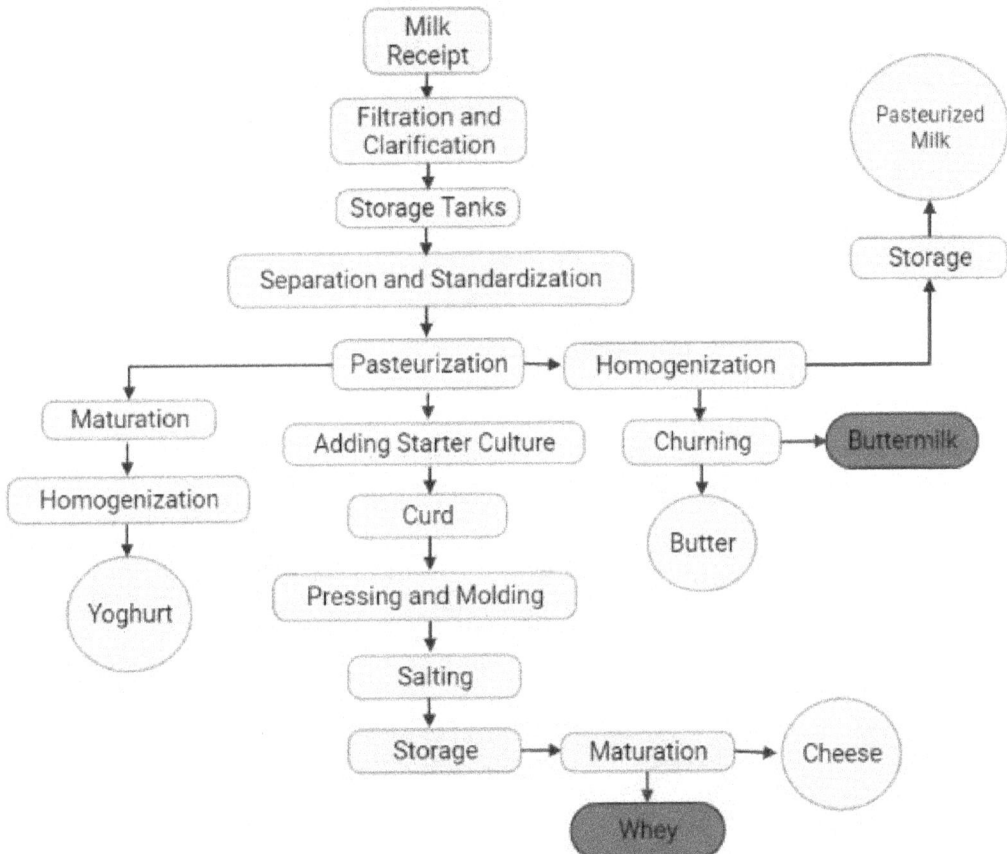

Figure 14. Dairy products flow chart.

Table 7. Nutritional composition of milk and dairy product/by-product.

Product/By-Product	Dry matter (%)	Fat (%)	Protein (%)	Lactose (%)	Ash (%)	References
Whole Milk	11.5–12.5	3.00–4.00	3.30	4.80	0.70	Kolev Slavov (2017), Wijesinha-Bettoni and Burlingame (2013)
Yogurt	12.1	3.30	3.50	4.70		Wijesinha-Bettoni and Burlingame (2013)
Cheese	41.0–47.0	17.0–21.0	17.5		5.80	Çelik and Uysal (2009), Demirci (1988)
Butter	84.1	81.1	0.90	0.10		Wijesinha-Bettoni and Burlingame (2013)
Buttermilk		0.10	3.50	5.10	0.80	Yıldırım and Güzeler (2013)
Whey	6.00–6.20	0.05–0.20	0.75–1.00	4.50–4.80		Kolev Slavov (2017)

sources of the bioactive compounds of buttermilk (Öğe and Yüceer 2021). Therefore, the disposal of buttermilk without any further treatment results in the loss of a valuable by-product containing high amounts of nutritive and functional compounds.

Egg is an important foodstuff in terms of high protein quality, digestibility, essential amino acid and essential fatty acid (especially linoleic and oleic acid) contents. It is also rich in fat-soluble and B complex vitamins. Egg and egg by-products are utilized for their techno-functional properties including gelling, foaming, binding, flavoring, swelling and emulsification. Liquid egg and egg powder production is becoming widespread due to their ease of use along with being microbiologically safe due to the applied pasteurization (Yüceer and Caner 2019). The main process steps of the egg products are given in Fig. 15.

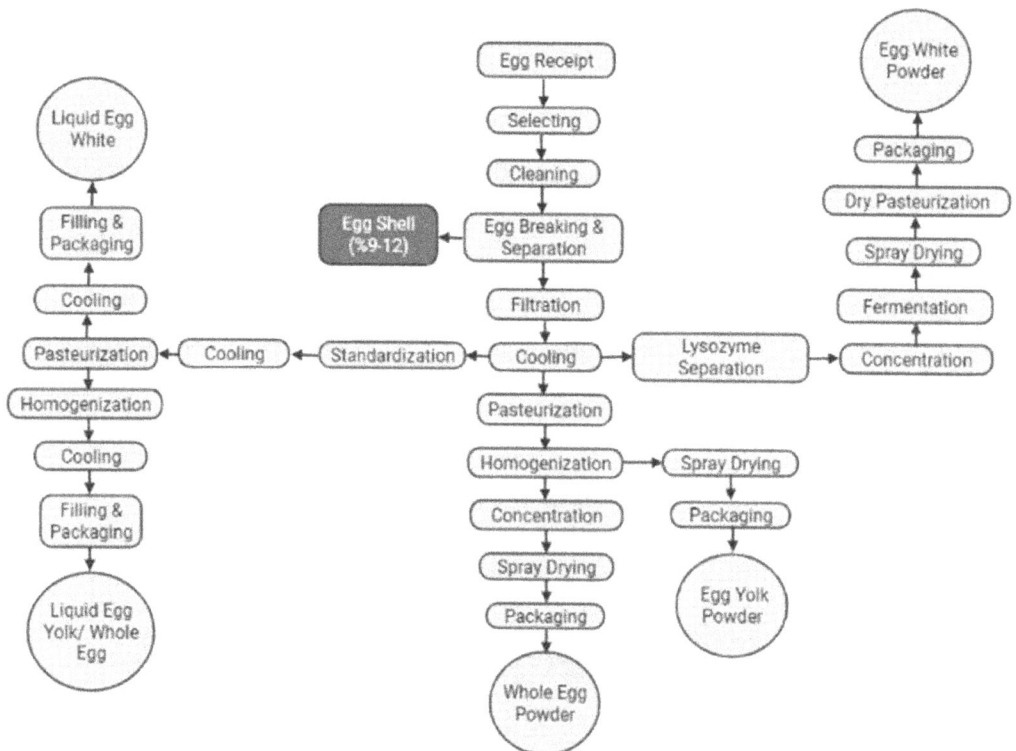

Figure 15. Egg products flow chart (Yüceer and Caner 2019).

Table 8. Nutritional composition of egg by-products (%).

By-product	Dry matter	Fat	Protein	Carbohydrate	Mineral	References
Egg yolk	45.0	26.7	15.5	1.00	1.60	Réhault-Godbert et al. (2019)
Egg white	12.3	0.10	10.8	0.80	0.40	
Egg shell	99.5	0.35	3.92	0.43	94.6	Ray et al. (2017)

Egg shells are the main waste generated during the production of egg products. Egg shells basically consist of a three-layer structure: Foamy cuticle outer layer, spongy middle layer and lamellar inner layer (Abdulrahman et al. 2014). However, 94–96% of an eggshell is calcium carbonate ($CaCO_3$), which represents 2.2–5.5 g of $CaCO_3$. The rest is composed of magnesium carbonate ($MgCO_3$), calcium phosphate [$Ca_3(PO_4)_2$] and protein fibers. It also contains trace amounts of sodium, potassium, zinc, iron and copper (Zaman et al. 2018). The nutritional composition of egg by-products was presented in Table 8.

6.2 Sustainable Valorization Options

The dairy industry releases toxic greenhouse gases such as carbon dioxide, methane and nitrous oxide into the atmosphere. Greenhouse gases cause ocean acidification, ozone depletion and air pollution as well as the global warming. In addition, these toxic gases affect the plant flora (Sinha et al. 2019). The greenhouse gas emission of cheese is 5.9 kg CO_2 eqv./kg, and it has been stated that the dairy sector contributes 2.7% of total greenhouse gas emissions (Milani et al. 2011).

Wastewater discharge, which has significant effects on the ecosystem, is the main source of pollution from industries. While the wastes generated in the dairy industry vary in concentration and composition depending on the type of product processed, operating methods and the type of water treatment applied affect the quality of air, water and soil (Adesra et al. 2021). Untreated dairy wastewater may affect the physical and chemical properties of the soil, thereby reducing the crop yield (Ahmad et al. 2019). The organic constituents, high content of nitrogen and phosphorous present in the dairy wastewater create intense foul odor by reducing the level of dissolved oxygen in the water (Sinha et al. 2019, Ahmad et al. 2019). The reduction of dissolved oxygen is one of the serious problems caused by wastewater discharge directly into the environment. Also, the oil and grease in the wastewater form a film on the water surface that prevents the transfer of oxygen. High concentrations of total soluble solids found in dairy wastes lead the fish gills to become clogged, and thus suffocating to the growth of some algae and bacteria that deplete the oxygen in the water. Wastewater may also become a suitable environment for flies that carry several diseases such as malaria (Ahmad et al. 2019). In addition, when wastewater which plays a significant role in the growth of microbes and the development of eutrophication is discharged into sewers, sewage fungus can develop due to the slime-producing bacterial colonies (Ahmad et al. 2019, Adesra et al. 2021). Since the temperature of dairy waste water is higher than that of municipal wastewater, biological degradation occurs faster and creates serious problems in the sewer system (Ahmad et al. 2019). Milk wastewater containing high organic matter concentration is cloudy and white in color (Ahmad et al. 2019). Carbohydrates, lactose, proteins and lipids contribute to the high nutrient levels of effluents from dairy processing plants. In addition to fats, 97.7% of the total COD in cheese wastewater consists of components such as lactose and protein (Goli et al. 2019). Lactose is the main carbohydrate in dairy waste (Goli et al. 2019), and is considered the most polluting component due to its high levels of BOD and COD (Ahmad et al. 2019). Nitrogen is also found in dairy by-products in the form of either organic compounds or ions (Goli et al. 2019). It has been stated that the discharge of wastewater containing phosphorus and nitrogen causes eutrophication of the water (Arvanitoyannis and Giakoundis 2006). High phosphorus level is stated as one of the most important quality indicators in fresh water (Osei et al. 2003).

Table 9. Composition of sweet and acidic whey (%) (Dinçoğlu and Ardıç 2012).

Composition	Sweet whey	Acidic whey
Dry matter	6.30–7.00	6.30–7.00
Lactose	4.60–5.20	4.40–5.20
Protein	0.60–1.00	0.60–0.80
Fat	0.30–0.35	0.20–0.25
Calcium	0.04–0.06	0.12–0.16
Phosphates	0.10–0.30	0.20–0.45
Lactate	0.20	0.64
Chlorides	0.11	0.11

Whey is defined as a by-product of cheese production that can be produced as sweet or acidic, as a result of the coagulation of casein obtained by separating from the curd after cutting process (Božanić et al. 2014, Yıldırım and Güzeler 2013). Whey basically contains about half of the dairy ingredients, such as lactose, whey proteins, minerals and a low amount of fat. Calcium, phosphate, lactic acid and lactate content of sweet whey is generally lower than those of acidic whey (Božanić et al. 2014). The nutritional composition of sweet and acidic whey is presented in Table 9. Whey proteins are stated to have a higher degree of bioavailability than other animal-derived proteins, including eggs. In addition, whey proteins contain bioactive peptides that have positive effects on health such as antimicrobial, antioxidant, antithrombotic and antihypertensive effects. In cheese production, most of the carbohydrates within the milk pass into the whey, 90% being lactose as the major component. Lactose facilitates the absorption of calcium and phosphorus in the body, and improves the digestion of fat and other nutrients. Also, some water-soluble vitamins can be transferred into whey. Among them, riboflavin gives whey its characteristic color (Božanić et al. 2014).

Another important by-product of the dairy industry is buttermilk (Skryplonek et al. 2019). Buttermilk contains water-soluble components (lactose, casein, whey protein) found in milk (Skryplonek et al. 2019, Ali 2019). In addition, whole milk contains about 0.12 mg/g of phospholipids, while buttermilk contains about seven times more (0.89 mg/g), as the milk fat globule membrane is rich in phospholipids (Barukčić et al. 2019). The cholesterol-lowering and anti-inflammatory properties of the polar lipids of the milk fat globule membrane have shown to have a positive effect on human health. Vitamin A, carotenoids, enzymes and fatty acid-binding proteins in buttermilk have also a beneficial effect on human health such as the anticancer properties (Skryplonek et al. 2019). Considering the valuable nutrients of egg and dairy by-products, whey, buttermilk and eggshell have been used in the food industry to produce value-added functional foods, thanks to the developing technology and elevated amounts of scientific studies.

Emulsification, gelation, foaming stability and viscosity are important functional properties of whey proteins. Accordingly, whey-based beverages or whey-added foods have been developed by using powdered, deproteinized or diluted whey (Barukčić et al. 2019). Some potential usage areas of whey include the enrichment of taste, the production of salami and sausages with fresh yeast-fermented whey proteins, and hardening the yogurt structure with whey protein concentrates or whey powder (Dinçoğlu and Ardıç 2012). In a study conducted by Pareek et al. (2014) on carbonated orange juice drink containing 70% orange juice and 30% whey, it was observed that there was an increase in vitamin C, protein, sodium and potassium levels compared to conventional orange juice. Yetim et al. (2001) stated that adding whey to the frankfurter type sausage formulation increased the emulsion stability, chemical and sensory properties. Bilgin et al. (2006) conducted a comparative study by adding different rates of pasteurized whey or buttermilk to the bread formulation. Authors reported that the buttermilk added bread had a higher water absorption capacity than whey containing bread. This may be associated with the increased hydration capacity of whey and buttermilk due to water-soluble components such as lactose, albumins and globulins. Breads containing buttermilk showed more browning in crust color, as more intense Maillard and caramelization reactions occur during baking due to higher lactose and protein content. Skryplonek et al. (2019) carried out a study using processed buttermilk as a raw material in the production of quark cheese. Although the hardness value of the produced cheese was improved, it showed similar physicochemical properties with classical Quark cheese. This study has shown that buttermilk is a successful alternative as a raw material for soft immature cheese production. Sheth and Hirdyani (2016) developed a buttermilk-based fermented beverage containing barley and fructooligosaccharides which was highly appreciated by the sensory analysis panel members. Also, Mudgil et al. (2016) increased the viscosity of the buttermilk by adding aloe vera juice and a new cultured buttermilk beverage was formulated using dairy by-products.

It is known that dairy by-products have a wide range of uses in different industries, apart from the food industry. Biogas, which is formed as a result of fermentation of organic wastes of whey, is an alternative energy source that can be used for heating or electricity generation (Sözer and Yaldız 2006). As in whey, biogas can be produced by anaerobic digestion of buttermilk and converted into renewable energy. In addition, the biomass resulting from biological digestion can also be used as fertilizer (TÜGİŞ 2016). Andrade et al. (2019) produced edible film from whey proteins and used them in sliced salami packaging, and reported that the whey protein film inhibited oxygen permeability and delayed lipid oxidation. Whey protein edible film was also applied to various foods such as chicken breast fillet (Fernández-Pan et al. 2014), salmon slice (Boyacı et al. 2016) and Ricotta cheese (Di Pierro et al. 2011). Results showed an inhibitory effect against microorganisms after incorporation of dairy by-products to the packaging films. The lactose obtained by crystallization of the remaining liquid after removing proteins, fats and minerals in the whey can be used in the production of special diet supplements, penicillin, lactose syrups or lactic acid. In addition, derivatives of lactose (acetic acid, citric acid), vitamin B_{12} and single cell protein are some examples that can be obtained from whey (Coton 1985, Dinçoğlu and Ardıç 2012).

The utilization of eggshells in various industries depends on its main component calcium carbonate, which contains about 40% calcium (Waheed et al. 2019, 2020b). Since the extraction of calcium from the eggshell is not challenging due to its high solubility, eggshells can be mainly used as a calcium supplement. In addition, eggshells can also be a good raw material for calcium-fortified functional foods. Daengprok et al. (2002) has developed a novel functional food by converting calcium carbonate in eggshell into calcium lactate powder, to use in fermented pork sausage formulation. In a study conducted by de Paula et al. (2014), calcium citrate powder obtained from eggshells was added to roasted and ground coffee. Authors observed no significant difference in the sensory properties of coffee, even an increase in calcium bioavailability was achieved. The studies regarding the calcium fortification in bakery products by adding eggshell powder directly to formulations include chocolate cake (Ray et al. 2017), muffins (Afzal et al. 2020), and white bread (Platon et al. 2020). Fina et al. (2016) produced a calcium-rich functional beverage suitable for lactose intolerant individuals. Eggshells can also be used as calcium supplement for animal feeds to meet the daily calcium requirement of animals. It is reported that eggshell powder used in poultry feeds plays an important role in providing calcium source to hens for laying eggs. Furthermore, calcium obtained from eggshell can also be used in the pharmaceutical industry for patients having calcium deficiency, since it is a cheap material unlike other calcium sources (Aditya et al. 2021).

7. Conclusions

Foods include a wide range of biological materials either of plant or animal origin. Therefore, food wastes discarded during processing or wasted streams are among the most important factors to be considered for environmental sustainability. The growing population and the awareness regarding to food-related subjects have led scientists and industries to focus on novel green production strategies. The valorization of food waste is not only beneficial in terms of sustainability, but also may open up new opportunities including the production of bioactive components having high economical value. Accordingly, the current chapter aimed to provide a detailed viewpoint on the utilization options for food waste and by-products, by interrelating among literature studies for each basic food groups. As a result, even considering the increasing attention to valorization attempts, especially in the last 10 years, further studies are still required for the discovery of new techniques for the exploitation of all types of waste streams.

References

Abbey, L., Glover-Amengor, M., Atikpo, M.O., Atter, A. and Toppe, J. 2017. Nutrient content of fish powder from low value fish and fish byproducts. Food Sci. Nutr. 5: 374–379.

Abdulrahman, I., Tijani, H.I., Mohammed, B.A., Saidu, H., Yusuf, H., Jibrin, M.N., et al. 2014. From garbage to biomaterials: an overview on egg shell based hydroxyapatite. J. Matter. 2014: 802467.

Adeleke, B.S. and Babalola, O.O. 2020. Oilseed crop sunflower (*Helianthus annuus*) as a source of food: Nutritional and health benefits. Food Sci. Nutr. 8: 4666–4684.

Adesra, A., Srivastava, V.K. and Varjani, S. 2021. Valorization of dairy wastes: integrative approaches for value added products. Indian J. Microbiol. 61: 270–278.

Aditya, S., Stephen, J. and Radhakrishnan, M. 2021. Utilization of eggshell waste in calcium-fortified foods and other industrial applications: A review. Trends Food Sci. Technol. 115: 422–432.

Afzal, F., Mueen-ud-Din, G., Nadeem, M., Murtaza, M.A. and Mahmood, S. 2020. Effect of eggshell powder fortification on the physicochemical and organoleptic characteristics of muffins. Pure Appl. Biol. 9: 1488–1496.

Ahmad, T. and Danish, M. 2018. Prospects of banana waste utilization in wastewater treatment: A review. J. Environ. Manage, 206: 330–348.

Ahmad, T., Aadil, R.M., Ahmed, H., ur Rahman, U., Soares, B.C.V., Souza, S.L.Q., et al. 2019. Treatment and utilization of dairy industrial waste: A review. Trends Food Sci. Technol. 88: 361–372.

Ahmad Khorairi, A.N.S., Sofian-Seng, N.-S., Othaman, R., Abdul Rahman, H., Mohd Razali, N.S., Lim, S.J., et al. 2021. A review on agro-industrial waste as cellulose and nanocellulose source and their potentials in food applications. Food Rev. Int. 1–26.

Ahmed, M.S., Nair, K.P., Khan, M.S., Algahtani, A. and Rehan, M. 2021. Evaluation of date seed (*Phoenix dactylifera* L.) oil as crop base stock for environment friendly industrial lubricants. Biomass Convers. Biorefin. 11: 559–568.

Ajila, C.M., Brar, S.K., Verma, M., Tyagi, R.D., Godbout, S. and Valéro, J.R. 2012. Bio-processing of agro-byproducts to animal feed. Crit. Rev. Biotechnol. 32: 382–400.

Ajmal, M., Butt, M.S., Sharif, K., Nasir, M. and Nadeem, M.T. 2006. Preparation of fiber and mineral enriched pan bread by using defatted rice bran. Int. J. Food Prop. 9: 623–636.

Akpabio, U.D., Akpakpan, A.E. and Enin, G.N. 2012. Evaluation of proximate compositions and mineral elements in the star apple peel, pulp and seed. J. Basic. Appl. Sci. Res. 2: 4839–4843.

Aktaş, K. and Akın, N. 2020. Influence of rice bran and corn bran addition on the selected properties of tarhana, a fermented cereal based food product. LWT 129: 109574.

Al-Abdallah, W. and Dahman, Y. 2013. Production of green biocellulose nanofibers by *Gluconacetobacter xylinus* through utilizing the renewable resources of agriculture residues. Bioprocess Biosyst. Eng. 36: 1735–1743.

Alexandri, M., López-Gómez, J.P., Olszewska-Widdrat, A. and Venus, J. 2020. Valorising agro-industrial wastes within the circular bioeconomy concept: The case of defatted rice bran with emphasis on bioconversion strategies. Fermentation 6: 42.

Ali, A.H. 2019. Current knowledge of buttermilk: Composition, applications in the food industry, nutritional and beneficial health characteristics. Int. J. Dairy Technol. 72: 169-182.

Alkaltham, M.S., Uslu, N., Özcan, M.M., Salamatullah, A.M., Ahmed, I.A.M. and Hayat, K. 2021. Effect of drying process on oil, phenolic composition and antioxidant activity of avocado (*cv.* Hass) fruits harvested at two different maturity stages. LWT 148: 111716.

Almasi, S., Najafi, G., Ghobadian, B. and Jalili, S. 2021. Biodiesel production from sour cherry kernel oil as novel feedstock using potassium hydroxide catalyst: Optimization using response surface methodology. Biocatal. Agric. Biotechnol. 35: 102089.

Almodares, A., Jafarinia, M. and Hadi, M.R. 2009. The effects of nitrogen fertilizer on chemical compositions in corn and sweet sorghum. Am. Eurasian J. Agric. Environ. Sci. 6: 441–446.

Alves, F.J.L., Ferreira, M.A., Urbano, S.A., Andrade, R.P.X., Silva, A.E.M., Siqueira, M.C.B., et al. Performance of lambs fed alternative protein sources to soybean meal. R. Bras. Zootec. 45: 145–150.

Amarowicz, R., Naczk, M. and Shahidi, F. 2000. Antioxidant activity of crude tannins of canola and rapeseed hulls. J. Am. Oil Chem. Soc. 77: 957.

Amin, M. and Bhatti, H.N. 2014. Effect of physicochemical parameters on lipase production by *Penicillium fellutanum* using canola seed oil cake as substrate. Int. J. Agric. Biol. 16: 118–124.

Anbesaw, M.S. 2021. Characterization and potential application of bromelain from pineapple (*Ananas comosus*) waste (peel) in recovery of silver from X-ray films. Int. J. Biomater. 2021.

Ancuţa, P. and Sonia, A. 2020. Oil press-cakes and meals valorization through circular economy approaches: A review. Applied Sciences 10: 7432.

Andrade, M.A., Ribeiro-Santos, R., Guerra, M. and Sanches-Silva, A. 2019. Evaluation of the oxidative status of salami packaged with an active whey protein film. Foods 8: 387.

Angin, D. 2014. Utilization of activated carbon produced from fruit juice industry solid waste for the adsorption of Yellow 18 from aqueous solutions. Bioresour. Technol. 168: 259–266.

Annamalai, N., Elayaraja, S., Vijayalakshmi, S. and Balasubramanian, T. 2011. Thermostable, alkaline tolerant lipase from *Bacillus licheniformis* using peanut oil cake as a substrate. Afr. J. Biochem. Res. 5: 176–181.

Anwar, B., Bundjali, B., Sunarya, Y. and Arcana, I.M. 2021. Properties of bacterial cellulose and its nanocrystalline obtained from pineapple peel waste juice. Fibers and Polymers 22: 1228–1236.

Anwar, F., Naseer, R., Bhanger, M.I., Ashraf, S., Talpur, F.N. and Aladedunye, F.A. 2008. Physico-chemical characteristics of citrus seeds and seed oils from Pakistan. J. Am. Oil Chem. Soc. 85: 321–330.

Araújo, C.S., Rodrigues, A.M.C., Peixoto Joele, M.R.S., Araújo, E.A.F. and Lourenço, L.F.H. 2018. Optimizing process parameters to obtain a bioplastic using proteins from fish byproducts through the response surface methodology. Food Packag. Shelf Life 16: 23–30.

Arvanitoyannis, I.S. and Giakoundis, A. 2006. Current strategies for dairy waste management: A review. Crit. Rev. Food Sci. Nutr. 46: 379–390.

Arzami, A.N., Ho, T.M. and Mikkonen, K.S. 2021. Valorization of cereal by-product hemicelluloses: Fractionation and purity considerations. Food. Res. Int. 151: 110818.

Ashraf, S.A., Adnan, M., Patel, M., Siddiqui, A.J., Sachidanandan, M., Snoussi, M. et al. 2020. Fish-based bioactives as potent nutraceuticals: Exploring the therapeutic perspective of sustainable food from the sea. Mar. drugs 18: 265.

Aslam, F., Ansari, A., Aman, A., Baloch, G., Nisar, G., Baloch, A.H. et al. 2020. Production of commercially important enzymes from *Bacillus licheniformis* KIBGE-IB3 using date fruit wastes as substrate. J. Genet. Eng. Biotechnol. 18.

Atchudan, R., Edison, T.N.J.I., Shanmugam, M., Perumal, S., Somanathan, T. and Lee, Y.R. 2021. Sustainable synthesis of carbon quantum dots from banana peel waste using hydrothermal process for *in vivo* bioimaging. Phys. E Low-dimens. Syst. Nanostruct. 126: 114417.

Atef, M. and Ojagh, S.M. 2017. Health benefits and food applications of bioactive compounds from fish byproducts: A review. J. Funct. Foods. 35: 673–681.

Awasthi, M.K., Ferreira, J.A., Sirohi, R., Sarsaiya, S., Khoshnevisan, B., Baladi, S. et al. 2021. A critical review on the development stage of biorefinery systems towards the management of apple processing-derived waste. Renew. Sust. Energ. Rev. 143: 110972.

Ayrılmış, N., Kaymakçı, A. and Özdemir, F. 2013. Sunflower seed cake as reinforcing filler in thermoplastic composites. J. Appl. Polym. Sci. 129: 1170–1178.

Bagheri, S., Alinejad, M., Nejad, M. and Aliakbarian, B. 2021. Creating incremental revenue from industrial cherry wastes. Chem. Eng. Trans. 87: 553–558.

Bakari, M. and Yusuf, H.O. 2018. Utilization of locally available binders for densification of rice husk for biofuel production. Banats J. Biotechnol. 9: 47–55.

Balcı, F. and Bayram, M. 2015. Recovery of waste-water from wheat washing operation. Academic Food J. 13: 15–21.

Balcı, F. and Bayram, M. 2020. Bulgur cooking process: Recovery of energy and wastewater. J. Food Eng. 269: 109734.

Banerjee, S., Ranganathan, V., Patti, A. and Arora, A. 2018. Valorisation of pineapple wastes for food and therapeutic applications. Trends Food. Sci. Technol. 82: 60–70.

Banik, S. and Nandi, R. 2004. Effect of supplementation of rice straw with biogas residual slurry manure on the yield, protein and mineral contents of oyster mushroom. Ind. Crop Prod. 20: 311–319.

Barukčić, I., Jakopović, K.L. and R. Božanić. 2019. Valorisation of whey and buttermilk for production of functional beverages–an overview of current possibilities. Food Technol. Biotechnol. 57: 448.

Bechaux, J., Gatellier, P., Le Page, J.F., Drillet, Y. and Sante-Lhoutellier, V. 2019. A comprehensive review of bioactive peptides obtained from animal by-products and their applications. Food Funct. 10: 6244–6266.

Bedin, S., Zanella, K., Bragagnolo, N. and Taranto, O.P. 2020. Implication of microwaves on the extraction process of rice bran protein. Braz. J. Chem. Eng. 36: 1653–1665.

Behera, S., Indumathi, K., Mahadevamma, S. and Sudha, M.L. 2013. Oil cakes–a by-product of agriculture industry as a fortificant in bakery products. Int. J. Food Sci. Nutr. 64: 806–814.

Belc, N., Mustatea, G., Apostol, L., Iorga, S., Vlăduţ, V.N. and Mosoiu, C. 2019. Cereal supply chain waste in the context of circular economy. EDP Sci. Romania 112: 03031.

Beres, C., Costa, G.N.S., Cabezudo, I., da Silva-James, N.K., Teles, A.S.C., Cruz, A.P.G., et al. 2017. Towards integral utilization of grape pomace from winemaking process: A review. Waste Manage 68: 581–594.

Bhattacharyya, P., Bisen, J., Bhaduri, D., Priyadarsini, S., Munda, S., Chakraborti, M., et al. 2021. Turn the wheel from waste to wealth: Economic and environmental gain of sustainable rice straw management practices over field burning in reference to India. Sci. Total Environ. 775: 145896.

Bhise, S., Kaur, A., Ahluwali, P. and Thind, S.S. 2014. Texturization of deoiled cake of sunflower, soybean and flaxseed into food grade meal and its utilization in preparation of cookies. Nutr. Food Sci. 44: 576–585.

Bhuvaneshwari, S., Hettiarachchi, H. and Meegoda, J.N. 2019. Crop residue burning in India: policy challenges and potential solutions. Int. J. Environ. Res. Public Health 16: 832.

Bilgiçli, N., Elgün, A., Herken, E.N., Türker, S., Ertaş, N. and İbanoğlu, Ş. 2006. Effect of wheat germ/bran addition on the chemical, nutritional and sensory quality of tarhana, a fermented wheat flour-yoghurt product. J. Food Eng. 77: 680–686.

Bilgin, B., Dağlıoğlu, O. and Konyalı, M. 2006. Functionality of bread made with pasteurized whey and/or buttermilk. Ital. J. Food Sci. 18: 277–286.

Bisht, T.S., Sharma, S.K., Rawat, L., Chakraborty, B. and Yadav, V. 2020. A novel approach towards the fruit specific waste minimization and utilization: A review. J. Pharmacogn. Phytochem. 9: 712–722.

Blanco, M., Vázquez, J.A., Pérez-Martín, R.I. and Sotelo, C.G. 2017. Hydrolysates of fish skin collagen: An opportunity for valorizing fish industry byproducts. Mar. Drugs 15: 131.

Bledzki, A.K., Mamun, A.A. and Volk, J. 2010. Barley husk and coconut shell reinforced polypropylene composites: the effect of fibre physical, chemical and surface properties. Compos. Sci. Technol. 70: 840–846.

Bordiga, M., Travaglia, F. and Locatelli, M. 2019. Valorisation of grape pomace: an approach that is increasingly reaching its maturity–a review. Int. J. Food Sci. Technol. 54: 933–942.

Bouaziz, F., Abdeddayem, A.B., Koubaa, M., Barba, F.J., Jeddou, K.B., Kacem, I., et al. 2020. Bioethanol production from date seed cellulosic fraction using *Saccharomyces cerevisiae*. Separations 7: 67.

Bouchet, A.S., Laperche, A., Bissuel-Belaygue, C., Snowdon, R., Nesi, N. and Stahl, A. 2016. Nitrogen use efficiency in rapeseed. A review. Agron. Sustain. Dev. 36: 1–20.

Boyacı, D., Korel, F. and Yemenicioğlu, A. 2016. Development of activate-at-home-type edible antimicrobial films: An example pH-triggering mechanism formed for smoked salmon slices using lysozyme in whey protein films. Food Hydrocoll. 60: 170–178.

Božanić, R., Barukčić, I., Lisak, K., Jakopovic, Ž. and Tratnik, L. 2014. Possibilities of whey utilisation. Austin J. Nutr. Food Sci. 2: 7.

Brandelli, A., Sala, L. and Kalil, S.J. 2015. Microbial enzymes for bioconversion of poultry waste into added-value products. Food Res. Int. 73: 3–12.

Bujak, J.W. 2015. New insights into waste management–Meat industry. Renew. Energy 83: 1174–1186.

Bulkan, G., Ferreira, J.A. and Taherzadeh, M.J. 2021. Retrofitting analysis of a biorefinery: Integration of 1st and 2nd generation ethanol through organosolv pretreatment of oat husks and fungal cultivation. Bioresour. Technol. Rep. 15: 100762.

Burton, A.D. 2007. Field plot conditions for the expression and selection of straw fibre concentration in oilseed flax. M.Sc Thesis, University of Saskatchewan, Saskatchewan, Canada.

Călinoiu, L.F. and Vodnar, D.C. 2020. Thermal processing for the release of phenolic compounds from wheat and oat bran. Biomolecules 10: 21.

Campanati, C., Willer, D., Schubert, J. and Aldridge, D.C. 2021. Sustainable intensification of aquaculture through nutrient recycling and circular economies: More fish, less waste, blue growth. Rev. Fish. Sci. Aquac. 1–50.

Campone, L., Celano, R., Piccinelli, A.L., Pagano, I., Carabetta, S., Di Sanzo, R., et al. 2018. Response surface methodology to optimize supercritical carbon dioxide/co-solvent extraction of brown onion skin by-product as source of nutraceutical compounds. Food Chem. 269: 495–502.

Cancalon, P. 1971. Chemical composition of sunflower seed hulls. J. Am. Oil Chem. Soc. 48: 629.

Carré, P., Citeau, M., Robin, G. and Estorges, M. 2016. Hull content and chemical composition of whole seeds, hulls and germs in cultivars of rapeseed (*Brassica napus*). OCL 23: A302.

Casabar, J.T., Unpaprom, Y. and Ramaraj, R. 2019. Fermentation of pineapple fruit peel wastes for bioethanol production. Biomass Convers. Biorefin. 9: 761–765.

Casabar, J.T., Ramaraj, R., Tipnee, S. and Unpaprom, Y. 2020. Enhancement of hydrolysis with *Trichoderma harzianum* for bioethanol production of sonicated pineapple fruit peel. Fuel 279: 118437.

Casoni, A.I., Gutierrez, V.S. and Volpe, M.A. 2019. Conversion of sunflower seed hulls, waste from edible oil production, into valuable products. J. Environ. Chem. Eng. 7: 102893.

Cassales, A., de Souza-Cruz, P.B., Rech, R. and Záchia Ayub, M.A. 2011. Optimization of soybean hull acid hydrolysis and its characterization as a potential substrate for bioprocessing. Biomass Bioenerg. 35: 4675–4683.

Castro-Muñoz, R. and Ruby-Figueroa, R. 2019. Membrane technology for the recovery of high-added value compounds from meat processing coproducts. pp. 127–143. *In*: Galanakis, C.M. (ed.). Sustainable Meat Production and Processing. Academic Press, London, UK.

Çavuşoğlu, Ş., Yılmaz, N. and İşlek, F. 2021. Effect of methyl jasmonate treatments on fruit quality and antioxidant enzyme activities of sour cherry (*Prunus cerasus* L.) during cold storage. J. Agric. Sci. 27: 460–468.

Celano, R., Docimo, T., Piccinelli, A.L., Gazzerro, P., Tucci, M., Di Sanzo, R., et al. 2021. Onion peel: Turning a food waste into a resource. Antioxidants: 10: 304.

Çelik, İ., Işık, F. and Yılmaz, Y. 2010. Chemical, rheological and sensory properties of tarhana with wheat bran as a functional constituent. Akademik Gıda 8: 11–17.

Çelik, Ş. and Uysal, Ş. 2009. Composition, Quality, Microflora and Ripening of Beyaz Cheese. Atatürk University Journal of Agricultural Faculty. 40: 141–151.

Cerveró, J.M., Skovgaard, P.A., Felby, C., Sørensen, H.R. and Jørgensen, H. 2010. Enzymatic hydrolysis and fermentation of palm kernel press cake for production of bioethanol. Enzyme Microb. Tech. 46: 177–184.

Chaipet, N., Raita, M., Siriwatwechakul, W. and Champreda, V. 2020. Efficiency of protein extraction from palm kernel cake via different chemical extraction methods. IOP Conf. Ser: Mater. Sci. Eng. Thailand 965: 012008.

Chand, P. and Pakade, Y.B. 2015. Utilization of chemically modified apple juice industrial waste for removal of Ni 2+ ions from aqueous solution. J. Matter. Cycles Waste Manage. 17: 163–173.

Chaouch, M.A. and Benvenuti, S. 2020. The role of fruit by-products as bioactive compounds for intestinal health. Foods 9: 1716.

Chauhan, V. 2021. Nutritional quality analysis of sunflower seed cake (SSC). J. Pharm. Innov. 10: 720–728.

Chen, M., Liu, S., Yuan, X., Li, Q.X., Wang, F., Xin, F., et al. 2021. Methane production and characteristics of the microbial community in the co-digestion of potato pulp waste and dairy manure amended with biochar. Renew. Energ. 163: 357–367.

Chen, X., Ma, X., Peng, X., Lin, Y., Wang, J. and Zheng, C. 2018. Effects of aqueous phase recirculation in hydrothermal carbonization of sweet potato waste. Bioresour. Technol. 267: 167–174.

Cheng, Y., Zhang, J., Luo, K. and Zhang, G. 2017. Oat bran β-glucan improves glucose homeostasis in mice fed on a high-fat diet. RSC Advances 7: 54717–54725.

Cheok, C.Y., Mohd Adzahan, N., Abdul Rahman, R., Zainal Abedin, N.H., Hussain, N., Sulaiman, R., et al. 2018. Current trends of tropical fruit waste utilization. Crit. Rev. Food Sci. Nutr. 58: 335–361.

Cherif, M.M., Grigorakis, S., Halahlah, A., Loupassaki, S. and Makris, D.P. 2020. High-efficiency extraction of phenolics from wheat waste biomass (bran) by combining deep eutectic solvent, ultrasound-assisted pretreatment and thermal treatment. Environ. Process. 7: 845–859.

Chintagunta, A.D., Jacob, S. and Banerjee, R. 2016. Integrated bioethanol and biomanure production from potato waste. Waste Manage. 49: 320–325.

Cho, E.J., Choic, Y.S. and Bae, H.J. 2021. Bioconversion of onion waste to valuable biosugar as an alternative feed source for honey bee. Waste Biomass Valor. 12: 4503–4512.

Chou, R.L., Her, B.Y., Su, M.S., Hwang, G., Wu, Y.H. and Chen, H.Y. 2004. Substituting fish meal with soybean meal in diets of juvenile cobia *Rachycentron canadum*. Aquaculture 229: 325–333.

Collazo-Bigliardi, S., Ortega-Toro, R. and Boix, A.C. 2018. Isolation and characterisation of microcrystalline cellulose and cellulose nanocrystals from coffee husk and comparative study with rice husk. Carbohydr. Polym. 191: 205–215.

Conesa, C., Seguí, L., Laguarda-Miró, N. and Fito, P. 2016. Microwaves as a pretreatment for enhancing enzymatic hydrolysis of pineapple industrial waste for bioethanol production. Food Bioprod. Process. 100: 203–213.

Costa, L.A., Fonseca, A.F., Pereira, F.V. and Druzian, J.I. 2015. Extraction and characterization of cellulose nanocrystals from corn stover. Cell Chem. Technol. 49: 127–133.

Coton, S.G. 1985. Whey resources and utilization. Int. J. Dairy Technol. 38: 97–100.

Cristóbal, J., Castellani, V., Manfredi, S. and Sala, S. 2018. Prioritizing and optimizing sustainable measures for food waste prevention and management. Waste Manage. 72: 3–16.

Črnivec, I.G.O., Skrt, M., Šeremet, D., Sterniša, M., Farčnik, D., Štrumbelj, E., et al. 2021. Waste streams in onion production: Bioactive compounds, quercetin and use of antimicrobial and antioxidative properties. Waste Manage. 126: 476–486.

Daengprok, W., Garnjanagoonchorn, W. and Mine, Y. 2002. Fermented pork sausage fortified with commercial or hen eggshell calcium lactate. Meat Sci. 62: 199–204.

Dalgaard, R., Schmidt, J., Halberg, N., Christensen, P., Thrane, M. and Pengue, W.A. 2008. LCA of soybean meal. Int. J. Life Cycle Assess 13: 240–254.

Dave, D. and Routray, W. 2018. Current scenario of Canadian fishery and corresponding underutilized species and fishery by-products: A potential source of omega-3 fatty acids. J. Clean. Prod. 180: 617–641.

de Erive, M.O., Wang, T., He, F. and Chen, G. 2020. Development of high-fiber wheat bread using microfluidized corn bran. Food Chem. 310: 125921.

de Leonardis, A., Macciola, V. and di Domenico, N. 2005. A first pilot study to produce a food antioxidant from sunflower seed shells (*Helianthus annuus*). Eur. J. Lipid Sci. Technol. 107: 220–227.

de Oliveira, J.P., Bruni, G.P., Lima, K.O., El Halal, S.L.M., da Rosa, G.S., Dias, A.R.G., et al. 2017. Cellulose fibers extracted from rice and oat husks and their application in hydrogel. Food Chem. 221: 153–160.

de Paula, L.N., de Souza, A.H.P., Moreira, I.C., Gohara, A.K., de Oliveira, A.F. and Dias, L.F. 2014. Calcium fortification of roasted and ground coffee with different calcium salts. Acta Sci. Technol. 36: 707–712.

Debiagi, F., Faria-Tischer, P.C.S. and Mali, S. 2021. A green approach based on reactive extrusion to produce nanofibrillated cellulose from oat hull. Waste Biomass Valor. 12: 1051–1060.

Del Castillo-Llamosas, A., Rodríguez-Martínez, B., del Rio, P.G., Eibes, G., Garrote, G. and Gullón, B. 2021. Hydrothermal treatment of avocado peel waste for the simultaneous recovery of oligosaccharides and antioxidant phenolics. Bioresour. Technol. 342: 125981.

Demir, G., Nemlioglu, S., Yazgic, U., Dogan, E.E. and Bayat, C. 2005. Determination of some important emissions of sunflower oil production industrial wastes incineration. J. Sci. Ind. Res. 64: 226–228.

Demir, H. and Tarı, C. 2014. Valorization of wheat bran for the production of polygalacturonase in SSF of *Aspergillus sojae*. Ind. Crop Prod. 54: 302–309.

Demirbas, A. 2017. Utilization of date biomass waste and date seed as bio-fuels source. Energy Sources Part A Recovery Util. Environ. Eff. 39: 754–760.

Demirci, M. 1988. Ülkemizin önemli peynir çeşitlerinin mineral madde düzeyi ve kalori değerleri. Food 13: 17–21.

Deshwal, G.K., Alam, T., Panjagari, N.R. and Bhardwaj, A. 2021. Utilization of cereal crop residues, cereal milling, sugarcane and dairy processing by-products for sustainable packaging solutions. J. Polym. Environ. 29: 2046–2061.

Despoudi, S., Bucatariu, C., Otles, S., Kartal, C., Otles, S., Despoudi, S. et al. 2021. Food waste management, valorization, and sustainability in the food industry. pp. 3–19. *In:* Galanakis, C.M. (ed.). Food Waste Recovery. Academic Press, MA, USA.

Dhillon, G.S., Kaur, S. and Brar, S.K. 2013. Perspective of apple processing wastes as low-cost substrates for bioproduction of high value products: A review. Renewable and Sustainable Energy Reviews 27: 789–805.

Di Pierro, P., Sorrentino, A., Mariniello, L., Giosafatto, C.V.L. and Porta, R. 2011. Chitosan/whey protein film as active coating to extend Ricotta cheese shelf-life. LWT-Food Sci. Technol. 44: 2324–2327.

Díaz-Calderón, P., MacNaughtan, B., Hill, S., Foster, T., Enrione, J. and Mitchell, J. 2018. Changes in gelatinisation and pasting properties of various starches (wheat, maize and waxy maize) by the addition of bacterial cellulose fibrils. Food Hydrocoll. 80: 274–280.

Dinçoğlu, A.H. and Ardıç, M. 2012. The importance of whey on nutrition and use possibilities. Harran University Journal of Veterinary Medicine 1: 54–60.

Doan, V.D., Luc, V.S., Nguyen, T.L.H., Nguyen, T.D. and Nguyen, T.D. 2020. Utilizing waste corn-cob in biosynthesis of noble metallic nanoparticles for antibacterial effect and catalytic degradation of contaminants. Environ. Sci. Pollut. Res. 27: 6148–6162.

Dogaris, I., Vakontios, G., Kalogeris, E., Mamma, D. and Kekos, D. 2009. Induction of cellulases and hemicellulases from *Neurospora crassa* under solid-state cultivation for bioconversion of sorghum bagasse into ethanol. Ind. Crop Prod. 29: 404–411.

Dołżyńska, M., Obidziński, S., Piekut, J. and Yildiz, G. 2020. The utilization of plum stones for pellet production and investigation of post-combustion flue gas emissions. Energies 13: 5107.

Donadelli, R.A., Jones, C.K. and Beyer, R.S. 2018. The amino acid composition and protein quality of various egg, poultry meal by-products, and vegetable proteins used in the production of dog and cat diets. Poultry Sci. 98: 1371–1378.

Doshi, P., Srivastava, G., Pathak, G. and Dikshit, M. 2014. Physicochemical and thermal characterization of nonedible oilseed residual waste as sustainable solid biofuel. Waste Manage. 34: 1836–1846.

Dourado, L.R.B., Pascoal, L.A.F., Sakomura, N.K., Costa, F.G.P. and Biagiotti, D. 2011. Soybeans (Glycine max) and soybean products in poultry and swine nutrition. Intech Publishing, Croatia.

Duan, Y., Mehariya, S., Kumar, A., Singh, E., Yang, J., Kumar, S., et al. 2021. Apple orchard waste recycling and valorization of valuable product-A review. Bioengineered 12: 476–495.

Eckey, E.W. 1954. Vegetable Fats and Oils. Reinhold Press, New York.

Edwiges, T., Frare, L., Mayer, B., Lins, L., Triolo, J.M., Flotats, X., et al. 2018. Influence of chemical composition on biochemical methane potential of fruit and vegetable waste. Waste Manage. 71: 618–625.

El Barnossi, A., Moussaid, F. and Iraqi Housseini, A. 2021. Tangerine, banana and pomegranate peels valorisation for sustainable environment: A review. Biotechnol. Rep. 29: e00574.

El-Hamidi, M. and Zaher, F.A. 2018. Production of vegetable oils in the world and in Egypt: an overview. Bull. Natl. Res. Cent. 42: 1–9.

ElMekawy, A., Diels, L., De Wever, H. and Pant, D. 2013. Valorization of cereal based biorefinery byproducts: reality and expectations. Environ. Sci. Technol. Environ. 47: 9014–9027.

Fathi, H.I., El-Shazly, A.H., El-Kady, M.F. and Madih, K. 2018. Assessment of new technique for production cellulose nanocrystals from agricultural waste. Mater. Sci. Forum 928: 83–88.

Enniya, I., Rghioui, L. and Jourani, A. 2018. Adsorption of hexavalent chromium in aqueous solution on activated carbon prepared from apple peels. Sustain. Chem. Pharm. 7: 9–16.

Eriksson, M. and Spångberg, J. 2017. Carbon footprint and energy use of food waste management options for fresh fruit and vegetables from supermarkets. Waste Manage. 60: 786–799.

Ermiş, E. and Karasu, E.N. 2020. Spray drying of de-oiled sunflower protein extracts: functional properties and characterization of the powder. Food 45: 39–49.

Esparza, I., Jiménez-Moreno, N., Bimbela, F., Ancín-Azpilicueta, C. and Gandía, L.M. 2020. Fruit and vegetable waste management: Conventional and emerging approaches. J. Environ. Manage. 265: 110510.

EU Waste Framework Directive. 2008. Directive 2008/98/EC of the European Parliament and the Council of 19 November 2008 on Waste and Repealing Certain Directives. EU.

Fafiolu, A.O., Oduguwa, O.O., Jegede, A.V., Tukura, C.C., Olarotimi, I.D., Teniola, A.A., et al. 2015. Assessment of enzyme supplementation on growth performance and apparent nutrient digestibility in diets containing undecorticated sunflower seed meal in layer chicks. Poultry Sci. 94: 1917–1922.

FAOSTAT. 2018. Dairy products production amount in 2018. Food and Agriculture Organization of the United Nations (FAO), Rome, Italy.

FAOSTAT. 2019. FAOSTAT database. Food and Agriculture Organization of the United Nations (FAO), Rome, Italy.

Faryar, R., Linares-Pastén, J.A., Immerzeel, P., Mamo, G., Andersson, M., Stålbrand, H., Mattiasson, B., et al. 2015. Production of prebiotic xylooligosaccharides from alkaline extracted wheat straw using the K80R-variant of a thermostable alkali-tolerant xylanase. Food Bioprod. Process 93: 1–10.

Fernández-Pan, I., Carrión-Granda, X. and Maté, J.I. 2014. Antimicrobial efficiency of edible coatings on the preservation of chicken breast fillets. Food Control 36: 69–75.

Figlas, N.D., Gonzalez Matute, R. and Curvetto, N.R. 2016. Sunflower seed hull: its value as a broad mushroom substrate. Ann. Food Process. Preserv. 1: 1002.

Fina, B.L., Brun, L.R. and Rigalli, A. 2016. Increase of calcium and reduction of lactose concentration in milk by treatment with kefir grains and eggshell. Int. J. Food Sci. 67: 133–140.

François, I.E., Lescroart, O., Veraverbeke, W.S., Marzorati, M., Possemiers, S., Evenepoel, P., et al. 2012. Effects of a wheat bran extract containing arabinoxylan oligosaccharides on gastrointestinal health parameters in healthy adult human volunteers: a double-blind, randomised, placebo-controlled, cross-over trial. Br. J. Nutr. 108: 2229–2242.

Fraterrigo Garofalo, S., Tommasi, T. and Fino, D. 2021. A short review of green extraction technologies for rice bran oil. Biomass Convers. Biorefin. 11: 569–587.

Fritsch, C., Staebler, A., Happel, A., Márquez, M.A.C., Aguiló-Aguayo, I., Abadias, M., et al. 2017. Processing, valorization and application of bio-waste derived compounds from potato, tomato, olive and cereals: A review. Sustainability 9: 1492.

Gaber, M.A.F.M., Tujillo, F.J., Mansour, M.P. and Juliano, P. 2018. Improving oil extraction from canola seeds by conventional and advanced methods. Food Eng. Rev. 10: 198–210.

Gamel, T.H., Abdel-Aal, E.S.M., Ames, N.P., Duss, R. and Tosh, S.M. 2014. Enzymatic extraction of beta-glucan from oat bran cereals and oat crackers and optimization of viscosity measurement. J. Cereal Sci. 59: 33–40.

Ganesh, K.S., Sridhar, A. and Vishali, S. 2022. Utilization of fruit and vegetable waste to produce value-added products: Conventional utilization and emerging opportunities-A review. Chemosphere 287: 132221.

García-Vargas, M.C., Contreras, M.D.M. and Castro, E. 2020. Avocado-derived biomass as a source of bioenergy and bioproducts. Appl. Sci. 10: 8195.

Gebrechristos, H.Y. and Chen, W. 2018. Utilization of potato peel as eco-friendly products: A review. Food Sci. Nutr. 6: 1352–1356.

George, T.S., Taylor, M.A., Dodd, I.C. and White, P.J. 2017. Climate change and consequences for potato production: A review of tolerance to emerging abiotic stress. Potato Res. 60: 239–268.

Ghosh, P.R., Fawcett, D., Sharma, S.B. and Poinern, G.E.J. 2016. Progress towards sustainable utilisation and management of food wastes in the global economy. Int. J. Food Sci. 2016: 1–22.

Ghosh, S., Chowdhury, R. and Bhattacharya, P. 2017. Sustainability of cereal straws for the fermentative production of second generation biofuels: A review of the efficiency and economics of biochemical pretreatment processes. Appl. Energy 198: 284–298.

Gil, L.S. and Maupoey, P.F. 2018. An integrated approach for pineapple waste valorisation. Bioethanol production and bromelain extraction from pineapple residues. J. Clean. Prod. 172: 1224–1231.

Goli, A., Shamiri, A., Khosroyar, S., Talaiekhozani, A., Sanaye, R. and Azizi, K. 2019. A review on different aerobic and anaerobic treatment methods in dairy industry wastewater. J. Environ. Treat. Tech. 6: 113–141.

González-García, E., García, M.C. and Marina, M.L. 2018. Capillary liquid chromatography-ion trap-mass spectrometry methodology for the simultaneous quantification of four angiotensin-converting enzyme-inhibitory peptides in Prunus seed hydrolysates. J. Chromatogr A 1540: 47–54.

Görgüç, A., Özer, P. and Yılmaz, F.M. 2020. Microwave-assisted enzymatic extraction of plant protein with antioxidant compounds from the food waste sesame bran: Comparative optimization study and identification of metabolomics using LC/Q-TOF/MS. J. Food Process. Preserv. 44: e14304.

Gowman, A.C., Picard, M.C., Rodriguez-Uribe, A., Misra, M., Khalil, H., Thimmanagari, M., et al. 2019. Physicochemical analysis of apple and grape pomaces. BioResources 14: 3210–3230.

Grippo, V., Romano, S. and Vastola, A. 2019. Multi-criteria evaluation of bran use to promote circularity in the cereal production chain. Nat. Resour. Res. 28: 125–137.

Grzelak-Błaszczyk, K., Milala, J., Kosmala, M., Kołodziejczyk, K., Sójka, M., Czarnecki, A., et al. 2018. Onion quercetin monoglycosides alter microbial activity and increase antioxidant capacity. J. Nutr. Biochem. 56: 81–88.

Gu, I., Lam, W.S., Marasini, D., Brownmiller, C., Savary, B.J., Lee, J.A., et al. 2021. *In vitro* fecal fermentation patterns of arabinoxylan from rice bran on fecal microbiota from normal-weight and overweight/obese subjects. Nutrients 13: 2052.

Gül, H., Özer, M.S. and Dizlek, H. 2009. Improvement of the wheat and corn bran bread quality by using glucose oxidase and hexose oxidase. J. Food Qual. 32: 209–223.

Gültekin Subaşı, B., Vahapoğlu, B., Capanoglu, E. and Mohammadifar, M.A. 2021. A review on protein extracts from sunflower cake: techno-functional properties and promising modification methods. Crit. Rev. Food Sci. Nutr. 61: 1–16.

Guo, C., Yang, J., Wei, J., Li, Y., Xu, J. and Jiang, Y. 2003. Antioxidant activities of peel, pulp and seed fractions of common fruits as determined by FRAP assay. Nutr. Res. 23: 1719–1726.

Gupta, A., Sharma, A., Pathak, R., Kumar, A. and Sharma, S. 2018. Solid state fermentation of non-edible oil seed cakes for production of proteases and cellulases and degradation of anti-nutritional factors. J. Food Biotechnol. Res. 2: 1–6.

Gupta, N., Poddar, K., Sarkar, D., Kumari, N., Padhan, B. and Sarkar, A. 2019. Fruit waste management by pigment production and utilization of residual as bioadsorbent. J. Environ. Manage. 244: 138–143.

Gutierrez, M.M. 2019. Ethanol production from cereal food waste: an enriched carbohydrate source. M.Sc Thesis, Kansas State University, Manhattan, Kansas.

Habib, H.M., Kamal, H., Ibrahim, W.H. and Al Dhaheri, A.S. 2013. Carotenoids, fat soluble vitamins and fatty acid profiles of 18 varieties of date seed oil. Ind. Crop Prod. 42: 567–572.

Hao, Y., Wang, Z., Zou, Y., He, R., Ju, X. and Yuan, J. 2020. Effect of static-state fermentation on volatile composition in rapeseed meal. J. Sci. Food Agric. 100: 2145–2152.

Hassannejad, H. and Nouri, A. 2018. Sunflower seed hull extract as a novel green corrosion inhibitor for mild steel in HCl solution. J. Mol. Liq. 254: 377–382.

Hayta, M., Benli, B., İşçimen, E.M. and Kaya, A. 2021. Antioxidant and antihypertensive protein hydrolysates from rice bran: optimization of microwave assisted extraction. J. Food Meas. Charact. 15: 2904–2914.

Heidarinejad, Z., Rahmanian, O., Fazlzadeh, M. and Heidari, M. 2018. Enhancement of methylene blue adsorption onto activated carbon prepared from date press cake by low frequency ultrasound. J. Mol. Liq. 264: 591–599.

Henrion, M., Francey, C., Lê, K.A. and Lamothe, L. 2019. Cereal B-glucans: The impact of processing and how it affects physiological responses. Nutrients 11: 1729.

Heraldy, E., Lestari, W.W., Permatasari, D. and Arimurti, D.D. 2018. Biosorbent from tomato waste and apple juice residue for lead removal. J. Environ. Chem. Eng. 6: 1201–1208.

Hicks, T.M. and Verbeek, C.J.R. 2016. Meat industry protein by-products: sources and characteristics. pp. 37–61. *In:* Dhillon, G.S. [ed.]. Protein By-Products. Academic Press, London, UK.

Hijosa-Valsero, M., Paniagua-García, A.I. and Díez-Antolínez, R. 2017. Biobutanol production from apple pomace: the importance of pretreatment methods on the fermentability of lignocellulosic agro-food wastes. Appl. Microbiol. Biotechnol. 101: 8041–8052.

Hikal, W.M., Mahmoud, A.A., Said-Al Ahl, H.A.H., Bratovcic, A., Tkachenko, K.G., Kačániová, M., et al. 2021. Pineapple (*Ananas comosus* L. Merr.), waste streams, characterisation and valorisation: An Overview. Open J. Ecol. 11: 610–634.

Hiwot, T. 2017. Determination of oil and biodiesel content, physicochemical properties of the oil extracted from avocado seed (*Persea americana*) grown in Wonago and Dilla (gedeo zone), southern Ethiopia. Chem. Int. 3: 311–319.

Ho, Y.S. and Ofomaja, A.E. 2006. Kinetic studies of copper ion adsorption on palm kernel fibre. J. Hazard. Mater. 137: 1796–1802.

Holloway, J.W. and Wu, J. 2019. Meat by-products. pp: 161–166. *In*: Holloway, J.W. and Wu, J. [eds.]. Red. Meat Science and Production. Science Press, Singapore.

Hossain, A.S., Alshammari, A.M., Alessa, M. and Khalil, M. 2015. Prospect of biofuel production and its application in engine emission and bioelectricity generation from fish, chicken and camel waste by-product as sustainable energy. Camel Publishing House 387–390.

Hossain, M.S., Norulaini, N.A.N., Naim, A.Y.A., Zulkhairi, A.R.M., Bennama, M.M. and Omar, A.K.M. 2016. Utilization of the supercritical carbon dioxide extraction technology for the production of deoiled palm kernel cake. J. CO_2 Util. 16: 121–129.

Hromádková, Z., Košt'Álová, Z. and Ebringerová, A. 2008. Comparison of conventional and ultrasound-assisted extraction of phenolics-rich heteroxylans from wheat bran. Ultrason. Sonochem. 15: 1062–1068.

Hu, Z., Qiu, L., Sun, Y., Xiong, H. and Ogra, Y. 2019. Improvement of the solubility and emulsifying properties of rice bran protein by phosphorylation with sodium trimetaphosphate. Food Hydrocoll. 96: 288–299.

Hussain, S., Jõudu, I. and Bhat, R. 2020. Dietary fiber from underutilized plant resources—a positive approach for valorization of fruit and vegetable wastes. Sustainability 12: 5401.

Inglett, G.E. and Chen, D. 2011. Antioxidant activity and phenolic content of air-classified corn bran. Cereal Chem. 88: 36–40.

Irshad, A. and Sharma, B.D. 2015. Abattoir by-product utilization for sustainable meat industry: a review. J. Anim. Prod. Adv. 5: 681–696.

Ivanov, D.S., Lević, J.D. and Sredanović, S.A. 2010. Fatty acid composition of various soybean products. Food Nutr. Res. 37: 65–70.

Izaguirre, J.K., da Fonseca, M.M.R., Castañón, S., Villarán, M.C. and Cesário, M.T. 2020. Giving credit to residual bioresources: from municipal solid waste hydrolysate and waste plum juice to poly (3-hydroxybutyrate). Waste Manage. 118: 534–540.

Izidoro, D.R., Scheer, A.P., Sierakowski, M.R. and Haminiuk, C.W. 2008. Influence of green banana pulp on the rheological behaviour and chemical characteristics of emulsions (mayonnaises). LWT 41: 1018–1028.

Jabar, J.M. 2016. Effect of chemical modification on physicochemical properties of coir, empty fruit bunch and palm kernel fibres. Appl. Tropical Agric. 21: 153–158.

Jalgaonkar, K., Mahawar, M.K., Bibwe, B. and Kannaujia, P. 2020. Postharvest profile, processing and waste utilization of dragon fruit (*Hylocereus* spp.): A Review. Food Rev. Int. 27: 1–27.

Jankowski, K. and Budzyński, W. 2003. Energy potential of oilseed crops. Electronic Journal of Polish Agricultural Universities 6: 31–38.

Javed, A., Ahmad, A., Tahir, A., Shabbir, U., Nouman, M. and Hameed, A. 2019. Potato peel waste-its nutraceutical, industrial and biotechnological applacations. AIMS Agric. Food 4: 807–823.

Jiménez, R., Sandoval-Flores, G., Alvarado-Reyna, S., Alemán-Castillo, S.E., Santiago-Adame, R. and Velázquez, G. 2021. Extraction of starch from Hass avocado seeds for the preparation of biofilms. Food Sci. Technol.

Jonathan, S.G., Okorie, A.N., Garuba, E.O. and Babayemi, O.J. 2012. Bioconversion of sorghum stalk and rice straw into value added ruminant feed using *Pleurotus pulmonarius*. Nat. Sci. 10: 10–16.

Kaewdiew, J., Ramaraj, R., Koonaphapdeelert, S. and Dussadee, N. 2019. Assessment of the biogas potential from agricultural waste in northern Thailand. Maejo Int. J. Energ. Environ. Comm. 1: 40–47.

Kalaydzhiev, H., Ivanova, P., Stoyanova, M., Pavlov, A., Rustad, T., Silva, C.L., et al. 2020. Valorization of rapeseed meal: influence of ethanol antinutrients removal on protein extractability, amino acid composition and fractional profile. Waste Biomass Valor. 11: 2709–2719.

Kale, M.S., Hamaker, B.R. and Campanella, O.H. 2013. Alkaline extraction conditions determine gelling properties of corn bran arabinoxylans. Food Hydrocoll. 31: 121–126.

Kao, T. H. and Chen, B.H. 2006. Functional components in soybean cake and their effects on antioxidant activity. J. Agric. Food Chem. 54: 7544–7555.

Kao, T.H., Wu, W.M., Hung, C.F., Wu, W.B. and Chen, B.H. 2007. Anti-inflammatory effects of isoflavone powder produced from soybean cake. J. Agric. Food Chem. 55: 11068–11079.

Kao, T.H., Chien, J.T. and Chen, B.H. 2008. Extraction yield of isoflavones from soybean cake as affected by solvent and supercritical carbon dioxide. Food Chem. 107: 1728–1736.

Karakaş, F.P., Keskin, C.N., Agil, F. and Zencirci, N. 2021. Profiles of vitamin B and E in wheat grass and grain of einkorn (*Triticum monococcum* spp. *monococcum*), emmer (*Triticum dicoccum* ssp. *dicoccum* Schrank.), durum (*Triticum durum* Desf.), and bread wheat (*Triticum aestivum* L.) cultivars by LC-ESI-MS/MS analysis. J. Cereal Sci. 98: 103177.

Karp, S., Wyrwisz, J. and Kurek, M.A. 2019. Comparative analysis of the physical properties of o/w emulsions stabilised by cereal β-glucan and other stabilisers. Int. J. Biol. Macromol. 132: 236–243.

Kart, M.Ç.Ö. and Demircan, V. 2014. Recent developments in production, consumption and marketing of milk and dairy products in the world and Turkey. Akademik Gıda 12: 78–96.

Kasapoğlu, K.N., Demircan, E., Eryılmaz, H.S., Karaça, A.C. and Özçelik, B. 2021. Sour cherry kernel as an unexploited processing waste: Optimisation of extraction conditions for protein recovery, functional properties and *in vitro* digestibility. Waste Biomass Valor. 12: 6685–6698.

Kaur, A., Singh, D., Kaur, H. and Sogi, D.S. 2018. Drying characteristics and antioxidant properties of Java plum seed and skin waste. J. Stored Prod. Postharvest 9: 36–43.

Kaur, G., Sharma, S., Nagi, H.P.S. and Dar, B.N. 2012. Functional properties of pasta enriched with variable cereal brans. J. Food Sci. Technol. 49: 467–474.

Kazempour-Samak, M., Rashidi, L., Ghavami, M., Sharifan, A. and Hosseini, F. 2021. Antibacterial and antioxidant activity of sour cherry kernel oil (*Cerasus vulgaris Miller*) against some food-borne microorganisms. J. Food Meas. Charact. 15: 4686–4695.

Kim, J., Oh, J., Averilla, J.N., Kim, H.J., Kim, J.S. and Kim, J.S. 2019a. Grape peel extract and resveratrol inhibit wrinkle formation in mice model through activation of Nrf2/HO-1 signaling pathway. J. Food Sci. 84: 1600–1608.

Kim, S. and Dale, B.E. 2004. Global potential bioethanol production from wasted crops and crop residues. Biomass Bioenerg. 26: 361–375.

Kim, S.W., Ko M.J. and Chung, M.S. 2019b. Extraction of the flavonol quercetin from onion waste by combined treatment with intense pulsed light and subcritical water extraction. J. Clean. Prod. 231: 1192–1199.

Kolev Slavov, A. 2017. General characteristics and treatment possibilities of dairy wastewater–a review. Food Technol. Biotechnol. 55: 14–28.

Koneva, S.I., Egorova, E.Y., Kozubaeva, L.A., Kuzmina, S.S. and Zakharova, A.S. 2018. Influence of flaxseed flour on dough rheology from wheat-flaxseed meal. Adv. Eng. Softw. 151: 370–377.

Kostić, M.D., Veličković, A.V., Joković, N.M., Stamenković, O.S. and Veljković, V.B. 2016. Optimization and kinetic modeling of esterification of the oil obtained from waste plum stones as a pretreatment step in biodiesel production. Waste Manage. 48: 619–629.

Kot, A.M., Pobiega, K., Piwowarek, K., Kieliszek, M., Błażejak, S., Gniewosz, M., et al. 2020. Biotechnological methods of management and utilization of potato industry waste—a review. Potato Res. 6: 431–447.

Kotecka-Majchrzak, K., Sumara, A., Fornal, E. and Montowska, M. 2020. Oilseed proteins–Properties and application as a food ingredient. Trends Food Sci. Technol. 106: 160–170.

Kouřimská, L., Pokhrel, K., Božik, M., Tilami, S.K. and Horčička, P. 2021. Fat content and fatty acid profiles of recently registered varieties of naked and hulled oats with and without husks. J. Cereal Sci. 99: 103216.

Kovalcik, A., Pernicova, I., Obruca, S., Szotkowski, M., Enev, V., Kalina, M., et al. 2020. Grape winery waste as a promising feedstock for the production of polyhydroxyalkanoates and other value-added products. Food Bioprod. Process. 124: 1–10.

Krishna, G.V., Bhagwan, A., Kumar, A.K., Girwani, A., Sreedhar, M., Reddy, S.N., et al. 2020. Influence of pre harvest application of plant bio regulators and chemicals on postharvest quality and shelf life of Mango (*Mangifera indica* L.). J. Pharmacogn. Phytochem. 9: 134–142.

Krishna, T.A., Pragalyaashree, M.M. and Balamurugan, P. 2018. Extraction and quantification of lutein from sweet potato leaves (*Ipomoea batatas*). Drug Invent. Today, 10: 2618–2621.

Kumar, M., Barbhai, M.D., Hasan, M., Punia, S., Dhumal, S., Rais, N., et al. 2022. Onion (*Allium cepa* L.) peels: A review on bioactive compounds and biomedical activities. Biomed. Pharmacother. 146: 112498.

Kumar, S.J., Kumar, N.S. and Chintagunta, A.D. 2020. Bioethanol production from cereal crops and lignocelluloses rich agro-residues: prospects and challenges. SN Appl. Sci. 2: 1–11.

Kumar, V., Kumar Biswas, A., Kumar Chatli, M. and Sahoo, J. 2011. Effect of banana and soybean hull flours on vacuum-packaged chicken nuggets during refrigeration storage. Int. J. Food Sci. 46: 122–129.

Kummu, M., De Moel, H., Porkka, M., Siebert, S., Varis, O. and Ward, P.J. 2012. Lost food, wasted resources: Global food supply chain losses and their impacts on freshwater, cropland, and fertiliser use. Sci. Total Environ. 438: 477–489.

Lagos, L.V. and Stein, H.H. 2017. Chemical composition and amino acid digestibility of soybean meal produced in the United States, China, Argentina, Brazil, or India. J. Anim. Sci. 95: 1626–1636.

Lammerskötter, A., Seggert, H., Matthäus, B., Raß, M., Bart, H.J. and Jordan, V. 2017. Rapeseed hull oil as a source for phytosterols and their separation by organic solvent nanofiltration. Eur. J. Lipid Sci. Technol. 119: 1600090.

Lapeña, D., Vuoristo, K.S., Kosa, G., Horn, S.J. and Eijsink, V.G. 2018. Comparative assessment of enzymatic hydrolysis for valorization of different protein-rich industrial by-products. J. Agric. Food Chem. 66: 9738–9749.

Lapo, B., Bou, J.J., Hoyo, J., Carrillo, M., Peña, K., Tzanov, T., et al. 2020. A potential lignocellulosic biomass based on banana waste for critical rare earths recovery from aqueous solutions. Environ. Pollut. 264: 114409.

Laso, J., Margallo, M., Celaya, J., Fullana, P., Bala, A., Gazulla, C., et al. 2016. Waste management under a life cycle approach as a tool for a circular economy in the canned anchovy industry. Waste Manag. Res. 34: 724–733.

Le Clef, E. and Kemper, T.. 2015. Sunflower seed preparation and oil extraction. pp. 187–226. *In:* Martínez-Force, E., Dunford, N.T. and Salas, J.J. [eds.]. Sunflower. AOCS Press, Amsterdam, The Netherlands.

Leal, F.H.P.N., Senna C.D.A, Kupski, L., Mendes, G.D.R.L. and Badiale-Furlong, E. 2021. Enzymatic and extrusion pretreatments of defatted rice bran to improve functional properties of protein concentrates. Int. J. Food Sci. 56: 5445–5451.

Lemaire, A. and Limbourg, S. 2019. How can food loss and waste management achieve sustainable development goals? J. Clean. Prod. 234: 1221–1234.

Leming, R. and Lember, A. 2005. Chemical composition of expeller-extracted and cold-pressed rapeseed cake. Agraarteadus 16: 96–109.

Lerma-García, M.J., Herrero-Martínez, J.M., Simó-Alfonso, E.F., Mendonça, C.R. and Ramis-Ramos, G. 2009. Composition, industrial processing and applications of rice bran γ-oryzanol. Food Chem. 115: 389–404.

Levent, H., Koyuncu, M., Bilgicli, N., Adıgüzel, E. and Dedeoğlu, M. 2020. Improvement of chemical properties of noodle and pasta using dephytinized cereal brans. LWT 128: 109470.

Li, P., Wang, Y., Hou, Q. and Li, X. 2018. Isolation and characterization of microfibrillated cellulose from agro-industrial soybean residue (okara). BioResources 13: 7944–7956.

Liaudanskas, M., Okulevičiūtė, R., Lanauskas, J., Kviklys, D., Zymonė, K., Rendyuk, T., et al. 2020. Variability in the content of phenolic compounds in plum fruit. Plants 9: 1611.

Liu, H., Li, Y., You, M. and Liu, X. 2021a. Comparison of physicochemical properties of β-glucans extracted from hull-less barley bran by different methods. Int. J. Biol. Macromol. 182: 1192–1199.

Liu, J., Xu, X., Zhao, P.F., Tian, Q.Y., Zhang, S., Li, P., et al. 2015. Evaluation of energy digestibility and prediction of digestible and metabolisable energy in sunflower seed meal fed to growing pigs. Ital. J. Anim. Sci. 14: 3533.

Liu, L., Wen, W., Zhang, R., Wei, Z., Deng, Y., Xiao, J., et al. 2017. Complex enzyme hydrolysis releases antioxidative phenolics from rice bran. Food Chem. 214: 1–8.

Liu, Q., He, W.Q., Aguedo, M., Xia, X., Bai, W.B., Dong, Y.Y., et al. 2021b. Microwave-assisted alkali hydrolysis for cellulose isolation from wheat straw: Influence of reaction conditions and non-thermal effects of microwave. Carbohydr. Polym. 253: 117170.

Liu, R., Liu, R., Shi, L., Zhang, Z., Zhang, T., Lu, M., et al. 2019. Effect of refining process on physicochemical parameters, chemical compositions and in vitro antioxidant activities of rice bran oil. LWT 109: 26–32.

Long, J.M. and Mohan, A. 2021. Development of meat powder from beef by-product as value-added food ingredient. LWT 146: 111460.

Łopusiewicz, Ł., Drozłowska, E., Siedlecka, P., Mężyńska, M., Bartkowiak, A., Sienkiewicz, M., et al. 2019. Development, characterization, and bioactivity of non-dairy kefir-like fermented beverage based on flaxseed oil cake. Foods 8: 544.

Louis, A.C.F. and Venkatachalam, S. 2020. Energy efficient process for valorization of corn cob as a source for nanocrystalline cellulose and hemicellulose production. Int. J. Biol. Macromol. 163: 260–269.

Lu, F., Liu, Y. and Li, B. 2013. Okara dietary fiber and hypoglycemic effect of okara foods. Bioact. Carbohydr. Diet. Fibre. 2: 126–132.

Lyu, F., Luiz, S.F., Azeredo, D.R.P., Cruz, A.G., Ajlouni, S. and Ranadheera, C.S. 2020. Apple pomace as a functional and healthy ingredient in food products: A review. Processes 8: 319.

Mahato, N., Sharma, K., Sinha, M., Baral, E.R., Koteswararao, R., Dhyani, A., et al. 2020. Bio-sorbents, industrially important chemicals and novel materials from citrus processing waste as a sustainable and renewable bioresource: A review. J. Adv. Res. 23: 61–82.

Mahmoudi, G., Sufimahmoudi, E. and Sajadi, S.M. 2020. Bioactive metal oxide nanoparticles from some common fruit wastes and Euphorbia condylocarpa plant. Food Sci. Nutr. 8: 5521–5531.

Maina, S., Kachrimanidou, V., Ladakis, D., Papanikolaou, S., de Castro, A.M. and Koutinas, A. 2019. Evaluation of 1, 3-propanediol production by two Citrobacter freundii strains using crude glycerol and soybean cake hydrolysate. Environ. Sci. Pollut. Res. 26: 35523–35532.

Majzoobi, M., Karambakhsh, G., Golmakani, M.T., Mesbahi, G.R. and Farahnaki, A. 2019. Chemical composition and functional properties of date press cake, an agro-industrial waste. J. Agric. Sci. Technol. 21: 1807–1817.

Mangia, M. 2010. Free and hidden fumonisins in maize and gluten-free products. Ph.D. Thesis, University of Parma, Parma, Italy.

Manhongo, T.T., Chimphango, A., Thornley, P. and Röder, M. 2021. Techno-economic and environmental evaluation of integrated mango waste biorefineries. J. Clean. Prod. 325: 129335.

Manzocchi, E., Guggenbühl, B., Kreuzer, M. and Giller, K. 2020. Effects of the substitution of soybean meal by spirulina in a hay-based diet for dairy cows on milk composition and sensory perception. Int. J. Dairy Sci. 103: 11349–11362.

Marcotuli, I., Hsieh, Y.S.Y., Lahnstein, J., Yap, K., Burton, R.A., Blanco, A., et al. 2016. Structural variation and content of arabinoxylans in endosperm and bran of durum wheat (*Triticum turgidum* L.). J. Agric. Food Chem. 64: 2883–2892.

Marenda, F.R.B., Colodel, C., Canteri, M.H.G., de Olivera Müller, C.M., Amante, E.R., de Oliveira Petkowicz, C.L., et al. 2019. Investigation of cell wall polysaccharides from flour made with waste peel from unripe banana (*Musa sapientum*) biomass. J. Sci. Food Agric. 99: 4363–4372.

Maroušek, J., Rowland, Z., Valášková, K. and Král, P. 2020. Techno-economic assessment of potato waste management in developing economies. Clean Technol. Environ. Policy. 22: 937–944.

Martí-Quijal, F.J., Tornos, A., Príncep, A., Luz, C., Meca, G., Tedeschi, P., et al. 2020. Impact of fermentation on the recovery of antioxidant bioactive compounds from sea bass by-products. Antioxidants 9: 239.

Marvizadeh, M.M. and Akbari, N. 2019. Development and Utilization of Rice Bran in Hamburger as a Fat Replacer. J. Chem. Health Risks. 9: 245–251.

Matthäus, B. 1998. Effect of dehulling on the composition of antinutritive compounds in various cultivars of rapeseed. Lipid/Fett 100: 295–301.

Maurício, E.M., Rosado, C., Duarte, M.P., Fernando, A.L. and Díaz-Lanza, A.M. 2020. Evaluation of industrial sour cherry liquor wastes as an ecofriendly source of added value chemical compounds and energy. Waste Biomass Valor. 11: 201–210.

McCance, K.R., Suarez, A., McAlexander, S.L., Davis, G., Blanchard, M.R. and Venditti, R.A. 2021. Modeling a biorefinery: converting pineapple waste to bioproducts and biofuel. J. Chem. Educ. 98: 2047–2054.

Mendes, J.A., Prozil, S.O., Evtuguin, D.V. and Lopes, L.P.C. 2013. Towards comprehensive utilization of winemaking residues: Characterization of grape skins from red grape pomaces of variety Touriga Nacional. Ind. Crops Prod. 43: 25–32.

Merino, D., Bertolacci, L., Paul, U.C., Simonutti, R. and Athanassiou, A. 2021. Avocado peels and seeds: Processing strategies for the development of highly antioxidant bioplastic films. ACS Appl. Mater. Interfaces. 13: 38688–38699.

Miedzińska, K., Członka, S., Strąkowska, A. and Strzelec, K. 2021. Biobased polyurethane composite foams reinforced with plum stones and silanized plum stones. Int. J. Mol. Sci. 22: 4757.

Mikušová, L., Gereková, P., Kocková, M., Šturdík, E., Valachovičová, M., A. Holubková, et al. 2013. Nutritional, antioxidant, and glycaemic characteristics of new functional bread. Chem. Pap. 67: 284–291.

Milani, F.X., Nutter, D. and Thoma, G. 2011. Invited review: Environmental impacts of dairy processing and products: A review. Int. J. Dairy Sci. 94: 4243–4254.

Milićević, N., Sakač, M., Hadnađev, M., Škrobot, D., Šarić, B., Hadnađev, T.D., et al. 2020. Physico-chemical properties of low-fat cookies containing wheat and oat bran gels as fat replacers. J. Cereal Sci. 95: 103056.

Mingyai, S., Kettawan, A., Srikaeo, K. and Singanusong, R. 2017. Physicochemical and antioxidant properties of rice bran oils produced from colored rice using different extraction methods. J. Oleo Sci. 66: 565–572.

Mironeasa, S., Leahu, A., Codina, G.G., Stroe, S.G. and Mironeasa, C. 2010. Grape Seed: physico-chemical, structural characteristics and oil content. J. Agroaliment. Processes Technol. 16: 1–6.

Mitaru, B.N., Blair, R., Reichert, R.D. and Roe, W.E. 1984. Dark and yellow rapeseed hulls, soybean hulls and a purified fiber source: Their effects on dry matter, energy, protein and amino acid digestibilities in cannulated pigs. J. Anim. Sci. 59: 1510–1518.

Mosenthin, R., Messerschmidt, U., Sauer, N., Carré, P., Quinsac, A. and Schöne, F. 2016. Effect of the desolventizing/toasting process on chemical composition and protein quality of rapeseed meal. J. Anim. Sci. Biotechnol. 7: 1–12.

Mosna, D., Bottani, E., Vignali, G. and Montanari, R. 2021. Environmental benefits of pet food obtained as a result of the valorisation of meat fraction derived from packaged food waste. Waste Manage. 125: 132–144.

Motaung, T.E. and Mtibe, A. 2015. Alkali treatment and cellulose nanowhiskers extracted from maize stalk residues. Int. J. Mater. Sci. Appl. 6: 1022.

Moure, A., Sineiro, J., Domínguez, H. and Parajó, J.C. 2006. Functionality of oilseed protein products: A review. Food Res. Int. 39: 945–963.

Mourtzinos, I., Prodromidis, P., Grigorakis, S., Makris, D.P., Biliaderis, C.G. and Moschakis, T. 2018. Natural food colorants derived from onion wastes: Application in a yoghurt product. Electrophoresis 39: 1975–1983.

Mrabet, A., Jiménez-Araujo, A., Guillén-Bejarano, R., Rodríguez-Arcos, R. and Sindic, M. 2020. Date seeds: A promising source of oil with functional properties. Foods 9: 787.

Muhammad, N., Maigandi, S.A., Hassan, W.A. and Daneji, A.I. 2008. Growth performance and economics of sheep production with varying levels of rice milling waste. Sokoto J. Vet. Sci. 7: 59–64.

Mullen, A.M., Álvarez, C., Zeugolis, D.I., Henchion, M., O'Neill, E. and Drummond, L. 2017. Alternative uses for co-products: Harnessing the potential of valuable compounds from meat processing chains. Meat Sci. 132: 90–98.

Mustafayev, S.K. and Smychagin, E.O. 2018. Organization of fodder production based on sunflower seed waste. Adv. Eng. Res. 157: 429–434.

Naczk, M. and Shahidi, F. 1990. Carbohydrates of canola and rapeseed. pp. 211–220. *In:* Shadidi, F. [ed.]. Canola and Rapeseed: Production, Chemistry, Nutrition, and Processing Technology. Springer Press, Boston, MA.

Nahman, A. and de Lange, W. 2013. Costs of food waste along the value chain: Evidence from South Africa. Waste Manage. 33: 2493–2500.

Narisetty, V., Cox, R., Willoughby, N., Aktas, E., Tiwari, B., Matharu, A.S., et al. 2021. Recycling bread waste into chemical building blocks using a circular biorefining approach. Sustain. Energy Fuels 5: 4842–4849.

Ng, H.S., Kee, P.E., Yim, H.S., Chen, P.T., Wei, Y.H. and Lan. J.C.W. 2020. Recent advances on the sustainable approaches for conversion and reutilization of food wastes to valuable bioproducts. Bioresour Technol. 302: 122889.

Nile, S.H., Nile, A.S., Keum, Y.S. and Sharma, K. 2017. Utilization of quercetin and quercetin glycosides from onion (*Allium cepa L.*) solid waste as an antioxidant, urease and xanthine oxidase inhibitors. Food Chem. 235: 119–126.

Niño-Medina, G., Carvajal-Millán, E., Lizardi, J., Rascon-Chu, A., Marquez-Escalante, J.A., Gardea, A., et al. 2009. Maize processing waste water arabinoxylans: Gelling capability and cross-linking content. Food Chem. 115: 1286–1290.

Noor, R.S., Hussain, F., Farooq, M.U. and Umair, M. 2020. Cost and profitability analysis of cherry production: The case study of district Quetta, Pakistan. Big Data Agric. (BDA) 2: 65–71.

Noorzai, S., Verbeek, C.J.R., Lay, M.C. and Swan, J. 2020. Collagen extraction from various waste bovine hide sources. Waste Biomass Valor. 11: 5687–5698.

Obibuzor, J.U., Okogbenin, E.A. and Abigor, R.D. 2012. Oil recovery from palm fruits and palm kernel. pp. 299-328 *In:* Lai, O.M., Tan, C.P. and Akoh, C.C. [eds.]. Palm Oil: Production, Processing, Characterization, and Uses. AOCS Press. Urbana, IL, USA.

Odebode, F.D., Ekeleme, O.T., Ijarotimi, O.S., Malomo, S.A., Idowu, A.O., Badejo, A.A., et al. 2018. Nutritional composition, antidiabetic and antilipidemic potentials of flour blends made from unripe plantain, soybean cake, and rice bran. J. Food Biochem. 42: e12447.

Ofomaja, E.A., Unuabonah, I.E. and Oladoja, N.A. 2005. Removal of lead from aqueous solution by palm kernel fibre. S. Afr. J. Chem. 58: 127–130.

Öğe, Ç. and Yüceer, Y.K. 2021. Determination of physicochemical and sensory properties and volatile compounds of buttermilk drink. Gıda/The Journal of Food 46: 1243–1255

Okoroigwe, E.C., Saffron, C.M. and Kamdem, P.D. 2014. Characterization of palm kernel shell for materials reinforcement and water treatment. J. Chem. Eng. Mater. Sci. 5: 1–6.

Oladipupo Kareem, M., Edathil, A.A., Rambabu, K., Bharath, G., Banat, F., Nirmala, G.S., et al. 2021. Extraction, characterization and optimization of high quality bio-oil derived from waste date seeds. Chem. Eng. Commun. 208: 801–811.

Oladzad, S., Fallah, N., Mahboubi, A., Afsham, N. and Taherzadeh, M.J. 2021. Date fruit processing waste and approaches to its valorization: A review. Bioresour. Technol. 340: 125625.

Oliveira, F., Souza, C.E., Peclat, V.R., Salgado, J.M., Ribeiro, B.D., Coelho, M.A., et al. 2017. Optimization of lipase production by Aspergillus ibericus from oil cakes and its application in esterification reactions. Food Bioprod. Process. 102: 268–277.

Omar, H. and Rohani, S. 2015. Treatment of landfill waste, leachate and landfill gas: A review. Front. Chem. Sci. and Eng. 9: 15–32.

Osei, E., Gassman, P.W., Hauck, L.M., Jones, R., Beran, L., Dyke, P.T., et al. 2003. Environmental benefits and economic costs of manure incorporation on dairy waste application fields. J. Environ. Manage. 68: 1–11.

Osma, J.F., Herrera, J.L.T. and Couto, S.R. 2007. Banana skin: A novel waste for laccase production by *Trametes pubescens* under solid-state conditions. Application to synthetic dye decolouration. Dyes and Pigments 75: 32–37.

Osorio, L.L.D.R., Flórez-López, E. and Grande-Tovar, C.D. 2021. The potential of selected agri-food loss and waste to contribute to a circular economy: Applications in the food, cosmetic and pharmaceutical industries. Molecules 26: 515.

Östbring, K., Malmqvist, E., Nilsson, K., Rosenlind, I. and Rayner, M. 2020. The effects of oil extraction methods on recovery yield and emulsifying properties of proteins from rapeseed meal and press cake. Foods 9: 19.

Otles, S., Despoudi, S., Bucatariu, C. and Kartal, C. 2015. Food waste management, valorization, and sustainability in the food industry. pp. 3–23 *In:* Galanakis, C.M. [ed.]. Food Waste Recovery: Processing Technologies, Industrial Techniques, and Applications. Academic Press. Waltham, MA, USA.

O'Toole, D.K. 1999. Characteristics and use of okara, the soybean residue from soy milk production a review. J. Agric. Food Chem. 47: 363–371.

Özkaya, B., Baumgartner, B. and Özkaya, H. 2018. Effects of concentrated and dephytinized wheat bran and rice bran addition on bread properties. J. Texture Stud. 49: 84–93.

Pal, D.B., Tiwari, A.K., Prasad, N., Srivastav, N., Almalki, A.H., Haque S., et al. 2021. Thermo-chemical potential of solid waste seed biomass obtained from plant *Phoenix Dactylifera* and *Aegle Marmelos* L. Fruit core cell. Bioresour. Technol. 345: 126441.

Palmarola-Adrados, B., Chotěborská, P., Galbe, M. and Zacchi, G. 2005. Ethanol production from non-starch carbohydrates of wheat bran. Bioresour. Technol. 96: 843–850.

Pap, S., Knudsen, T.Š., Radonić, J., Maletić, S., Igić, S.M. and Sekulić, M.T. 2017. Utilization of fruit processing industry waste as green activated carbon for the treatment of heavy metals and chlorophenols contaminated water. J. Clean. Prod. 162: 958–972.

Papademas, P. and Kotsaki, P. 2019. Technological utilization of whey towards sustainable exploitation. Adv. Dairy Res. 7: 231.

Papageorgiou, M. and Skendi, A. 2018. Introduction to cereal processing and by-products. pp. 1–25 *In:* Galanakis, C.M. [ed.]. Sustainable Recovery and Reutilization of Cereal Processing By-Products. Woodhead Publishing, Amsterdam, The Netherlands.

Pareek, N., Gupta, A. and Sengar, R. 2014. Preparation of healthy fruit based carbonated whey beverages using whey and orange juice. Asian J. Dairy. Food Res. 33: 5–8.

Parfitt, J., Barthel, M. and Macnaughton, S. 2010. Food waste within food supply chains: quantification and potential for change to 2050. Philos. Trans. R. Soc. Lond., B, Biol. Sci. 365: 3065–3081.

Parihar, D.K. 2012. Production of lipase utilizing linseed oilcake as fermentation substrate. Int. J. Sci. Environ. Technol. 1: 135–143.

Paritosh, K., Kushwaha, S.K., Yadav, M., Pareek, N., Chawade, A. and Vivekanand, V. 2017. Food waste to energy: an overview of sustainable approaches for food waste management and nutrient recycling. BioMed Res. Int. 2017: 2370927.

Patiño-Rodríguez, O., Bello-Pérez, L.A., Agama-Acevedo, E. and Pacheco-Vargas, G. 2020. Pulp and peel of unripe stenospermocarpic mango (*Mangifera indica* L. cv *Ataulfo*) as an alternative source of starch, polyphenols and dietary fibre. Food Res. Int. 138: 109719.

Paynor, K.A., David, E.S. and Valentino, M.J.G. 2016. Endophytic fungi associated with bamboo as possible sources of single cell protein using corn cob as a substrate. Mycosphere 7: 139–147.

Peanparkdee, M., Patrawart, J. and Iwamoto, S. 2019. Effect of extraction conditions on phenolic content, anthocyanin content and antioxidant activity of bran extracts from Thai rice cultivars. J. Cereal Sci. 86: 86–91.

Peña-Saldarriaga, L.M., Pérez-Alvarez, J.A. and Fernández-López, J. 2020. Quality properties of chicken emulsion-type sausages formulated with chicken fatty by-products. Foods 9: 507.

Pereira, G.S., Cipriani, M., Wisbeck, E., Souza, O., Strapazzon, J.O. and Gern, R.M. 2017. Onion juice waste for production of Pleurotus sajor-caju and pectinases. Food Bioprod. Process. 106: 11–18.

Permal, R., Chang, W.L., Seale, B., Hamid, N. and Kam, R. 2020. Converting industrial organic waste from the cold-pressed avocado oil production line into a potential food preservative. Food Chem. 306: 125635.

Phongthai, S., Lim, S.T. and Rawdkuen, S. 2017. Ultrasonic-assisted extraction of rice bran protein using response surface methodology. J. Food Biochem. 41: e12314.

Pińkowska, H., Wolak, P. and Oliveros, E. 2013. Application of Doehlert matrix for determination of the optimal conditions of hydrothermolysis of rapeseed meal in subcritical water. Fuel 106: 258–264.

Platon, N., Arus, V.A., Georgescu, A.M., Nistor, I.D. and Barsan, N. 2020. White bread fortified with calcium from eggshell powder. Rev. Chim. 71: 299–306.

Pontonio, E., Lorusso, A., Gobbetti, M. and Rizzello, C.G. 2017. Use of fermented milling by-products as functional ingredient to develop a low-glycaemic index bread. J. Cereal Sci. 77: 235–242.

Principato, L., Ruini, L., Guidi, M. and Secondi, L. 2019. Adopting the circular economy approach on food loss and waste: The case of Italian pasta production. Resour. Conserv. Recycl. 144: 82–89.

Prueckler, M., Siebenhandl-Ehn, S., Apprich, S., Hoeltinger, S., Haas, C., Schmid, E., et al. 2014. Wheat bran-based biorefinery 1: Composition of wheat bran and strategies of functionalization. LWT. 56: 211–221.

Pycia, K., Kapusta, I. and Jaworska, G. 2020. Walnut oil and oilcake affect selected the physicochemical and antioxidant properties of wheat bread enriched with them. J. Food Process. Preserv. 44: e14573.

Qing, Q., Guo, Q., Zhou, L., Gao, X., Lu, X. and Zhang, Y. 2017. Comparison of alkaline and acid pretreatments for enzymatic hydrolysis of soybean hull and soybean straw to produce fermentable sugars. Ind. Crop. Prod. 109: 391–397.

Rajesh Kana, S. and Shaija, A. 2020. Performance, combustion and emission characteristics of a diesel engine using waste avocado biodiesel with manganese-doped alumina nanoparticles. Int. J. Ambient Energy 1–8.

Rajmalwar, S. and Dabholkar, P.S. 2009. Production of protease by *Aspergillus* sp. using solid-state fermentation. Afr. J. Biotechnol. 8: 4197–4198.

Rambabu, K., Bharath, G., Banat, F. and Show, P.L. 2021a. Green synthesis of zinc oxide nanoparticles using Phoenix dactylifera waste as bioreductant for effective dye degradation and antibacterial performance in wastewater treatment. J. Hazard. Mater. 402: 123560.

Rambabu, K., Bharath, G., Banat, F., Hai, A., Show, P.L. and Nguyen, T.H.P. 2021b. Ferric oxide/date seed activated carbon nanocomposites mediated dark fermentation of date fruit wastes for enriched biohydrogen production. Int. J. Hydrog. Energy. 46: 16631–16643.

Ray, S., Barman, A.K., Roy, P.K. and Singh, B.K. 2017. Chicken eggshell powder as dietary calcium source in chocolate cakes. The Pharma Innovation 6: 1–4.

Reddy, N. and Yang, Y. 2009. Natural cellulose fibers from soybean straw. Bioresour. Technol. 100: 3593–3598.

Redondo-Cuenca, A., Villanueva-Suárez, M.J. and Mateos-Aparicio, I. 2008. Soybean seeds and its by-product okara as sources of dietary fibre. Measurement by AOAC and Englyst methods. Food Chem. 108: 1099–1105.

Réhault-Godbert, S., Guyot, N. and Nys, Y. 2019. The golden egg: nutritional value, bioactivities, and emerging benefits for human health. Nutrients 11: 684.

Reyna-Bensusan, N., Wilson, D.C., Davy, P.M., Fuller, G.W., Fowler, G.D. and Smith, S.R. 2019. Experimental measurements of black carbon emission factors to estimate the global impact of uncontrolled burning of waste. Atmos. Environ. 213: 629–639.

Ribeiro, R.D.X., Medeiros, A.N., Oliveira, R.L., de Araújo, G.G.L., Queiroga, R.D.C.D.E., Ribeiro, M.D., et al. 2018. Palm kernel cake from the biodiesel industry in goat kid diets. Part 2: Physicochemical composition, fatty acid profile and sensory attributes of meat. Small Rumin. Res. 165: 1–7.

Roda, A. and Lambri, M. 2019. Food uses of pineapple waste and by-products: a review. Int. J. Food Sci. Technol. 54: 1009–1017.

Rosas-Mendoza, E.S., Méndez-Contreras, J.M., Aguilar-Lasserre, A.A., Vallejo-Cantú, N.A. and Alvarado-Lassman, A. 2020. Evaluation of bioenergy potential from citrus effluents through anaerobic digestion. J. Clean. Prod. 254: 120128.

Rosicka-Kaczmarek, J., Komisarczyk, A., Nebesny, E. and Makowski, B. 2016. The influence of arabinoxylans on the quality of grain industry products. Eur. Food Res. Technol. 242: 295–303.

Roye, C., Henrion, M., Chanvrier, H., De Roeck, K., De Bondt, Y., Liberloo, I., et al. 2020. Extrusion-cooking modifies physicochemical and nutrition-related properties of wheat bran. Foods 9: 738.

Ruini, L., Marino, M., Pignatelli, S., Laio, F. and Ridolfi, L. 2013. Water footprint of a large-sized food company: The case of *Barilla* pasta production. Water Resour. Ind. 1: 7–24.

Saeed, A., Hanif, M.A., Nawaz, H. and Qadri, R.W.K. 2021. The production of biodiesel from plum waste oil using nano-structured catalyst loaded into supports. Sci. Rep. 11: 1–18.

Saeed, M., Khan, M.I., Butt, M.S. and Riaz, F. 2020. Characterization of peptides fractions produced through enzymatic hydrolysis of meat by-products for their antihypertensive and antioxidant activities. Pak. J. Agric. Sci. 57: 545–551.

Saeed, S., Aslam, S., Mehmood, T., Naseer, R., Nawaz, S., Mujahid, H., et al. 2021. Production of gallic acid under solid-state fermentation by utilizing waste from food processing industries. Waste Biomass Valor. 12: 155–163.

Sagar, N.A., Pareek, S., Sharma, S., Yahia, E.M. and Lobo, M.G. 2018. Fruit and vegetable waste: Bioactive compounds, their extraction, and possible utilization. Compr. Rev. Food Sci. Food Saf. 17: 512–531.

Şahin, S. and Elhussein, E.A.A. 2018. Valorization of a biomass: phytochemicals in oilseed by-products. Phytochem. Rev. 17: 657–668.

Salem, I.B., Saleh, M.B., Iqbal, J., El Gamal, M. and Hameed, S. 2021. Date palm waste pyrolysis into biochar for carbon dioxide adsorption. Energy Reports 7: 152–159.

Sánchez, H., Ponce, W., Brito, B., Viera, W., Baquerizo, R. and Riera, M.A. 2021. Biofilms production from avocado waste. Ing. Univ. 25: 1–16.

Sanchez-Camargo, A.P., Gutiérrez, L.F., Vargas, S.M., Martinez-Correa, H.A., Parada-Alfonso, F. and Narváez-Cuenca, C.-E. 2019. Valorisation of mango peel: Proximate composition, supercritical fluid extraction of carotenoids, and application as an antioxidant additive for an edible oil. J. Supercrit. Fluids 152: 104574.

Sánchez-Quezada, V., Campos-Vega, R. and Loarca-Piña, G. 2021. Prediction of the physicochemical and nutraceutical characteristics of 'Hass' avocado seeds by correlating the physicochemical avocado fruit properties according to their ripening state. Plant Foods Hum. Nutr. 76: 311–318.

Santiago, B., Calvo, A.A., Gullon, B., Feijoo, G., Moreira, M.T. and Gonzalez-Garcia, S. 2020. Production of flavonol quercetin and fructooligosaccharides from onion (*Allium cepa* L.) waste: An environmental life cycle approach. Chem. Eng. J. 392: 123772.

Sari, N.H., Wardana, I.N.G., Irawan, Y.S. and Siswanto, E. 2018. Characterization of the chemical, physical, and mechanical properties of NaOH-treated natural cellulosic fibers from corn husks. J. Nat. Fibers 15: 545–558.

Satari, B. and Karimi, K. 2018. Citrus processing wastes: Environmental impacts, recent advances, and future perspectives in total valorization. Resour. Conserv. Recycl. 129: 153–167.

Satya, C.V., Anuradha, C. and Reddy, D.S.R. 2014. Palm oil cake: A potential substrate for L-asparaginase production. Int. J. Innov. Res. Sci. Eng. Technol. 3: 14627–14632.

Savic, I.M. and Gajic, I.M.S. 2021. Optimization study on extraction of antioxidants from plum seeds (*Prunus domestica* L.). Optim. Eng. 22: 141–158.

Schmidt, S. and Pokorný, J. 2005. Potential application of oilseeds as sources of antioxidants for food lipids–a review. Czech J. Food Sci. 23: 93–102.

Seidavi, A., Zaker-Esteghamati, H. and Scanes, C.G. 2019. Poultry by-products. pp 123–146. *In:* Simpson, B.K., Aryee, A.N. and Toldrá, F. (eds.). By-products from Agriculture and Fisheries: Adding Value for Food, Feed, Pharma, and Fuels. John Wiley & Sons, Hoboken, NJ, USA.

Selani, M.M., Brazaca, S.G.C., dos Santos Dias, C.T., Ratnayake, W.S., Flores, R.A. and Bianchini, A. 2014. Characterisation and potential application of pineapple pomace in an extruded product for fibre enhancement. Food Chem. 163: 23–30.

Serrem, C.A., de Kock, H.L. and Taylor, J.R. 2011. Nutritional quality, sensory quality and consumer acceptability of sorghum and bread wheat biscuits fortified with defatted soy flour. Int. J. Food Sci. Technol. 46: 74–83.

Sezer, D.B., Ahmed, J., Sumnu, G. and Sahin, S. 2021. Green processing of sour cherry (*Prunus cerasus* L.) pomace: process optimization for the modification of dietary fibers and property measurements. J. Food Meas. Charact. 15: 3015–3025.

Shaikhiev, I.G., Kraysman, N.V. and Sverguzova, S.V. 2022. Onion (*Allium Cepa*) processing waste as a sorption material for removing pollutants from aqueous media. Biointerface Res. Appl. Chem. 12: 3173–3185.

Sharma, K., Mahato, N., Cho, M.H. and Lee, Y.R. 2017. Converting citrus wastes into value-added products: Economic and environmently friendly approaches. Nutrition 34: 29–46.

Sharma, P., Gaur, V.K., Kim, S.H. and Pandey, A. 2020. Microbial strategies for bio-transforming food waste into resources. Bioresour. Technol. 299: 122580.

Shehu, I., Akanbi, T.O., Wyatt, V. and Aryee, A.N. 2019. Fruit, nut, cereal, and vegetable waste valorization to produce biofuel. pp. 665–684. *In:* Simpson, B.K., Aryee, A.N. and Toldrá, F. (eds.). By-products from Agriculture and Fisheries: Adding Value for Food, Feed, Pharma, and Fuels. John Wiley & Sons, Hoboken, NJ, USA.

Shen, X., Zhang, M., Bhandari, B. and Gao, Z. 2019. Novel technologies in utilization of by-products of animal food processing: A review. Crit. Rev. Food Sci. Nutr. 59: 3420–3430.

Sheth, M. and Hirdyani, H. 2016. Development and sensory analysis of a buttermilk based fermented drink using barley and fructooligosaccharide as functional ingredients. Int. J. Home Sci. 2: 235–239.

Shi, Z., Zhang, Y., Phillips, G.O. and Yang, G. 2014. Utilization of bacterial cellulose in food. Food Hydrocolloids 35: 539–545.

Sielicka-Różyńska, M. and Gwiazdowska, D. 2020. Antioxidant and antibacterial properties of lemon, sweet, and cereal grasses. J. Food Process. Preserv. 44: e14984.

Singh, B. and Satyanarayana, T. 2006. Phytase production by thermophilic mold *Sporotrichum thermophile* in solid-state fermentation and its application in dephytinization of sesame oil cake. Appl. Biochem. Biotechnol. 133: 239–250.

Singh, S. and Immanuel, G. 2014. Extraction of antioxidants from fruit peels and its utilization in paneer. J. Food Process. Technol. 5: 1–5.

Sinha, S., Srivastava, A., Mehrotra, T. and Singh, R. 2019. A review on the dairy industry waste water characteristics, its impact on environment and treatment possibilities. pp. 73–84 *In:* Jindal, T. (ed.). Emerging Issues in Ecology and Environmental Science. Springer, Cham, Switzerland.

Skendi, A. and Biliaderis, C.G. 2016. Gelation of wheat arabinoxylans in the presence of Cu+ 2 and in aqueous mixtures with cereal β-glucans. Food Chem. 203: 267–275.

Skendi, A., Zinoviadou, K.G., Papageorgiou, M. and Rocha, J.M. 2020. Advances on the valorisation and functionalization of by-products and wastes from cereal-based processing industry. Foods 9: 1243.

Skryplonek, K., Dmytrów, I. and Mituniewicz-Małek, A. 2019. The use of buttermilk as a raw material for cheese production. Int. J. Dairy Technol. 72: 610–616.

Smets, K., Adriaensens, P., Reggers, G., Schreurs, S., Carleer, R. and Yperman, J. 2011. Flash pyrolysis of rapeseed cake: Influence of temperature on the yield and the characteristics of the pyrolysis liquid. J. Anal. Appl. Pyrolysis 90: 118–125.

Smets, K., Roukaerts, A., Czech, J., Reggers, G., Schreurs, S., Carleer, R., et al. 2013. Slow catalytic pyrolysis of rapeseed cake: Product yield and characterization of the pyrolysis liquid. Biomass and Bioenergy 57: 180–190.

Solarte-Toro, J.C., Ortiz-Sanchez, M., Restrepo-Serna, D.L., Piñeres, P.P., Cordero, A.P. and Alzate, C.A.C. 2021. Influence of products portfolio and process contextualization on the economic performance of small-and large-scale avocado biorefineries. Bioresour. Technol. 342: 126060.

Son, S.W., Nam, Y.J., Lee, S.H., Lee, S.M., Lee, S.H., Kim, M.J., et al. 2011. Toxigenic fungal contaminants in the 2009-harvested rice and its milling-by products samples collected from rice processing complexes in Korea. Research in Plant Disease 17: 280–287.

Sousa, D., Salgado, J.M., Cambra-López, M., Dias, A.C. and Belo, I. 2021. Degradation of lignocellulosic matrix of oilseed cakes by solid-state fermentation: fungi screening for enzymes production and antioxidants release. J. Sci. Food Agric.

Sözer, S. and Yaldız, O. 2006. A research on biogas production from cattle manure and cheese whey mixtures. Mediterr. Agric. Sci. 19: 179–183.

Srilatha, K. and Krishnakumari, K. 2003. Proximate composition and protein quality evaluation of recipes containing sunflower cake. Plant Foods Hum. Nutr. 58: 1–11.

Srivastava, N., Srivastava, M., Alhazmi, A., Kausar, T., Haque, S., Singh, R., et al. 2021. Technological advances for improving fungal cellulase production from fruit wastes for bioenergy application: A review. Environ. Pollut. 287: 117370.

Stephenson, A.L., Dennis, J.S. and Scott, S.A. 2008. Improving the sustainability of the production of biodiesel from oilseed rape in the UK. Process Saf. Environ. Prot. 86: 427–440.

Storino, F., Arizmendiarrieta, J.S., Irigoyen, I., Muro, J. and Aparicio-Tejo, P.M. 2016. Meat waste as feedstock for home composting: Effects on the process and quality of compost. Waste Manage. 56: 53–62.

Suárez, S., Mu, T., Sun, H. and Añón, M.C. 2020. Antioxidant activity, nutritional, and phenolic composition of sweet potato leaves as affected by harvesting period. Int. J. Food Prop. 23: 178–188.

Sudha, M.L., Ramasarma, P.R. and Venkateswara Rao, G. 2011. Wheat bran stabilization and its use in the preparation of high-fiber pasta. Food Sci. Technol. Int. 17: 47–53.

Sue, T.T. 2004. Quality and characteristics of Malaysian palm kernel cakes/expellers. Palm Oil Development 34: 1–3.

Summo, C., De Angelis, D., Difonzo, G., Caponio, F. and Pasqualone, A. 2020. Effectiveness of oat-hull-based ingredient as fat replacer to produce low fat burger with high beta-glucans content. Foods 9: 1057.

Sun, X., Wei, X., Zhang, J., Ge, Q., Liang, Y., Ju, Y., et al. 2020. Biomass estimation and physicochemical characterization of winter vine prunings in the Chinese and global grape and wine industries. Waste Manage. 104: 119–129.

Syaftika, N. and Matsumura, Y. 2018. Comparative study of hydrothermal pretreatment for rice straw and its corresponding mixture of cellulose, xylan, and lignin. Bioresour. Technol. 255: 1–6.

Syazili, A., Ahmad, K. and Umakaapa, I. 2021, October. Using tuna fish bone waste as mineral sources in feed formulation of tilapia (*Oreochromis niloticus*). IOP Conf. Ser. Earth Environ. Sci. 890: 012026.

Szydłowska-Czerniak, A., Amarowicz, R. and Szłyk, E. 2010. Antioxidant capacity of rapeseed meal and rapeseed oils enriched with meal extract. Eur. J. Lipid Sci. Technol. 112: 750–760.

Tagirova, P.R., Khasikhanov, M.S., Kasyanov, G.I., Saidulaev, S.S., Masaeva, L.M., Erzhapova, R.S., et al. 2018. Food and ecological safety of grape by-products. Adv. Eng. Res. 151: 941–945.

Tan, S.H., Mailer, R.J., Blanchard, C.L. and Agboola, S.O. 2011. Extraction and characterization of protein fractions from Australian canola meals. Food Res. Int. 44: 1075–1082.

Teh, S.S., Bekhit, A.E.D. and Birch, J. 2014. Antioxidative polyphenols from defatted oilseed cakes: Effect of solvents. Antioxidants 3: 67–80.

Terpinc, P., Čeh, B., Ulrih, N.P. and Abramovič, H. 2012. Studies of the correlation between antioxidant properties and the total phenolic content of different oil cake extracts. Ind. Crop. Prod. 39: 210–217.

Tesfaye, T., Ayele, M., Ferede, E., Gibril, M., Kong, F. and Sithole, B. 2021. A techno-economic feasibility of a process for extraction of starch from waste avocado seeds. Clean Technol. Environ. Policy 23: 581–595.

Thakur, S.A., Nemade, S.N. and Sharanappa, A. 2015. Solid state fermentation of overheated soybean meal (Waste) for production of protease using *Aspergillus Oryzae*. Int. J. Innov. Res. Sci. Eng. Technol. 4: 18456–18461.

Thanikachalam, J. and Ganesh, N. 2020, September. Exploration of properties of meat waste incinerated fly ash/cement/sand dust associated fly ash bricks. IOP Conf. Ser. Mater. Sci. Eng. 923: 012040.

Toldrá, F., Mora, L. and Reig, M. 2016. New insights into meat by-product utilization. Meat Sci. 120: 54–59.

Toldrá-Reig, F., Mora, L. and Toldrá, F. 2020. Trends in biodiesel production from animal fat waste. Appl. Sci. 10: 3644.

Torres, M.D. and Domínguez, H. 2020. Valorisation of potato wastes. Int. J. Food Sci. Technol. 55: 2296–2304.

Torres-León, C., Ramírez-Guzmán, N., Ascacio-Valdes, J., Serna-Cock, L., dos Santos Correia, M.T., Contreras-Esquivel, J.C., et al. 2019. Solid-state fermentation with Aspergillus niger to enhance the phenolic contents and antioxidative activity of Mexican mango seed: A promising source of natural antioxidants. LWT. 112: 108236.

Torres-León, C., Ramírez-Guzman, N., Londoño-Hernandez, L., Martinez-Medina, G.A., Díaz-Herrera, R., Navarro-Macias, V., et al. 2018a. Food waste and byproducts: An opportunity to minimize malnutrition and hunger in developing countries. Front. Sustain. Food Syst. 2: 52.

Torres-León, C., Vicente, A.A., Flores-López, M.L., Rojas, R., Serna-Cock, L., Alvarez-Pérez, O.B., et al. 2018b. Edible films and coatings based on mango (var. Ataulfo) by-products to improve gas transfer rate of peach. LWT. 97: 624–631.

[TÜGİŞ] Turkish Food and Beverage Industry Association. 2016. Gıdalarda Atıkların Azaltılması ve Geri Kazanımı Projesi. Süt ve Süt Ürünleri Sektöründe Atık Yönetimi. Turkey.

Tuncel, N.Y., Kaya, E. and Karaman, M. 2017. Rice bran substituted Turkish noodles (erişte): textural, sensorial, and nutritional properties. Cereal Chem. 94: 903–908.

Uluata, S. and Özdemir, N. 2017. Evaluation of chemical characterization, antioxidant activity and oxidative stability of some waste seed oil. Turkish J. Agric. Food Sci. Technol. 5: 48–53.

Upadhyay, A., Lama, J.P. and Tawata, S. 2010. Utilization of pineapple waste: a review. J. Food Sci. Technol. Nepal 6: 10–18.

USDA. 2021. Oilseeds: World Markets and Trade. World Production, Markets, and Trade Report. United States Department of Agriculture, USDA.

Valensya, D., Rozalia, I. and Syamsuddin, Y. 2020. Utilization of avocado seed waste as raw material for producing biodiesel with CaO catalyst from eggshell. IOP Conf. Ser.: Mater. Sci. Eng. 845: 012020.

Vanara, F., Scarpino, V. and Blandino, M. 2018. Fumonisin distribution in maize dry-milling products and by-products: Impact of two industrial degermination systems. Toxins 10: 357.

Venkatesagowda, B., Ponugupaty, E., Barbosa, A.M. and Dekker, R.F. 2015. Solid-state fermentation of coconut kernel-cake as substrate for the production of lipases by the coconut kernel-associated fungus *Lasiodiplodia theobromae* VBE-1. Ann. Microbiol. 65: 129–142.

Vieira, I.M.M., Santos, B.L.P., Silva, L.S., Ramos, L.C., de Souza, R.R., Ruzene, D.S., et al. 2021. Potential of pineapple peel in the alternative composition of culture media for biosurfactant production. Environ. Sci. and Pollut. Res. 28: 1–15.

Vilariño, M.V., Franco, C. and Quarrington, C. 2017. Food loss and waste reduction as an integral part of a circular economy. Front. Environ. Sci. 5: 1–5.

Vishkarma, K., Suman, R., Sharanagat, V.S. and Singh, M. 2015. Modeling of hot air drying kinetics of tofu. Agric. Eng. Today 39: 47–56.

Vitez, T., Dokulilová, T., Vitezova, M., Elbl, J., Kintl, A., Kynicky, J., et al. 2020. The digestion of waste from vegetables and maize processing. Waste Biomass Valor. 11: 2467–2473.

Vladic, J., Gavaric, A., Jokic, S., Pavlovic, N., Moslavac, T., Popovic, L., et al. 2020. Alternative to conventional edible oil sources: cold pressing and supercritical co2 extraction of plum (*Prunus Domestica* L.) kernel seed. Acta Chimica Slovenica 67: 778–784.

Vong, W.C. and Liu, S.Q. 2019. The effects of carbohydrase, probiotic *Lactobacillus paracasei* and yeast *Lindnera saturnus* on the composition of a novel okara (soybean residue) functional beverage. LWT. 100: 196–204.

Vu, H.T., Scarlett, C.J. and Vuong, Q.V. 2020. Encapsulation of phenolic-rich extract from banana (Musa cavendish) peel. J. Food Sci. Technol. 57: 2089–2098.

Waheed, A., Goyal, M., Gupta, D., Khanna, A., Hassanien, A.E. and Pandey, H.M. 2020a. An optimized dense convolutional neural network model for disease recognition and classification in corn leaf. Comput. Electron. Agric. 175: 105456.

Waheed, M., Butt, M.S., Shehzad, A., Adzahan, N.M., Shabbir, M.A., Suleria, H.A.R., et al. 2019. Eggshell calcium: A cheap alternative to expensive supplements. Trends Food Sci. Technol. 91: 219–230.

Waheed, M., Yousaf, M., Shehzad, A., Inam-Ur-Raheem, M., Khan, M.K.I., Khan, M.R., et al. 2020b. Channelling eggshell waste to valuable and utilizable products: a comprehensive review. Trends Food Sci. Technol. 106: 78–90.

Walton, G.E., Lu, C., Trogh, I., Arnaut, F. and Gibson, G.R. 2012. A randomised, double-blind, placebo controlled cross-over study to determine the gastrointestinal effects of consumption of arabinoxylan-oligosaccharides enriched bread in healthy volunteers. Nutr. J. 11: 1–11.

Wan, C., Zhou, Y. and Li, Y. 2011. Liquid hot water and alkaline pretreatment of soybean straw for improving cellulose digestibility. Bioresour. Technol. 102: 6254–6259.

Wang, Q., Nnanna, P.C., Shen, F., Huang, M., Tian, D., Hu, J., et al. 2021. Full utilization of sweet sorghum for bacterial cellulose production: A concept of material crop. Ind. Crop Prod. 162: 113256.

Wang, Y., Chen, Y.J., Cho, J.H., Yoo, J.S., Huang, Y., Kim, H.J., et al. 2009. Effect of soybean hull supplementation to finishing pigs on the emission of noxious gases from slurry. Anim. Sci. J. 80: 316–321.

Wang, Y., Kong, L.J., Li, C. and Bureau, D.P. 2006. Effect of replacing fish meal with soybean meal on growth, feed utilization and carcass composition of cuneate drum (*Nibea miichthioides*). Aquaculture 261: 1307–1313.

Wang, Z., He, X., Yan, L., Wang, J., Hu, X., Sun, Q., et al. 2020. Enhancing enzymatic hydrolysis of corn stover by twin-screw extrusion pretreatment. Ind. Crop Prod. 143: 111960.

Wanyo, P., Meeso, N. and Siriamornpun, S. 2014. Effects of different treatments on the antioxidant properties and phenolic compounds of rice bran and rice husk. Food Chem. 157: 457–463.

Weber, C.T., Trierweiler, L.F. and Trierweiler, J.O. 2020. Food waste biorefinery advocating circular economy: Bioethanol and distilled beverage from sweet potato. J. Clean. Prod. 268: 121788.

Wianowska, D., Olszowy-Tomczyk, M. and Garbaczewska, S. 2021. A central composite design in increasing the quercetin content in the aqueous onion waste isolates with antifungal and antioxidant properties. Eur. Food Res. Technol. 1–9.

Wijesinha-Bettoni, R. and Burlingame, B. 2013. Milk and dairy product composition. pp. 41–102. *In:* Muehlhoff, E., Bennett, A. and McMahon, D. [Eds.]. Milk and Dairy Products in Human Nutrition. Food and Agriculture Organisation, Rome, Italy.

Wikandari, R., Millati, R., Cahyanto, M.N. and Taherzadeh, M.J. 2014. Biogas production from citrus waste by membrane bioreactor. Membranes 4: 596–607.

Wu, X., Li, F. and Wu, W. 2020. Effects of rice bran rancidity on the oxidation and structural characteristics of rice bran protein. LWT. 120: 108943.

Xiao, Y., Liu, Y., Wang, X., Li, M., Lei, H. and Xu, H. 2019. Cellulose nanocrystals prepared from wheat bran: Characterization and cytotoxicity assessment. Int. J. Biol. Macromol. 140: 225–233.

Xu, J., Krietemeyer, E.F., Boddu, V.M., Liu, S.X. and Liu, W.C. 2018. Production and characterization of cellulose nanofibril (CNF) from agricultural waste corn stover. Carbohydr. Polym. 192: 202–207.

Yahaya, Y., Uauri, U.A.B. and Bagudo, B.U. 2010. Study of nutrient content variation in bulb and stalk of onions (*Allium Sepa*) cultivated in Aliero, Aliero, Kebbi State, Nigeria. Nig. J. Basic Appl. Sci. 18: 83–89.

Yan, M., Hantoko, D., Susanto, H., Ardy, A., Waluyo, J., Weng, Z., et al. 2019. Hydrothermal treatment of empty fruit bunch and its pyrolysis characteristics. Biomass Convers. Biorefin. 9: 709–717.

Yang, T., Yan, H.L. and Tang, C.H. 2021. Wet media planetary ball milling remarkably improves functional and cholesterol-binding properties of okara. Food Hydrocoll. 111: 106386.

Ye, G., Wu, Y., Wang, L., Tan, B., Shen, W., Li, X., et al. 2021. Comparison of six modification methods on the chemical composition, functional properties and antioxidant capacity of wheat bran. LWT. 149: 111996.

Yetim, H., Müller, W.D. and Eber, M. 2001. Using fluid whey in comminuted meat products: effects on technological, chemical and sensory properties of frankfurter-type sausages. Food Res. Int. 34: 97–101.

Yıldırım, Ç. and Güzeler, N. 2013. Evaluation of whey and buttermilk as powder. J. Agric. Fac. Çukurova Univ. 28: 11–20.

Yılmaz, E., Keskin, O. and Ok, S. 2020. Valorization of sour cherry and cherry seeds: Cold press oil production and characterization. J. Agroaliment. Process. Technol. 26: 228–240.

Yılmaz, F.M., Görgüç, A., Karaaslan, M., Vardin, H., Ersus Bilek, S., Uygun, Ö., et al. 2019. Sour cherry by-products: Compositions, functional properties and recovery potentials–a review. Crit. Rev. Food Sci. Nutr. 59: 3549–3563.

Yılmaz, I. 2004. Effects of rye bran addition on fatty acid composition and quality characteristics of low-fat meatballs. Meat Sci. 67: 245–249.

Yu, S., Wang, J., Li, Y., Wang, X., Ren, F. and Wang, X. 2021. Structural studies of water-insoluble β-Glucan from oat bran and its effect on improving lipid metabolism in mice fed high-fat diet. Nutrients 13: 3254.

Yüceer, M. and Caner, C. 2019. An assessment and review of the halal food certification process requirements of industrial egg products. J. Halal Life Style 1: 23–34.

Zago, E., Lecomte, J., Barouh, N., Aouf, C., Carré, P., Fine, F. and Villeneuve, P. 2015. Influence of rapeseed meal treatments on its total phenolic content and composition in sinapine, sinapic acid and canolol. Ind. Crop Prod. 76: 1061–1070.

Zaman, T., Mostari, M., Mahmood, M.A.A. and Rahman, M.S. 2018. Evolution and characterization of eggshell as a potential candidate of raw material. Cerâmica 64: 236–241.

Zandona, E., Blažić, M. and Režek Jambrak, A. 2021. Whey utilization: Sustainable uses and environmental approach. Food Technol. Biotechnol. 59: 147–161.

Zang, X., Yue, C., Wang, Y., Shao, M. and Yu, G. 2019. Effect of limited enzymatic hydrolysis on the structure and emulsifying properties of rice bran protein. J. Cereal Sci. 85: 168–174.

Zardo, I., de Espíndola Sobczyk, A., Marczak, L.D.F. and Sarkis, J. 2019. Optimization of ultrasound assisted extraction of phenolic compounds from sunflower seed cake using response surface methodology. Waste Biomass Valor. 10: 33–44.

Zardo, I., Rodrigues, N.P., Sarkis, J.R. and Marczak, L.D. 2020. Extraction and identification by mass spectrometry of phenolic compounds from canola seed cake. J. Sci. Food Agric. 100: 578–586.

Zarubin, N.Y., Kharenko, E.N., Bredikhina, O.V., Arkhipov, L.O., Zolotarev, K.V., Mikhailov, A.N., et al. 2021. Application of the *Gadidae* fish processing waste for food grade gelatin production. Marine Drugs, 19: 455.

Zema, D.A., Calabrò, P.S., Folino, A., Tamburino, V., Zappia, G. and Zimbone, S.M. 2018. Valorisation of citrus processing waste: A review. Waste Manage. 80: 252–273.

Zen, C.K., Sartor, K.B., Silva, R.V., Colla, L.M. and Reinehr, C.O. 2020. Improvement of nutritional quality of ruminant feed composed of soybean straw and ryegrass hay. Sci. Plena 16: 12303.

Zeyada, N.N., Zeitoum, M.A.M. and Barbary, O.M. 2008. Utilization of some vegetables and fruit waste as natural antioxidants. Alex. J. Food Sci. Technol. 5: 1–11.

Zhao, Z., Xu, Z., Le, K., Azordegan, N., Riediger, N.D. and Moghadasian, M.H. 2009. Lack of evidence for antiatherogenic effects of wheat bran or corn bran in apolipoprotein E-knockout mice. J. Agric. Food Chem. 57: 6455–6460.

Zhivkova, V. 2020. Characterization of nutritional and mineral content of plum and cherry waste. Calitatea 21: 141–144.

Zhu, S.M., Lin, S.L., Ramaswamy, H.S., Yu, Y. and Zhang, Q.T. 2017. Enhancement of functional properties of rice bran proteins by high pressure treatment and their correlation with surface hydrophobicity. Food Bioproc. Tech. 10: 317–327.

Zou, Y., Gao, Y., He, H. and Yang, T. 2018. Effect of roasting on physico-chemical properties, antioxidant capacity, and oxidative stability of wheat germ oil. LWT 90: 246–253.

Index

For Product Safety Concerns and Information please contact our EU
representative GPSR@taylorandfrancis.com
Taylor & Francis Verlag GmbH, Kaufingerstraße 24, 80331 München, Germany

9 781032 138664